EPIC: EXPLORATION PROGRAMS IN CALCULUS, 2/E

by James W. Burgmeier and Larry L. Kost

This interactive software package for the IBM Personal Computer is a collection of programs that allows users to explore a variety of calculus concepts, including: equations of lines, intersections of curves, maxima and minima, differentiation, and anti-derivatives. The hardware requirements include a 256K memory, one disk drive, and a graphics card. The software also offers the following features:

1. *Users may superimpose graphs of functions with parameters.*
2. *Users have complete control over the range used for the x and y variables on graphs.*
3. *Functions may include parameters that users can alter (e.g., users may enter* asin bx *and specify a,b later).*
4. *Users can invoke options or gain information by a single keystroke (e.g., they can change the range along the x axis, get the max and min of the function, or trace polar curves).*

A departmental version of this software package is available and offers the same features listed above, as well as:

1. *A built-in math calculator with 25 memory addresses and 10 formula saving registers*
2. *Help screens and on-screen reminders*
3. *Extended options for: function plotting, integration, and parametric/polar equations*

ORDER YOUR COPY TODAY!

Please send me:

☐ **EPIC: EXPLORATION PROGAMS IN CALCULUS, 2/E**
(student version) (Title Code #28325–9) . . . . . $ 13.00

☐ **EPIC: EXPLORATION PROGRAMS IN CALCULUS, 2/E**
(departmental version) (Title Code #28323–4) $225.00

QUANTITY	TITLE/AUTHOR	TITLE CODE	PRICE	TOTAL
(Sample Entry) 2	**Burgmeier/EPIC (Student)**	28325–9	$13.00	$26.00
			TOTAL PRICE:	

SAVE!
If payment accompanies order, plus your state's sales tax where applicable, Prentice Hall pays postage and handling charges. Same return privilege refund guaranteed.

☐ PAYMENT ENCLOSED—shipping and handling to be paid by publisher (please include your state's tax where applicable).

MAIL TO:
PRENTICE HALL
Book Distribution Center
Route 59 at Brook Hill Drive
West Nyack, NY 10995

NAME _____

ORGANIZATION _____

STREET ADDRESS _____

CITY _____ STATE _____ ZIP _____

I Prefer to charge my ☐ VISA   ☐ MasterCard

Card Number _____ Expiration Date _____

Signature _____

All prices in this catalog are subject to change without notice. Prices not valid outside the U.S.

Dept. 1

D–EPSD–FS(0)

CALCULUS
WITH APPLICATIONS
to the Management, Social,
Behavioral, and Biomedical Sciences

Regarding the front cover illustration:

I do not know what I may appear to the world; but to myself I seem to have been only like a boy playing on the seashore, and diverting myself in now and then finding a smoother pebble or a prettier shell than ordinary, whilst the great ocean of truth lay all undiscovered before me.

ISAAC NEWTON

In Latin the word ''calculus'' means ''small stone.''

CALCULUS

WITH APPLICATIONS
to the Management, Social,
Behavioral, and Biomedical Sciences

Geoffrey C. Berresford

Department of Mathematics
C.W. Post College
Long Island University

PRENTICE HALL
Englewood Cliffs, New Jersey 07632

Library of Congress Cataloging-in-Publication Data

Berresford, Geoffrey C.,
  Calculus, with applications to the management, social, behavioral, and biomedical sciences.

  Includes index.
  1. Calculus.   I. Title.
QA303.B53   1989   515   88-19646
ISBN 0-13-110628-7

Editorial/production supervision: Nicholas Romanelli
Cover design: Maureen Eide
Manufacturing buyer: Paula Massenara
Photo research: Lorinda Morris-Nantz and Page Poore
Cover art: © J. DiMaggio/J. Kalish, Peter Arnold, Inc.

Photo credits: 1, 41, H.E. Edgerton, M.I.T.; 65, New York Public Library; 99, The New York Times; 120, UPI/Bettman Newsphotos; 122, Convention and Visitor's Bureau of Greater St. Louis; 261, Fred Lombardi; 279, Kennecott Copper Corporation; 333, Frank Hoffman; 335, Grant White; 349, Laimute E. Druskis; 404, American Airlines; 411, R.G. Kessel and R. H. Kardon; 420, Florida State News Bureau.

© 1989 by Prentice-Hall, Inc.
A Division of Simon & Schuster
Englewood Cliffs, New Jersey 07632

Printed in the United States of America

10 9 8 7 6 5 4 3 2 1

ISBN 0-13-110628-7

Prentice-Hall International (UK) Limited, *London*
Prentice-Hall of Australia Pty. Limited, *Sydney*
Prentice-Hall Canada Inc., *Toronto*
Prentice-Hall Hispanoamericana, S.A., *Mexico*
Prentice-Hall of India Private Limited, *New Delhi*
Prentice-Hall of Japan, Inc., *Tokyo*
Prentice-Hall of Southeast Asia Pte. Ltd., *Singapore*
Editora Prentice-Hall do Brasil, Ltda., *Rio de Janeiro*

CONTENTS

■ CONTENTS

PREFACE

TO THE STUDENT

Calculus has an enormous variety of useful and interesting applications. In this book we use calculus to predict the national debt, maximize longevity, and estimate the dangers of cigarette smoking. Even beyond its utility, however, there is a beauty to calculus, and I hope that while studying its many important applications you will appreciate its conceptual unity and simplicity.

Prerequisites The only prerequisite for this course is some knowledge of algebra and graphing, and these are reviewed in Chapter 1. Chapter 1 also covers the preliminary material on functions that is necessary to begin the study of calculus.

Practice Exercises Learning calculus requires your active participation ("mathematics is not a spectator sport"). As you read you should use pencil and paper to solve the "practice exercises" placed throughout the reading. These practice exercises will help you consolidate your understanding before moving on to new material. Complete solutions for all practice exercises are given at the end of each section. Experience has shown that solving the practice exercises substantially increases your understanding of, and success with, calculus.

Exercises Each section contains numerous applied exercises. These demonstrate that calculus is not just the manipulation of abstract symbols but is deeply connected with everyday life. Anyone who ever learned calculus did so by solving many, many problems, and the exercises are the most essential part of learning calculus. Answers to all odd-numbered exercises are given at the end of the book. Exercises marked with the symbol ▦ require a calculator with $\boxed{\ln x}$ and $\boxed{e^x}$ keys for natural logarithms and exponentiation and a $\boxed{y^x}$ key for powers.

Margin Notes The notes in the right-hand margins restate most of the mathematical results in words. They show how to "read mathematics."

Student Solutions Manual and EPIC Software The Student Solutions Manual, written by Gloria Langer, contains completely worked-out solutions to every other odd-numbered exercise. EPIC (Exploration Programs in Calculus) is a set of interactive computer programs for exploring a variety of calculus concepts. Both are available through your bookstore or from the publisher.

TO THE INSTRUCTOR

This is a textbook on calculus and its applications to the management, social, behavioral, and biomedical sciences.

Treatment The basic nature of such a course should be very "applied." Therefore, this book contains an unusually large number of applications, and intuitive explanations have been preferred to formal proofs.

I wrote this book with several principles in mind. One is that to learn something, it is best to begin doing it as soon as possible. Therefore, the preliminary Chapter 1 has been kept short, so that students begin taking derivatives as soon as possible. This early start allows more time during the course for useful applications and review. As another example, in the first section on graphing (Section 3.1) students are graphing polynomials instead of just finding critical values.

Another principle is that the mathematics should be done together with the applications. Consequently, there are applications in every section (there are no "pure math" sections).

Organization Whenever possible, review material has been put close to where it is needed. Explicit page references are used throughout the book, and references to earlier material have been minimized by restating results wherever they are used.

Many students will be able to skip parts of Chapter 1. The following sections are optional in that they are not prerequisites for later material: 3.3 (graphing rational functions), 3.6 (lot size and harvest size), 3.7 (implicit differentiation and related rates), 3.8 (chapter review), 4.4 (related rates and elasticity of demand), 4.5 (chapter review), 5.5 (consumers' surplus and income distribution), 5.7 (chapter review), all of Chapter 6 (integration techniques and differential equations) except that 6.6 depends on 6.5, and 7.4 through 7.7 (least squares, Lagrange multipliers, multiple integrals, and chapter review).

Supplements and Other Teaching Aids Several supplements (besides the Student Solutions Manual) are available from the publisher. The Instructor's Manual consists of answers to all even-numbered exercises, and a large collection of sample test questions. The sample test questions (with a test-generating program developed by Bibliogem, Inc.) are also available on computer disk. Individual tests consisting of any of these questions may also be ordered by telephone from the publisher. A set of transparencies allows overhead projection of some of the more complicated diagrams and tables from the text. EPIC (Exploration Programs in Calculus), the interactive software package for students, is available in an institutional version for site licensing. Finally, an Instructor's Edition of the text contains answers to *all* exercises (rather than just the odd-numbered answers). All of these are available from the publisher.

I first wrote this book for my own students. I then rewrote it to include applications from a wider variety of fields. This made the book too long, so I wrote it a third time. With the knowledge that it will never be completely finished, I invite corrections, criticism, and suggestions from any reader.

ACKNOWLEDGMENTS

I would like to thank Andrew Rockett for suggesting that I write this book, and all of my "Math 6" students for serving as proofreaders over several years. Susan Halter expertly solved every problem and checked all the answers, for which I am very grateful. Theodore Faticoni class-tested and proofread the final draft at Fordham University and made numerous very useful suggestions. I am also indebted to Ed Immergut, Ethel Matin, John Stevenson, and many others for many useful conversations and correspondence.

The following reviewers have contributed greatly to the progression from first draft to finished book:

R.B. Campbell, University of Northern Iowa
Edward A. Connors, University of Massachusetts at Amherst
W.E. Conway, University of Arizona
Bob C. Denton, Orange Coast College
Joseph Elich, Utah State University
Theodore G. Faticoni, Fordham College
Catherine Folio, Brookdale Community College
Howard Frisinger, Colorado State University
James W. Newsom, Tidewater Community College
Charles A. Nicol, University of South Carolina
Martha H. Pratt, Mississippi State University
Daniel Shea, University of Wisconsin
Clifton T. Whyburn, University of Houston
Melvin R. Woodard, Indiana University of Pennsylvania

Much useful advice came from my editors at Prentice Hall, Bob Sickles and David Ostrow, who helped beyond words, as did Nicholas Romanelli, who expertly supervised all aspects of production. To all others at Prentice

Hall who gave their time (Page Poore, Thom Moore, Bruce Colin, and many more unknown to me), I express my gratitude.

Finally and most importantly, I thank my wife, Barbara, and my children, Lee and Christopher, for their (almost) inexhaustible patience while "the thing" was being finished.

Chilmark, 1989   **G.C.B.**

CALCULUS
WITH APPLICATIONS
to the Management, Social,
Behavioral, and Biomedical Sciences

1

FUNCTIONS

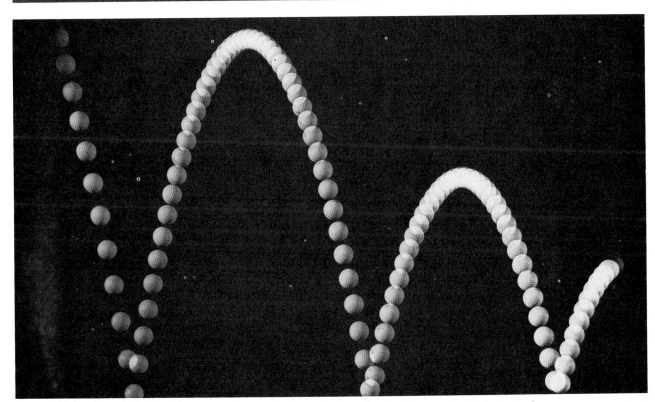

Parabolas described by a bouncing ball

1.1 EXPONENTS

Introduction

Calculus is, quite simply, the study of rates of change. The two graphs below show how the United States inflation rate and the world population have changed over the years. Calculus enables us to analyze these and other rates of change with great precision.

U.S. Inflation Rate

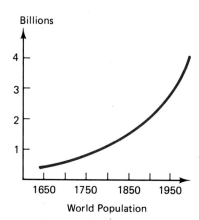

World Population

Practice Exercises

This book has many "Practice Exercises" placed throughout the reading. These are short exercises to check your understanding before moving on to new material. Full solutions are given at the end of each section.

Calculators

Some exercises require a calculator with $\boxed{\ln x}$ and $\boxed{e^x}$ keys for natural logarithms and exponentiation, and a $\boxed{y^x}$ key for powers. Such calculators are quite inexpensive ($10 to $20) and are well worth this small investment.

Positive Integer Exponents

Before studying how quantities change, we review some of the ways in which numerical quantities can be expressed. We begin with exponential notation. For a positive integer n,

$$x^n \quad \text{means} \quad \overbrace{x\cdot x \cdots x}^{n}$$

product of n x's

The number being raised to the power is called the *base*:

$$x^n \quad \text{exponent} \quad \text{base}$$

Laws of Exponents

There are several laws of exponents.

Addition Law of Exponents

$$x^m \cdot x^n = x^{m+n}$$

to multiply powers of the same base, add the exponents

Subtraction Law of Exponents

$$\frac{x^m}{x^n} = x^{m-n}$$

to divide powers of the same base, subtract the exponents (top exponent minus bottom exponent)

Example 1

(a) $x^2 \cdot x^3 = x^5$ $2 + 3$

(b) $\dfrac{x^5}{x^3} = x^2$ $5 - 3$

(c) $x^5 \cdot x^2 \cdot x = x^8$ $5 + 2 + 1$

(d) $\dfrac{x^{11}}{x^2 x^5} = \dfrac{x^{11}}{x^7} = x^4$

by the addition law of exponents   by the subtraction law of exponents ▪

These rules for adding and subtracting exponents apply only when the bases are the same. For example, $x^3 y^4$ cannot be simplified.

Multiplication Law of Exponents

$$(x^m)^n = x^{m \cdot n}$$

to raise a power to a power, multiply the powers

Example 2

(a) $(x^2)^3 = x^6$ $2 \cdot 3$

(b) $((x^2)^3)^4 = x^{24}$ $2 \cdot 3 \cdot 4$ ▪

Remember that for exponents in the form $x^2 \cdot x^3 = x^5$, *add* exponents, but for exponents in the form $(x^2)^3 = x^6$, *multiply* exponents.

Solve the following practice exercise. Then check your answer on page 11.

Practice Exercise 1

Simplify

(a) $\dfrac{x^5 x}{x^2}$

(b) $((x^3)^2)^2$

(solutions on page 11)

There are two more laws of exponents:

$$(xy)^m = x^m \cdot y^m$$

to raise a product to a power, raise each factor to the power

$$\left(\frac{x}{y}\right)^m = \frac{x^m}{y^m}$$

to raise a fraction to a power, raise the numerator and denominator to the power

Example 3

(a) $(2w)^3 = 2^3 w^3 = 8w^3$

(b) $\left(\dfrac{x}{3}\right)^4 = \dfrac{x^4}{3^4} = \dfrac{x^4}{81}$ ▪

Zero and Negative Exponents

For any number x other than zero, we define

$$x^0 = 1$$

x to the power zero is 1

$$x^{-1} = \frac{1}{x}$$

x to the power −1 is 1 over x

$$x^{-2} = \frac{1}{x^2}$$

x to the power −2 is 1 over x squared

and in general

$$x^{-n} = \frac{1}{x^n}$$

x to the negative n is 1 over x to the positive n

Example 4

(a) $5^0 = 1$

(b) $7^{-1} = \dfrac{1}{7}$

(c) $3^{-2} = \dfrac{1}{3^2} = \dfrac{1}{9}$

(d) $(-2)^{-3} = \dfrac{1}{(-2)^3} = \dfrac{1}{-8} = -\dfrac{1}{8}$

(e) 0^0 and 0^{-3} are undefined. ▪

Practice Exercise 2

Evaluate

(a) 2^0

(b) 2^{-4}

(solutions on page 11)

The definitions of x^0 and x^{-n} are motivated by the subtraction law of exponents:

$$1 = \frac{x^2}{x^2} = x^{2-2} = x^0, \quad \text{therefore} \quad x^0 = 1$$

$$\frac{1}{x^n} = \frac{x^0}{x^n} = x^{0-n} = x^{-n}, \quad \text{therefore} \quad x^{-n} = \frac{1}{x^n}$$

For a fraction to a negative power, we use the fact that to divide by a fraction we "invert and multiply."

$$\left(\frac{a}{b}\right)^{-1} = \frac{1}{\frac{a}{b}} = 1 \cdot \frac{b}{a} = \frac{b}{a}$$

↳ reciprocal of the original fraction

Therefore, for $a \neq 0$ and $b \neq 0$,

$$\left(\frac{a}{b}\right)^{-1} = \frac{b}{a}$$

a fraction to the power -1 is the reciprocal of the fraction

and similarly,

$$\left(\frac{a}{b}\right)^{-n} = \left(\frac{b}{a}\right)^{n}$$

a fraction to a negative power is the reciprocal of the fraction to the positive power

Example 5
(a) $\left(\frac{3}{2}\right)^{-1} = \frac{2}{3}$

(b) $\left(\frac{1}{2}\right)^{-3} = \left(\frac{2}{1}\right)^3 = \frac{2^3}{1^3} = 8$ ■

Practice Exercise 3
Simplify $\left(\frac{2}{3}\right)^{-2}$.

(solution on page 11)

Roots
We may take the square root of any *nonnegative* number, and the cube root of *any* number.

Example 6
(a) $\sqrt{9} = 3$

(b) $\sqrt{-9}$ is undefined

(c) $\sqrt[3]{8} = 2$

(d) $\sqrt[3]{-8} = -2$

(e) $\sqrt[3]{\frac{27}{8}} = \frac{\sqrt[3]{27}}{\sqrt[3]{8}} = \frac{3}{2}$ ■

square roots of negative numbers are not defined

since $(-2)^3 = -8$

There are *two* square roots of 9, namely 3 and -3, but the radical sign $\sqrt{}$ means just the *positive* one (called the *principal* square root). In general,

$$\sqrt[n]{a} \text{ means the principal } n\text{th root of } a.$$

where the word "principal" means that if there is a positive and a negative root, take the positive one. We may take *odd* roots of *any* number, but *even* roots only if the number is positive or zero.

Example 7

(a) $\sqrt[4]{81} = 3$

(b) $\sqrt[5]{-32} = -2$

since $3^4 = 81$ and $(-2)^5 = -32$ ■

Fractional Exponents

Fractional exponents are defined as follows.

$$x^{1/2} = \sqrt{x}$$
$$x^{1/3} = \sqrt[3]{x}$$

the power 1/2 means the principal square root
the power 1/3 means the cube root

In general, for any positive integer n,

$$x^{1/n} = \sqrt[n]{x}$$

the power 1/n means the principal nth root

Example 8

(a) $9^{1/2} = \sqrt{9} = 3$

(b) $\left(\dfrac{4}{9}\right)^{1/2} = \sqrt{\dfrac{4}{9}} = \dfrac{\sqrt{4}}{\sqrt{9}} = \dfrac{2}{3}$

(c) $8^{1/3} = \sqrt[3]{8} = 2$

(d) $(-125)^{1/3} = \sqrt[3]{-125} = -5$

(e) $\left(\dfrac{8}{27}\right)^{1/3} = \sqrt[3]{\dfrac{8}{27}} = \dfrac{\sqrt[3]{8}}{\sqrt[3]{27}} = \dfrac{2}{3}$

(f) $81^{1/4} = \sqrt[4]{81} = 3$

(g) $(-32)^{1/5} = \sqrt[5]{-32} = -2$ ■

Practice Exercise 4

Evaluate

(a) $(-27)^{1/3}$

(b) $\left(\dfrac{16}{81}\right)^{1/4}$

(solutions on page 11)

The definition of $x^{1/2}$ is motivated by the multiplication law of exponents:

$$(x^{1/2})^2 = x^1 = x$$

and taking square roots gives

$$x^{1/2} = \sqrt{x}$$

More General Fractional Exponents

To define $x^{m/n}$ for positive integers m and n, the exponent m/n must be fully reduced (for example, 4/6 must be reduced to 2/3).   Then

$$x^{m/n} = (x^{1/n})^m = (x^m)^{1/n}$$

leading to the definition

$$x^{m/n} = (\sqrt[n]{x})^m = \sqrt[n]{x^m}$$

$x^{m/n}$ means the mth power of the nth root, or the nth root of the mth power

Both expressions $(\sqrt[n]{x})^m$ and $\sqrt[n]{x^m}$ will give the same answer. In either case, the numerator determines the power and the denominator determines the root.

$$x^{m/n}$$

power/root

Example 9

(a) $8^{2/3} = \sqrt[3]{8^2} = \sqrt[3]{64} = 4$ *the power and then the root*

(b) $8^{2/3} = (\sqrt[3]{8})^2 = (2)^2 = 4$ *the root and then the power*

(c) $25^{3/2} = (\sqrt{25})^3 = 5^3 = 125$

(d) $\left(\dfrac{-27}{8}\right)^{2/3} = \left(\sqrt[3]{\dfrac{-27}{8}}\right)^2 = \left(\dfrac{-3}{2}\right)^2 = \dfrac{9}{4}$ ▪

Practice Exercise 5

Evaluate

(a) $16^{3/2}$

(b) $\left(\dfrac{8}{27}\right)^{2/3}$

(solutions on page 12)

Negative Fractional Exponents

A negative fractional exponent, $x^{-m/n}$, means the reciprocal of the number to the positive exponent:

$$x^{-m/n} = \frac{1}{x^{m/n}}$$

Example 10

$$8^{-2/3} = \frac{1}{8^{2/3}} = \frac{1}{(\sqrt[3]{8})^2} = \frac{1}{2^2} = \frac{1}{4}$$ ▪

Practice Exercise 6

Evaluate $25^{-3/2}$.   *(solution on page 12)*

As before, a negative exponent inverts a fraction:

$$\left(\frac{a}{b}\right)^{-m/n} = \left(\frac{b}{a}\right)^{m/n}$$

Example 11

$$\left(\frac{9}{4}\right)^{-3/2} = \left(\frac{4}{9}\right)^{3/2} = \left(\sqrt{\frac{4}{9}}\right)^3 = \left(\frac{2}{3}\right)^3 = \frac{8}{27}$$ ▪

Decimal Exponents

For a *decimal* exponent, such as $x^{.7}$, we express the decimal as a fraction.

Example 12

(a) $x^{.7} = x^{7/10} = \sqrt[10]{x^7}$ (b) $x^{.2} = x^{2/10} = x^{1/5} = \sqrt[5]{x}$ ▪
$$\hspace{6.5cm} \underset{\text{reduced!}}{\uparrow}$$

Evaluating such powers, of course, is most easily done using a calculator.

Practice Exercise 7

Use a calculator with a $\boxed{y^x}$ key to evaluate $5^{1.3}$. (*Hint*: For most calculators, press 5, then $\boxed{y^x}$, then 1.3, then $\boxed{=}$.)   *(solution on page 12)*

A Warning About Roots and Powers

The square root of a sum is *not* equal to the sum of the square roots,

$$\sqrt{a + b} \quad \text{is } not \text{ equal to} \quad \sqrt{a} + \sqrt{b}$$

For example,

$$\underbrace{\sqrt{9 + 16}}_{\sqrt{25}} \neq \underbrace{\sqrt{9}}_{3} + \underbrace{\sqrt{16}}_{4}$$

and the two sides, 5 and 7, are *not* the same. (The corresponding property for products *is* true: $\sqrt{ab} = \sqrt{a} \cdot \sqrt{b}$, but not for sums or differences.) Therefore, do not "simplify" $\sqrt{x^2 + 9}$ into $x + 3$. The expression $\sqrt{x^2 + 9}$ cannot be simplified. To state this warning in terms of fractional exponents,

$$(a + b)^{1/2} \quad \text{is } not \text{ equal to} \quad a^{1/2} + b^{1/2}$$

The same warning holds for all exponents. For example,

$$(x + y)^2 \quad \text{is } not \text{ equal to} \quad x^2 + y^2$$

The expression $(x + y)^2$ means $(x + y)$ times itself,

$$(x + y)^2 = (x + y)(x + y) = x^2 + xy + yx + y^2 = x^2 + 2xy + y^2$$

and the middle term, $2xy$, must not be forgotten. This result is worth remembering, since we will use it frequently.

$$(x + y)^2 = x^2 + 2xy + y^2$$

$(x + y)^2$ is the first number squared plus twice the product of the numbers plus the second number squared

Application of Fractional Exponents: Allometry

The study of size and shape is called *allometry*. Many laws of allometry involve fractional exponents. For example, among all four-legged animals, from mice to elephants, (average) leg width and body length are governed (approximately) by the law

$$(\text{leg width}) = (\text{a constant})(\text{body length})^{3/2}$$

The constant depends on the units (inches, centimeters, etc.), but the exponent is always 3/2. This relationship is an example of what is called the *law of simple allometry*, which says that two different measurements x and y will be related by a *power law*,

$$y = bx^a$$

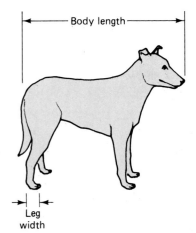

with constants *a* and *b*. If animals grew stricly in proportion, leg width and body length would obey the law

$$\text{(leg width)} = \text{(a constant)}\text{(body length)}^1$$

with exponent 1. The exponent 3/2 in the actual relationship means that for larger animals, leg width increases *faster* than body length. This exponent 3/2 is the reason nature cannot build a land animal very much larger than an elephant—its oversized legs would get in each other's way. (Some dinosaurs were larger, but they spent much of their time in water.*)

Another application of fractional exponents applies to maps and illustrations. Studies have shown that when people are asked to estimate the size of objects, the perceived size and the actual size are related by a power law

$$\text{(perceived size)} = \text{(a constant)}\text{(actual size)}^a$$

with exponent $a < 1$. The value of the exponent a depends on whether the object is one-, two-, or three-dimensional, but the exponent is always less than 1. This means that if you were using a geometric symbol on a map to indicate rainfall, to suggest that one country's rainfall was twice as great as another's, you should use a symbol *more* than twice as large.

Application: The Learning Curve in Airplane Production

It is a truism that the more you practice a task, the faster you can do it. Successive repetitions generally take less time, following a *learning curve* like that shown on the right. Learning curves are widely used in industrial production. For example, it took 150,000 hours to build the first Boeing 707 airliner, whereas later planes ($n = 2, 3, \ldots, 300$) took less time:

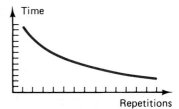

$$\text{(time to build plane number } n) = 150n^{-.322} \text{ thousand hours}$$

The time for the tenth Boeing 707 is found by substituting $n = 10$,

$$\text{(time to build plane 10)} = 150(10)^{-.322} \approx 71.46 \text{ thousand hours}$$

(the $\approx$ sign means "is approximately equal to"). This shows that building the tenth Boeing 707 took about 71,460 hours, which is less than half of the 150,000 hours needed for the first. For the 100th Boeing 707 we substitute $n = 100$,

$$\text{(time to build plane 100)} = 150(100)^{-.322} \approx 34.05 \text{ thousand hours}$$

or about 34,050 hours, which is less than half the time needed to build the

---

* For further information on allometry, see D'Arcy Wentworth Thompson, *On Growth and Form* (Cambridge University Press, Cambridge, 1942); Stephen Jay Gould, "Allometry and Size in Ontogeny and Philogeny," *Biol. Rev.*, **41** (1966), 587–640.

tenth plane. Such learning curves are found by curve-fitting techniques (discussed in Chapter 7), and are used for estimating the cost of building several planes.*

Summary

We defined zero, negative, and fractional exponents as follows:

$$x^0 = 1$$ *for x ≠ 0*

$$x^{-n} = \frac{1}{x^n}$$ *for x ≠ 0*

$$x^{m/n} = \sqrt[n]{x^m} = (\sqrt[n]{x})^m$$ *for m/n fully reduced, m > 0, n > 0*

With these definitions, the following laws of exponents hold for *all* exponents, whether integral or fractional, positive or negative.

$$x^m \cdot x^n = x^{m+n} \qquad (x \cdot y)^m = x^m \cdot y^m$$

$$\frac{x^m}{x^n} = x^{m-n} \qquad \left(\frac{x}{y}\right)^m = \frac{x^m}{y^m}$$

$$(x^m)^n = x^{m \cdot n}$$

SOLUTIONS TO PRACTICE EXERCISES

1 **(a)** $\dfrac{x^5 x}{x^2} = \dfrac{x^6}{x^2} = x^4$ **(b)** $(x^3)^2)^2 = x^{3 \cdot 2 \cdot 2} = x^{12}$

2 **(a)** $2^0 = 1$ **(b)** $2^{-4} = \dfrac{1}{2^4} = \dfrac{1}{16}$

3 $\left(\dfrac{2}{3}\right)^{-2} = \left(\dfrac{3}{2}\right)^2 = \dfrac{9}{4}$

4 **(a)** $(-27)^{1/3} = \sqrt[3]{-27} = -3$

(b) $\left(\dfrac{16}{81}\right)^{1/4} = \sqrt[4]{\dfrac{16}{81}} = \dfrac{2}{3}$

---

* For more information on learning curves in industrial production, see J. M. Dutton et al., ''The History of Progress Functions as a Manageral Technology,'' *Bus. Hist. Rev.*, 58:204–33, Sum. 84.

5 (a) $16^{3/2} = (\sqrt{16})^3 = 4^3 = 64$

   (b) $\left(\dfrac{8}{27}\right)^{2/3} = \left(\sqrt[3]{\dfrac{8}{27}}\right)^2 = \left(\dfrac{2}{3}\right)^2 = \dfrac{4}{9}$

6 $25^{-3/2} = \dfrac{1}{25^{3/2}} = \dfrac{1}{(\sqrt{25})^3} = \dfrac{1}{(5)^3} = \dfrac{1}{125}$

7 $5^{1.3} \approx 8.10$ (the $\approx$ sign means "is approximately equal to")

EXERCISES 1.1

*Evaluate each expression **without** using a calculator.*

1 $(2^2 \cdot 2)^2$ **2** $(5^2 \cdot 4)^2$ **3** 2^{-4} **4** 3^{-3}

5 $\left(\dfrac{1}{2}\right)^{-3}$ **6** $\left(\dfrac{1}{3}\right)^{-2}$ **7** $\left(\dfrac{5}{8}\right)^{-1}$ **8** $\left(\dfrac{3}{4}\right)^{-1}$

9 $4^{-2} \cdot 2^{-1}$ **10** $3^{-2} \cdot 9^{-1}$ **11** $\left(\dfrac{3}{2}\right)^{-3}$ **12** $\left(\dfrac{2}{3}\right)^{-3}$

13 $\left(\dfrac{1}{3}\right)^{-2} - \left(\dfrac{1}{2}\right)^{-3}$ **14** $\left(\dfrac{1}{3}\right)^{-2} - \left(\dfrac{1}{2}\right)^{-2}$ **15** $\left(\left(\dfrac{2}{3}\right)^{-2}\right)^{-1}$ **16** $\left(\left(\dfrac{2}{5}\right)^{-2}\right)^{-1}$

17 $25^{1/2}$ **18** $36^{1/2}$ **19** $25^{3/2}$ **20** $16^{3/2}$

21 $16^{3/4}$ **22** $27^{2/3}$ **23** $(-8)^{2/3}$ **24** $(-27)^{2/3}$

25 $(-8)^{5/3}$ **26** $(-27)^{5/3}$ **27** $\left(\dfrac{25}{36}\right)^{3/2}$ **28** $\left(\dfrac{16}{25}\right)^{3/2}$

29 $\left(\dfrac{27}{125}\right)^{2/3}$ **30** $\left(\dfrac{125}{8}\right)^{2/3}$ **31** $\left(\dfrac{1}{32}\right)^{2/5}$ **32** $\left(\dfrac{1}{32}\right)^{3/5}$

33 $4^{-1/2}$ **34** $9^{-1/2}$ **35** $4^{-3/2}$ **36** $9^{-3/2}$

37 $8^{-2/3}$ **38** $16^{-3/4}$ **39** $(-8)^{-1/3}$ **40** $(-27)^{-1/3}$

41 $(-8)^{-2/3}$ **42** $(-27)^{-2/3}$ **43** $\left(\dfrac{25}{16}\right)^{-1/2}$ **44** $\left(\dfrac{16}{9}\right)^{-1/2}$

45 $\left(\dfrac{25}{16}\right)^{-3/2}$ **46** $\left(\dfrac{16}{9}\right)^{-3/2}$ **47** $\left(\dfrac{-1}{27}\right)^{-5/3}$ **48** $\left(\dfrac{-1}{8}\right)^{-5/3}$

Use a calculator with a $\boxed{y^x}$ key to evaluate each expression. Round answers to two decimal places.

49 $7^{.39}$ **50** $5^{.47}$ **51** $8^{2.7}$ **52** $5^{3.9}$

Simplify.

53 $(x^3 \cdot x^2)^2$ **54** $(x^4 \cdot x^3)^2$ **55** $(z^2(z \cdot z^2)^2 z)^3$ **56** $(z(z^3 \cdot z)^2 z^2)^2$

57 $((x^2)^2)^2$ **58** $((x^3)^3)^3$ **59** $\dfrac{(ww^2)^3}{w^3 w}$ **60** $\dfrac{(ww^3)^2}{w^3 w^2}$

61 $\dfrac{(5xy^4)^2}{25x^3y^3}$

62 $\dfrac{(4x^3y)^2}{8x^2y^3}$

63 $\dfrac{(9xy^3z)^2}{3(xyz)^2}$

64 $\dfrac{(5x^2y^3z)^2}{5(xyz)^2}$

65 $\dfrac{(2u^2vw^3)^2}{4(uw^2)^2}$

66 $\dfrac{(u^3vw^2)^2}{9(u^2w)^2}$

Express without using negative exponents.

67 $(x^2 + 3)^{-1}$

68 $(x^3 + 2)^{-1}$

69 $(x + 1)^{-5}$

70 $(x^2 + 2)^{-4}$

Express in simplest form using negative exponents.

71 $\dfrac{1}{(x^2 + 4)^3}$

72 $\dfrac{1}{(x^2 + 2)^5}$

73 $\dfrac{1}{(x^2 + 1)^2(x^2 + 5)^3}$

74 $\dfrac{1}{(x + 1)^3(x + 2)^2}$

Express in simplest form using fractional exponents instead of radicals.

75 $\sqrt{x^2 + 1}$

76 $\sqrt[3]{(x + 4}$

77 $\sqrt[3]{(x + 3)^2}$

78 $\sqrt{(x + 1)^3}$

Express using radicals instead of fractional exponents

79 $(x^3 + 1)^{1/4}$

80 $(x^2 + 1)^{1/3}$

81 $(2x + 3)^{5/4}$

82 $(3x - 4)^{7/3}$

Express in simplest form using negative fractional exponents.

83 $\dfrac{1}{\sqrt[3]{x + 1}}$

84 $\dfrac{1}{\sqrt[5]{x^4 - 1}}$

85 $\dfrac{1}{\sqrt[5]{(3x + 2)^7}}$

86 $\dfrac{1}{\sqrt[3]{(2x + 1)^2}}$

Express in simplest form using radicals instead of fractional exponents.

87 $(x^3 + 1)^{-1/4}$

88 $(x^2 + 1)^{-1/3}$

89 $(x + 1)^{-3/4}$

90 $(x + 5)^{-7/2}$

■   **APPLIED EXERCISES**

▥ **91–92** (*Allometry*) Most four-legged animals, from mice to elephants, obey (approximately) the following power law:

$$\left(\begin{array}{c}\text{average body}\\ \text{thickness}\end{array}\right) = .4 \text{ (hip-to-shoulder length)}^{3/2}$$

where body thickness is measured vertically and all measurements are in feet. Assuming that this same relationship held for dinosaurs, find the average body thickness of the following dinosaurs, whose hip-to-shoulder length can be measured from their skeletons.

91 Diplodocus, whose hip-to-shoulder length was 16 feet

92 Triceratops, whose hip-to-shoulder length was 14 feet

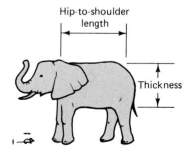

Hip-to-shoulder length

Thickness

93–94 (*Business—The Rule of .6 in Industrial Production*) Many chemical and refining companies use the *rule of point six* to estimate the cost of new equipment. According to this rule, if a piece of equipment (such as a storage tank) originally cost C dollars, then the cost of similar equipment which is x times as large will be approximately $x^{.6}C$ dollars. For example, if the original equipment cost C dollars, new equipment with twice ($x = 2$) the capacity of the old equipment would cost $2^{.6}C \approx 1.516C$ dollars, that is, about 1.5 times as much. Therefore, to increase capacity by 100% costs only about 50% more. Although the rule of .6 is only a rough "rule of thumb," it can be somewhat justified on the basis that the equipment of such industries consists mainly of containers, and the cost of a container depends on its surface area (square units), which increases more slowly than its capacity (cubic units).

93 Use the rule of .6 to find how costs change if a company wants to triple ($x = 3$) its capacity.

94 Use the rule of .6 to find how costs change if a company wants to quadruple ($x = 4$) its capacity.

95–96 (*Biomedical—Heart Rate*) It is well known that the hearts of smaller animals beat faster than the hearts of larger animals. The actual relationship is approximately

$$(\text{heart rate}) = 250\,(\text{weight})^{-1/4}$$

where the heart rate is in beats per minute and the weight is in pounds. Use this relationship to estimate the heart rate of

95 a 16-pound dog

96 a 625-pound grizzly bear

97–98 (*Business and Psychology—Learning Curves in Airplane Production*) Recall that the learning curve for the production of Boeing 707 airplanes is $150x^{-.322}$ (in thousands of hours). Find how many hours it took to build

97 the 50th Boeing 707

98 the 250th Boeing 707

1.2 FUNCTIONS

Introduction

In Section 1.1 we saw that the time required to build a Boeing 707 airplane depends upon the number that have already been built. Such mathematical relationships, in which one quantity depends upon another, are called *functions*. In this section we define and give some examples of functions. First some preliminaries.

Inequalities

In the following inequalities, the letters a and b stand for any (real) numbers.

The Inequality:	Is Read:
$x < a$	x is less than a
$x \le a$	x is less than or equal to a
$x > a$	x is greater than a
$x \ge a$	x is greater than or equal to a

A *double inequality*, such as $a \le x < b$, means that *both* inequalities $a \le x$ and $x < b$ hold.

Sets

Braces { } are read "the set of all," and a vertical bar | is read "such that."

Example 1

(a) $\{x \mid x > 3\}$ means "the set of all x such that x is greater than 3."

the set such
of all that

(b) $\{x \mid x \le -5\}$ means "the set of all x such that x is less than or equal to -5."   ■

Practice Exercise 1

(a) Write in set notation: the set of all x such that x is less than 7.

(b) Express in words: $\{x \mid x \ge -1\}$.   *(solutions on page 26)*

The symbol $\mathbb{R}$ denotes the set of *all* (real) numbers.

Functions

A function of a variable x is a rule f that assigns to each value of x a unique number $f(x)$ (read "f of x"). The number $f(x)$ is called *the value of the function at x*. A function is often defined by an algebraic formula for calculating $f(x)$ from x.

Example 2   For the function defined by $f(x) = 2x^2 + 4x - 5$, find $f(3)$ and $f(-3)$.

Solution $f(3)$ is simply $f(x) = 2x^2 + 4x - 5$ with each occurrence of x replaced by 3,

$$f(3) = 2(3)^2 + 4 \cdot 3 - 5 = 18 + 12 - 5 = 25$$

$$f(-3) = 2(-3)^2 + 4 \cdot (-3) - 5 = 18 - 12 - 5 = 1$$

$f(x) = x^2 + 4x - 5$ with each x replaced by 3

$f(x) = x^2 + 4x - 5$ with each x replaced by -3

■

A function f may be thought of as a numerical "machine" that takes an "input" number x and produced an "output" number $f(x)$.

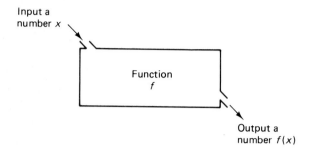

Any letters may be used for defining functions.

Example 3   For $g(z) = \sqrt{z - 1}$, find $g(10)$.

Solution

$$g(10) = \sqrt{10 - 1} = \sqrt{9} = 3 \qquad\qquad \text{\textit{g(z)} = √z − 1 with z replaced by 10}$$

■

Example 4   For $h(w) = w^{1/3}$, find $h(-8)$.

Solution

$$h(-8) = (-8)^{1/3} = \sqrt[3]{-8} = -2 \qquad\qquad \text{\textit{since } w^{1/3} = \sqrt[3]{w}}$$

■

   The *domain* of a function f is the set of all numbers for which the function is defined. Formally,

> A function f is a rule that assigns to each number x in the domain of the function one and only one number $f(x)$.

If the domain of a function is not stated, then the domain is the *largest* set of values of the variable for which the function is defined (sometimes called the *natural domain* of the function).

Example 5

(a)   The domain of $f(x) = \dfrac{1}{x}$ is $\{x \mid x \neq 0\}$.   *all numbers except 0 (since 1/x is not defined at x = 0)*

(b)   The domain of $g(z) = \sqrt{z - 1}$ is $\{z \mid z \geq 1\}$.   *since we cannot take square roots of negative numbers*

(c)   The domain of $h(w) = w^{1/3}$ is $\mathbb{R}$.   *since cube roots are defined for all numbers*

■

Practice Exercise 2

Find the domain of

(a) $f(x) = \dfrac{1}{x + 2}$

(b) $g(y) = \sqrt{y - 4}$

(solutions on page 27)

Graphs of Functions

To graph a function f we set $y = f(x)$ and plot points (x, y) for values of x in the domain. However, not every graph is the graph of a *function*. A function f must have a *unique* value $y = f(x)$ for each x in the domain, so its graph cannot have two y-coordinates (two distinct points) with the same x-coordinate. This leads to the following condition for a curve to be the graph of a function:

> **Vertical Line Test**
>
> A curve in the x–y plane is the graph of a *function* if and only if no vertical line intersects the curve at more than one point.

Example 6

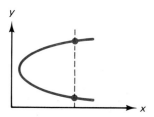

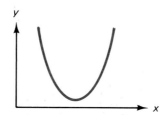

This is <u>not</u> the graph of a function of **x**, because there is a vertical line (shown dashed) that intersects the curve at more than one point.

This <u>is</u> the graph of a function of **x**, because no vertical line intersects the curve at more than one point. ▪

In this book we will be concerned almost exclusively with functions, so we will often use the terms "function," "graph," and "curve" interchangeably.

Straight Lines

The symbol Δ (read "delta," the Greek letter D) means "the change in." For two points (x_1, y_1) and (x_2, y_2) we define

$$\Delta y = y_2 - y_1$$

the change in y

and

$$\Delta x = x_2 - x_1$$

the change in x

The *slope, m,* of a nonvertical line is defined as *the change in y over the change in x* for any two points on the line:

$$m = \frac{\Delta y}{\Delta x} = \frac{y_2 - y_1}{x_2 - x_1}$$

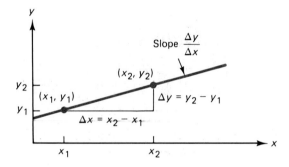

Any nonvertical line can be written in the form

$$y = mx + b$$

where *m* is the slope of the line and *b* is the *y*-intercept (where the line crosses the *y*-axis).

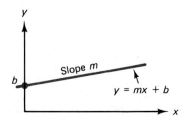

A line with *positive* slope *rises* as you move to the right, a line with *negative* slope *falls* as you move to the right, and a line with *zero* slope remains level (a horizontal line).

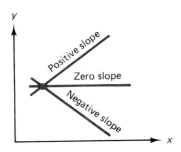

A *vertical* line can be written in the form

$$x = a$$

where a is a constant. A vertical line has no slope (its slope is undefined).

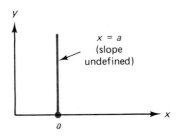

Linear Functions

We may write the line $y = mx + b$ in function notation:

$f(x) = mx + b$ is called a *linear function*. Its graph is a straight line with slope m and y-intercept b.

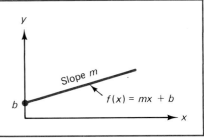

Example 7   Graph the linear function $f(x) = -4x + 3$.

Solution   Since the function is linear, only two points are needed to graph the line. We may choose *any* two x-values, calculating the y-values from $y = f(x)$:

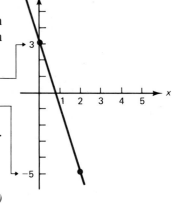

Choosing $x = 0$: $f(0) = -4\cdot 0 + 3 = 3$ point: $(0, 3)$

Choosing $x = 2$: $f(2) = -4\cdot 2 + 3 = -5$ point: $(2, -5)$

The line through these two points is the graph of the function $f(x) = -4x + 3$.

Practice Exercise 3

Graph the function $f(x) = 2x - 6$.   *(solution on page 27)*

Many important relationships are linear.

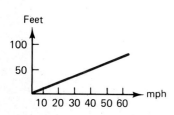

The stopping time for a car traveling at speed x.

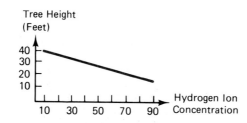

Average tree height versus sulphur dioxide pollution.

Example 8   An electronics company manufactures pocket calculators at a cost of $9 each, and the company's fixed costs (such as rent) amount to $500 per week. Find a function $C(x)$ that gives the total cost of producing x pocket calculators per week.

Solution   Each calculator costs $9 to produce, so x calculators will cost $9x$ dollars, to which we must add the fixed costs of 500, giving

$$C(x) = 9x + 500$$

total cost | unit cost | number of units | fixed cost

Notice that the *slope* of the cost function $C(x) = 9x + 500$ is 9, which is also the *rate of change* of the cost as production increases (that is, costs increase at the rate of $9 per additional calculator). This same 9 is also the company's *marginal cost* (the cost of producing one more calculator). That is, the *slope,* the *rate of change,* and the *marginal* cost are all identical. Calculus will show that for *any* function, these three ideas are the same.

Mathematical Models

A mathematical description of a real-world situation is called a *mathematical model*. For example, the cost function $C(x) = 9x + 500$ is a mathematical model for the cost of manufacturing calculators. In some models, variables may take on fractional values (half a ton of grain), while in others (pocket calculators) it makes no sense to speak of half a unit. In the latter case, we should restrict the variable to integer values ($x = 1, 2, 3, \ldots$), and our graphs should consist of isolated dots rather than continuous curves. Instead, however, we will take the easier course of allowing the variables in our mathematical models to take *any* value, and then round up or down as necessary at the end.

Quadratic Functions

For constants a, b, and c, with $a \neq 0$,

$f(x) = ax^2 + bx + c$ is called a *quadratic* function. Its graph is a *parabola*.

(The condition $a \neq 0$ keeps the function from becoming $f(x) = bx + c$, which would be linear.) Quadratic functions are sometimes called *quadratics*.

Many familiar curves are parabolas.

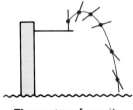

The center of gravity of a diver describes a parabola.

A stream of water from a hose takes the shape of a parabola.

The parabola $f(x) = ax^2 + bx + c$ opens *upward* if the constant a is *positive,* and *downward* if the constant a is *negative.*

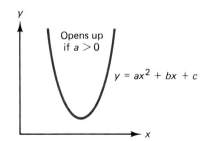

Opens up if $a > 0$ $y = ax^2 + bx + c$

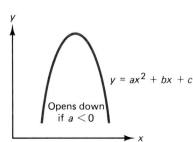

$y = ax^2 + bx + c$ Opens down if $a < 0$

The *vertex* of a parabola is the lowest point if the parabola opens upward, and the highest point if the parabola opens downward. There is a formula for the *x*-coordinate of the vertex of a parabola:

For the parabola $f(x) = ax^2 + bx + c$, the *x*-coordinate of the vertex is $x = \dfrac{-b}{2a}$.

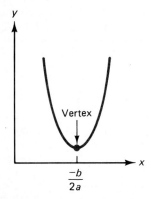 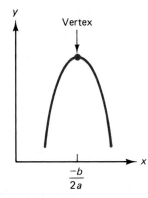

We will not prove this formula here, since in Chapter 2 we can give a very simple derivation using calculus.

We graph a parabola by plotting the vertex and one or more points on either side.

Example 9 Graph $f(x) = -2x^2 + 12x - 3$.

Solution The vertex formula $x = \dfrac{-b}{2a}$ gives

$$x = \frac{-b}{2a} = \frac{-12}{2(-2)} = \frac{-12}{-4} = 3$$

Therefore, the vertex is at $x = 3$. We choose two *x*-values on either side of $x = 3$, calculating the *y*-values from $y = f(x)$:

$x = 0$: $y = f(0) = -2(0)^2 + 12(0) - 3 = -3$ point: $(0, -3)$

$x = 2$: $y = f(2) = -2(2)^2 + 12(2) - 3 = 13$ point: $(2, 13)$

$x = 3$: $y = f(3) = -2(3)^2 + 12(3) - 3 = 15$ point: $(3, 15)$

$x = 4$: $y = f(4) = -2(4)^2 + 12(4) - 3 = 13$ point: $(4, 13)$

$x = 5$: $y = f(5) = -2(5)^2 + 12(5) - 3 = 7$ point: $(5, 7)$

Plotting these points and drawing a smooth curve through them gives the graph shown on the right.

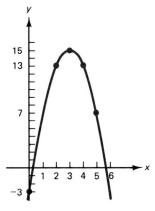

The graph of
$f(x) = -2x^2 + 12x - 3$ ■

A quadratic function such as this might be used to model daily sales for a product. Sales increase while the product is being advertised, then level off, and finally decrease after the advertising has stopped. The vertex formula $x = -b/(2a)$ gives the day on which sales were greatest.

Factoring Quadratic Functions

To *factor* an expression means to write it as a product. To see how to factor a quadratic, we multiply out $(x + d)(x + e)$, obtaining

$$(x + d)(x + e) = x^2 + \underbrace{(d + e)}x + \underbrace{d \cdot e}$$

$$\text{\textit{sum} of} \qquad \text{\textit{product} of}$$
$$d \text{ and } e \qquad d \text{ and } e$$

Therefore, to reverse the process and *factor* the quadratic $x^2 + bx + c$, we want to find two numbers (for the boxes below) that add to b and multiply to c:

$$x^2 + bx + c = (x + \square)(x + \square)$$

product / sum

Example 10   Factor $x^2 + 8x + 12$.

Solution   Trying various factors of 12, only 2 and 6 add up to 8, so

$$x^2 + 8x + 12 = (x + 2)(x + 6)$$

product / sum

as may be checked by multiplication

■

Example 11   Factor $x^2 - 9x + 18$.

Solution   We want numbers whose *product is 18* and whose *sum is* -9. Both numbers must have the same sign (since the product is positive), and at least one must be negative (since the sum is negative), therefore *both* must be negative. Trying the (negative) factors of 18, only -3 and -6 add to -9, so

$$x^2 - 9x + 18 = (x - 3)(x - 6)$$ ■

Example 12   Factor $x^2 - 4x - 12$.

Solution   We want numbers whose *product is* -12 (so they must have opposite signs) and whose *sum is* -4. Trying the factors of -12, only -6 and 2 add to -4, so

$$x^2 - 4x - 12 = (x - 6)(x + 2)$$ ■

Practice Exercise 4

Factor $x^2 - 2x - 8$. *(solution on page 27)*

There is a useful formula for factoring the difference of two squares:

$$x^2 - a^2 = (x + a)(x - a)$$

which may be checked by multiplying out the right-hand side.

Example 13 Factor $x^2 - 16$.

Solution

$$x^2 - 16 = (x + 4)(x - 4)$$ *since $16 = 4^2$*

■

Solving Quadratic Equations

We will use factoring to solve equations. The procedure is as follows:

(i) Express the equation with zero on the right-hand side.

(ii) Factor the expression on the left-hand side.

(iii) The solutions are the values that make one or more of the factors equal zero.

Example 14 Solve $x^2 - 2x - 3 = 0$.

Solution

$$x^2 - 2x - 3 = 0$$ *the equation already has zero on the right*

$$(x - 3)\,(x + 1) = 0$$ *factoring*

$\underbrace{}$ $\underbrace{}$ *finding x-values that make each factor zero*
equals 0   equals 0
at $x = 3$ at $x = -1$

Therefore, the solutions are $x = 3$ and $x = -1$ (as may be checked by substitution into the original equation). Geometrically, these are the numbers at which the curve crosses the x-axis (see graph on right).

For the next example, remember the first rule of factoring: **take out common factors.**

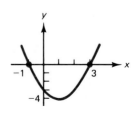

$f(x) = x^2 - 2x - 3$

■

Example 15   Solve $2x^2 + 4x = 48$.

Solution

$$2x^2 + 4x - 48 = 0$$

subtracting 48 to get zero on the right

$$2(x^2 + 2x - 24) = 0$$

taking out the common factor 2

$$2(x + 6)(x - 4) = 0$$

factoring $x^2 + 2x - 24$

equals 0   equals 0
at $x = -6$ at $x = 4$

finding x-values that make each factor zero

Therefore, the solutions are $x = -6$ and $x = 4$.   ■

Example 16   Solve $9x - 3x^2 = -30$.

Solution

$$-3x^2 + 9x + 30 = 0$$

reordering, with zero on the right

$$-3(x^2 - 3x - 10) = 0$$

taking out the common factor -3

$$3(x - 5)(x + 2) = 0$$

factoring $x^2 + 3x - 10$

equals 0   equals 0
at $x = 5$ at $x = -2$

finding x-values that make each factor zero

Therefore, the solutions are $x = 5$ and $x = -2$.   ■

Practice Exercise 5

Solve $2x^2 - 2x = 40$.

(solution on page 27)

Quadratic equations can also be solved by the *quadratic formula,*

Quadratic Formula

The solutions to $ax^2 + bx + c = 0$ are

$$x = \frac{-b \pm \sqrt{b^2 - 4ac}}{2a}$$

The sign ± ("plus or minus") means that there may be two solutions, one from the + sign and one from the − sign.

Example 17   Solve $x^2 - 2x - 3 = 0$.   *same as example 14*

Solution   The quadratic formula with $a = 1$, $b = -2$, and $c = -3$ gives

$$x = \frac{2 \pm \sqrt{4 - 4(1)(-3)}}{2(1)} = \frac{2 \pm \sqrt{4 + 12}}{2} = \frac{2 \pm \sqrt{16}}{2}$$

$$= \frac{2 \pm 4}{2} = \frac{6}{2} \text{ or } \frac{-2}{2} = 3 \text{ or } -1 \qquad ■$$

Example 18   Solve $\frac{1}{2}x^2 - 3x + 5 = 0$.

Solution   Using the quadratic formula with $a = \frac{1}{2}$, $b = -3$, and $c = 5$,

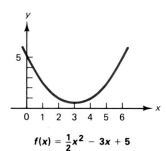

$$x = \frac{3 \pm \sqrt{9 - 4(\frac{1}{2})(5)}}{2(\frac{1}{2})} = \frac{3 \pm \sqrt{9 - 10}}{1} = 3 \pm \sqrt{-1}$$

The square root of -1 is undefined, so there are no solutions to the equation. The reason for this can be seen from the graph on the right: The curve never touches the x-axis.

$f(x) = \frac{1}{2}x^2 - 3x + 5$ ■

   It is usually easier to solve quadratic equations by factoring, using the quadratic formula only if you cannot find the factors.

Derivation of the Quadratic Formula

$$ax^2 + bx + c = 0 \qquad \text{\textit{setting the quadratic equal to zero}}$$

$$ax^2 + bx = -c \qquad \text{\textit{subtracting c}}$$

$$4a^2x^2 + 4abx = -4ac \qquad \text{\textit{multiplying by 4a}}$$

$$4a^2x^2 + 4abx + b^2 = b^2 - 4ac \qquad \text{\textit{adding } b^2}$$

$$(2ax + b)^2 = b^2 - 4ac \qquad \text{\textit{since } 4a^2x^2 + 4abx + b^2 = (2ax + b)^2}$$

$$2ax + b = \pm \sqrt{b^2 - 4ac} \qquad \text{\textit{taking square roots}}$$

$$2ax = -b \pm \sqrt{b^2 - 4ac} \qquad \text{\textit{subtracting b}}$$

$$x = \frac{-b \pm \sqrt{b^2 - 4ac}}{2a} \qquad \text{\textit{dividing by 2a gives the quadratic formula}}$$

SOLUTIONS TO PRACTICE EXERCISES

1. **(a)** $\{x \mid x < 7\}$

    (b) The set of all x such that x is greater than or equal to -1

2. **(a)** $\{x \mid x \neq -2\}$

    (b) $\{y \mid y \geq 4\}$

3.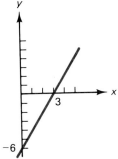

4. $(x - 4)(x + 2)$

5. $2x^2 - 2x - 40 = 0$
$2(x^2 - x - 20) = 0$
$2(x - 5)(x + 4) = 0$
$x = 5$ or $x = -4$

■ EXERCISES 1.2

Find the domain of the function and evaluate the given expression.

1 $g(w) = \sqrt{w - 1}$ and find $g(10)$

2 $g(w) = \sqrt{w - 4}$ and find $g(40)$

3 $h(z) = \dfrac{1}{z + 4}$ and find $h(-5)$

4 $h(z) = \dfrac{1}{z + 7}$ and find $h(-8)$

5 $h(x) = x^{1/4}$ and find $h(81)$

6 $h(x) = x^{1/6}$ and find $h(64)$

7 $h(x) = x^{-1/2}$ and find $h(81)$

8 $h(x) = x^{-2/5}$ and find $h(32)$

9 $h(x) = x^{-4/3}$ and find $h(27)$

10 $h(x) = x^{-5/4}$ and find $h(81)$

Determine whether each graph is the graph of a function of x.

11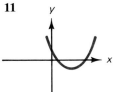

12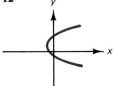

13

14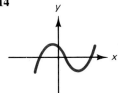

15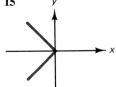

16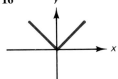

17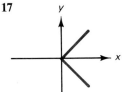

18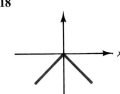

Graph each function.

19 $f(x) = 6x - 2$

20 $f(x) = 8x - 4$

21 $f(x) = -x + 1$

22 $f(x) = -4x + 5$

23 $f(x) = 2x^2 + 4x - 16$

24 $f(x) = 3x^2 - 6x - 9$

25 $f(x) = -3x^2 + 6x + 9$

26 $f(x) = -2x^2 + 4x + 16$

Factor each expression.

27 $x^2 + 7x + 12$

28 $x^2 + 9x + 20$

29 $x^2 - 6x + 8$

30 $x^2 - 5x + 4$

31 $x^2 - 2x - 15$

32 $x^2 - x - 12$

33 $x^2 - 7x$

34 $x^2 - 5x$

35 $x^2 - 81$

36 $x^2 - 144$

37 $3x^2 + 9x - 12$

38 $5x^2 + 5x - 30$

39 $5x^2 - 45$

40 $6x^2 - 150$

41 $3x^2 - 12x + 12$

42 $4x^2 + 8x + 4$

43 $-3x^2 + 24x$

44 $-4x^2 + 24x$

Solve each equation.

45 $x^2 - 6x - 7 = 0$

46 $x^2 - x - 20 = 0$

47 $x^2 + 2x = 15$

48 $x^2 - 3x = 54$

49 $2x^2 + 40 = 18x$

50 $3x^2 + 18 = 15x$

51 $5x^2 - 50x = 0$

52 $3x^2 - 36x = 0$

53 $2x^2 - 50 = 0$

54 $3x^2 - 27 = 0$

55 $4x^2 + 24x + 40 = 4$

56 $3x^2 - 6x + 9 = 6$

57 $-4x^2 + 12x = 8$

58 $-3x^2 + 6x = -24$

59 $2x^2 - 12x + 20 = 0$

60 $2x^2 - 8x + 10 = 0$

61 $3x^2 + 12 = 0$

62 $5x^2 + 20 = 0$

▪ APPLIED EXERCISES

63 (*Business–Cost Functions*) A lumberyard will deliver wood for $4 per board foot plus a delivery charge of $20. Find a function $C(x)$ for the cost of having x board feet of lumber delivered.

64 (*Business–Cost Functions*) A company manufactures bicycles at a cost of $55 each. If the company's fixed costs are $900, express the company's cost as a linear function of x, the number of bicycles produced.

65 (*General–Water Pressure*) At a depth of d feet underwater, the water pressure is $p(d) = .45d + 15$ pounds per square inch. Find the pressure at

(a) the bottom of a 6-foot-deep swimming pool

(b) the maximum ocean depth of 35,000 feet

66 (*General–Boiling Point*) At higher altitudes water boils at lower temperatures. This is why at high altitudes foods must be boiled for longer times—the lower boiling point imparts less heat to the food. At an altitude of h thousand feet above sea level, the boiling point of water (in degrees Fahrenheit) is $B(h) = -1.8h + 212$ (h is in *thousands* of feet). Find

the boiling point at 63 thousand feet above sea level. Your answer will show that at a high enough altitude, water boils at normal body temperature. This is why airplane cabins must be pressurized—at high enough altitudes one's blood would boil.

67 (*Biomedical–Cell Growth*) The number of cells in a culture after t days is given by $N(t) = 200 + 50t^2$. Find the size of the culture after

(a) 2 days

(b) 10 days

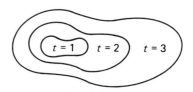

68 (*General–Juggling*) If you toss a ball h feet straight up, it will return to your hand after $T(h) = .5\sqrt{h}$ seconds. This leads to the "juggler's dilemma." Juggling more balls means tossing them higher. However, the square root in the above formula means that tossing twice as high does not gain twice as much time, but only $\sqrt{2} \approx 1.4$ as much time. Because of this there is a limit to the number of balls that a person can juggle, which seems to be about 10. Use this formula to find how long a ball will spend in the air if it is tossed to a height of

(a) 4 feet   (b) 8 feet

69 (*General–Impact Velocity*) If a marble is dropped from a height of x feet, it will hit the ground with velocity $v(x) = 5.45 \sqrt{x}$ miles per hour (neglecting air resistance). Use this formula to find the velocity with which a marble will strike the ground if it is dropped from the top of the world's tallest building, the 1454-foot Sears Tower in Chicago.

70 (*General–Tsunamis*) The speed of a tsunami (popularly known as a *tidal wave*, although it has nothing whatever to do with tides) depends on the depth of the water through which it is traveling. At a depth of d feet, the speed of a tsunami will be $s(d) = 3.86 \sqrt{d}$ miles per hour. Find the speed of a tsunami in the Pacific basin, where the average depth is 15,000 feet.

71–72 (*Business–Straight-Line Depreciation*) Straight-line depreciation is a method for calculating the value of a piece of equipment or some other asset. According to this method, given the original *price* of the asset, its *useful lifetime*, and its *scrap value* (its value at the end of its useful lifetime), the value of the asset after t years is

$$V(t) = (\text{price}) - \frac{(\text{price}) - (\text{scrap value})}{(\text{useful lifetime})} \cdot t$$

for $0 \leq t \leq$ (useful lifetime). The name derives from the fact that this function is a linear function of t. Note that the fraction in the formula is the amount by which the value will decrease each year.

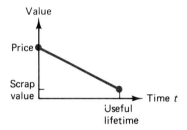

71 A company buys a computer for $500,000 and estimates its useful life to be 10 years, at the end of which its scrap value will be $10,000.

(a) Use the formula above to find a function for its value $V(t)$ after t years, for $0 \leq t \leq 10$.

(b) Use this function to find its value after 5 years.

72 A company buys a printing press for $800,000 and estimates its useful life to be 20 years, at the end of which its scrap value will be $20,000.

(a) Use the formula above to find a function for its value $V(t)$ after t years, for $0 \leq t \leq 20$.

(b) Use this function to find its value after 6 years.

1.3 FUNCTIONS, CONTINUED

Introduction

In this section we continue our study of functions, introducing other types of functions and some useful algebraic operations.

Polynomial Functions

A *polynomial function* (or simply a *polynomial*) is a function of the form

$$f(x) = a_n x^n + a_{n-1} x^{n-1} + \cdots + a_2 x^2 + a_1 x + a_0$$

where n is a nonnegative integer and $a_0, a_1, \ldots, a_n$ are (real) numbers. The *degree* of a polynomial is the highest power of the variable. For example, the following are polynomials:

$$f(x) = 2x^8 - 3x^7 + 4x^5 - 5$$ *polynomial of degree 8*

$$f(x) = -4x^2 - \frac{1}{3}x + 19$$ *quadratic polynomial*

$$f(x) = x - 1$$ *linear polynomial*

$$f(x) = 6$$ *constant polynomial*

The domain of a polynomial is $\mathbb{R}$, the set of all (real) numbers.

The polynomial graphed on the right might be used to model the total cost of manufacturing x units of a product. At first costs rise quite steeply, due to high startup expenses, and then more slowly as the economies of mass production come into play, and finally more steeply as original equipment needs to be replaced.

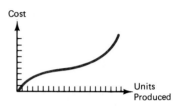

A cost function may increase at different rates at different production levels.

Solving Polynomial Equations

We will solve polynomial equations by factoring, just as we did with quadratic equations. Always remember to remove common factors first.

Example 1   Solve $3x^4 - 6x^3 = 24x^2$.

Solution

$$3x^4 - 6x^3 - 24x^2 = 0$$ *written with all terms on the left side*

$$3x^2(x^2 - 2x - 8) = 0$$ *factoring out $3x^2$*

$$3x^2\ (x - 4)\ (x + 2) = 0$$ *factoring further*

equals equals   equals
zero   zero   zero
at   at   at
$x = 0$ $x = 4$ $x = -2$

Therefore, the solutions are $x = 0$, $x = 4$, and $x = -2$.   ■

Example 1 shows that if a power of x can be factored out of a polynomial, then $x = 0$ is one of the solutions.

Practice Exercise 1

Solve $2x^3 - 4x^2 - 48x = 0$.   *(solution on page 38)*

Rational Functions

The word *ratio* means fraction or quotient, and a *rational function* is a quotient of two polynomials. The following are rational functions:

$$f(x) = \frac{3x^5 - 2x + 1}{x^2 + 1} \qquad g(x) = \frac{1}{x^3}$$

The domain of a rational function is all numbers except those that make the denominator zero. For example, the domain of the function on the left above is $\mathbb{R}$ (since $x^2 + 1$ is never zero), and the domain of the function on the right is $\{x \mid x \neq 0\}$.
   The rational function

$$f(x) = \frac{1000}{100 - x}$$

might be used to model the cost of removing successive amounts of iron from a mining deposit. Initially, costs are low (for iron near the surface), but they rise steeply as the mine becomes deeper.

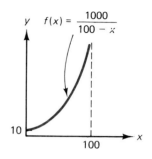

Exponential Functions

The function $f(x) = 2^x$ is called an *exponential function* because the variable appears in the exponent. Its graph may be found by plotting the points in the table on the right.

x	$y = 2^x$
2	$2^2 = 4$
1	$2^1 = 2$
0	$2^0 = 1$
-1	$2^{-1} = 1/2$
-2	$2^{-2} = 1/4$

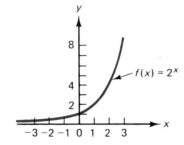

Exponential functions are used to model population growth and decline.

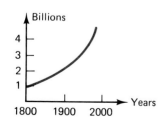

World population since the year 1800 can be approximated by an exponential function.

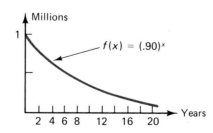

A population that declines by 10% each year is modeled by an exponential function.

Piecewise Linear Functions

The rule for calculating the values of a function may be given in several parts.

Example 2　The Tax Reform Act of 1986.　The 1988 federal income tax for a single taxpayer is calculated by a two-part rule: for incomes below $17,850 the tax is 15% of the income, and for higher incomes the tax is $2677.50 plus 28% of the income in excess of $17,850. For an income of x dollars this tax function can be expressed

$$f(x) = \begin{cases} .15x & \text{if } x < 17{,}850 \\ 2677.50 + .28(x - 17{,}850) & \text{if } x \geq 17{,}850 \end{cases}$$

This notation tells us to use the *upper* formula for incomes $x <$ $17,850 and the *lower* formula for incomes $x \geq$ $17,850. For example, the tax on an income of $4000 would be

$$f(4000) = .15(4000) = \$600$$

using the upper formula (since 4,000 < 17,850)

while the tax on an income of $25,000 would be

$$f(25000) = 2677.50 + .28(7150) = \$4679.50$$

$$\underbrace{\quad\quad}_{25{,}000 \,-\, 17{,}850}$$

using the lower formula (since 25,000 ≥ 17,850)

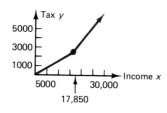

$$f(x) = \begin{cases} 0.15x & x < 17{,}850 \\ 2677.50 + 0.28(x - 17{,}850) & x \geq 17{,}850 \end{cases}$$

Such a function is called *piecewise linear* because its graph consists of "pieces" of straight lines.　■

The next example shows how to graph a piecewise linear function.

Example 3　Graph the piecewise linear function

$$f(x) = \begin{cases} 5 - 2x & \text{if } x \geq 2 \\ x + 3 & \text{if } x < 2 \end{cases}$$

Solution　We graph one "piece" at a time.

Step 1: To graph the first part, $f(x) = 5 - 2x$ if $x \geq 2$, plot points at the "endpoint" $x = 2$ and at $x = 4$ (or any other x-value satisfying $x \geq 2$). These points are $(2, 1)$ and $(4, -3)$, with the y-coordinates calculated from $f(x) = 5 - 2x$. Draw the line through these two points, but only for $x \geq 2$ (from $x = 2$ to the *right*).

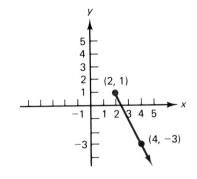

Step 2: For the second part, $f(x) = x + 3$ if $x < 2$, the restriction $x < 2$ means that the line ends just *before* $x = 2$. We mark this "missing point" $(2, 5)$ by an "open circle" (○) to indicate that it is *not* included in the graph [the y-coordinate comes from $f(x) = x + 3$]. For a second point, choose $x = 0$ (or any other $x < 2$), giving $(0, 3)$. Draw the line through these two points, but only for $x < 2$ (to the *left* of $x = 2$), completing the graph of the function.

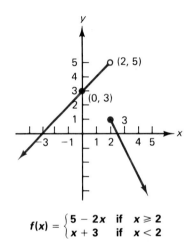

$$f(x) = \begin{cases} 5 - 2x & \text{if } x \geq 2 \\ x + 3 & \text{if } x < 2 \end{cases}$$

In accordance with the definition of function, no two points on the graph have the same x-coordinate [since the point $(2, 5)$ is not included the graph].

In Example 3 the "pieces" of the graph were not connected, while in the following example they are.

Example 4 Graph

$$f(x) = \begin{cases} x & \text{if } x \geq 0 \\ -x & \text{if } x < 0 \end{cases}$$

line $y = x$ *for* $x \geq 0$

line $y = -x$ *for* $x < 0$

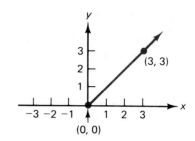

Solution

Step 1: To graph $f(x) = x$ if $x \geq 0$, use the endpoint $x = 0$, along with $x = 3$ (or any other $x \geq 0$). Calculating y-values from $f(x) = x$ gives the points $(0, 0)$ and $(3, 3)$. Draw the line through these points, but only for $x \geq 0$ (from $x = 0$ to the *right*).

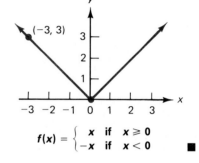

Step 2: For the other "piece," $f(x) = -x$ if $x < 0$, use the endpoint $x = 0$ (but plotting it with an "open circle" since it is excluded from this part). However, the resulting point $(0, 0)$ is already on the graph (from step 1), so it *is* included on the graph. For a second point, use $x = -3$ (or any other $x < 0$), obtaining $f(-3) = -(-3) = 3$, for the point $(-3, 3)$. Draw the line through these two points, but only for $x < 0$ (to the *left* of $x = 0$), completing the graph. Note that the two lines meet at the origin.

$$f(x) = \begin{cases} x & \text{if } x \geq 0 \\ -x & \text{if } x < 0 \end{cases}$$ ▪

This example showed that if one line of a piecewise linear function *does* include a point but the other line does not, that point *is* on the graph. For the piecewise linear function above, we found

$$f(3) = 3 \qquad\qquad f(x) = x \text{ with } x = 3$$

and

$$f(-3) = -(-3) = 3 \qquad\qquad f(x) = -x \text{ with } x = -3$$

showing that the function, when applied to either 3 or -3, gives *positive* 3. This function simply makes negative numbers positive. It is called the *absolute value function*.

The Absolute Value Function

$$f(x) = |x| = \begin{cases} x & \text{if } x \geq 0 \\ -x & \text{if } x < 0 \end{cases}$$

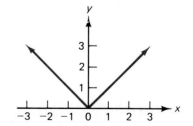

The absolute value function $f(x) = |x|$ has a "corner" at the origin.

Functions of Algebraic Expressions

In the same way that we substituted *numbers* into funcitons, we may substitute *algebraic expressions* into functions. We simply replace each occurrence of x by the algebraic expression.

Example 5   If $f(x) = x^2 - 5x$, find $f(3z)$.

Solution

$$f(3z) = (3z)^2 - 5(3z)$$

$$= 9z^2 - 15z$$

$f(x) = x^2 - 5x$ with each x replaced by 3z

multiplying out

▪

Notice that when substituting a quantity into x^2, the *whole quantity* is squared.

Example 6   If $f(x) = x^2 - 5x$, find $f(x + h)$.

Solution

$$f(x + h) = (x + h)^2 - 5(x + h)$$

$$= \underbrace{x^2 + 2xh + h^2}_{(x + h)^2} - 5x - 5h$$

$f(x) = x^2 - 5x$ with each x replaced by x + h

expanding

▪

Practice Exercise 2

For $f(x) = x^2 + 7x$, find $f(x + h)$.

(solution on page 38)

Example 7   Given $f(x) = 3x^2$, find and simplify $f(x + h) - f(x)$.

Solution $f(x + h) - f(x)$ means first find $f(x + h)$, then subtract $f(x)$:

$$f(x + h) - f(x) = \underbrace{3(x + h)^2}_{f(x + h)} - \underbrace{3x^2}_{f(x)}$$

$$= 3(x^2 + 2xh + h^2) - 3x^2$$

$$= 3x^2 + 6xh + 3h^2 - 3x^2$$

$$= \cancel{3x^2} + 6xh + 3h^2 - \cancel{3x^2}$$

$$= 6xh + 3h^2$$

expanding

multiplying out

canceling

▪

The following calculation will be important in Chapter 2.

Example 8 If $f(x) = x^2 - 4x + 1$, find and simplify $\dfrac{f(x + h) - f(x)}{h}$.

Solution

$$\frac{f(x + h) - f(x)}{h} = \frac{\overbrace{(x + h)^2 - 4(x + h) + 1}^{f(x+h)} - \overbrace{[x^2 - 4x + 1]}^{f(x)}}{h}$$

using $f(x) = x^2 - 4x + 1$ and
$f(x + h) = (x + h)^2 - 4(x + h) + 1$

$$= \frac{x^2 + 2xh + h^2 - 4x - 4h + 1 - x^2 + 4x - 1}{h}$$

expanding

$$= \frac{\cancel{x^2} + 2xh + h^2 - \cancel{4x} - 4h + \cancel{1} - \cancel{x^2} + \cancel{4x} - \cancel{1}}{h}$$

canceling

$$= \frac{2xh + h^2 - 4h}{h} = \frac{h(2x + h - 4)}{h}$$

factoring an h from the top

$$= \frac{\cancel{h}(2x + h - 4)}{\cancel{h}} = 2x + h - 4$$

dividing top and bottom by h

■

Practice Exercise 3

For $f(x) = 3x^2 - 2x + 1$, find and simplify $\dfrac{f(x + h) - f(x)}{h}$.

(solution on page 38)

Example 9 If $f(x) = \dfrac{1}{x}$, find and simplify $\dfrac{f(x + h) - f(x)}{h}$.

Solution

$$\frac{f(x + h) - f(x)}{h} = \frac{\overbrace{\dfrac{1}{x + h}}^{f(x+h)} - \overbrace{\dfrac{1}{x}}^{f(x)}}{h}$$

$$= \frac{1}{h}\left[\frac{1}{x + h} - \frac{1}{x}\right]$$

dividing by h is the same as multiplying by 1/h

$$= \frac{1}{h}\left[\frac{x}{(x + h)x} - \frac{x + h}{(x + h)x}\right]$$

using the common denominator $(x + h)x$

$$= \frac{1}{h}\frac{x - (x + h)}{(x + h)x} = \frac{1}{h}\frac{-h}{(x + h)x}$$

subtracting the fractions

$$= \frac{1}{\cancel{h}}\frac{\overset{-1}{\cancel{-h}}}{(x + h)x} = \frac{-1}{(x + h)x} \quad \text{answer}$$

canceling h

■

A Gallery of Functions

This chapter has introduced a variety of functions. Examples of these curves are shown below. In Chapter 3 we use calculus to develop efficient techniques for graphing such curves.

POLYNOMIALS

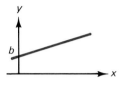

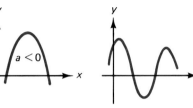

Linear function
$f(x) = mx + b$

Quadratic functions
$f(x) = ax^2 + bx + c$

$f(x) = ax^4 + bx^3 + cx^2 + dx + e$

RATIONAL FUNCTIONS

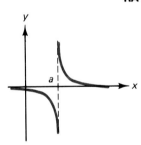

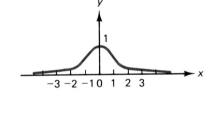

$f(x) = \dfrac{1}{x - a}$

$f(x) = \dfrac{1}{x^2 + 1}$

EXPONENTIAL FUNCTIONS

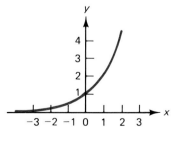

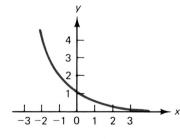

$f(x) = 2^x$

$f(x) = \left(\dfrac{1}{2}\right)^x$

PIECEWISE LINEAR FUNCTIONS

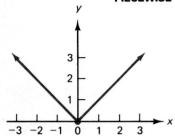

The absolute value function
$$f(x) = |x|$$

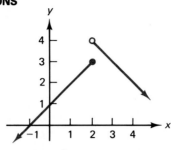

$$f(x) = \begin{cases} x + 1 & \text{if } x \le 2 \\ 6 - x & \text{if } x > 2 \end{cases}$$

SOLUTIONS TO PRACTICE EXERCISES

1. $2x(x^2 - 2x - 24) = 0$

$2x(x + 4)(x - 6) = 0$

Solutions: $x = 0$, $x = -4$, and $x = 6$

2. $f(x + h) = (x + h)^2 + 7(x + h) = x^2 + 2xh + h^2 + 7x + 7h$

3. $\dfrac{f(x + h) - f(x)}{h} = \dfrac{3(x + h)^2 - 2(x + h) + 1 - [3x^2 - 2x + 1]}{h}$

$$= \frac{\cancel{3x^2} + 6xh + 3h^2 - \cancel{2x} - 2h + \cancel{1} - \cancel{3x^2} + \cancel{2x} - \cancel{1}}{h}$$

$$= \frac{h(6x + 3h - 2)}{h} = 6x + 3h - 2$$

▨ EXERCISES 1.3

Find the domain of the function and evaluate the given expression.

1 $f(x) = \dfrac{x + 5}{x - 7}$ and find $f(6)$

2 $f(x) = \dfrac{x - 4}{x + 6}$ and find $f(-7)$

3 $f(x) = \dfrac{x + 5}{x(x + 3)}$ and find $f(-1)$

4 $f(x) = \dfrac{x + 12}{x(x + 6)}$ and find $f(-3)$

5 $g(x) = 4^x$ and find $g(-1/2)$

6 $g(x) = 8^x$ and find $g(-1/3)$

Solve each equation.

7 $x^5 + 2x^4 - 3x^3 = 0$

8 $x^6 - x^5 - 6x^4 = 0$

9 $5x^3 - 20x = 0$

10 $2x^5 - 50x^3 = 0$

11 $2x^3 + 18x = 12x^2$

13 $6x^5 + 30x^4 = 0$

12 $3x^4 + 12x^2 = 12x^3$

14 $5x^4 + 20x^3 = 0$

Graph each function.

15 $f(x) = 3^x$

16 $f(x) = (1/3)^x$

17 $f(x) = \begin{cases} 2x - 7 & \text{if } x \ge 4 \\ 2 - x & \text{if } x < 4 \end{cases}$

18 $f(x) = \begin{cases} 2 - x & \text{if } x \ge 3 \\ 2x - 4 & \text{if } x < 3 \end{cases}$

19 $f(x) = \begin{cases} 3 - x & \text{if } x > 3 \\ 3x - 6 & \text{if } x \le 3 \end{cases}$

20 $f(x) = \begin{cases} 2x - 1 & \text{if } x > 1 \\ 1 - x & \text{if } x \le 1 \end{cases}$

21 $f(x) = \begin{cases} 8 - 2x & \text{if } x \ge 2 \\ x + 2 & \text{if } x < 2 \end{cases}$

22 $f(x) = \begin{cases} 2x - 4 & \text{if } x > 3 \\ 5 - x & \text{if } x \le 3 \end{cases}$

23 $f(x) = \begin{cases} x & \text{if } x \le 0 \\ -x & \text{if } x > 0 \end{cases}$

24 $f(x) = \begin{cases} 1 - x & \text{if } x \le 1 \\ x - 1 & \text{if } x > 1 \end{cases}$

For each function, find and simplify (a) $f(x + h)$ and   (b) $f(x + h) - f(x)$.

25 $f(x) = 5x^2$

27 $f(x) = 2x^2 - 5x + 1$

26 $f(x) = 3x^2$

28 $f(x) = 3x^2 - 5x + 2$

For each function, find and simplify $\dfrac{f(x + h) - f(x)}{h}$.

29 $f(x) = 5x^2$

31 $f(x) = 2x^2 - 5x + 1$

33 $f(x) = 7x^2 - 3x + 2$

35 $f(x) = x^3$ [*Hint:* Use $(x + h)^3 = x^3 + 3x^2h + 3xh^2 + h^3$.]

36 $f(x) = x^4$ [*Hint:* Use $(x + h)^4 = x^4 + 4x^3h + 6x^2h^2 + 4xh^3 + h^4$.]

30 $f(x) = 3x^2$

32 $f(x) = 3x^2 - 5x + 2$

34 $f(x) = 4x^2 - 5x + 3$

37 $f(x) = \dfrac{2}{x}$ **38** $f(x) = \dfrac{3}{x}$ **39** $f(x) = \dfrac{1}{x^2}$

40 $f(x) = \sqrt{x}$
[*Hint:* Multiply top and bottom of the fraction by $(\sqrt{x + h} + \sqrt{x})$.]

▦ APPLIED EXERCISES

41–42 (*General–World Population*) The world population (in millions) since the year 1700 is approximated by the exponential function $P(x) = 514(1.007)^x$, where x is the number of years since 1700. Use a calculator with a $\boxed{y^x}$ key to estimate the world population in the year

41 1750   **42** 1800

43 (*Economics–Income Tax*) The following function expresses an income tax which is 10% of income up to $5000 plus 30% of income in excess of $5000.

$$f(x) = \begin{cases} .10x & \text{if } x \le 5000 \\ 500 + .30(x - 5000) & \text{if } x > 5000 \end{cases}$$

(a) Calculate the tax on an income of $3000.

(b) Calculate the tax on an income of $5000.

(c) Calculate the tax on an income of $10,000.

(d) Graph the function.

44 (*Economics–Income Tax*) The following function expresses an income tax which is 15% of income up to $6000 plus 40% of income in excess of $6000.

$$f(x) = \begin{cases} .15x & \text{if } x < 6000 \\ 900 + .40(x - 6000) & \text{if } x \geq 6000 \end{cases}$$

(a) Calculate the tax on an income of $3000.

(b) Calculate the tax on an income of $6000.

(c) Calculate the tax on an income of $10,000.

(d) Graph the function.

2

DERIVATIVES
AND
THEIR USES

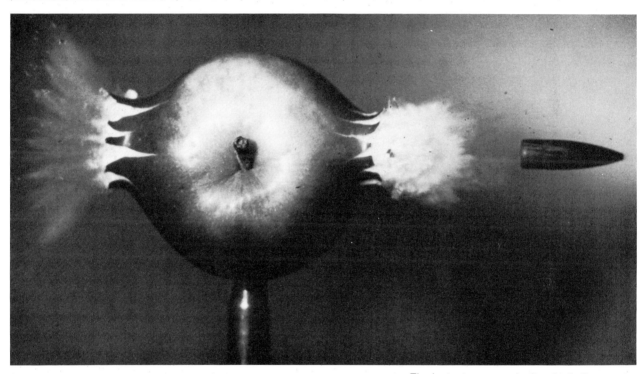

The instantaneous velocity of a bullet may be found by using calculus

2.1 LIMITS AND CONTINUITY

Introduction

While much of the material in Chapter 1 was review, in this section we introduce two new concepts, *limits* and *continuity*. We first discuss continuity, then limits, and then the relationship between them.

Continuous and Discontinuous Change

Some things in life change "continuously" (gradually), while others change "discontinuously" (abruptly). For example, changes in height are continuous (that is, you grow gradually rather than several inches in an instant), but changes in your bank balance are discontinuous, in jumps, whenever you make a deposit or a withdrawal. The difference between continuous and discontinuous change is like the difference between a wristwatch with hands, which registers time change continuously, and a digital watch, which registers time changes discontinuously, in "jumps" of 1 second.

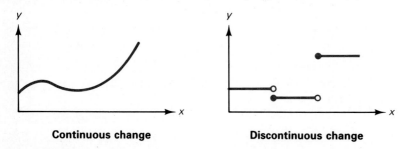

Continuous change   Discontinuous change

Some things change both continuously and discontinuously. For example, your understanding of calculus will increase continuously as you study, but will occasionally take sudden jumps, as you suddenly "see" something that you have been thinking about. The graph on the right shows continuous change, then a sudden "jump" of insight, and then more continuous change.

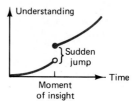

Continuous and Discontinuous Functions

Based on these ideas, a function whose graph has no jumps or gaps is called a *continuous function*. The following graphs show continuous functions.

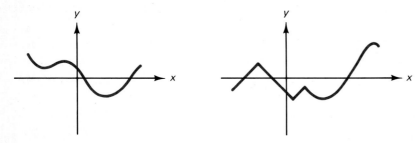

A function whose graph has at least one jump or gap is called a *discontinuous function*. The following graphs show discontinuous functions.

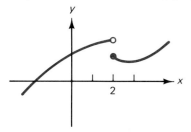

A function discontinuous at 2   **A function discontinuous at −1 and at 3**

Graphing a discontinuous function requires lifting your pencil to move from one part of the graph to another. If you must lift your pencil at $x = a$, we say that the function is *discontinuous at a*. For example, the function on the left above is discontinuous at 2, and the function on the right is discontinuous at −1 and at 3.

Practice Exercise 1

At which *x*-values is the function below discontinuous?   *(solution on page 51)*

Practice Exercise 2

Is the absolute value function $f(x) = |x|$ continuous or discontinuous?   *(solution on page 51)*

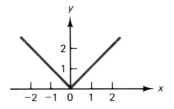

Which Functions Are Continuous?

Most of the functions that we encounter in applications are continuous. For example, *linear* functions (straight lines) are continuous, and *quadratic* functions (parabolas) are continuous, since their graphs have no jumps or gaps. Furthermore, sums and products of continuous functions are continuous. Therefore,

> All polynomial functions are continuous.

A rational function (the quotient of two polynomials) is continuous at all x-values except those at which its denominator is zero (since division by zero is undefined).

Example 1

(a)

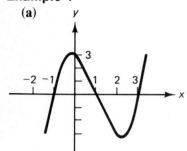

(b)

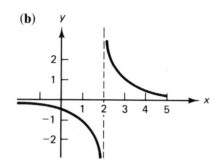

(a) The polynomial
$f(x) = x^3 - 3x^2 - x + 3$
is continuous.

(b) The rational function

$f(x) = \dfrac{1}{x - 2}$

is discontinuous at $x = 2$
(where its denominator is zero).

(c)

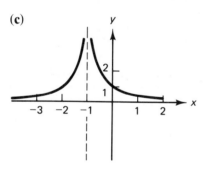

(c) The rational function

$f(x) = \dfrac{1}{(x + 1)^2}$

is discontinuous at $x = -1$.   ■

From this knowledge of polynomials and rational functions we can recognize many continuous and discontinuous functions from their algebraic formulas (that is, without looking at their graphs).

Example 2

(a) $f(x) = x^5 - 4x^3 + x^2 - 8x + 6$ is continuous (since it is a polynomial).

(b) $f(x) = \dfrac{x^2 - 3x + 17}{x - 3}$ is discontinuous at $x = 3$ (at which the denominator is zero).   ■

Limits

The word *limit* is used in everyday conversation to describe the "ultimate" behavior of something, as in the "limit of one's endurance" or the "limit of one's patience." In mathematics, the word "limit" has a similar but more precise meaning. When we speak of the *limit of a function* we mean a number that the function approaches. The following is an example from the "real world."

Absolute Zero Temperature

It has long been known that there is a coldest possible temperature (known as *absolute zero*), a temperature that can be approached but never actually reached. Absolute zero is the temperature of an object when all heat has been removed from it. This temperature is 460 degrees below zero in our "everyday" Fahrenheit temperature scale, or −273 in the centigrade (Celsius) system. We will use another temperature scale, the "absolute" or "Kelvin" scale (named after the nineteenth century scientist Lord Kelvin), in which absolute zero temperature is assigned the number 0.

   Certain types of electrical machinery, such as high-speed trains and computers, use temperatures near absolute zero for conducting electricity. It is impossible to measure the electrical conductivity *at* absolute zero, since absolute zero is unattainable, but one can measure conductivity as temperatures *approaches* absolute zero. The graph on the right shows the electrical conductivity of aluminum at temperatures near absolute zero. It shows the remarkable fact that as temperatures approach absolute zero, *the conductivity of aluminum approaches 100%.*

   In other words, at temperatures approaching absolute zero, aluminum loses all resistance to electricity and becomes "superconducting." (Other recently discovered substances become superconducting at higher temperatures.)

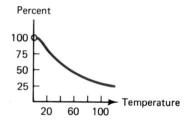

The conductivity of aluminum at temperatures near absolute zero.

Limit Notation

We will use a horizontal arrow, →, to mean "approaches," so that $t \to 0$ means "t approaches zero." If $f(t)$ stands for the conductivity of aluminum at temperature t, the statement that the conductivity approaches 100 percent as t approaches zero may be written

$$\lim_{t \to 0} f(t) = 100$$

"the limit of f(t) as t approaches 0 is 100"

This limit statement does *not* say what happens *at* $t = 0$ (since absolute zero is unattainable). Rather, it means that as t *approaches* (becomes closer and closer to) zero, $f(t)$ approaches (becomes closer and closer to) 100.

   Limits can be defined formally*, but we will treat them intuitively. For a

---

   * Formally, limits are defined as follows: $\lim_{x \to a} f(x) = L$ if and only if for every number $\varepsilon > 0$ there is a number $\delta > 0$ such that $|f(x) - L| < \varepsilon$ whenever $0 < |x - a| < \delta$.

function $f(x)$ and a number a,

$$\lim_{x \to a} f(x)$$

is the number (if it exists) that $f(x)$ approaches as x approaches a.

limit of $f(x)$ as x approaches a

Note that if the limit exists, it must be a *single* number.

Evaluating Limits

We may evaluate limits intuitively.

Example 3   Find $\lim_{x \to 3} (2x + 4)$.

limit of $2x + 4$ as x approaches 3

Solution   If x is very close to 3, then $2x$ will be very close to two times 3, which is 6, and then adding 4 gives a number very close to 10, so that

$$\lim_{x \to 3} (2x + 4) = 10$$

■

We may check this answer with a calculator by evaluating $2x + 4$ at x-values near 3. Using 2.99 and 3.01 (we must choose numbers on both sides of 3) gives the following:

At $x = 2.99$: $2x + 4 = 2(2.99) + 4 = 5.98 + 4 = 9.98$ ⟵ values

At $x = 3.01$: $2x + 4 = 2(3.01) + 4 = 6.02 + 4 = 10.02$ ⟵ close to 10

For x-values even *closer* to 3, $2x + 4$ will be even *closer* to 10.

At $x = 2.999$: $2x + 4 = 2(2.999) + 4 = 5.998 + 4 = 9.998$ ⟵ values even

At $x = 3.001$: $2x + 4 = 2(3.001) + 4 = 6.002 + 4 = 10.002$ ⟵ closer to 10

agreeing with our limit statement $\lim_{x \to 3} 2x + 4 = 10$.

We could continue choosing x-values even closer to 3 (like 2.9999 and 3.0001), and the results of substituting them into $2x + 4$ would indeed be even closer to 10. However, as useful as these numerical calculations are for verifying a limit, they are no substitute for the simple intuitive analysis that says:

If x is close to 3, then $2x$ will be close to 6,
and 4 more, $2x + 4$, will be close to 10, so
$$\lim_{x \to 3} (2x + 4) = 10$$

We may, of course, use other letters for the variables in limit problems.

Practice Exercise 3

Find $\lim_{h \to 6} (4h - 8)$.

(solution on page 51)

Evaluating Limits by Direct Substitution

In each of these examples, the correct limit could also have been found by *direct substitution* of the limiting number into the function. That is, in our first example,

$$\lim_{x \to 3} (2x + 4) = 10$$

simply substituting $x = 3$ into the function $2x + 4$ gives $2 \cdot 3 + 4 = 10$, which was also the correct *limit* of the function as x *approached* 3. Similarly, in Practice Exercise 3, evaluating the function $4h - 8$ at $h = 6$ gives $4 \cdot 6 - 8 = 16$, which is also the limit as h *approaches* 6. Since direct substitution is much easier than reasoning through the limit operation, why do we need limits at all? The answer is that sometimes direct substitution will not give the same answer as the limit. In fact, sometimes the limiting number cannot even be substituted into the function—it may make the function undefined.

Example 4   Direct substitution into the limit $\lim\limits_{h \to 0} \dfrac{h^2 + 2h}{h}$ gives the un-

defined quantity 0/0, showing that this limit cannot be evaluated by direct substitution. However, we will later be able to show that this limit *does* exist and equals 2.   ■

Rules of Limits

Many limits *can* be evaluated by direct substitution, using the following *rules of limits*. In these rules, the letters c and n stand for constants, with $n > 0$.

1. $\lim\limits_{x \to a} c = c$

the limit of a constant is just the constant

2. $\lim\limits_{x \to a} x^n = a^n$

the limit of a power is the power of the limit

3. $\lim\limits_{x \to a} \sqrt[n]{x} = \sqrt[n]{a}$

the limit of a root is the root of the limit

4. If $\lim\limits_{x \to a} f(x)$ and $\lim\limits_{x \to a} g(x)$ both exist, then

(a) $\lim\limits_{x \to a} [f(x) + g(x)] = \lim\limits_{x \to a} f(x) + \lim\limits_{x \to a} g(x)$

the limit of a sum is the sum of the limits

(b) $\lim\limits_{x \to a} [f(x) - g(x)] = \lim\limits_{x \to a} f(x) - \lim\limits_{x \to a} g(x)$

the limit of a difference is the difference of the limits

(c) $\lim\limits_{x \to a} [f(x)g(x)] = [\lim\limits_{x \to a} f(x)] \cdot [\lim\limits_{x \to a} g(x)]$

the limit of a product is the product of the limits

(d) $\lim\limits_{x \to a} \dfrac{f(x)}{g(x)} = \dfrac{\lim\limits_{x \to a} f(x)}{\lim\limits_{x \to a} g(x)}$

the limit of a quotient is the quotient of the limits [provided that $\lim\limits_{x \to a} g(x) \neq 0$]

These rules, saying that many limits can be evaluated by direct substitution, may be summarized as follows.

> For functions composed only of the operations of addition, subtraction, multiplication, division, powers, and roots, limits may be evaluated by direct substitution (provided that the resulting expression is defined):
>
> $$\lim_{x \to a} f(x) = f(a)$$

Example 5

(a) $\lim_{x \to 2} 7 = 7$ *rule 1*

(b) $\lim_{x \to 2} x^3 = 2^3 = 8$ *rule 2, direct substitution of $x = 2$*

(c) $\lim_{x \to 4} \sqrt{x} = \sqrt{4} = 2$ *rule 3, direct substitution of $x = 4$*

(d) $\lim_{x \to 2} (3x^2 + 4x - 7) = 3 \cdot 2^2 + 4 \cdot 2 - 7 = 13$ *rules 4, 2, and 1, direct substitution of $x = 2$*

(e) $\lim_{x \to 6} \dfrac{x^2}{x + 3} = \dfrac{6^2}{6 + 3} = \dfrac{36}{9} = 4$ *rules 4, 2, and 1, direct substitution of $x = 6$*

■

Practice Exercise 4

Find

    (a) $\lim_{x \to 3} (2x^2 - 4x + 1)$ (b) $\lim_{x \to 2} \dfrac{x^3 + 2}{x^2 - 2}$ *(solutions on page 51)*

Limits with Two Variables

Some problems involve two variables, with only one variable approaching a limit.

Example 6   Find $\lim_{h \to 0} (x^2 + xh + h^2)$.

Solution   Notice only h approaches zero, so that x is to be left alone. Since the function involves only powers of h, we may evaluate the limit by direct substitution of $h = 0$:

$$\lim_{h \to 0} (x^2 + xh + h^2) = x^2 + \underbrace{0 \cdot h}_{0} + \underbrace{0^2}_{0} = x^2$$

■

Practice Exercise 5

Find $\lim_{h \to 0} (3x^2 + 3xh + h^2)$.   *(solution on page 51)*

Algebraic Simplification of Limits

If direct substitution into a quotient gives the undefined expression 0/0, try factoring and simplifying.

Example 7   Find $\lim\limits_{h \to 0} \dfrac{2xh + h^2}{h}$.

Solution   Direct substitution of $h = 0$ gives 0/0, which is undefined. But simplifying yields

$$\lim\limits_{h \to 0} \frac{2xh + h^2}{h} = \lim\limits_{h \to 0} \frac{h(2x + h)}{h} = \lim\limits_{h \to 0} \frac{\cancel{h}(2x + h)}{\cancel{h}} = \lim\limits_{h \to 0} 2x + h = 2x$$

factoring ⌐   canceling   now use direct   limit
out an h the h substitution   *does* exist   ■

Example 7 illustrates once again that limits are *not* the same as direct substitution, since direct substitution into the original gave the undefined 0/0, but the limit *did* exist.

Practice Exercise 6

Find $\lim\limits_{h \to 0} \dfrac{5xh + 3h^2}{h}$.

(solution on page 51)

Limits May Not Exist

Sometimes there will be no single number that the function approaches. In that case we say that the limit *does not exist*.

Example 8   Find $\lim\limits_{x \to 2} \dfrac{1}{x - 2}$.

Solution   Substituting $x = 2$ gives the undefined expression 1/0, which means that we cannot use direct substitution (remember that we can use direct substitution on a quotient only if the denominator does not become zero). Substituting numbers close to 2 gives the following:

At $x - 1.99$: $\dfrac{1}{x - 2} = \dfrac{1}{1.99 - 2} = \dfrac{1}{-.01} = -100$ values very
                                                                                       far apart

At $x = 2.01$: $\dfrac{1}{x - 2} = \dfrac{1}{2.01 - 2} = \dfrac{1}{.01} = 100$

For x-values even *closer* to 2, the results are even farther apart:

At $x = 1.999$: $\dfrac{1}{x - 2} = \dfrac{1}{1.999 - 2} = \dfrac{1}{-.001} = -1000$ values even
                                                                                             farther apart

At $x = 2.001$: $\dfrac{1}{x - 2} = \dfrac{1}{2.001 - 2} = \dfrac{1}{.001} = 1000$

It is clear that as x approaches 2, the function $1/(x - 2)$ does not approach *any* single number (remember that the limit must be *one* number). Therefore, the limit $\lim\limits_{x \to 2} \dfrac{1}{x - 2}$ *does not exist*.   ■

The graph of the function $f(x) = 1/(x - 2)$ shows why this is so. Near $x = 2$ the graph becomes arbitrarily high and low (on either side of $x = 2$), not approaching any single value. Therefore, we see again that the limit as x approaches 2 *does not exist*.

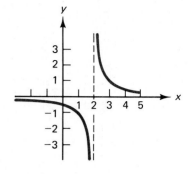

The graph of $f(x) = \dfrac{1}{x - 2}$

Limits and Continuity

Interpreted geometrically, the limit statement $\lim\limits_{x \to a} f(x) = f(a)$ says that as x approaches a, the graph of $f(x)$ approaches $f(a)$, the point on the graph *at* $x = a$. This means, however, that we do not have to lift the pencil to graph the function at $x = a$, which means that the function is *continuous at $x = a$.* Therefore, continuity may be defined* in terms of *limits*:

$f(x)$ is continuous at $x = a$ if and only if

$$\lim_{x \to a} f(x) = f(a)$$

We may see this result in pictures.

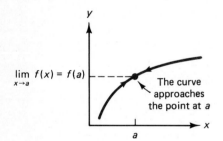

A function that is continuous at $x = a$, showing that $\lim\limits_{x \to a} f(x) = f(a)$.

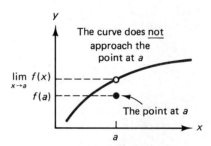

A function that is discontinuous at $x = a$, showing that $\lim\limits_{x \to a} f(x) \neq f(a)$.

Summary

A function is *continuous* if it has no jumps or gaps and can be graphed in one connected curve. A function is *discontinuous* if it has at least one jump or

---

* When we write $\lim\limits_{x \to a} f(x) = f(a)$, we of course mean that the quantities on both sides of the equation are defined. We may make this explicit by writing the definition of continuity in the longer but equivalent form: A function $f(x)$ is continuous at $x = a$ if and only if the following three conditions hold:

   1. $\lim\limits_{x \to a} f(x)$ exists.   **2.** $f(a)$ is defined.   **3.** $\lim\limits_{x \to a} f(x) = f(a)$.

gap. If you must lift your pencil at a particular x-value to graph a function, then the function is discontinuous *at that x-value*.

We defined the limit

$$\lim_{x \to a} f(x)$$

as the number that the function $f(x)$ approaches as x approaches a (if such a number exists). We found that for functions composed only of the operations of addition, subtraction, multiplication, division, roots, and powers, limits could be evaluated by *direct substitution* (provided that the resulting expression is defined). If direct substitution makes a quotient into 0/0, simplifying and *then* using direct substitution may work.

In general, the limit of a function as x approaches a number is not the same as the value of the function *at* the number (recall our very first example of absolute zero temperature—we can approach absolute zero but we can never reach it).

More could be said about limits, and we will return to the subject in Chapter 3 when we use limits for curve sketching.

SOLUTIONS TO PRACTICE EXERCISES

1. Discontinuous at -2 and at 3 (because of the jumps at these values).

2. The absolute value function $f(x) = |x|$ is continuous. (It has a corner at $x = 0$ but no jumps.)

3. When h is close to 6, $4h$ will be close to 24, so $4h - 8$ will be close to $24 - 8$, or 16. Therefore,

$$\lim_{h \to 6} (4h - 8) = 16$$

4. Using direct substitution,

   (a) $\lim_{x \to 3} (2x^2 - 4x + 1) = 2 \cdot 3^2 - 4 \cdot 3 + 1 = 18 - 12 + 1 = 7$

   (b) $\lim_{x \to 2} \dfrac{x^3 + 2}{x^2 - 2} = \dfrac{2^3 + 2}{4 - 2} = \dfrac{10}{2} = 5$

5. Using direct substitution,

$$\lim_{h \to 0} (3x^2 + 3xh + h^2) = 3x^2 + \underbrace{3x \cdot 0}_{0} + \underbrace{0^2}_{0} = 3x^2$$

6. Direct substitution of $h = 0$ gives 0/0 (undefined), but simplifying,

$$\lim_{h \to 0} \frac{5xh + 3h^2}{h} = \lim_{h \to 0} \frac{h(5x + 3h)}{h} = \lim_{h \to 0} \frac{\cancel{h}(5x + 3h)}{\cancel{h}} = \lim_{h \to 0} 5x + 3h = 5x$$

■ EXERCISES 2.1

For each of the following graphs, state whether the function is continuous or discontinuous. If it is discontinuous, state the x-values at which it is discontinuous.

1
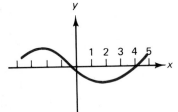

2
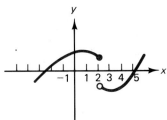

3
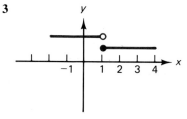

4
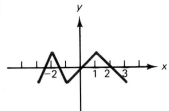

5
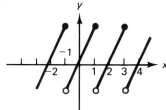

6
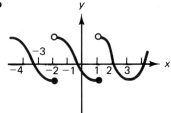

State whether each function is continuous or discontinuous. If it is discontinuous, state the values of x at which it is discontinuous.

7 $f(x) = 7x - 5$

8 $f(x) = 5x^3 - 6x^2 + 2x - 4$

9 $f(x) = 8x^4 + 4x^3 - 7x + 1$

10 $f(x) = 4x + 15$

11 $f(x) = \dfrac{x + 1}{x - 1}$

12 $f(x) = \dfrac{x - 1}{x + 1}$

13 $f(x) = \dfrac{x^2}{(x - 3)(x + 4)}$

14 $f(x) = \dfrac{x^3}{(x + 7)(x - 2)}$

Find the indicated limits.

15 $\lim\limits_{x \to 4} (2x^2 - 5x - 3)$

16 $\lim\limits_{x \to 3} (4x^2 - 10x + 2)$

17 $\lim\limits_{x \to 5} \dfrac{3x^2 - 5x}{7x - 10}$

18 $\lim\limits_{x \to 7} \dfrac{x^2 - x}{2x - 7}$

19 $\lim\limits_{r \to 8} \dfrac{r^2}{r - 2\sqrt[3]{r}}$

20 $\lim\limits_{r \to 8} \dfrac{r}{r^2 - 30\sqrt[3]{r}}$

21 $\lim\limits_{s \to 9} (s^{3/2} - 4s^{1/2})$

22 $\lim\limits_{s \to 4} (s^{3/2} - 3s^{1/2})$

23 $\lim\limits_{t \to 25} [(t + 5)t^{-1/2}]$

24 $\lim\limits_{t \to 36} [(t - 6)t^{-1/2}]$

25 $\lim\limits_{x \to 4} \sqrt{x^2 + x + 5}$

26 $\lim\limits_{x \to 3} \sqrt[3]{x^2 + x - 4}$

27 $\lim\limits_{h \to 0} (4x^2 + 3h)$

28 $\lim\limits_{h \to 0} (3x^2 - 4h)$

29 $\lim\limits_{h \to 0} (2x^2 + 4xh + h^2)$

30 $\lim\limits_{h \to 0} (5x^2 + 2xh - h^2)$

31 $\lim\limits_{h \to 0} 7$

32 $\lim\limits_{h \to 0} 9$

33 $\lim\limits_{h \to 0} \dfrac{2xh - 3h^2}{h}$

34 $\lim\limits_{h \to 0} \dfrac{6xh - 2h^2}{h}$

35 $\lim\limits_{h \to 0} \dfrac{7x^3h - 3xh}{h}$

36 $\lim\limits_{h \to 0} \dfrac{5x^4h - 9xh}{h}$

37 $\lim\limits_{h \to 0} \dfrac{4x^2h + xh^2 - h^2}{h}$

38 $\lim\limits_{h \to 0} \dfrac{x^2h - xh^2 + h^2}{h}$

Determine whether the limit exists. If it does, then evaluate it. If the limit does not exist, show this by substituting some appropriate numbers.

39 $\lim\limits_{x \to 3} \dfrac{1}{x - 4}$ **40** $\lim\limits_{x \to 5} \dfrac{1}{x - 5}$ **41** $\lim\limits_{x \to 4} \dfrac{1}{x - 4}$ **42** $\lim\limits_{x \to 4} \dfrac{1}{x - 5}$

43 $\lim\limits_{x \to 2} \dfrac{x}{x - 2}$ **44** $\lim\limits_{x \to 2} \dfrac{x}{x - 1}$ **45** $\lim\limits_{h \to 0} \dfrac{h}{h + 1}$ **46** $\lim\limits_{h \to 0} \dfrac{h + 1}{h}$

▨ APPLIED EXERCISES

47–48 (*General*) Experiments have shown that the length to which a crystal fiber will grow depends on the strength of gravity where it is growing. If the crystal's length (in inches) is given by the following functions, where g is the strength of gravity, find the limiting length that would be approached under the weightless conditions of space. That is, find the limit of the function as g approaches zero.

47 $\dfrac{6}{\sqrt{g + 4}}$ **48** $\dfrac{12}{\sqrt{g + 9}}$

49–50 (*General*) According to Einstein's theory of relativity, under certain conditions a 1-foot ruler moving with velocity v will appear to a stationary observer to have length given by the following functions (in which c is the speed of light). Find the limiting value of the apparent length as the velocity approaches the speed of light c. That is, find the limit of the function as v approaches c.

49 $\sqrt{1 - \left(\dfrac{v}{c}\right)^2}$ **50** $\sqrt{\dfrac{c^2 - v^2}{c^2}}$

2.2 INSTANTANEOUS RATES OF CHANGE AND SLOPES

Introduction

Of all the mathematical discoveries of the last 1000 years, the one that has had the widest range of applications is the *derivative*. Quite simply, derivatives are *rates of change*. (The word "rate" means one unit per another unit, such as miles per hour, dollars per employee, or people per square mile.) For example, the rate at which you grow (inches per year) varies with your age.

1 year old growth rate: 3 inches per year.

5 years old growth rate: 2.5 inches per year.

10 years old growth rate: 2.3 inches per year.

15 years old growth rate: 1.8 inches per year.

20 years old growth rate: 0.2 inches per year.

These and other rates of change may be expressed as derivatives.

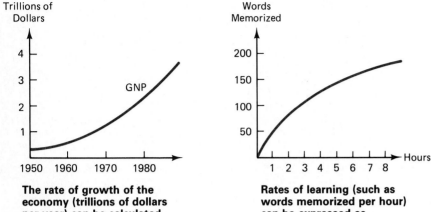

The rate of growth of the economy (trillions of dollars per year) can be calculated with derivatives.

Rates of learning (such as words memorized per hour) can be expressed as derivatives.

In this chapter we define derivatives, learn how to calculate them, and interpret them geometrically.

Rates of Change: An Example

Refining crude oil into gasoline involves several stages of heating and cooling. Controlling the process requires knowing how the temperature is changing (degrees per hour) at any particular moment. Suppose that the temperature of the oil after x hours of refining is given by a function $f(x)$,

$$f(x) = \begin{pmatrix} \text{temperature} \\ \text{at time } x \end{pmatrix}$$

Two problems might arise.

Problem 1 Find the *average rate of change of temperature* over the 4-hour time period from time x to time $x + 4$.

Solution The temperatures at times x and $x + 4$ are $f(x)$ and $f(x + 4)$. We subtract them, $f(x + 4) - f(x)$, to find the *change* during the 4 hours, and then divide by 4 to find the average change *per hour*:

$$\begin{pmatrix} \text{average rate of change of temperature} \\ \text{from time } x \text{ to time } x + 4 \end{pmatrix} = \frac{f(x + 4) - f(x)}{4} \qquad ■$$

Problem 2 Now find the rate of change of temperature *at the particular instant* x (the "instantaneous" rate of change).

Solution To calculate the *instantaneous* rate of change at time x we begin with the formula above for the average rate of change over 4 hours, but replacing the 4 by an arbitrary length of time h:

$$\begin{pmatrix} \text{average rate of change of temperature} \\ \text{from time } x \text{ to time } x + h \end{pmatrix} = \frac{f(x + h) - f(x)}{h}$$

We then let this time period shrink to an "instant" by letting the time period h approach zero (using the limit notation),

$$\begin{pmatrix} \textit{instantaneous} \text{ rate of change} \\ \text{of temperature at time } x \end{pmatrix} = \lim_{h \to 0} \frac{f(x + h) - f(x)}{h}$$

This gives a formula for calculating the instantaneous rate of change of temperature at any time x.   ■

Instantaneous Rates of Change in General

Note that this formula is not limited to temperature but can be used to find instantaneous rates of change of *any* function. We derived the formula as follows:

$$(\text{change in } f) = f(x + h) - f(x)$$

$$\begin{pmatrix} \text{average rate} \\ \text{of change of } f \end{pmatrix} = \frac{f(x + h) - f(x)}{h}$$

$$\begin{pmatrix} \text{instantaneous rate} \\ \text{of change of } f \end{pmatrix} = \lim_{h \to 0} \frac{f(x + h) - f(x)}{h}$$

$f(x + h) - f(x)$ gives the change in the function when x changes by an amount h, from x to x + h

dividing this total change by h gives the average rate of change of f(x) per unit change in x

taking the limit as h approaches zero gives the instantaneous rate of change at x

If this limit exists, we call the result *the derivative of the function f at x,* and symbolize it by $f'(x)$ (read "*f* prime of *x*").

The derivative of a function f at x is defined as

$$f'(x) = \lim_{h \to 0} \frac{f(x + h) - f(x)}{h}$$

(provided that the limit exists). It gives the instantaneous rate of change of f at x.

The expression $\frac{f(x + h) - f(x)}{h}$ (without the $\lim_{h \to 0}$) is called the *difference quotient* because it is a quotient (a fraction), and its numerator is a difference. It gives the *average* rate of change of the function over an interval of length h. It is the *limit* which makes the interval shrink to an instant, giving the *instantaneous* rate of change.

Example 1   Suppose that the function $f(x) = x^2$ gives the temperature of the oil after x hours. Find the instantaneous rate of change of the temperature at time $x = 3$ hours.

Solution   The instantaneous rate of change means the derivative. To find $f'(3)$, the derivative of $f(x) = x^2$ at $x = 3$, we begin with the definition of the derivative.

$$f'(x) = \lim_{h \to 0} \frac{f(x + h) - f(x)}{h}$$
definition of the derivative

$$f'(3) = \lim_{h \to 0} \frac{f(3 + h) - f(3)}{h}$$
substituting x = 3

$$= \lim_{h \to 0} \frac{(3 + h)^2 - (3)^2}{h}$$
f(x) = x² means f(3 + h) = (3 + h)² and f(3) = 3²

$$= \lim_{h \to 0} \frac{9 + 6h + h^2 - 9}{h}$$
expanding

$$= \lim_{h \to 0} \frac{\cancel{9} + 6h + h^2 - \cancel{9}}{h} = \lim_{h \to 0} \frac{6h + h^2}{h}$$
simplifying

$$= \lim_{h \to 0} \frac{h(6 + h)}{h} = \lim_{h \to 0} \frac{\cancel{h}(6 + h)}{\cancel{h}}$$
factoring out h and canceling

$$= \lim_{h \to 0} (6 + h) = 6$$
evaluating the limit by direct substitution

The answer, 6, means that at time $x = 3$ hours, the temperature is increasing at the rate of 6 degrees per hour.   ■

Practice Exercise 1

Given the same temperature function as in Example 1, $f(x) = x^2$, find the instantaneous rate of change of the temperature at time $x = 4$.

(*Hint*: Follow the steps of Example 1, but now with $x = 4$.)   *(solution on page 63)*

It is important that you try to solve this practice problem before reading any further, as its answer will be used in the rest of the section.

Example 1 showed that at time 3 hours the temperature was increasing at the rate of 6 degrees per hour, and Practice Exercise 1 showed that an hour later, at time $x = 4$, the temperature was increasing even faster, at the rate of 8 degrees per hour. Knowing how rapidly the temperature is increasing would be critical in controlling the refining process.

The Derivative As Slope

The derivative also has a geometric meaning. It gives the *slope of the graph* of a function.

At any point on a curve like the one shown here, we may draw the straight line that most closely matches the steepness of the curve at that point. Such a line is called the *tangent line* to the curve at the point. When we speak of the slope of a curve at a point, we mean the slope of the tangent line to the curve at that point.

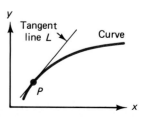

The slope of the curve at P is the slope of the line L.

To find slope of the tangent line at P, let two points P and Q on the graph of $f(x)$ have coordinates as follows.

P has coordinates $(x, f(x))$

Q has coordinates $(x + h, f(x + h))$

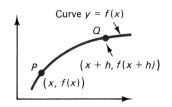

From the diagram on the right, we see that the slope of the line through P and Q is

$$\frac{\Delta y}{\Delta x} = \frac{f(x + h) - f(x)}{h}$$

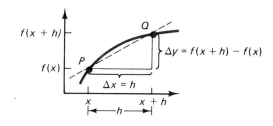

The points P and Q are a distance h apart (measured along the x-axis).

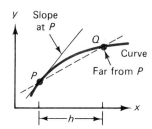

If h becomes smaller, the point Q moves toward the point P, making the slope of the (dashed) line through P and Q closer to the slope of the (solid) tangent line at P.

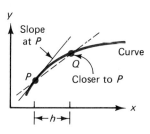

If h becomes even smaller, the slope of the (dashed) line through P and Q becomes even closer to the slope of the curve at P.

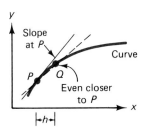

These pictures show that the slope of the line through P and Q is just the difference quotient $\dfrac{f(x + h) - f(x)}{h}$, and that as $h \to 0$ the difference quotient approaches the slope of the curve at P, giving

$$\begin{pmatrix} \text{slope of the} \\ \text{curve at } P \end{pmatrix} = \lim_{h \to 0} \frac{f(x + h) - f(x)}{h}$$

Since the right-hand side of the equation above is just the definition of the derivative, we see that *the derivative $f'(x)$ gives the slope of the graph of $f(x)$.*

The derivative

$$f'(x) = \lim_{h \to 0} \frac{f(x + h) - f(x)}{h}$$

gives the slope of the graph at x, in addition to giving the instantaneous rate of change of the function f at x.

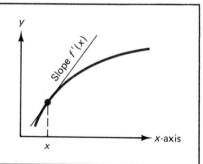

Therefore, the derivative has a *geometric* interpretation, the *slope* of the graph, as well as an "analytic" interpretation (that is, in terms of functions), the *instantaneous rate of change* of the function.

Earlier we found that the derivative of $f(x) = x^2$ at $x = 3$ is 6. Interpreted geometrically, this says that at the point on the graph where $x = 3$, the slope of the curve $y = x^2$ is 6, as shown here on the graph on the right. Therefore, we can now "see" derivatives as slopes of curves.

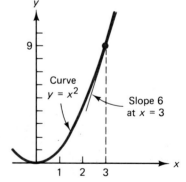

The Derivative as a Function

Calculating the derivative as a *function* (that is, with x left as a variable) is no more difficult than calculating the derivative at a particular number.

Example 2   For the same temperature function $f(x) = x^2$, find the instantaneous rate of change of the temperature at *any* time x [or equivalently, find the slope of the graph of $f(x) = x^2$ for *any* value of x].

Solution   The derivative is

$$f'(x) = \lim_{h \to 0} \frac{f(x + h) - f(x)}{h} \qquad \text{\textit{definition of} } f'(x)$$

$$= \lim_{h \to 0} \frac{(x + h)^2 - x^2}{h} \qquad \textit{since } f(x + h) = (x + h)^2 \textit{ and } f(x) = x^2$$

$$= \lim_{h \to 0} \frac{\cancel{x^2} + 2xh + h^2 - \cancel{x^2}}{h} \qquad \text{\textit{expanding and canceling}}$$

$$= \lim_{h \to 0} \frac{2xh + h^2}{h} = \lim_{h \to 0} \frac{\cancel{h}(2x + h)}{\cancel{h}} \qquad \text{\textit{factoring and canceling the h}}$$

$$= \lim_{h \to 0} (2x + h) = 2x \qquad \text{\textit{evaluating the limit by direct substitution}}$$

∎

Therefore, for any value of x, the derivative of $f(x) = x^2$ is $f'(x) = 2x$. This formula for the derivative enables us to find the derivative at *any* value of x merely by substituting numbers into $f'(x) = 2x$. For example:

At $x = 2$:	the derivative is $f'(2)$	$= 2 \cdot 2$	$= 4$	*$f'(x) = 2x$ with $x = 2$*
At $x = 3$:	the derivative is $f'(3)$	$= 2 \cdot 3$	$= 6$	*$f'(x) = 2x$ with $x = 3$*
At $x = 4$:	the derivative is $f'(4)$	$= 2 \cdot 4$	$= 8$	*$f'(x) = 2x$ with $x = 4$*
At $x = -1$:	the derivative is $f'(-1) = 2(-1) = -2$			*$f'(x) = 2x$ with $x = -1$*

The middle two results in this list are just the results that we found in Example 1 and Practice Exercise 1. The fact that the slope of the graph of $f(x) = x^2$ is $f'(x) = 2x$ is illustrated in the diagram below.

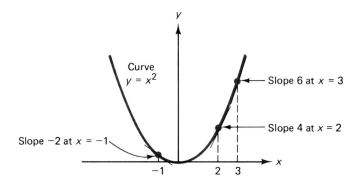

From now on we will calculate derivatives in this way, keeping x as a variable. This gives the derivative as a *function*, and the derivative at any particular x-value can then be found by substituting that value into the derivative function.

We should now view the operation of differentiation as an operation on *functions*, an operation that takes one function [such as $f(x) = x^2$] and gives another function [$f'(x) = 2x$]. The resulting function is called the "derivative" because it is "derived" from the first, and the process of obtaining it is called "differentiation."

Units of the Derivative

The units of the derivative come directly from the units of the original function. For example, if $f(x)$ gives the temperature (in degrees) at hour x, the derivative is measured in *degrees per hour,* and if $f(x)$ gives the height (in inches) of a child at age x years, $f'(x)$ is measured in *inches per year.*

The Derivative of a More Complicated Function

Example 3   Find the derivative of $f(x) = x^2 - 5x + 3$.

Solution

$$f'(x) = \lim_{h \to 0} \frac{f(x+h) - f(x)}{h}$$ *definition of the derivative*

$$= \lim_{h \to 0} \frac{\overbrace{(x+h)^2 - 5(x+h) + 3}^{f(x+h)} - \overbrace{[x^2 - 5x + 3]}^{f(x)}}{h}$$ *using $f(x) = x^2 - 5x + 3$*

$$= \lim_{h \to 0} \frac{x^2 + 2xh + h^2 - 5x - 5h + 3 - x^2 + 5x - 3}{h}$$ *expanding*

$$= \lim_{h \to 0} \frac{2xh + h^2 - 5h}{h} = \lim_{h \to 0} \frac{\cancel{h}[2x + h - 5]}{\cancel{h}}$$ *simplifying*

$$= \lim_{h \to 0} (2x + h - 5) = 2x - 5$$ *evaluating the limit by direct substitution*

Therefore, the derivative of $f(x) = x^2 - 5x + 3$ is $f'(x) = 2x - 5$.   ▪

Interpreting the Sign of the Derivative

Notice that substituting $x = 1$ into the derivative $f'(x) = 2x - 5$ gives $f'(1) = -3$. What does the negative sign mean? If we interpret the derivative as *slope,* it means that the slope is negative, so that the curve slopes *downward* at $x = 1$. If we interpret the derivative as an instantaneous rate of change, it means that the rate of change is negative, so that the function is *decreasing.* For example, if $f(x)$ gives the temperature at time x, then $f'(1) = -3$ means that at time $x = 1$ the temperature is *decreasing* at the rate of 3 degrees per hour.

Calculating Derivatives from the Definition

The general procedure for calculating derivatives is as follows.

(a)   Write the definition

$$f'(x) = \lim_{h \to 0} \frac{f(x+h) - f(x)}{h}$$

(b)   Replace $f(x + h)$ and $f(x)$ by expressions using the given function.

(c)   Simplify.

(d)   Evaluate the limit as $h \to 0$.

Practice Exercise 2

Find $f'(x)$ if $f(x) = 2x^2 + 3x - 7$.   *(solution on page 63)*

Final Example: The Derivative of 1/x

Example 4   Find the derivative of $f(x) = 1/x$.

Solution

$$f'(x) = \lim_{h \to 0} \frac{f(x + h) - f(x)}{h}$$ *definition of $f'(x)$*

$$= \lim_{h \to 0} \frac{\frac{1}{x + h} - \frac{1}{x}}{h}$$ *since $f(x + h) = \frac{1}{x + h}$ and $f(x) = \frac{1}{x}$*

$$= \lim_{h \to 0} \frac{1}{h}\left(\frac{1}{x + h} - \frac{1}{x}\right)$$ *since dividing by h is the same as multiplying by 1/h*

$$= \lim_{h \to 0} \frac{1}{h}\left(\frac{x - (x + h)}{(x + h)x}\right)$$ *subtracting the fractions, using common denominator $(x + h)x$*

$$= \lim_{h \to 0} \frac{1}{h}\frac{-h}{(x + h)x}$$ *simplifying the numerator*

$$= \lim_{h \to 0} \frac{1}{\cancel{h}} \cdot \frac{\overset{-1}{\cancel{-h}}}{(x + h)x} = \lim_{h \to 0} \frac{-1}{(x + h)x}$$ *canceling*

$$= \frac{-1}{(x)x} = \frac{-1}{x^2}$$ *evaluating the limit by direct substitution of $h = 0$*

Therefore, the derivative of $f(x) = 1/x$ is

$$f'(x) = \frac{-1}{x^2}.$$ ▪

Rates of Change Versus Actual Change

The derivative gives the rate of change *at a particular instant,* not an actual change over a period of time. Instantaneous rates of change are like the speeds shown on an automobile speedometer. The fact that at one moment you are driving at 50 miles per hour does not mean that an hour later you will have traveled exactly 50 miles, since the total distance depends on your speed during the entire hour. The derivative gives the *instantaneous* rate of change at a particular moment, and this gives the actual change over a period of time only if the rate remains steady.

We will, however, sometimes interpret derivatives as giving *approximate* changes. For example, we might say that if your speedometer now reads 50 miles per hour, you will travel *about* 50 miles during the next hour, meaning that this will be true provided that your speed remains steady throughout the hour. (In Chapter 5 we will see how to calculate actual changes from rates of change that do not stay constant.)

Summary

The derivative

$$f'(x) = \lim_{h \to 0} \frac{f(x + h) - f(x)}{h}$$

is the single most important concept in all of calculus. If the derivative $f'(x)$ exists at a given x-value, we say that the original function $f(x)$ is *differentiable* at that x-value. The derivative has two interpretations: *slope* and *instantaneous rate of change* (sometimes called simply the "rate of change," with "instantaneous" understood). In the coming sections we will find other interpretations for the derivative.

We now have three different ways of asking the same question:

(a)   Find the instantaneous rate of change of a function.

(b)   Find the slope of the graph of a function.

(c)   Find the derivative of a function.

These three ways are merely the analytic, the geometric, and the mathematical interpretations of differentiation.

The geometric relationship between a function and its derivative can be seen from the graphs of a function and its derivative shown below.

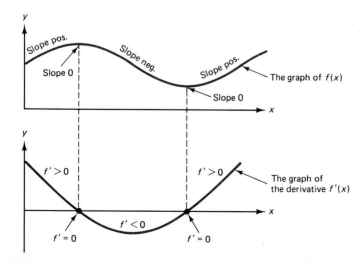

SOLUTIONS TO PRACTICE EXERCISES

1. $f'(x) = \lim_{h \to 0} \dfrac{f(x + h) - f(x)}{h}$ and so at $x = 4$

$$f'(4) = \lim_{h \to 0} \frac{f(4 + h) - f(4)}{h} = \lim_{h \to 0} \frac{(4 + h)^2 - (4)^2}{h}$$

$$= \lim_{h \to 0} \frac{16 + 8h + h^2 - 16}{h} = \lim_{h \to 0} \frac{8h + h^2}{h}$$

$$= \lim_{h \to 0} \frac{h(8 + h)}{h} = \lim_{h \to 0} (8 + h) = 8$$

2. $f'(x) = \lim_{h \to 0} \dfrac{f(x + h) - f(x)}{h}$

$$= \lim_{h \to 0} \frac{2(x + h)^2 + 3(x + h) - 7 - [2x^2 + 3x - 7]}{h}$$

$$= \lim_{h \to 0} \frac{2(x^2 + 2xh + h^2) + 3x + 3h - 7 - 2x^2 - 3x + 7}{h}$$

$$= \lim_{h \to 0} \frac{2x^2 + 4xh + 2h^2 + 3x + 3h - 7 - 2x^2 - 3x + 7}{h}$$

$$= \lim_{h \to 0} \frac{h(4x - 2h + 3)}{h} = \lim_{h \to 0} (4x - 2h + 3) = 4x + 3$$

▪ EXERCISES 2.2 APPLIED EXERCISES

Find all derivatives by the methods of this section.

1 (*General–Temperature*) The temperature of a piece of machinery is $f(x) = x^2 - 8x + 110$ degrees after x minutes.

(a) Find the instantaneous rate of change of the temperature at any time x.

(b) Use your answer to part (a) to find the instantaneous rate of change of the temperature after 2 minutes. Be sure to interpret the sign of your answer.

(c) Use the answer to part (a) to find the instantaneous rate of change after 5 minutes.

2 (*General–Population*) The population of a town is $f(x) = 3x^2 - 12x + 200$ people after x weeks.

(a) Find the instantaneous rate of change of the population for any value of x.

(b) Use your answer to part (a) to find the instantaneous rate of change of the population after 1 week. Be sure to interpret the sign of your answer.

(c) Use your answer to part (a) to find the instantaneous rate of change of the population after 5 weeks.

3 (*Behavioral Science–Learning Theory*) In a psychology experiment a person could memorize x words in $f(x) = 2x^2 - x$ seconds.

(a) Find the instantaneous rate of change of this memorization time.

(b) Evaluate your answer at $x = 5$ and interpret your answer in terms of the proper units.

4 (*Business–Advertising*) An automobile dealership finds that the number of cars that it sells on day x of an advertising campaign is $S(x) = -x^2 + 10x$ (for $x \leq 7$).

(a) Find the instantaneous rate of change of the sales.

(b) Use your answer to part (a) to find the instantaneous rate of change on day $x = 3$.

(c) Use your answer to part (a) to find the instantaneous rate of change on day $x = 6$.

Be sure to interpret the signs of your answers.

5 (*Biomedical–Temperature*) The temperature of a patient in a hospital on day x of an illness is given by $T(x) = -x^2 + 5x + 100$ (for $1 < x < 5$).

(a) Find the instantaneous rate of change of the patient's temperature.

(b) Use your answer to part (a) to find the rate of change on day 2.

(c) Use your answer to part (a) to find the rate of change on day 3.

(d) What do your answers tell you about the patient's health on those two days?

6 (*Biomedical–Bacteria*) The number of bacteria in a culture x hours after treatment with an antibiotic is given by $f(x) = -x^2 + 12x + 1000$ (for $0 < x < 30$).

(a) Find the instantaneous rate of change of the bacteria population after x hours.

(b) Use your answer to part (a) to find the rate of change after 2 hours.

(c) Use your answer to part (a) to find the rate of change after 20 hours.

Be sure to interpret the signs of your answers.

For each function:
  (a) *Find the slope of the graph at x.*
  (b) *Use your answer to part (a) to find the slope at $x = 2$.*
  (c) *Use your answer to part (a) to find the slope at $x = 4$.*
  (d) *Sketch the graph of the function $f(x)$. (Use the technique that we used in Section 1.2.) Your graph should agree with the slopes found in parts (b) and (c).*

7 $f(x) = x^2 - 6x + 10$ **8** $f(x) = 2x^2 - 12x + 20$

For each function:
  (a) *Calculate $f'(x)$ using the methods of this section.*
  (b) *Interpret geometrically your answer to part (a).*

9 $f(x) = 3x - 4$ **10** $f(x) = 2x - 9$

11 $f(x) = 5$ **12** $f(x) = 12$

13 $f(x) = mx + b$ **14** $f(x) = c$
  (m and b are constants)   (c is a constant)

Find the derivative of each function.

15 $f(x) = ax^2 + bx + c$ (a, b, and c are constants)

16 $f(x) = \frac{1}{2}x^2 + 1$

17 $f(x) = x^3$
  [*Hint*: Use $(x + h)^3 = x^3 + 3x^2h + 3xh^2 + h^3$.]

18 $f(x) = x^4$
  [*Hint*: Use $(x + h)^4 = x^4 + 4x^3h + 6x^2h^2 + 4xh^3 + h^4$.]

19 $f(x) = \frac{2}{x}$ **20** $f(x) = \frac{3}{x}$

21 $f(x) = \frac{1}{x^2}$ **22** $f(x) = \frac{2}{x^2}$

23 $f(x) = \sqrt{x}$ (*Hint*: Multiply the top and bottom of the difference quotient by $\sqrt{x + h} + \sqrt{x}$ and then simplify.)

24 $f(x) = \frac{1}{\sqrt{x}}$ (*Hint*: Multiply the top and bottom of the difference quotient by $\sqrt{x} + \sqrt{x + h}$ and then simplify.)

2.3 SOME DIFFERENTIATION FORMULAS

Introduction

In Section 2.2 we defined the *derivative* of a function and used it to calculate instantaneous rates of change and slopes. Even for a function as simple as $f(x) = x^2$, however, calculating the derivative from the definition was rather involved. Calculus would be of limited usefulness if all derivatives had to be calculated in this way.

In this section we will learn several *rules of differentiation* which will simplify differentiation. The rules are derived from the definition of the derivative, which is why we studied the definition first. We will also learn another important use for differentiation: calculating "marginals" (marginal revenue, marginal cost, and marginal profit).

Leibniz's Notation for the Derivative

Calculus was developed independently by two people, Isaac Newton (1642–1727) and Gottfried Wilhelm von Leibniz (1646–1716). Newton denoted derivatives by a dot over the function, $\dot{f}$, a notation that has been largely replaced by the "prime" that we have been using. Leibniz wrote the derivative of $f(x)$ by writing a $\frac{d}{dx}$ in front of the function: $\frac{d}{dx} f(x)$. That is,

read: "*d* over *dx* of *f* of *x*"

Isaac Newton

$$\frac{d}{dx} f(x) \quad \text{means} \quad f'(x)$$

For example, in Section 2.2 we saw that the derivative of $f(x) = x^2$ is $f'(x) = 2x$. In Leibniz's notation this would be written

$$\text{if} \quad f(x) = x^2 \quad \text{then} \quad \frac{d}{dx} f(x) = 2x$$

or more briefly,

$$\frac{d}{dx} x^2 = 2x$$

Gottfried Wilhelm Leibniz

the derivative of x^2 is $2x$

The following table shows equivalent statements in the two notations.

Prime Notation	Leibniz's Notation	
$f'(x)$	$\dfrac{d}{dx} f(x)$	
y'	$\dfrac{dy}{dx}$	*for y a function of x*
$g'(w)$	$\dfrac{d}{dw} g(w)$	*use d/dw to differentiate a function of w*
$h'(t)$	$\dfrac{d}{dt} h(t)$	*use d/dt to differentiate a function of t*

Each notation has its own advantages, and we will use both.*
Leibniz's notation comes from writing the derivative of $y = f(x)$ with Δx in place of h,

$$\frac{dy}{dx} = \lim_{\Delta x \to 0} \frac{f(x + \Delta x) - f(x)}{\Delta x}$$

The numerator $f(x + \Delta x) - f(x)$ is just the change in y (shown in the diagram below).

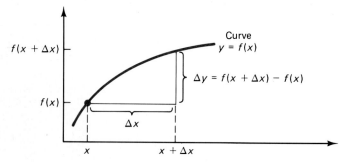

Therefore, $f(x + \Delta x) - f(x)$ may be replaced by Δy, giving

$$\frac{dy}{dx} = \lim_{\Delta x \to 0} \frac{\Delta y}{\Delta x}$$

In this equation it is as if the limiting operation turns each Δ (the Greek letter D) into a d, changing $\Delta y/\Delta x$ into dy/dx. In other words, Leibniz's dy/dx notation for the derivative reminds us that the derivative is the limit of the slope $\Delta y/\Delta x$.

---

* Other notations for the derivative are $Df(x)$ and $D_x f(x)$, but we shall not use them.

The Derivative of a Constant Function is Zero

The first rule of differentiation shows how to differentiate a constant function. For any constant c,

$$\frac{d}{dx} c = 0$$

the derivative of a constant is zero

Example 1

$$\frac{d}{dx} 7 = 0$$

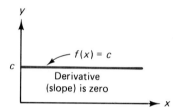

This rule is obvious geometrically. The graph of a constant function $f(x) = c$ is the horizontal line $y = c$. Since the slope of a horizontal line is zero, the derivative of $f(x) = c$ is zero.

A constant function (a horizontal line) has derivative (slope) zero.

This rule follows immediately from the definition of the derivative. The constant function $f(x) = c$ has the same value c for *any* value of x, so in particular, $f(x + h) = c$ and $f(x) = c$. Substituting these into the definition of the derivative gives

$$f'(x) = \lim_{h \to 0} \frac{\overbrace{f(x+h)}^{c} - \overbrace{f(x)}^{c}}{h} = \lim_{h \to 0} \frac{c - c}{h} = \lim_{h \to 0} \frac{\overbrace{0}^{0}}{h} = \lim_{h \to 0} 0 = 0$$

Therefore, the derivative of a constant function $f(x) = c$ is $f'(x) = 0$.

Power Rule

One of the most useful differentiation formulas in all of calculus is called the *power rule*. It tells how to differentiate powers like x^7 or x^{100}. For any constant exponent n,

Power Rule

$$\frac{d}{dx} x^n = n \cdot x^{n-1}$$

to differentiate x^n, bring down the exponent as a multiplier and then decrease the exponent by 1

A derivation of the power rule is given at the end of this section.

Example 2

(a) $\dfrac{d}{dx} x^7 = 7x^{7-1} = 7x^6$

bring down↗ ↖decrease the
the exponent   exponent by 1

(b) $\dfrac{d}{dx} x^{100} = 100x^{100-1} = 100x^{99}$

(c) $\dfrac{d}{dx} x^2 = 2x^{2-1} = 2x^1 = 2x$ *as we showed in a longer way on pages 58–59*

(d) $\dfrac{d}{dx} x^{-2} = -2x^{-2-1} = -2x^{-3}$ *the power rule holds for negative exponents*

(e) $\dfrac{d}{dx} x^{1/2} = \dfrac{1}{2} x^{(1/2)-1} = \dfrac{1}{2} x^{-1/2}$ *and for fractional exponents*

(f) $\dfrac{d}{dx} x^1 = 1x^{1-1} = x^0 = 1$ ■

This last result should be remembered separately since it is used so frequently.

$$\frac{d}{dx} x = 1$$

the derivative of x is 1

From now on we will skip the middle step in these examples, differentiating powers in one step:

$$\frac{d}{dx} x^{50} = 50x^{49}$$

$$\frac{d}{dx} x^{2/3} = \frac{2}{3} x^{-1/3}$$
$$\longleftarrow 2/3 - 1$$

Practice Exercise 1

Find

(a) $\dfrac{d}{dx} x^3$ (b) $\dfrac{d}{dx} x^{-5}$ (c) $\dfrac{d}{dx} x^{1/4}$ *(solutions on page 77)*

Constant Multiple Rule

The power rule shows how to differentiate a power like x^3. The *constant multiple rule* extends this result to functions like $5x^3$, a constant *times* a function. Briefly, to differentiate a constant times a function we simply "carry along" the constant and differentiate the function. For any constant c,

---

Constant Multiple Rule

$$\frac{d}{dx}[c \cdot f(x)] = c \cdot f'(x)$$

the derivative of a constant times a function is the constant times the derivative of the function

---

[provided, of course, that the derivative $f'(x)$ exists]. A derivation of this rule is given at the end of this section.

Example 3

(a) $\dfrac{d}{dx} 5x^3 = 5 \cdot 3x^2 = 15x^2$ (b) $\dfrac{d}{dx} 3x^{-4} = 3(-4)x^{-5} = -12x^{-5}$

  carry along⌐ ⌐derivative
  the constant   of x^3

(c) $\dfrac{d}{dx} 6x^{1/3} = 6 \cdot \dfrac{1}{3} x^{-2/3} = 2x^{-2/3}$ ▪

Again we will skip the middle step, bringing down the exponent and immediately multiplying it by the number in front of the x.

Example 4

(a) $\dfrac{d}{dx} 8x^{-1/2} = -4x^{-3/2}$ (b) $\dfrac{d}{dx} 7x = 7 \cdot 1 = 7$

    └─ $8(-1/2)$ └ derivative of x ▪

This example, showing that the derivative of $7x$ is just 7, leads to a very useful general rule. For any constant c,

---

$$\frac{d}{dx}(cx) = c$$

the derivative of a constant times x is just the constant

Example 5

(a) $\dfrac{d}{dx}(-5x) = -5$ (b) $\dfrac{d}{dx} 7 = 0$

for a constant alone the derivative is zero

(c) $\dfrac{d}{dx}(7x^2) = 7 \cdot 2x = 14x$

but for a constant times a function the derivative is the constant times the derivative of the function

▪

Sum Rule

The *sum rule* extends differentiation to *sums* of functions. Briefly, to differentiate a *sum* of two functions, just differentiate the functions separately and add the results.

Sum Rule

$$\frac{d}{dx}[f(x) + g(x)] = f'(x) + g'(x)$$

the derivative of a sum is the sum of the derivatives

[provided, of course, that $f'(x)$ and $g'(x)$ exist]. A derivation of the sum rule is given at the end of this section. For example, the derivative of the sum

$$x^3 + x^5$$

is just the sum of the derivatives

$$3x^2 + 5x^4$$

differentiating each separately

A similar rule holds for the *difference* of two functions,

$$\frac{d}{dx}[f(x) - g(x)] = f'(x) - g'(x)$$

the derivative of a difference is the difference of the derivatives

---

[provided that $f'(x)$ and $g'(x)$ exist]. These two rules may be written together:

$$\frac{d}{dx}[f(x) \pm g(x)] = f'(x) \pm g'(x)$$

the $\pm$ sign means read both $+$ signs or both $-$ signs

---

and the same rule holds for sums and differences of any number of terms.

Using these rules we may differentiate any polynomial, or for that matter, functions with variables raised to *any* constant powers.

Example 6

(a) $\dfrac{d}{dx}(x^3 - x^5) = 3x^2 - 5x^4$

derivatives taken separately

(b) $\dfrac{d}{dx}(5x^{-2} - 6x^{1/3} + 7x) = -10x^{-3} - 2x^{-2/3} + 7$

(c) $\dfrac{d}{dx}(2x^4 - x^2 + 4x - 1) = 8x^3 - 2x + 4$

the constant -1 has derivative 0

■

Evaluation of Derivatives

The notation $f'(2)$ means differentiate the function and then evaluate at 2.

```
                    ┌── derivative
                    │
                    ↓
              f'(2)
                ↑
                └── evaluated at x = 2
```

In Leibniz's notation this would be written

```
                    ┌────── derivative
                    │
                    ↓
          df │
          ── │
          dx │x=2
                 ↑
                 └── evaluated at x = 2
```

(The bar | means "evaluated at.") Both notations mean *first* differentiate and *then* evaluate.

$f'(c)$ and $\dfrac{df}{dx}\Big|_{x=c}$ mean the derivative evaluated at $x = c$.

Example 7 If $f(x) = x^4$, find $f'(2)$.

Solution

$$f'(x) = 4x^3 \qquad\qquad\qquad\qquad\qquad \textit{first differentiate}$$

$$f'(2) = 4 \cdot 2^3 = 4 \cdot 8 = 32 \qquad\qquad \textit{then evaluate } f' \textit{ at } x = 2$$

▪

Practice Exercise 2

If $f(x) = 2x^3$, find $f'(-2)$.   *(solution on page 77)*

Example 8 If $f(x) = 16x^{-1/2}$, find $\dfrac{df}{dx}\Big|_{x=4}$.

Solution

$$\frac{df}{dx} = -8x^{-3/2} \qquad\qquad\qquad\qquad \textit{first differentiate}$$

$$\frac{df}{dx}\Big|_{x=4} = -8(4)^{-3/2} = -8\frac{1}{(\sqrt{4})^3} = -8\frac{1}{8} = -1 \qquad \textit{then evaluate at } x = 4$$

▪

Example 9 If $h(t) = t^2 - \dfrac{64}{t} - 24\sqrt[3]{t} + 7$, find $\dfrac{dh}{dt}\Big|_{t=8}$.

Solution

$$h(t) = t^2 - 64t^{-1} - 24t^{1/3} + 7 \qquad \textit{writing } h(t) \textit{ in exponential form}$$

$$\frac{dh}{dt} = 2t + 64t^{-2} - 8t^{-2/3} \qquad \textit{differentiating}$$

$$\left.\frac{dh}{dt}\right|_{t=8} = 2(8) + 64(8)^{-2} - 8(8)^{-2/3} \qquad \textit{evaluating at } t = 8$$

$$= 16 + \frac{64}{64} - 8\left(\frac{1}{4}\right) = 16 + 1 - 2 = 15 \qquad ■$$

Practice Exercise 3

If $f(x) = 4\sqrt{x} + \dfrac{32}{x} - 3$, find $\left.\dfrac{df}{dx}\right|_{x=4}$. *(solution on page 77)*

Derivatives in Business and Economics: Marginals

There is another interpretation for the derivative, one which is particularly important in business and economics. Suppose that a company has calculated the revenue, cost, and profit functions for one of its products,

$R(x)$ = the *total revenue* (income) from selling x units of the product

$C(x)$ = the *total cost* of producing x units of the product

$P(x)$ = the *total profit* from producing and selling x units of the product

The term *marginal cost* means the additional cost of producing one more unit. Marginal cost is measured in dollars per unit, and is therefore the *rate* at which cost changes as production increases. If we calculate rates of change as *instantaneous* rates of change (that is, as derivatives), we see that the marginal cost is the *derivative* of the cost function:

The marginal cost function $MC(x)$ is the derivative of the cost function,

$$MC(x) = C'(x)$$

Similarly for revenue and profit:

The marginal revenue function $MR(x)$ is the derivative of the revenue function,

$$MR(x) = R'(x)$$

> The marginal profit function MP(x) is the derivative of the profit function,
>
> $$MP(x) = P'(x)$$

All of this can be summarized very briefly: "marginal" means "derivative of."

We now have three interpretations for the derivative: *slopes*, *instantaneous rates of change*, and *marginals*.

Example 10   A company manufactures cordless telephones and finds that its cost function (the total cost of manufacturing x telephones) is

$$C(x) = 400\sqrt{x} + 500$$

dollars, where x is the number of telephones produced.

(**a**)   Find the marginal cost function MC(x).

(**b**)   Find the marginal cost when 100 telephones have been produced, and interpret your answer.

Solution

(**a**)   The marginal cost function is the derivative of the cost function $C(x) = 400x^{1/2} + 500$,

$$MC(x) = 200x^{-1/2} = \frac{100}{\sqrt{x}} \qquad \text{\textit{derivative of }} C(x)$$

(**b**)   To find the marginal cost when 100 telephones have been produced, we evaluate the marginal cost function at $x = 100$,

$$MC(100) = \frac{200}{\sqrt{100}} = \frac{200}{10} = \$20$$

Interpretation: When 100 telephones have been produced, the marginal cost is \$20, meaning that to produce one more telephone costs about \$20.

■

Practice Exercise 4

An electronics company finds that its total profit from selling x computer chips is $P(x) = .02x^{3/2} - 200$ dollars.

(**a**)   Find the company's marginal profit function.

(**b**)   Find the marginal profit when 10,000 units have been produced and interpret your answer.

(solutions on page 78)

Application to Behavioral Science

Example 11   A psychology researcher finds that the number of names that a person can memorize in x minutes is approximately $f(x) = 6\sqrt[3]{x^2}$. Find the instantaneous rate of change of this function after 8 minutes and interpret your answer.

Solution

$$f(x) = 6x^{2/3}$$

function in exponential form

$$f'(x) = 6 \cdot \frac{2}{3} x^{-1/3} = 4x^{-1/3}$$

the instantaneous rate of change is $f'(x)$

$$f'(8) = 4(8)^{-1/3} = 4\left(\frac{1}{\sqrt[3]{8}}\right) = 4\left(\frac{1}{2}\right) = 2$$

evaluating at $x = 8$

Interpretation of $f'(8) = 2$: After 8 minutes the person can memorize about two additional names per minute.   ■

Overview

You may have noticed that calculus requires a more abstract point of view than precalculus mathematics. In earlier courses you looked at functions and graphs as collections of individual points, to be plotted one at a time. Now, however, we are operating on *whole functions* all at once (for example, differentiating the function x^3 to obtain the function $3x^2$). In calculus, the basic objects of interest are *functions*, and a function should be thought of as a *single* object.

This is in keeping with a trend toward increasing abstraction as you learn mathematics. You first studied single numbers, then points (pairs of numbers), then functions (collections of points) and now collections of functions (polynomials, differentiable functions, and so on). Each stage has been a generalization of the previous stage as you reach higher levels of sophistication. This process of generalization or "chunking" of knowledge enables you to express ideas of wider applicability and power.

Summary

The developments of this chapter have followed two quite different lines— one technical (the *rules* for derivatives) and the other conceptual (the *meaning* of derivatives). We will first review the technical developments.

Although we have learned several differentiation rules, we really only know how to differentiate one kind of function, *x to a power:*

$$\frac{d}{dx} x^n = n x^{n-1}$$

The other rules,

$$\frac{d}{dx}[c{\cdot}f(x)] = c{\cdot}f'(x)$$

$$\frac{d}{dx}[f(x) \pm g(x)] = f'(x) \pm g'(x)$$

simply extend the power rule to sums, differences, and constant multiples of such powers. Therefore, any function to be differentiated must first be expressed in terms of powers. This is why we reviewed exponential notation so carefully in Chapter 1.

Besides these differentiation techniques, we have three interpretations for the derivative: instantaneous rates of change, slopes, and marginals.

Verification of the Power Rule for Positive Integer Exponents

We know that $(x + h)^2 = x^2 + 2xh + h^2$, and multiplying this by $(x + h)$ would give

$$(x + h)^3 = x^3 + 3x^2h + 3xh^2 + h^3$$

In general, by repeatedly multiplying by $(x + h)$ we could show that for any positive integer n,

$$(x + h)^n = x^n + nx^{n-1}h + n(n - 1)x^{n-2}h^2 + \cdots + nxh^{n-1} + h^n$$
$$= x^n + nx^{n-1}h + h^2[n(n - 1)x^{n-2} + \cdots + nxh^{n-3} + h^{n-2}] \qquad \textit{factoring out } h^2$$
$$= x^n + nx^{n-1}h + h^2{\cdot}P$$

where P stands for the polynomial in the square brackets in the preceding line. This formula,

$$(x + h)^n = x^n + nx^{n-1}h + h^2{\cdot}P$$

will be useful below.

To prove the power rule for any positive integer n we use the definition of the derivative to differentiate $f(x) = x^n$.

$$f'(x) = \lim_{h \to 0} \frac{f(x+h) - f(x)}{h}$$ *definition of the derivative*

$$= \lim_{h \to 0} \frac{(x+h)^n - x^n}{h}$$ *since $f(x+h) = (x+h)^n$ and $f(x) = x^n$*

$$= \lim_{h \to 0} \frac{x^n + nx^{n-1}h + h^2 \cdot P - x^n}{h}$$ *expanding, using the formula derived above*

$$= \lim_{h \to 0} \frac{nx^{n-1}h + h^2 \cdot P}{h}$$ *canceling the x^n and the $-x^n$*

$$= \lim_{h \to 0} \frac{h(nx^{n-1} + h \cdot P)}{h}$$ *factoring out an h*

$$= \lim_{h \to 0} (nx^{n-1} + h \cdot P)$$ *canceling the h*

$$= nx^{n-1}$$ *evaluating the limit by direct substitution of $h = 0$*

This shows that the derivative of x^n is nx^{n-1}.

Verification of the Constant Multiple Rule

For a constant c and a function f, let $g(x) = c \cdot f(x)$. If $f'(x)$ exists, we may calculate the derivative $g'(x)$ as follows:

$$g'(x) = \lim_{h \to 0} \frac{g(x+h) - g(x)}{h}$$ *definition of the derivative*

$$= \lim_{h \to 0} \frac{c \cdot f(x+h) - c \cdot f(x)}{h}$$ *since $g(x+h) = c \cdot f(x+h)$ and $g(x) = c \cdot f(x)$*

$$= \lim_{h \to 0} \frac{c \cdot [f(x+h) - f(x)]}{h}$$ *factoring out the c*

$$= c \cdot \lim_{h \to 0} \frac{f(x+h) - f(x)]}{h}$$ *taking c outside the limit (using limit rules 1 and 4c on page 47)*

$$= c \cdot f'(x)$$ *constant multiple rule*

where the last step comes from recognizing that the limit expression in the previous line is just the definition of the derivative $f'(x)$. This shows that the derivative of a constant times a function, $c \cdot f(x)$, is the constant times the derivative of the function, $c \cdot f'(x)$.

Verification of the Sum Rule

For two functions f and g, let their sum be $s(x) = f(x) + g(x)$. If $f'(x)$ and $g'(x)$ exist, we may calculate $s'(x)$ as follows:

$$s'(x) = \lim_{h \to 0} \frac{s(x + h) - s(x)}{h}$$

definition of the derivative

$$= \lim_{h \to 0} \frac{[f(x + h) + g(x + h)] - [f(x) + g(x)]}{h}$$

since $s(x) = f(x) + g(x)$ and $s(x + h) = f(x + h) + g(x + h)$

$$= \lim_{h \to 0} \frac{f(x + h) + g(x + h) - f(x) - g(x)}{h}$$

dropping the brackets

$$= \lim_{h \to 0} \frac{f(x + h) - f(x) + g(x + h) - g(x)}{h}$$

rearranging the numerator

$$= \lim_{h \to 0} \left[\frac{f(x + h) - f(x)}{h} + \frac{g(x + h) - g(x)}{h} \right]$$

separating the fraction into two parts

$$= \lim_{h \to 0} \frac{f(x + h) - f(x)}{h} + \lim_{h \to 0} \frac{g(x + h) - g(x)}{h}$$

using limit rule 4a on page 47

$$= f'(x) + g'(x)$$

sum rule

where the last step comes from recognizing that the limit expressions in the previous line are just the definitions of the derivatives $f'(x)$ and $g'(x)$. This shows that the derivative of a sum $f(x) + g(x)$ is the sum of the derivatives $f'(x) + g'(x)$.

SOLUTIONS TO PRACTICE EXERCISES

1. **(a)** $\dfrac{d}{dx} x^3 = 3x^{3-1} = 3x^2$

(b) $\dfrac{d}{dx} x^{-5} = -5x^{-5-1} = -5x^{-6}$

(c) $\dfrac{d}{dx} x^{(1/4)} = \dfrac{1}{4} x^{(1/4)-1} = \dfrac{1}{4} x^{-3/4}$

2. $f'(x) = 6x^2$

$f'(-2) = 6(-2)^2 = 6 \cdot 4 = 24$

3. $f(x) = 4x^{1/2} + 32x^{-1} - 3$

$\dfrac{df}{dx} = 2x^{-1/2} - 32x^{-2}$

$\left. \dfrac{df}{dx} \right|_{x=4} = 2(4)^{-1/2} - 32(4)^{-2} = \dfrac{2}{\sqrt{4}} - \dfrac{32}{4^2} = \dfrac{2}{2} - \dfrac{32}{16} = 1 - 2 = -1$

4. **(a)** $MP(x) = P'(x) = .02(3/2)x^{1/2} = .03x^{1/2}$

(b) $MP(10,000) = .03(10,000)^{1/2} = .03\sqrt{10,000} = .03 \cdot 100 = 3$
Interpretation: After 10,000 chips, the profit on each additional chip is about $3.

EXERCISES 2.3

Find the derivative of each function.

1 $f(x) = x^4$ **2** $f(x) = x^5$ **3** $f(x) = x^{500}$ **4** $f(x) = x^{1000}$

5 $f(x) = x^{1/2}$ **6** $f(x) = x^{1/3}$ **7** $g(x) = \dfrac{1}{2}x^4$ **8** $f(x) = \dfrac{1}{3}x^9$

9 $g(w) = 6\sqrt[3]{w}$ **10** $g(w) = 12\sqrt{w}$ **11** $h(x) = \dfrac{3}{x^2}$ **12** $h(x) = \dfrac{4}{x^3}$

13 $f(x) = 4x^2 - 3x + 2$ **14** $f(x) = 3x^2 - 5x + 4$

15 $g(x) = \sqrt{x} - \dfrac{1}{x}$ **16** $g(x) = \sqrt[3]{x} - \dfrac{1}{x}$

17 $h(x) = 6\sqrt[3]{x^2} - \dfrac{12}{\sqrt[3]{x}}$ **18** $h(x) = 8\sqrt{x^3} - \dfrac{8}{\sqrt[4]{x}}$

19 $f(x) = \dfrac{10}{\sqrt{x}} - 9\sqrt[3]{x^5} + 17$ **20** $f(x) = \dfrac{9}{\sqrt[3]{x}} - 16\sqrt{x^5} - 14$

Find the indicated derivatives.

21 (a) Find the derivative of $f(x) = 2$.

    (b) Interpret your answer in terms of slope.

    (c) Interpret your answer in terms of instantaneous rates of change.

22 (a) Find the derivative of $f(x) = 3x$.

    (b) Interpret your answer in terms of slope.

    (c) Interpret your answer in terms of instantaneous rates of change.

23 If $f(x) = x^5$, find $f'(-2)$.

24 If $f(x) = x^4$, find $f'(-3)$.

25 If $g(w) = 12w^{2/3}$, find $g'(8)$.

26 If $g(w) = 36w^{1/3}$, find $g'(8)$.

27 If $h(t) = t^4 + 2t - \dfrac{1}{t}$, find $h'(-1)$.

28 If $h(t) = t^6 - 5t + \dfrac{1}{t}$, find $h'(-1)$.

29 If $f(x) = 6\sqrt[3]{x^2} - \dfrac{48}{\sqrt[3]{x}}$, find $f'(8)$.

30 If $f(x) = 12\sqrt[3]{x^2} + \dfrac{48}{\sqrt[3]{x}}$, find $f'(8)$.

31 If $f(x) = x^3$, find $\dfrac{df}{dx}\Big|_{x=-3}$.

32 If $f(x) = x^4$, find $\dfrac{df}{dx}\Big|_{x=-2}$.

33 If $g(w) = 120w^{2/5}$, find $\dfrac{dg}{dw}\Big|_{w=32}$.

34 If $g(w) = 60w^{3/5}$, find $\dfrac{dg}{dw}\Big|_{w=32}$.

35 If $h(t) = t^3 - t^2 + \dfrac{1}{t^2}$, find $\dfrac{dh}{dt}\Big|_{t=-1}$.

36 If $h(t) = t^4 - t^3 + \dfrac{1}{t^3}$, find $\dfrac{dh}{dt}\Big|_{t=-1}$.

37 If $f(x) = \dfrac{16}{\sqrt{x}} + 8\sqrt{x}$, find $\dfrac{df}{dx}\Big|_{x=4}$.

38 If $f(x) = \dfrac{54}{\sqrt{x}} + 12\sqrt{x}$, find $\dfrac{df}{dx}\Big|_{x=9}$.

▨ APPLIED EXERCISES

39 (*Business–Marginal Profit*) A company's profit function is

$$P(x) = -\frac{1}{8}x^4 + x^3 - 6x^2 + 28x - 400$$

dollars, where x is the quantity of good produced each day.

(a) Find the marginal profit function.

(b) Find the marginal profit at $x = 2$ and interpret your answer.

(c) Find the marginal profit at $x = 4$ and interpret your answer.

40 (*Business–Marginal Cost*) A steel mill finds that its cost function is

$$C(x) = 8000\sqrt{x} - 6000\sqrt[3]{x} + 300$$

dollars, where x is the (daily) production of steel (in tons).

(a) Find the marginal cost function.

(b) Find the marginal cost when 64 tons of steel are produced.

41 (*General–Population*) A company that makes games for teenage children forecasts that the teenage population in the United States x years from now will be

$$P(x) = 12,000,000 - 12,000x + 600x^2 + 100x^3$$

Find the rate of change of the teenage population

(a) x years from now

(b) 1 year from now and interpret your answer

(c) 10 years from now and interpret your answer

42 (*Biomedical–Flu Epidemic*) The number of people newly infected on day t of a flu epidemic is $f(t) = 13t^2 - t^3$ (for $t \le 13$). Find the instantaneous rate of change of this number on

(a) day 5 and interpret your answer

(b) day 10 and interpret your answer

43 (*Business–Advertising*) It has been found that the number of people who will see a newspaper advertisement obeys a law of the form

$$N(x) = T - \frac{T/2}{x} \qquad \text{for } x \ge 1$$

where T is the total readership of the newspaper and x is the number of days that the ad runs. If a newspaper has a circulation of 400,000, an ad that runs for x days will be seen by

$$N(x) = 400,000 - \frac{200,000}{x}$$

people. Find how fast this pool of potential customers is growing when this ad has run for 5 days.

44 (*General–Pollution*) An electrical generating plant burns high-sulfur oil, and the amount of sulfur dioxide pollution x miles downwind of the plant is $f(x) = 108x^{-2}$ parts per million (ppm). Find the instantaneous rate of change of the pollution level 2 miles from the source. Interpret your answer in the proper units.

45 (*Biomedical*) Nitroglycerin is often prescribed to enlarge blood vessels that have become too constricted. If the cross-sectional area of a blood vessel t hours after nitroglycerine is administered is $A(t) = .01t^2$ square centimeters (for $1 \le t \le 5$), find the instantaneous rate of change of the cross-sectional area 4 hours after the administration of nitroglycerine.

46 (*General–Hailstones*) Hailstones are frozen raindrops that increase in size as long as the updrafts keep them in the clouds. The weight of a typical hailstone that remains in a cloud for t minutes is $W(t) = .05t^3$ ounces. Find the instantaneous rate of change of the weight after 2 minutes.

47 (*Psychology–Learning Rates*) A language school has found that its students can memorize $p(t) = 24\sqrt{t}$ phrases in t hours of class (for $1 \le t \le 10$). Find the instantaneous rate of change of this quantity after 4 hours of class.

48 (*General–Ecology*) Downstream from a waste treatment plant the amount of dissolved oxygen in the water usually decreases for some distance (due to bacteria consuming the oxygen) and then increases (due to natural purification). A graph of the dissolved oxygen at various distances downstream looks like the curve below (known as the oxygen sag). The amount of dis-

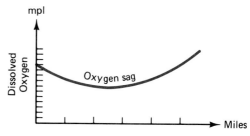

Distance Downstream from Treatment Plant

solved oxygen is usually taken as a measure of the health of the river. Suppose that the amount of dissolved oxygen x miles downstream is $D(x) = .2x^2 - 2x + 10$ mpl (milligrams per liter). Use this formula to find the instantaneous rate of change of the dissolved oxygen

(a) 1 mile downstream

(b) 10 miles downstream

and interpret the signs of your answers.

49–50 (*Economics–Marginal   Utility*) Generally,   the more you have of something, the less valuable each additional unit becomes. For example, a dollar is less valuable to a millionaire than to a beggar. Economists define a person's "utility function" $U(x)$ for a product as the "perceived value" of having x units of that product. The *derivative* of $U(x)$ is called the *marginal utility function*, $MU(x) = U'(x)$. Suppose that a person's utility function for money is given by the function below. That is, $U(x)$ is the utility (perceived value) of x dollars.

(a) Find the marginal utility function $MU(x)$.

(b) Find $MU(1)$, the marginal utility of the first dollar.

(c) Find $MU(1,000,000)$, the marginal utility of the millionth dollar.

49 $U(x) = 100\sqrt{x}$ **50** $U(x) = 12\sqrt[3]{x}$ dollar

2.4 PRODUCT AND QUOTIENT RULES

Introduction

In Section 2.3 we learned how to differentiate the sum and difference of two functions—we simply take the sum or difference of the derivatives. In this section we learn how to differentiate the *product* and *quotient* of two functions. Unfortunately, we do not simply take the product or quotient of the derivatives. Matters are a little more complicated.

Product Rule

To differentiate the product of two functions, $f(x) \cdot g(x)$, we use the *product rule,*

Product Rule

$$\frac{d}{dx}[f(x) \cdot g(x)] = f'(x) \cdot g(x) + f(x) \cdot g'(x)$$

---

the derivative of a product is the derivative of the first function times the second function plus the first function times the derivative of the second function

[provided, of course, that the derivatives $f'(x)$ and $g'(x)$ both exist]. The formula is clearer if we write the functions simply as f and g.

$$\frac{d}{dx} [f \cdot g] = f' \cdot g + f \cdot g'$$

derivative   second   first   derivative
of the first function function of the second

A derivation of the product rule is given at the end of this section.

Example 1   Use the product rule to calculate $\dfrac{d}{dx} [x^3 \cdot x^5]$.

Solution

$$\frac{d}{dx} [x^3 \cdot x^5] = 3x^2 \cdot x^5 + x^3 \cdot 5x^4 = 3x^7 + 5x^7 = 8x^7 \qquad ■$$

derivative   second first derivative
of the first   of the second

We may check this answer by simplifying the original product, $x^3 \cdot x^5 = x^8$, and then differentiating,

$$\frac{d}{dx} [x^3 \cdot x^5] = \frac{d}{dx} x^8 = 8x^7$$

x^8

which agrees with our answer using the product rule.
    Notice that the derivative of a product is *not* the product of the derivatives: $(f \cdot g)' \neq f' \cdot g'$. For $x^3 \cdot x^5$ the product of the derivatives is $3x^2 \cdot 5x^4 = 15x^6$, which is *not* the correct answer $8x^7$ that we found above. The product rule shows the correct way to differentiate a product.

Example 2   Use the product rule to find $\dfrac{d}{dx} [(x^2 - x + 2)(x^3 + 3)]$.

Solution

$$\frac{d}{dx} [(x^2 - x + 2)(x^3 + 3)] = (2x - 1)(x^3 + 3) + (x^2 - x + 2)(3x^2)$$

derivative   derivative
of $x^2 - x + 2$ of $x^3 + 3$

$$= 2x^4 + 6x - x^3 - 3 + 3x^4 - 3x^3 + 6x^2 \qquad \textit{multiplying out}$$

$$= 5x^4 - 4x^3 + 6x^2 + 6x - 3 \qquad \textit{simplifying}$$

■

This problem could also have been done by first multiplying out the original function and then differentiating.

Practice Exercise 1

Use the product rule to find $\dfrac{d}{dx}[x^3(x^2 - x)]$.

(solution on page 89)

Quotient Rule

The *quotient rule* shows how to differentiate a quotient of two functions.

Quotient Rule

$$\frac{d}{dx}\left(\frac{f(x)}{g(x)}\right) = \frac{g(x) \cdot f'(x) - g'(x) \cdot f(x)}{[g(x)]^2}$$

the bottom times the derivative of the top, minus the derivative of the bottom times the top

the bottom squared

[provided that the derivatives $f'(x)$ and $g'(x)$ both exist and that $g(x) \neq 0$].
A derivation of the quotient rule is given at the end of this section.

The quotient rule looks less formidable if we write the functions simply as f and g,

$$\frac{d}{dx}\left(\frac{f}{g}\right) = \frac{g \cdot f' - g' \cdot f}{g^2}$$

or even as

$$\frac{d}{dx}\left(\frac{\text{numerator}}{\text{denominator}}\right) = \frac{(\text{denominator}) \cdot \left(\dfrac{d}{dx}\text{ numerator}\right) - \left(\dfrac{d}{dx}\text{ denominator}\right) \cdot (\text{numerator})}{(\text{denominator})^2}$$

Example 3 Use the quotient rule to find $\dfrac{d}{dx}\left(\dfrac{x^9}{x^3}\right)$.

Solution

$$\frac{d}{dx}\left(\frac{x^9}{x^3}\right) = \frac{(x^3)(9x^8) - (3x^2)(x^9)}{(x^3)^2} = \frac{9x^{11} - 3x^{11}}{x^6} = \frac{6x^{11}}{x^6} = 6x^5$$

bottom — derivative of the top — derivative of the bottom — top

bottom squared ■

We may check this answer by simplifying the original quotient and then differentiating,

$$\frac{d}{dx}\left(\frac{x^9}{x^3}\right) = \frac{d}{dx}\,x^6 = 6x^5$$

x^6

which agrees with our answer using the quotient rule.

Notice that the derivative of a quotient is *not* the quotient of the derivatives:

$$\left(\frac{f}{g}\right)' \neq \frac{f'}{g'}$$

For the quotient x^9/x^3, taking the quotient of the derivatives gives $(9x^8)/(3x^2) = 3x^6$, which is *not* the answer correct $6x^5$ that we found above. The quotient rule shows the correct way to differentiate a quotient.

Example 4   Find $\dfrac{d}{dx}\left(\dfrac{x^2}{x+1}\right)$.

Solution   The quotient rule gives

$$\frac{d}{dx}\left(\frac{x^2}{x+1}\right) = \frac{(x+1)(2x) - (1)(x^2)}{(x+1)^2} = \frac{2x^2 + 2x - x^2}{(x+1)^2} = \frac{x^2 + 2x}{(x+1)^2}$$

with the labels: bottom, derivative of the top, derivative of the bottom, top, and bottom squared.   ■

Practice Exercise 2

Find $\dfrac{d}{dx}\left(\dfrac{2x^2}{x^2+1}\right)$.   *(solution on page 89)*

Application: The Cost of Clean Water

The cost of purifying water increases rapidly for higher and higher degrees of purity. That is, achieving 99% purity is usually far more expensive than achicving only 95% purity.

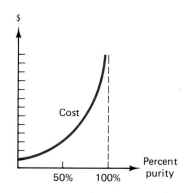

The cost of water purification increases rapidly for degrees of purity approaching 100%.

Example 5 For some methods, the cost of purifying a gallon of water to a purity of x percent is

$$C(x) = \frac{2}{100 - x} \qquad x < 100$$

dollars. Find the rate of change of this cost and evaluate it for

(a) 90% purity

(b) 98% purity

and interpret your answers.

Solution The instantaneous rate of change of costs is found by differentiating the cost function,

$$C'(x) = \frac{d}{dx}\left(\frac{2}{100 - x}\right) = \frac{\overset{\text{derivative of 2}}{(100 - x)(0)} - \overset{\text{derivative of } 100 - x}{(-1)(2)}}{(100 - x)^2}$$

$$= \frac{0 + 2}{(100 - x)^2} = \frac{2}{(100 - x)^2}$$

(a) For 90% purity we substitute $x = 90$.

$$C'(90) = \frac{2}{(100 - 90)^2} = \frac{2}{10^2} = \frac{2}{100}$$

Interpretation: At 90% purity, further purification of a gallon of water costs about 2/100 dollars, or *2 cents for each additional percent of purity*.

(b) For 98% purity, we substitute $x = 98$.

$$C'(98) = \frac{2}{(100 - 98)^2} = \frac{2}{2^2} = \frac{2}{4} = \frac{1}{2}$$

Interpretation: At 98% purity, further purification costs about half a dollar, or 50 cents, per additional percentage point, which is 25 times as costly as at the 90% level. ■

Application to Business and Economics: Average Cost per Unit

It is often useful to calculate not just the *total* cost of producing x units of some product, but also the *average* cost *per unit*. The average cost per unit, denoted $AC(x)$, is found by dividing the total cost, $C(x)$, by the number of units, x.

$$AC(x) = \frac{C(x)}{x}.$$

average cost is total cost divided by the number of units

Average revenue (AR) and average profit (AP) are defined similarly.

$$AR(x) = \frac{R(x)}{x}.$$

average revenue is total revenue divided by the number of units

$$AP(x) = \frac{P(x)}{x}.$$

average profit is total profit divided by the number of units

The derivative of the average cost function is called the *marginal average cost* (MAC).

$$MAC(x) = \frac{d}{dx}\left(\frac{C(x)}{x}\right).$$

marginal average cost

Marginal average revenue (MAR) and marginal average profit (MAP) are defined similarly.

$$MAR(x) = \frac{d}{dx}\left(\frac{R(x)}{x}\right).$$

marginal average revenue

$$MAP(x) = \frac{d}{dx}\left(\frac{P(x)}{x}\right).$$

marginal average profit

Example 6 It costs a book publisher $12 to produce each book, and fixed costs are $1500. Therefore, the company's cost function is

$$C(x) = 12x + 1500$$

total cost of producing x books

(a) Find the average cost function.

(b) Find the marginal average cost function.

(c) Find the marginal average cost at $x = 100$ and interpret your answer.

Solution

(a) The average cost function is

$$AC(x) = \frac{12x + 1500}{x}$$

total cost divided by number of units

(b) The *marginal average cost* is

$$MAC(x) = \frac{d}{dx}\left(\frac{12x + 1500}{x}\right)$$

derivative of AC(x)

$$= \frac{x \cdot 12 - 1(12x + 1500)}{x^2}$$

using the quotient rule

$$= \frac{12x - 12x - 1500}{x^2} = \frac{-1500}{x^2}$$

simplifying

(c) Evaluate $MAC(x)$ at $x = 100$.

$$MAC(100) = \frac{-1500}{100^2} = \frac{-1500}{10,000} = -.15$$

Interpretation: When 100 books have been produced, the average cost per book is decreasing at about 15 cents per additional book produced. This reflects the fact that while *total* costs will rise when you produce more, the *average* cost *per unit* will decrease, due to the economy of mass production. ■

Practice Exercise 3

Do you see an easier way to calculate the derivative in Exercise 6, $\frac{d}{dx}\left(\frac{12x + 1500}{x}\right)$? (*Hint*: Simplify the function before differentiating.)

(solution on page 90)

Application: Safe Driving

Example 7 A certain mathematics professor drives 25 miles to his office every day, mostly on highways. If he drives at constant speed v, his travel time (distance divided by speed) is

$$T(v) = \frac{25}{v}$$

Find $T'(55)$ and interpret this number.

Solution Since $T(v) = 25/v$ is a quotient, we could differentiate it by the quotient rule. However, it is easier to write $25/v$ as a *power*,

$$T(v) = 25v^{-1}$$

and differentiate using the power rule,

$$T'(v) = -25v^{-2}$$

This gives the rate of change of the travel time with respect to driving speed. $T'(v)$ is negative, showing that as speed increases, travel time *decreases*. Evaluating this at speed $v = 55$ gives

$$T(55) = -25(55)^{-2} = \frac{-25}{(55)^2} \approx -.00826 \qquad \textit{using a calculator}$$

This number, the rate of change of travel time with respect to driving speed, means that when driving at 55 miles per hour, you save only .00826 hour for each extra mile per hour of speed. Multiplying by 60 gives the saving in *minutes*,

$$(-.00826)(60) \approx -.50 = -\frac{1}{2}$$

That is, each extra mile per hour of speed saves only about half a minute, or 30 seconds. One must then decide whether this slight difference is worth the risk of an accident or a speeding ticket. ▪

Even though the function in this example was a quotient, we did not differentiate it by the quotient rule. The quotient rule is the most complicated of the differentiation rules, and should be avoided if there is an easier way. In this case it was easier to write the function as a power and use the power rule.

Summary

The following is a list of the differentiation formulas that we have learned so far. The letters c and n stand for constants, and f and g stand for functions of x.

$$\frac{d}{dx}c = 0$$

$$\frac{d}{dx}x^n = nx^{n-1} \qquad \left(\text{special case: } \frac{d}{dx}x = 1\right)$$

$$\frac{d}{dx}(c\cdot f) = c\cdot f' \qquad \left(\text{special case: } \frac{d}{dx}(cx) = c\right)$$

$$\frac{d}{dx}(f \pm g) = f' \pm g'$$

$$\frac{d}{dx}(f \cdot g) = f' \cdot g + f \cdot g'$$

$$\frac{d}{dx}\left(\frac{f}{g}\right) = \frac{g \cdot f' - g' \cdot f}{g^2}$$

These formulas are used extensively throughout calculus, and you should not proceed to Section 2.5 until you have mastered them.

Verification of the Product Rule

For two functions f and g let their product be $p(x) = f(x) \cdot g(x)$. If $f'(x)$ and $g'(x)$ exist, we may calculate $p'(x)$ as follows.

$$p'(x) = \lim_{h \to 0} \frac{p(x + h) - p(x)}{h} \qquad \text{\textit{definition of the derivative}}$$

$$= \lim_{h \to 0} \frac{f(x + h)g(x + h) - f(x)g(x)}{h} \qquad \text{\textit{since } } p(x + h) = f(x + h) \cdot g(x + h) \text{ and } p(x) = f(x) \cdot g(x)$$

$$= \lim_{h \to 0} \frac{f(x + h)g(x + h) - f(x)g(x + h) + f(x)g(x + h) - f(x)g(x)}{h} \qquad \text{\textit{subtracting and adding } } f(x)g(x + h)$$

$$= \lim_{h \to 0} \left[\frac{f(x + h)g(x + h) - f(x)g(x + h)}{h} + \frac{f(x)g(x + h) - f(x)g(x)}{h} \right] \qquad \text{\textit{separating the fraction into two parts}}$$

$$= \lim_{h \to 0} \frac{[f(x + h) - f(x)]g(x + h)}{h} + \lim_{h \to 0} \frac{f(x)[g(x + h) - g(x)]}{h} \qquad \text{\textit{using limit rule 4a on page 47 and factoring}}$$

$$= \underbrace{\lim_{h \to 0} \frac{[f(x + h) - f(x)]}{h}}_{f'(x)} \underbrace{\lim_{h \to 0} g(x + h)}_{g(x)} + \underbrace{f(x)}_{f(x)} \underbrace{\lim_{h \to 0} \frac{[g(x + h) - g(x)]}{h}}_{g'(x)} \qquad \text{\textit{using limit rule 4c on page 47}}$$

$$= f'(x)g(x) + f(x)g'(x) \qquad \text{\textit{product rule}}$$

The final step comes from recognizing the definitions of the derivatives $f'(x)$ and $g'(x)$, and that $g(x + h)$ approaches $g(x)$ as $h \to 0$. This shows that the derivative of a product $f(x)g(x)$ is $f'(x)g(x) + f(x)g'(x)$.

Verification of the Quotient Rule

For two functions f and g let the quotient be $q(x) = f(x)/g(x)$. If $f'(x)$ and $g'(x)$ exist, we may calculate $q'(x)$ as follows.

$$q'(x) = \lim_{h \to 0} \frac{q(x + h) - q(x)}{h}$$ *definition of the derivative*

$$= \lim_{h \to 0} \frac{\dfrac{f(x + h)}{g(x + h)} - \dfrac{f(x)}{g(x)}}{h}$$ $q(x + h) = \dfrac{f(x + h)}{g(x + h)}$ and $q(x) = \dfrac{f(x)}{g(x)}$

$$= \lim_{h \to 0} \frac{1}{h} \left[\frac{f(x + h)}{g(x + h)} - \frac{f(x)}{g(x)} \right]$$ *since dividing by h is equivalent to multiplying by 1/h*

$$= \lim_{h \to 0} \left(\frac{1}{h} \cdot \frac{g(x)f(x + h) - g(x + h)f(x)}{g(x + h)g(x)} \right)$$ *subtracting the fractions in the bracket, using common denominator* $g(x + h)g(x)$

$$= \lim_{h \to 0} \left(\frac{1}{h} \cdot \frac{g(x)f(x + h) - g(x)f(x) - [g(x + h)f(x) - g(x)f(x)]}{g(x + h)g(x)} \right)$$ *subtracting and adding* $g(x)f(x)$

$$= \lim_{h \to 0} \left(\frac{1}{g(x + h)g(x)} \cdot \frac{g(x)[f(x + h) - f(x)] - [g(x + h) - g(x)]f(x)}{h} \right)$$ *factoring in the numerator and switching the denominators*

$$= \lim_{h \to 0} \left(\frac{1}{\underbrace{g(x+h)g(x)}_{\substack{\text{approaches} \\ g(x)}}} \left[g(x) \underbrace{\lim_{h \to 0} \frac{f(x+h) - f(x)}{h}}_{f'(x)} - \underbrace{\lim_{h \to 0} \frac{g(x+h) - g(x)}{h}}_{g'(x)} f(x) \right] \right)$$ *using limit rules 4b and c on page 47*

$$= \frac{1}{[g(x)]^2} [g(x)f'(x) - g'(x)f(x)]$$ *using limit rules 1 and 4d*

$$= \frac{g(x)f'(x) - g'(x)f(x)}{[g(x)]^2}$$ *quotient rule*

This shows that the derivative of a quotient $f(x)/g(x)$ is

$$\frac{d}{dx} \left[\frac{f(x)}{g(x)} \right] = \frac{g(x)f'(x) - g'(x)f(x)}{[g(x)]^2}$$

SOLUTIONS TO PRACTICE EXERCISES

1. $\dfrac{d}{dx} [x^3(x^2 - x)] = 3x^2(x^2 - x) + x^3(2x - 1) = 3x^4 - 3x^3 + 2x^4 - x^3$

     $= 5x^4 - 4x^3$

2. $\dfrac{d}{dx} \left(\dfrac{2x^2}{x^2 + 1} \right) = \dfrac{(x^2 + 1)4x - 2x \cdot 2x^2}{(x^2 + 1)^2} = \dfrac{4x^3 + 4x - 4x^3}{(x^2 + 1)^2} = \dfrac{4x}{(x^2 + 1)^2}$

3. $\dfrac{d}{dx}\left(\dfrac{12x+1500}{x}\right) = \dfrac{d}{dx}\left(\dfrac{12x}{x}+\dfrac{1500}{x}\right) = \dfrac{d}{dx}(12+1500x^{-1}) = -1500x^{-2}$

EXERCISES 2.4

Find the derivative of each function in two ways:
*(a) Using the **product** rule.*
*(b) Multiplying out the function and then using the **power** rule.*
Your answers to parts (a) and (b) should agree.

1 $x^4 \cdot x^6$ **2** $x^7 \cdot x^2$ **3** $x^4(x^5+1)$ **4** $x^5(x^4+1)$

*Show that the derivative of a product is **not** the product of the derivatives by calculating the product of the derivatives, $f' \cdot g'$, for the following products and verifying that the answer does **not** agree with your answer to the corresponding exercise above.*

5 $x^4 \cdot x^6$ **6** $x^7 \cdot x^2$ **7** $x^4(x^5+1)$ **8** $x^5(x^4+1)$

Find the derivative of each function using the product rule.

9 $f(x)=x^2(x^3+1)$ **10** $f(x)=x^3(x^2+1)$

11 $f(x)=(x^2+1)(x^2-1)$ **12** $f(x)=(x^3-1)(x^3+1)$

13 $f(x)=(x^2+x)(3x+1)$ **14** $f(x)=(x^2+2x)(2x+1)$

15 $f(x)=(\sqrt{x}-1)(\sqrt{x}+1)$ **16** $f(x)=(\sqrt{x}+2)(\sqrt{x}-2)$

Find the derivative of each function in two ways:
*(a) Using the **quotient** rule.*
*(b) Simplifying the quotient and then using the **power** rule.*
Your answers to parts (a) and (b) should agree.

17 $\dfrac{x^8}{x^2}$ **18** $\dfrac{x^9}{x^3}$ **19** $\dfrac{1}{x^3}$ **20** $\dfrac{1}{x^4}$

*Show that the derivative of a quotient is **not** the quotient of the derivatives by calculating the quotient of the derivatives, f'/g', for the following quotients and verifying that the answer does **not** agree with your answer to the corresponding exercise above.*

21 $\dfrac{x^8}{x^2}$ **22** $\dfrac{x^9}{x^3}$ **23** $\dfrac{1}{x^3}$ **24** $\dfrac{1}{x^4}$

Find the derivative of each function by the quotient rule.

25 $f(x)=\dfrac{x^4+1}{x^3}$ **26** $f(x)=\dfrac{x^5-1}{x^2}$ **27** $f(x)=\dfrac{x+1}{x-1}$

28 $f(x) = \dfrac{x - 1}{x + 1}$ **29** $f(t) = \dfrac{t^2 - 1}{t^2 + 1}$ **30** $f(t) = \dfrac{t^2 + 1}{t^2 - 1}$

Find the instantaneous rate of change of the function at the given value of x.

31 $f(x) = \dfrac{x^2}{x - 1}$ at $x = 2$ **32** $f(x) = \dfrac{x^2}{x - 2}$ at $x = 4$

33 If $f(x) = \dfrac{x}{\sqrt{x}}$ at $x = 4$ **34** If $f(x) = \dfrac{x}{\sqrt{x}}$ at $x = 16$

---

35 (*The Product Rule for Three Functions*) Show that if f, g, and h are differentiable functions of x, then

$$\frac{d}{dx} [f{\cdot}g{\cdot}h] = f'{\cdot}g{\cdot}h + f{\cdot}g'{\cdot}h + f{\cdot}g{\cdot}h'$$

(*Hint*: Write the function as $f{\cdot}[g{\cdot}h]$ and apply the product rule twice.)

36 (*The Product Rule for Four Functions*) Based on your work in Exercise 35, find and prive a product rule for the product of *four functions*.

37 Find a formula for $\dfrac{d}{dx} [f(x)]^2$ by writing it as $\dfrac{d}{dx} [f(x)f(x)]$ and using the product rule. Be sure to simplify your answer.

38 Find a formula for $\dfrac{d}{dx} [f(x)]^{-1}$ by writing it as $\dfrac{d}{dx} \left(\dfrac{1}{f(x)}\right)$ and using the quotient rule. Be sure to simplify your answer.

▨   **APPLIED EXERCISES**

39 (*Economics–Marginal Average Revenue*) Use the quotient rule to find a general expression for the marginal average revenue. That is, calculate $\dfrac{d}{dx} \left(\dfrac{R(x)}{x}\right)$ and simplify your answer.

40 (*Economics–Marginal Average Profit*) Use the quotient rule to find a general expression for the marginal average profit. That is, calculate $\dfrac{d}{dx} \left(\dfrac{P(x)}{x}\right)$ and simplify your answer.

41 (*General–Water Purification*) If the cost of purifying a gallon of water to a purity of x percent is

$$C(x) = \frac{100}{100 - x} \quad \text{cents} \qquad \text{for } 50 \le x < 100$$

(a) Find the instantaneous rate of change of the cost with respect to purity.

(b) Evaluate this rate of change for a purity of 95% and interpret your answer.

(c) Evaluate this rate of change for a purity of 98% and interpret your answer.

42 (*Business–Marginal Average Cost*) A toy company can produce plastic trucks at a cost of $8 each, while fixed costs are $1200 per day. Therefore, the company's cost function is $C(x) = 8x + 1200$.

(a) Find the average cost function $AC(x) = C(x)/x$.

(b) Find the marginal average cost function $MAC(x)$.

(c) Evaluate $MAC(x)$ at $x = 200$ and interpret your answer.

43 (*Business–Marginal Average Profit*) A company's profit function is $P(x) = 12x - 1800$.

(a) Find the average profit function $AP(x) = P(x)/x$.

(b) Find the marginal average profit function $MAP(x)$.

(c) Evaluate $MAP(x)$ at $x = 300$ and interpret your answer.

44 (*Business–Sales*) The number of bottles of whiskey that a store will sell at a price of p dollars per bottle is

$$N(p) = \frac{2250}{p + 7} \quad (p \geq 5)$$

Find the rate of change of this quantity when the price is $8 and interpret your answer.

45 (*General–Body Temperature*) If a person's temperature after x hours of strenuous exercise is $T(x) = x^3(4 - x^2) + 98.6$ (for $x \leq 2$), find the rate of change of the temperature after 1 hour.

46 (*General–Record Sales*) After x months, monthly sales of a record are predicted to be $S(x) = x^2(8 - x^3)$ thousand (for $x \leq 2$). Find the rate of change of the sales after 1 month.

Find the derivative of each function.

47 $(x^3 + 2)\dfrac{x^2 + 1}{x + 1}$

48 $(x^5 + 1)\dfrac{x^3 + 2}{x + 1}$

49 $\dfrac{(x^2 + 3)(x^3 + 1)}{x^2 + 2}$

50 $\dfrac{(x^3 + 2)(x^2 + 2)}{x^3 + 1}$

51 $\dfrac{\sqrt{x} - 1}{\sqrt{x} + 1}$

52 $\dfrac{\sqrt{x} + 1}{\sqrt{x} - 1}$

2.5 HIGHER–ORDER DERIVATIVES

Repeated Differentiation

We have seen that from one function we can calculate a new function, the *derivative* of the original function. This new function, however, can itself be differentiated, giving what is called the *second derivative* of the original function. Differentiating again gives the *third derivative* of the original function, and so on.

Example 1   From $f(x) = x^3 - 6x^2 + 2x - 7$ we calculate $f'(x)$:

$$f'(x) = 3x^2 - 12x + 2 \qquad\qquad \text{"first" derivative}$$

Now *this* function can itself be differentiated, giving the *second* derivative of the original function, indicated by *two* primes on the f,

$$f''(x) = 6x - 12 \qquad\qquad \text{f double prime}$$

Differentiating again gives the *third* derivative of the original,

$$f'''(x) = 6 \qquad\qquad \text{f triple prime}$$

Differentiating again gives the *fourth* derivative,

$$f''''(x) = 0$$

All further derivatives of this function will, of course, be zero.   ■

The prime notation rapidly becomes awkward for higher derivatives. For example, the tenth derivative would require 10 primes. Instead, we may replace the primes by the *number* of derivatives *in parentheses*. For example, the tenth derivative is written $f^{(10)}(x)$. The parentheses around the 10 are important, for without them this would mean the tenth power instead of the tenth derivative.

Practice Exercise 1

If $f(x) = x^3 - x^2 + x - 1$, find

    (a) $f'(x)$ **(b)** $f''(x)$ **(c)** $f'''(x)$ **(d)** $f^{(4)}(x)$ *(solutions on page 100)*

Example 1 showed that differenting a third-degree polynomial four times gives zero. This suggests (and it is true) that any polynomial, differentiated often enough, gives zero. The same, however, is *not* true for nonpolynomial functions, as the following example shows.

Example 2 Calculate the first five derivatives of the rational function $f(x) = \dfrac{1}{x}$.

Solution

$$f(x) = x^{-1} \qquad\qquad\qquad \textit{f(x) in exponential form}$$

$$f'(x) = -x^{-2} \qquad\qquad\qquad \textit{first derivative}$$

$$f''(x) = 2x^{-3} \qquad\qquad\qquad \textit{second derivative}$$

$$f'''(x) = -6x^{-4} \qquad\qquad\qquad \textit{third derivative}$$

$$f^{(4)}(x) = 24x^{-5} \qquad\qquad\qquad \textit{fourth derivative}$$

$$f^{(5)}(x) = -120x^{-6} \qquad\qquad\qquad \textit{fifth derivative}$$

                                                            ▪

Clearly, we will never get zero, no matter how many times we differenti- ate this function.

Practice Exercise 2

If $f(x) = 16x^{-1/2}$, find

    (a) $f'(x)$ **(b)** $f''(x)$ **(c)** $f'''(x)$ **(d)** $f^{(4)}(x)$ *(solutions on page 100)*

Higher-Order Derivatives in Leibniz's Notation

In the d/dx notation, the second derivative $\dfrac{d}{dx}\left(\dfrac{d}{dx}f(x)\right)$ is written $\dfrac{d^2}{dx^2}f(x)$. The following table shows equivalent statements in the two notations.

Prime Notation	d/dx Notation	
$f''(x)$	$\dfrac{d^2}{dx^2} f(x)$	*second derivative*
y''	$\dfrac{d^2y}{dx^2}$	
$f'''(x)$	$\dfrac{d^3}{dx^3} f(x)$	*third derivative*
y'''	$\dfrac{d^3y}{dx^3}$	
$f^{(n)}(x)$	$\dfrac{d^n}{dx^n} f(x)$	*n^{th} derivative*
$y^{(n)}$	$\dfrac{d^ny}{dx^n}$	

Note that in Leibniz's notation the superscript comes after the d on the top and after the x on the bottom.

Calculating higher derivatives merely requires repeated use of the differentiation rules.

Example 3   Find $\dfrac{d^2}{dx^2} \left(\dfrac{x^2 + 1}{x} \right)$.

Solution

$$\frac{d}{dx} \left(\frac{x^2 + 1}{x} \right) = \frac{x\,(2x) - (x^2 + 1)}{x^2}$$

the first derivative, using the quotient rule

$$= \frac{2x^2 - x^2 - 1}{x^2} = \frac{x^2 - 1}{x^2}$$

simplifying

Differentiating this answer gives the *second* derivative of the original,

$$\frac{d}{dx} \left(\frac{x^2 - 1}{x^2} \right) = \frac{x^2(2x) - 2x(x^2 - 1)}{x^4}$$

differentiating the first derivative

$$= \frac{2x^3 - 2x^3 + 2x}{x^4} = \frac{2x}{x^4} = \frac{2}{x^3}$$ ■

The function in the previous example was a quotient, so it was perhaps natural to use the quotient rule. It is easier, however, to simplify the original function first,

$$\frac{x^2 + 1}{x} = \frac{x^2}{x} + \frac{1}{x} = x + x^{-1}$$

and then differentiate by the power rule. The first derivative of $x + x^{-1}$ is $1 - x^{-2}$, and differentiating again gives $2x^{-3}$, agreeing with the answer found by the quotient rule. *Moral*: **Always simplify before differentiating.**

Practice Exercise 3

Find $f''(x)$ if $f(x) = \dfrac{x + 1}{x}$.   *(solution on page 100)*

Evaluation of Higher Derivatives

As before, expressions like $f''(3)$ or $\dfrac{d^2}{dx^2} f(x)\Big|_{x=3}$, involving differentiation *and* evaluation, mean *differentiate first, then evaluate.*

Example 4   Find $f''\left(\dfrac{1}{4}\right)$ if $f(x) = \dfrac{1}{\sqrt{x}}$.

Solution

$$f(x) = x^{-1/2} \qquad\qquad\qquad\qquad \textit{f(x) in exponential form}$$

$$f'(x) = -\frac{1}{2} x^{-3/2} \qquad\qquad\qquad \textit{differentiating once}$$

$$f''(x) = \frac{3}{4} x^{-5/2} \qquad\qquad\qquad\quad \textit{differentiating again}$$

$$f''\left(\frac{1}{4}\right) = \frac{3}{4} \left(\frac{1}{4}\right)^{-5/2} = \frac{3}{4} (4)^{5/2} = \frac{3}{4} (\sqrt{4})^5 = \frac{3}{4} (2)^5 = \frac{3}{4} (32) = 24 \qquad \textit{evaluating } f''(x) \textit{ at } \frac{1}{4}$$ ■

Practice Exercise 4

Find $\dfrac{d^2}{dx^2} (x^4 + x^3 + 1)\Big|_{x=-1}$ *(solution on page 101)*

Derivatives As Velocities

There is another important interpretation for the derivative, an interpretation that also gives a meaning to the *second* derivative. Imagine that you are driving along a straight road, and let $s(t)$ stand for your distance (in miles) from your starting point after t hours of driving. Then the derivative $s'(t)$ gives the instantaneous rate of change of distance with respect to time (miles per hour). However, "miles per hour" means speed or velocity, so the derivative of the *distance* function $s(t)$ is just the *velocity* function $v(t)$, giving your velocity at any time t.

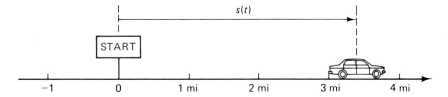

In general, for an object moving along a straight line, measuring distance from some fixed point,

> If $s(t) = distance$ at time t
>
> then $s'(t) = velocity$ at time t.

or using $v(t)$ for velocity at time t,

> $$v(t) = s'(t)$$

the velocity function is the derivative of the distance function

The units of velocity come directly from the distance and time units of $s(t)$. For example, if distance is in feet and time t is in seconds, then the velocity $s'(t)$ is in *feet per second,* while if distance is in miles and time is in hours, then $s'(t)$ is in *miles per hour.*

Accelerations Are Second Derivatives

The word "acceleration" is used in everyday speech to mean an increase in speed (when you are "accelerating" you are "going faster"). Since rates of increase are just derivatives, *acceleration is the derivative of velocity.* Since velocity is itself the derivative of distance, $v(t) = s'(t)$, acceleration is the *second* derivative of distance, $s''(t)$.

Using $s(t)$ to stand for distance, $v(t)$ for velocity, and $a(t)$ for acceleration at time t, all of this may be summarized very briefly:

> $$v(t) = s'(t)$$
> $$a(t) = v'(t) = s''(t)$$

velocity is the derivative of distance

acceleration is the derivative of velocity, which is the second derivative of distance

Therefore, we have the following interpretations for the first and second derivatives. For distance measured along a straight line from some fixed point, if $s(t) = distance$ at time t

then $s'(t) = velocity$ at time t

and $s''(t) = acceleration$ at time t.

Example 5   A delivery truck is driving east along a straight road, and after t hours it is

$$s(t) = 24t^2 - 4t^3$$

miles east of its starting point (for $t \le 6$).

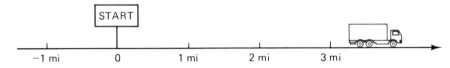

(a)   Find the velocity of the truck after 2 hours.

(b)   Find the velocity of the truck after 5 hours.

(c)   Find the acceleration of the truck after 1 hour.

Solution

(a)   To find velocity we differentiate distance,

$$v(t) = 48t - 12t^2 \qquad \text{\textit{differentiating } } s(t) = 24t^2 - 4t^3$$

and the velocity after 2 hours is

$$v(2) = 48 \cdot 2 - 12 \cdot (2)^2 \qquad \text{\textit{substituting } } t = 2$$

$$= 96 - 48 = 48 \text{ miles per hour}$$

(b)   The velocity after 5 hours is

$$v(5) = 48 \cdot 5 - 12 \cdot (5)^2 \qquad \text{\textit{evaluating } } v(t) \text{\textit{ at } } t = 5$$

$$= 240 - 300 = -60 \text{ miles per hour}$$

What does the negative sign mean? Since distances are measured *eastward* (according to the original problem), the "positive" direction is east, so a negative velocity means a *westward* velocity. Therefore, at time $t = 5$ the truck is driving *westward at 60 miles per hour* (that is, returning to its starting point).

(c)   The acceleration is

$$a(t) = 48 - 24t \qquad \text{\textit{differentiating } } v(t) = 48t - 12t^2$$

and at time $t = 1$,

$$a(1) = 48 - 24 = 24$$

Therefore, after 1 hour the acceleration of the truck is 24 mi/hr². ■

Units of Acceleration

In Example 5 velocity was in miles per hour (mi/hr), so acceleration, the rate of change of this with respect to time, is in miles per hour *per hour*, written mi/hr². In general, the units of acceleration are distance/time².

Practice Exercise 5

A helicopter rises vertically, and after t seconds its height above the ground is $s(t) = 6t^2 - t^3$ feet (for $t \le 6$).

(a)   Find its velocity after 2 seconds.

(b)   Find its velocity after 5 seconds.

(c)   Find its acceleration after 1 second.

(*Hint*: Distances are measured *upward*, so a negative velocity means *downward*.)

(solutions on page 101)

Other Interpretations of Second Derivatives

The second derivative has other meanings besides acceleration. In general, second derivatives measure how the rate of change is itself changing. That is, the second derivative tells whether growth is speeding up or slowing down.

Example 6   Demographers predict that t years from now the population of a city will be

$$P(t) = 2,000,000 + 28,800t^{1/3}$$

Find $P'(8)$ and $P''(8)$ and interpret these answers.

Solution   The derivative is

$$P'(t) = 9600t^{-2/3} \qquad \text{derivative of } P(t)$$

so

$$P'(8) = 9600(8)^{-2/3} = 9600\left(\frac{1}{4}\right) = 2400 \qquad \text{evaluating at } t = 8$$

Interpretation of $P'(8) = 2400$: Eight years from now the population will be growing at the rate of 2400 people per year.

The second derivative is

$$P''(t) = -6400t^{-5/3}$$

derivative of P'(t) = 9600t⁻²/³

derivative of $P'(t) = 9600t^{-2/3}$

so

$$P''(8) = -6400(8)^{-5/3} = -6400\left(\frac{1}{32}\right) = -200$$

The second derivative (the rate of change of the rate of change) being *negative* means that the growth rate is slowing.

Interpretation of P''(8) = −200: After 8 years the growth rate is *decreasing* by about 200 people per year per year. In other words, in the following year the population will continue to grow, but at a slower rate, about 2400 − 200 = 2200 people per year.   ■

Note that the first derivative being positive means that the population is growing, and the second derivative being negative means that the rate of growth is slowing. If the second derivative had been positive, the growth would have been *accelerating*.

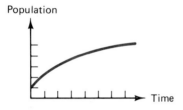

A population continuing to grow, but at a slower rate

Statements about second derivatives occur frequently in everyday events. For example, "the economy is growing more slowly" (its second derivative is negative), and "crime is increasing at a faster rate" (its second derivative is positive).

Practice Exercise 6

Sociologists predict that *t* years from now the annual number of crimes committed in a certain city will be

$$N(t) = 72,000 + 800t^{3/2}$$

Find $N'(4)$ and $N''(4)$ and interpret your answers.

APRIL PRICES UP 0.4% BUT PACE OF CLIMB CONTINUES TO SLOW

BANKS' PRIME RATE HITS 20.5%

Transportation, Food, Natural Gas and Electricity Costs Rise in the Metropolitan Area

By ROBERT D. HERSHEY Jr.
Special to The New York Times

WASHINGTON, May 22 — In yet another piece of surprising economic news, the Labor Department reported today

Second derivatives make the front page! The headline announces that the first derivative of prices is positive, but the second derivative is negative.

(solutions on page 101)

Summary

By simply repeating the process of differentiation we can calculate second, third, and even higher derivatives. We also have another interpretation for the derivative, one that gives an interpretation for the second derivative as well. For distance measured along a straight line from some fixed point,

---

> If $s(t) = distance$ at time t
>
> then $s'(t) = velocity$ at time t
>
> and $s''(t) = acceleration$ at time t.

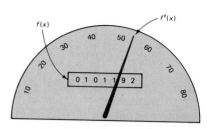

Therefore, whenever you are driving along a straight road, your speedometer reading is the derivative of your odometer reading.

We now have four interpretations for the derivatives: instantaneous rates of change, slopes, marginals, and velocity. It has been said that science is at its best when it unifies, and the derivative, unifying these four different concepts, is one of the most important ideas in all of science. We also saw that second derivatives, which measure the rate of change of the rate of change, can show whether growth is speeding up or slowing down.

Remember, however, that derivatives measures just what an automobile speedometer measures, the velocity at a particular *instant*. Although this statement may be obvious for velocities, it is easy to forget when dealing with marginals. For example, suppose that the marginal cost for a product is $15 when 100 units have been produced [which may be written $C'(100) = 15$]. Therefore, costs are increasing at the rate of $15 per additional unit, but only at the instant when $x = 100$. Although this may be used to *estimate* future costs (*about* $15 for each additional unit), it does not mean that one additional unit will increase costs by exactly $15, two more by exactly $30, and so on, since the marginal rate may change at any time. That is, $C'(100) = 15$ does not give the exact cost of the next few units any more than your velocity at one moment gives your exact mileage over the next few hours.

SOLUTIONS TO PRACTICE EXERCISES

1. **(a)** $f'(x) = 3x^2 - 2x + 1$ **(b)** $f''(x) = 6x - 2$

    (c) $f'''(x) = 6$ **(d)** $f^{(4)}(x) = 0$

2. **(a)** $f'(x) = -8x^{-3/2}$ **(b)** $f''(x) = 12x^{-5/2}$

    (c) $f'''(x) = -30x^{-7/2}$ **(d)** $f^{(4)}(x) = 105x^{-9/2}$

3. $f(x) = \dfrac{x}{x} + \dfrac{1}{x} = 1 + x^{-1}$ *simplifying first*

    $f'(x) = -x^{-2}$

    $f''(x) = 2x^{-3}$

4. $\dfrac{d}{dx}(x^4 + x^3 + 1) = 4x^3 + 3x^2$

$\dfrac{d^2}{dx^2}(x^4 + x^3 + 1) = \dfrac{d}{dx}(4x^3 + 3x^2) = 12x^2 + 6x$

$(12x^2 + 6x)\big|_{x=-1} = 12 - 6 = 6$

5. (a) $v(t) = 12t - 3t^2$

  $v(2) = 24 - 12 = 12$ ft/sec

 (b) $v(5) = 60 - 75 = -15$ ft/sec or 15 ft/sec *downward*

 (c) $a(t) = 12 - 6t$

  $a(1) = 12 - 6 = 6$ ft/sec^2

6. $N'(t) = 1200t^{1/2}$

$N'(4) = 1200(4)^{1/2} = 1200(2) = 2400$

Interpretation: Crime is increasing at the rate of 2400 crimes per year.

$N''(t) = 600t^{-1/2}$

$N''(4) = 600(4)^{-1/2} = 600\left(\dfrac{1}{2}\right) = 300$

Interpretation: The growth in crime is accelerating, with the rate of growth increasing by about 300 crimes per year each year.

EXERCISES 2.5

For each function, find (a) $f''(x)$, (b) $f''(2)$, (c) $f'''(x)$, (d) $f'''(2)$, (e) $f^{(4)}(x)$, and (f) $f^{(5)}(x)$.

1 $f(x) = x^4 - 2x^3 - 3x^2 + 5x - 7$

2 $f(x) = x^4 - 3x^3 + 2x^2 - 8x + 4$

3 $f(x) = 1 + x + \dfrac{1}{2}x^2 + \dfrac{1}{6}x^3 + \dfrac{1}{24}x^4 + \dfrac{1}{120}x^5$

4 $f(x) = 1 + x + \dfrac{1}{2}x^2 + \dfrac{1}{6}x^3 + \dfrac{1}{24}x^4$

For each function, find (a) $f''(x)$, (b) $f''(4)$, (c) $f'''(x)$, (d) $f'''(4)$, (e) $f^{(4)}(x)$, and (f) $f^{(5)}(x)$.

5 $f(x) = \sqrt{x^5}$

6 $f(x) = \sqrt{x^3}$

For each function, find (a) $f''(x)$ and (b) $f''(3)$.

7 $f(x) = \dfrac{x-1}{x}$

8 $f(x) = \dfrac{x+2}{x}$

9 $f(x) = \dfrac{x+1}{2x}$

10 $f(x) = \dfrac{x-2}{4x}$

11 $f(x) = \dfrac{\infty}{6x^2}$

12 $f(x) = \dfrac{\infty}{12x^3}$

*Find the **second** derivative of each function.*

13 $f(x) = (x^2 - 2)(x^2 + 3)$

14 $f(x) = (x^2 - 1)(x^2 + 2)$

15 $f(x) = \dfrac{27}{\sqrt[3]{x}}$

16 $f(x) = \dfrac{32}{\sqrt[5]{x}}$

17 $f(x) = \dfrac{x}{x - 1}$

18 $f(x) = \dfrac{x}{x - 2}$

Evaluate each expression.

19 $\dfrac{d^2}{dr^2}(\pi r^2)$

20 $\dfrac{d^2}{dr^2}\left(\dfrac{4}{3}\pi r^3\right)$

21 $\dfrac{d^2}{dx^2} x^{10}\Big|_{x=-1}$

22 $\dfrac{d^2}{dx^2} x^{11}\Big|_{x=-1}$

23 $\dfrac{d^3}{dx^3} x^{10}\Big|_{x=-1}$

24 $\dfrac{d^3}{dx^3} x^{11}\Big|_{x=-1}$

25 $\dfrac{d^2}{dx^2} \sqrt{x^3}\,\Big|_{x=1/16}$

26 $\dfrac{d^2}{dx^2} \sqrt[3]{x^4}\,\Big|_{x=1/27}$

---

27 Find $\dfrac{d^{100}}{dx^{100}}(x^{99} - 4x^{98} + 3x^{50} + 6)$.

(*Hint:* No calculation is necessary. Think of what happens when an *n*th-degree polynomial is differentiated $n + 1$ times. For example, try differentiating x^3 four times.)

28 Find a general formula for $\dfrac{d^n}{dx^n} x^{-1}$.

[*Hint:* Calculate the first few derivatives and look for a pattern. You may use the "factorial" notation: $n! = n(n - 1) \cdots 1$. For example, $3! = 3 \cdot 2 \cdot 1 = 6$.]

29 Verify the following formula for the *second* derivative of a product, where *f* and *g* are differentiable functions of *x*:

$$\frac{d^2}{dx^2}[f \cdot g] = f'' \cdot g + 2f' \cdot g' + f \cdot g''$$

(*Hint:* Use the product rule repeatedly.)

30 Verify the following formula for the *third* derivative of a product, where *f* and *g* are differentiable functions of *x*:

$$\frac{d^3}{dx^3}[f \cdot g] = f''' \cdot g + 3f'' \cdot g' + 3f' \cdot g'' + f \cdot g'''.$$

(*Hint:* Differentiate the formula in Exercise 29 by the product rule.)

APPLIED EXERCISES

31 (*General–Velocity*) After t hours a freight train is $s(t) = 18t^2 - 2t^3$ miles due north of its starting point (for $t \le 9$).

(a) Find its velocity at time $t = 3$ hours.

(b) Find its velocity at time $t = 7$ hours.

(c) Find its acceleration at time $t = 1$ hour.

32 (*General–Velocity*) After t hours a passenger train is $s(t) = 24t^2 - 2t^3$ miles due west of its starting point (for $t \le 12$).

(a) Find its velocity at time $t = 4$ hours.

(b) Find its velocity at time $t = 10$ hours.

(c) Find its acceleration at time $t = 1$ hour.

33 (*General–Velocity*) A rocket can rise to a height of $h(t) = t^3 + .5t^2$ feet in t seconds. Find its velocity and acceleration 10 seconds after it is launched.

34 (*General–Velocity*) After t hours a car is a distance $s(t) = 60t + \dfrac{100}{t + 3}$ miles from its starting point. Find the velocity after 2 hours.

35 (*General–Impact Velocity*) A penny dropped from a building will fall a distance $s(t) = 16t^2$ feet in t seconds (neglecting air resistance).

(a) With what velocity will it hit the ground if it does so 5 seconds after it is dropped? (This is called the *impact velocity*.)

(b) Find the acceleration at any time t. (This number is called the *acceleration due to gravity*.)

36 (*General–Impact Velocity*) If a marble is dropped from the top of the Sears Tower in Chicago, its height above the ground t seconds after it is dropped will be $s(t) = 1454 - 16t^2$ (neglecting air resistance).

(a) How long will it take to reach the ground? (*Hint*: Find when the height equals zero.)

(b) Use your answer to part (a) to find the velocity with which it will strike the ground.

37 (*General–Maximum Height*) A bullet is shot straight up so that its height (in feet) above the ground t seconds after it is fired is $s(t) = -16t^2 + 1280t$.

(a) Find the velocity function.

(b) Find the time t when the bullet will be at its maximum height.
  [*Hint*: At its maximum height the bullet is moving neither up nor down, and has velocity zero. Therefore, find the time when the velocity $v(t)$ equals zero.]

(c) Find the maximum height the bullet will reach. [*Hint*: Use the time found in part (b) together with the height function $s(t)$.]

38 (*General–Maximum Height*) A frog can jump to a height of $s(t) = -4.9t^2 + 4.41t$ meters above the ground in t seconds. Find the maximum height that the frog will reach.

39 (*Economics–National Debt*) The national debt of a South American country t years from now is predicted to be $D(t) = 65 + 9t^{4/3}$ billion dollars. Find $D'(8)$ and $D''(8)$ and interpret your answers.

40 (*General–Earth Temperature*) If the average temperature of the earth t years from now is predicted to be $T(t) = 65 - 4/t$ (for $t \geq 8$), find $T'(10)$ and $T''(10)$ and interpret your answers.

41–42 (*General–Sea Level*) The burning of fossil fuels (such as coal and oil) generates carbon dioxide which traps heat in the atmosphere, thereby increasing in the temperature of the earth (the "greenhouse effect").

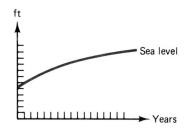

The higher temperature in turn melts the polar icecaps, raising the sea level. The precise results are very difficult to predict, but suppose that

41 in t years the average sea level on the east coast will be $L(t) = 30 - 4t^{-1/2}$ feet (for $t > 2$). Find $L'(4)$ and $L''(4)$ and interpret your answers.

42 in t years the average sea level on the west coast will be $L(t) = 35 - 8t^{-1/2}$ feet (for $t > 2$). Find $L'(4)$ and $L''(4)$ and interpret your answers.

Find the second derivative of each function.

43 $(x^2 - x + 1)(x^3 - 1)$

44 $(x^3 + x - 1)(x^3 + 1)$

45 $\dfrac{x}{x^2 + 1}$

46 $\dfrac{x}{x^2 - 1}$

47 $\dfrac{2x - 1}{2x + 1}$

48 $\dfrac{3x + 1}{3x - 1}$

2.6 THE CHAIN RULE AND THE GENERALIZED POWER RULE

Introduction

In this section we define composite functions (functions of functions) and learn the last of the general rules of differentiation, the chain rule and the generalized power rule.

Composite Functions

In Section 1.3, given a function $f(x)$ we found $f(x + h)$ simply by replacing each occurrence of x by $x + h$. Similarly, given two functions $f(x)$ and $g(x)$, to find $f(g(x))$ we take $f(x)$ and replace each occurrence of x by $g(x)$. The resulting function is called the *composition** of $f(x)$ and $g(x)$.

Composite Functions

The *composition* of two functions f and g is the function $f(g(x))$.

Example 1 If $f(x) = x^7$ and $g(x) = x^3 - 2x$, find the composition $f(g(x))$.

Solution

$$f(g(x)) = [g(x)]^7 \qquad\qquad f(x) = x^7 \text{ with } x \text{ replaced by } g(x)$$

$$= [x^3 - 2x]^7 \qquad\qquad \text{using } g(x) = x^3 - 2x$$

■

Example 2 If $f(x) = \dfrac{x + 1}{x - 1}$ and $g(x) = x^5$, find $f(g(x))$.

Solution

$$f(g(x)) = \frac{g(x) + 1}{g(x) - 1} \qquad\qquad f(x) = \frac{x + 1}{x - 1} \text{ with } x \text{ replaced by } g(x)$$

$$= \frac{x^5 + 1}{x^5 - 1} \qquad\qquad \text{using } g(x) = x^5$$

■

If we think of $f(x)$ and $g(x)$ as "numerical machines," then the composition $f(g(x))$ may be thought of as a combined machine in which the output of $g(x)$ is connected to the input of $f(x)$.

---

* The composition $f(g(x))$ is also written $f \circ g(x)$, although we will not use this notation here.

104

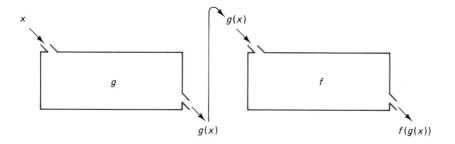

The order of composition is important: $f(g(x))$ is not the same as $g(f(x))$.

Example 3   If $f(x) = \sqrt{x}$ and $g(x) = 2x + 1$, find $f(g(x))$ and $g(f(x))$.

Solution

$$f(g(x)) = \sqrt{g(x)} \qquad = \sqrt{2x + 1}$$

$f(x) = \sqrt{x}$ with x replaced by $g(x) = 2x + 1$

$$g(f(x)) = 2f(x) + 1 = 2\sqrt{x} + 1$$

$g(x) = 2x + 1$ with x replaced by $f(x) = \sqrt{x}$

To emphasize that $f(g(x))$ and $g(f(x))$ are *not* the same, we evaluate each at $x = 4$:

$$f(g(4)) = \sqrt{2\cdot4 + 1} = \sqrt{9} \qquad = 3$$
$$g(f(4)) = 2\sqrt{4} + 1 = 2\cdot2 + 1 = 5$$

different answers

Practice Exercise 1

If $f(x) = x^5 + x^2$ and $g(x) = \sqrt[3]{x}$, find
  (a) $f(g(x))$ **(b)** $g(f(x))$ *(solutions on page 112)*

Besides building compositions out of simpler functions, we also want to "decompose" functions into compositions of simpler functions.

Example 4   Find $f(x)$ and $g(x)$ such that $(x^2 - 5x - 1)^{10}$ is the composition $f(g(x))$.

Solution   Think of $(x^2 - 5x - 1)^{10}$ as an "inside" function $x^2 - 5x - 1$ and an "outside" function $(x)^{10}$.

inside
function

outside function
(raising to the 10th power)

Therefore, for $(x^2 - 5x - 1)^{10} = f(g(x))$, we take

inside   inside

$$g(x) = x^2 - 5x - 1$$
$$f(x) = x^{10}$$

Notice that expressing a function as a composition involves thinking of the function in terms of "blocks," an inside block on which a further operation is performed.

Practice Exercise 2

Find $f(x)$ and $g(x)$ such that $\sqrt{x^5 - 7x + 1}$ is the composition $f(g(x))$.   *(solutions on page 112)*

The Chain Rule

The *chain rule* shows how to differentiate a composite function $f(g(x))$.

Chain Rule

$$\frac{d}{dx} f(g(x)) = f'(g(x)) \cdot g'(x)$$

---

to differentiate $f(g(x))$, differentiate $f(x)$, then replace each x by $g(x)$, and finally, multiply by the derivative of $g(x)$

[provided that the derivatives $f'(x)$ and $g'(x)$ both exist]. The name comes from thinking of compositions as "chains" of functions. A verification of the chain rule is given at the end of this section.

Example 5   Use the chain rule to find $\dfrac{d}{dx} (x^2 - 5x + 1)^{10}$.

Solution   In Example 4 we expressed $(x^2 - 5x + 1)^{10}$ as $f(g(x))$ with

$$g(x) = x^2 - 5x + 1$$
$$f(x) = x^{10}$$

From these,

$$g'(x) = 2x - 5$$
$$f'(x) = 10x^9$$

so

$$f'(g(x)) = 10(g(x))^9 = 10(x^2 - 5x + 1)^9$$

Substituting these into the chain rule

$$\frac{d}{dx} f(g(x)) = f'(g(x))g'(x) \qquad \text{\textit{chain rule}}$$

gives

$$\frac{d}{dx} (x^2 - 5x + 1)^{10} = 10(x^2 - 5x + 1)^9 (2x - 5) \qquad ■$$

This result says that to differentiate $(x^2 - 5x + 1)^{10}$, we bring down the exponent 10, reduce the exponent to 9 (steps familiar from the power rule), and finally multiply by the derivative of the inside function.

$$\frac{d}{dx}\,(x^2 - 5x + 1)^{10} = 10(x^2 - 5x + 1)^9(2x - 5)$$

| inside | bring down | power | derivative of |
| function | the power n | $n - 1$ | the inside function |

Generalized Power Rule

Example 5 leads to a general rule for differentiating a function to a power.

Generalized Power Rule

$$\frac{d}{dx}\,[g(x)]^n = n \cdot [g(x)]^{n-1} \cdot g'(x)$$

---

to differentiate a function to a power, bring down the power as a multiplier, reduce the exponent by 1, and then multiply by the derivative of the inside function

[provided, of course, that the derivative $g'(x)$ exists]. The generalized power rule follows from the chain rule by reasoning similar to that of Example 5: The derivative of $f(x) = x^n$ is $f'(x) = nx^{n-1}$, so

$$\frac{d}{dx}\,f(g(x)) = f'(g(x))g'(x) \qquad \textit{chain rule}$$

gives

$$\frac{d}{dx}\,[g(x)]^n = n[g(x)]^{n-1}g'(x) \qquad \textit{generalized power rule}$$

Example 6   Find $\dfrac{d}{dx}\sqrt{x^4 - 3x^3 - 4}$.

Solution

$$\frac{d}{dx}\,(x^4 - 3x^3 - 4)^{1/2} = \tfrac{1}{2}(x^4 - 3x^3 - 4)^{-1/2}(4x^3 - 9x^2)$$

| inside | bring down | power | derivative of |
| function | the n | $n - 1$ | the inside function ▪ |

Think of the generalized power rule "from the outside in." That is, first bring down the outer exponent and then reduce it by 1, and only then multiply by the derivative of the inside function.

Be careful—it is the *original function* (not differentiated) that is raised to the power $n - 1$. Only at the end do you multiply by the derivative of the inside function.

Example 7   Find $\dfrac{d}{dx}\left(\dfrac{1}{x^2 + 1}\right)^3$.

Solution   Writing the function as $(x^2 + 1)^{-3}$ gives

$$\frac{d}{dx}(x^2 + 1)^{-3} = -3(x^2 + 1)^{-4}(2x) = -6x(x^2 + 1)^{-4} = -\frac{6x}{(x^2 + 1)^4}$$

inside
function   bring down
the n power
$n - 1$ derivative of
the inside function   ■

Practice Exercise 3

Find $\dfrac{d}{dx}(x^3 - x)^{-1/2}$.

(solution on page 112)

Application to Oil Pollution

Example 8   An oil tanker hits a reef, and after t days the radius of the oil slick is $r(t) = \sqrt{4t + 1}$ miles. How fast is the radius of the oil slick expanding after 2 days?

Solution   To find the rate of change of the radius, we differentiate.

$$\frac{d}{dt}(4t + 1)^{1/2} = \frac{1}{2}(4t + 1)^{-1/2}(4) = 2(4t + 1)^{-1/2}$$

derivative of $4t + 1$

and at $t = 2$ this is

$$2(4\cdot2 + 1)^{-1/2} = 2\cdot9^{-1/2} = 2\cdot\frac{1}{3} = \frac{2}{3}$$

That is, after 2 days the oil slick is growing at the rate of 2/3 of a mile per day.

■

Further Examples

Some problems require the generalized power rule in combination with another differentiation rule, such as the product or quotient rule.

Example 9   Find $\dfrac{d}{dx}[(5x - 2)^4(9x + 2)^7]$.

Solution   Since this is a product of powers, $(5x - 2)^4$ times $(9x + 2)^7$, we use the product rule together with the generalized power rule.

$$\frac{d}{dx}[(5x - 2)^4(9x + 2)^7] = 4(5x - 2)^3(5)(9x + 2)^7 + (5x - 2)^4 7(9x + 2)^6(9)$$

$$\underbrace{\qquad\qquad}_{\substack{\text{derivative of} \\ (5x - 2)^4}} \qquad\qquad \underbrace{\qquad\qquad}_{\substack{\text{derivative of} \\ (9x + 2)^7}}$$

$$= 20(5x - 2)^3(9x + 2)^7 + 63(5x - 2)^4(9x + 2)^6$$

$$\underset{4 \cdot 5}{\uparrow} \qquad\qquad\qquad \underset{7 \cdot 9}{\uparrow} \qquad\qquad\qquad ■$$

Example 10 Find $\dfrac{d}{dx}\left(\dfrac{x}{x + 1}\right)^4$.

Solution Since the function is a quotient raised to a power, we use the quotient rule together with the generalized power rule. Working from the outside in, we obtain

$$\frac{d}{dx}\left(\frac{x}{x + 1}\right)^4 = 4\left(\frac{x}{x + 1}\right)^3 \underbrace{\frac{(x + 1)(1) - (1)(x)}{(x + 1)^2}}_{\substack{\text{derivative of the} \\ \text{inside function } x/(x + 1)}}$$

$$= 4\left(\frac{x}{x + 1}\right)^3 \frac{x + 1 - x}{(x + 1)^2} = 4\left(\frac{x}{x + 1}\right)^3 \frac{1}{(x + 1)^2}$$

$$= 4 \frac{x^3}{(x + 1)^3} \frac{1}{(x + 1)^2} = \frac{4x^3}{(x + 1)^5} \qquad\qquad ■$$

Example 11 Find $\dfrac{d}{dz}[z^2 + (z^2 - 1)^3]^5$.

Solution Since this is a function to a power, where the inside also contains a function to a power, we must use the generalized power rule *twice*.

$$\frac{d}{dz}[z^2 + (z^2 - 1)^3]^5 = 5[z^2 + (z^2 - 1)^3]^4 \underbrace{[2z + 3(z^2 - 1)^2(2z)]}_{\substack{\text{derivative of} \\ z^2 + (z^2 - 1)^3}}$$

$$= 5[z^2 + (z^2 - 1)^3]^4[2z + 6z(z^2 - 1)^2] \qquad\qquad ■$$

Chain Rule in Leibniz's Notation

If we write the composition $y = f(g(x))$ as $y = f(u)$ and $u = g(x)$, then y may be considered either as a function of u or as a function of x. From these we get

$$\frac{dy}{du} = f'(u) \quad \text{and} \quad \frac{du}{dx} = g'(x)$$

and the chain rule $\dfrac{d}{dx}f(g(x)) = f'(g(x))g'(x)$

$$\underbrace{}_{y} \quad \underbrace{}_{\frac{dy}{du}} \cdot \underbrace{}_{\frac{du}{dx}}$$

becomes

$$\frac{dy}{dx} = \frac{dy}{du}\cdot\frac{du}{dx}$$

In this form the chain rule is easy to remember since it looks as if the *du* in the top and bottom cancel:

$$\frac{dy}{dx} = \frac{dy}{\cancel{du}}\cdot\frac{\cancel{du}}{dx}$$

product of the derivatives

However, since derivatives are not really fractions (they are *limits* of fractions), this is only a convenient device for remembering the chain rule.

In our discussion of the product rule in Section 2.4, we found that the derivative of a product is *not* the product of the derivatives. We now see where the product of the derivatives *does* appear. It appears in the chain rule, when differentiating composite functions. In other words, the product of the derivatives comes not from *products* but from *compositions* of functions.

A Simple Example of the Chain Rule

The derivation of the chain rule is rather technical, but we can show the basic idea in a simple example. Suppose that your company produces steel, and you want to calculate your company's total revenue in dollars per year. You would take the output of your company in tons of steel per year, and multiply by the value of a ton of steel. In symbols,

$$\frac{\$}{\text{year}} = \frac{\$}{\text{ton}}\cdot\frac{\text{ton}}{\text{year}}$$

note that "ton" cancels

If we were to express these rates as derivatives, the equation above would become the chain rule.

Summary

To differentiate a composite function (a function of a function), we have the chain rule,

$$\frac{d}{dx}f(g(x)) = f'(g(x))g'(x)$$

or in Leibniz's notation, writing $y = f(g(x))$ as $y = f(u)$ and $u = g(x)$, this becomes

$$\frac{dy}{dx} = \frac{dy}{du} \cdot \frac{du}{dx}$$

To differentiate a function to a power, $[f(x)]^n$, we have the *generalized power rule* (a special case of the chain rule),

$$\frac{d}{dx} [f(x)]^n = n[f(x)]^{n-1} f'(x)$$

At the moment the generalized power rule is more useful than the chain rule, but in Chapter 5 we will make important use of the chain rule.

Verification of the Chain Rule

Let $f(x)$ and $g(x)$ be two differentiable functions. We define k by

$$k = g(x + h) - g(x)$$

or equivalently,

$$g(x + h) = g(x) + k$$

On page 88 we saw that for a differentiable function g we have

$$\lim_{h \to 0} \underbrace{[g(x + h) - g(x)]}_{k} = \lim_{h \to 0} k = 0$$

showing that $h \to 0$ implies $k \to 0$. With these relations we may calculate the derivative of the composition $f(g(x))$.

$$\frac{d}{dx} f(g(x)) = \lim_{h \to 0} \frac{f(g(x + h)) - f(g(x))}{h} \qquad \text{\textit{definition of the derivative of }} f(g(x))$$

$$= \lim_{h \to 0} \left[\frac{f(g(x + h)) - f(g(x))}{g(x + h) - g(x)} \cdot \frac{g(x + h) - g(x)}{h} \right] \qquad \text{\textit{dividing and multiplying by }} g(x + h) - g(x)$$

$$= \lim_{h \to 0} \frac{f(g(x + h)) - f(g(x))}{g(x + h) - g(x)} \lim_{h \to 0} \frac{g(x + h) - g(x)}{h} \qquad \text{\textit{using limit rule 4c on page 47}}$$

$$= \lim_{k \to 0} \underbrace{\frac{f(g(x) + k) - f(g(x))}{k}}_{f'(g(x))} \lim_{h \to 0} \underbrace{\frac{g(x + h) - g(x)}{h}}_{g'(x)} \qquad \begin{array}{l} \textit{using the relations} \\ \quad g(x + h) = g(x) + k \textit{ and} \\ \quad k = g(x + h) - g(x) \\ \textit{and that } h \to 0 \textit{ implies } k \to 0 \end{array}$$

$$= f'(g(x)) g'(x) \qquad \text{\textit{chain rule}}$$

The last step comes from recognizing the two limit expressions as derivatives. This verifies the chain rule,

$$\frac{d}{dx} f(g(x)) = f'(g(x))g'(x)$$

Strictly speaking, this verification requires an additional assumption, that the denominator $g(x + h) - g(x)$ is never zero. This assumption can be avoided by a more technical argument, which we omit.

SOLUTIONS TO PRACTICE EXERCISES

1. **(a)** $f(g(x)) = [g(x)]^5 + [g(x)]^2 = [\sqrt[3]{x}]^5 + [\sqrt[3]{x}]^2$

   (b) $g(f(x)) = \sqrt[3]{f(x)} = \sqrt[3]{x^5 + x^2}$

2. $f(x) = \sqrt{x},\ g(x) = x^5 - 7x + 1$

3. $\dfrac{d}{dx}(x^3 - x)^{-1/2} = -\dfrac{1}{2}(x^3 - x)^{-3/2}(3x^2 - 1)$

▨ EXERCISES 2.6

For each f(x) and g(x), find   (a) f(g(x))   and   (b) g(f(x)).

1 $f(x) = x^5,\ g(x) = 7x - 1$

2 $f(x) = x^8,\ g(x) = 2x + 5$

3 $f(x) = \dfrac{1}{x},\ g(x) = x^2 + 1$

4 $f(x) = \sqrt{x},\ g(x) = x^3 - 1$

5 $f(x) = x^3 - x^2,\ g(x) = \sqrt{x} - 1$

6 $f(x) = x - \sqrt{x},\ g(x) = x^2 + 1$

7 $f(x) = \dfrac{x^3 - 1}{x^3 + 1},\ g(x) = x^2 - x$

8 $f(x) = \dfrac{x^4 + 1}{x^4 - 1},\ g(x) = x^3 + x$

Find f(x) and g(x) such that the given function is the composition f(g(x)).

9 $\sqrt{x^2 - 3x + 1}$

10 $(5x^2 - x + 2)^4$

11 $(x^2 - x)^{-3}$

12 $\dfrac{1}{x^2 + x}$

13 $\dfrac{x^3 + 1}{x^3 - 1}$

14 $\dfrac{\sqrt{x} - 1}{\sqrt{x} + 1}$

15 $\left(\dfrac{x + 1}{x - 1}\right)^4$

16 $\sqrt{\dfrac{x - 1}{x + 1}}$

17 $\sqrt{x^2 - 9} + 5$

18 $\sqrt[3]{x^3 + 8} - 5$

Use the generalized power rule to find the derivative of each function.

19 $f(x) = (x^2 + 1)^3$

20 $f(x) = (x^3 + 1)^4$

21 $g(x) = (9x - 4)^5$

22 $g(x) = (4x - 5)^4$

23 $h(z) = (3z^2 - 5z + 2)^4$

24 $h(z) = (5z^2 + 3z - 1)^3$

25 $f(x) = \sqrt{x^4 - 5x + 1}$

26 $f(x) = \sqrt{x^6 + 3x - 1}$

27 $w(z) = \sqrt[3]{9z - 1}$

28 $w(z) = \sqrt[5]{10z - 4}$

29 $y = (x^2 + 9)^{200}$

30 $y = (x^2 - 1)^{100}$

31 $y = (4 - x^2)^4$

32 $y = (9 - x^2)^3$

33 $y = (2 - x)^{40}$

34 $y = (1 - x)^{50}$

35 $y = \left(\dfrac{1}{x^3 - 1}\right)^4$

36 $y = \left(\dfrac{1}{x^4 + 1}\right)^5$

37 $y = x^4 + (1 - x)^4$

38 $y = x^5 + (1 - x)^5$

39 $f(x) = (x^2 + 1)^4 - (x^2 + 1)^3$

40 $f(x) = (x^2 + 4)^3 - (x^2 + 4)^2$

41 $G(w) = \sqrt{2w^3 + 3w^2}$

42 $G(w) = \sqrt{w^4 + 2w^5}$

43 $f(x) = \dfrac{1}{\sqrt[3]{(9x + 1)^2}}$

44 $f(x) = \dfrac{1}{\sqrt[3]{(3x - 1)^2}}$

45 $f(x) = [(x^2 + 1)^3 + x]^3$

46 $f(x) = [(x^3 + 1)^2 - x]^4$

47 $f(x) = (2x + 1)^3(2x - 1)^4$

48 $f(x) = (2x - 1)^3(2x + 1)^4$

49 $f(x) = \dfrac{(x + 1)^2}{x^2}$

50 $f(x) = \dfrac{(x - 1)^2}{x^2}$

51 $f(x) = \left(\dfrac{x + 1}{x - 1}\right)^3$

52 $f(x) = \left(\dfrac{x - 1}{x + 1}\right)^5$

53 $f(x) = \dfrac{\sqrt{x} - 1}{\sqrt{x} + 1}$

54 $f(x) = \dfrac{\sqrt{x} + 1}{\sqrt{x} - 1}$

55 $f(x) = x^2 \sqrt{1 + x^2}$

56 $f(x) = x^2 \sqrt{x^2 - 1}$

57 $f(x) = \sqrt{1 + \sqrt{x}}$

58 $f(x) = \sqrt[3]{1 + \sqrt[3]{x}}$

---

59 Find the derivative of $(x^2 + 1)^2$ in two ways:

(a) by the generalized power rule

(b) by "squaring out" the original expression and then differentiating

Your answers should agree.

60 Find the derivative of $1/x^2$ in three ways:

(a) by the quotient rule

(b) by writing $1/x^2$ as $(x^2)^{-1}$ and using the generalized power rule

(c) by writing $1/x^2$ as x^{-2} and using the (ordinary) power rule

Your answers should agree.

61 Find the derivative of $1/(3x + 1)$ in two ways:

(a) by the quotient rule

(b) by writing it as $(3x + 1)^{-1}$ and using the generalized power rule

Your answers should agree. Which way was easiest? Remember this for the future.

62 Find an expression for the derivative of the composition of *three* functions, $\dfrac{d}{dx} f(g(h(x)))$. (*Hint:* Use the chain rule repeatedly.)

*Find the **second** derivative of each function.*

63 $f(x) = (x^2 + 1)^{10}$

64 $f(x) = (x^3 - 1)^5$

APPLIED EXERCISES

65 (*Business–Cost*) A company's cost function is $C(x) = \sqrt{4x^2 + 900}$ dollars. Find the marginal cost function and evaluate it at $x = 20$.

66 (*Sociology–Educational Status*) A 1971 study* estimated how a person's social status (rated on a scale of 0 to 100) depended on years of education. Based on this study, with e years of education, a person's status would be $S(e) = .22(e + 4)^{2.1}$. Find $S'(12)$ and interpret your answer.

67 (*Sociology–Income Status*) A 1971 study* estimated how a person's social status (rated on a scale of 0 to 100) depended upon income. Based on this study, with an income of i thousand dollars, a person's status would be $S(i) = .45(i - 1)^{.53}$. Find $S'(25)$ and interpret your answer.

68 (*Economics–Compound Interest*) If $1000 is deposited in a bank paying $r\%$ interest compounded annually, 5 years later its value will be

$$V(r) = 1000(1 + .01r)^5 \text{ dollars}$$

Find $V'(5)$ and interpret your answer. (*Hint*: $r = 5$ corresponds to 5% interest.)

69 (*Biomedical–Drug Sensitivity*) The strength of a patient's reaction to a dose of x milligrams of a certain drug is $R(x) = 4x\sqrt{11 + 0.5x}$. The derivative $R'(x)$ is called the *sensitivity* to the drug. Find $R'(50)$, the sensitivity to a dose of 50 mg.

70 (*Biomedical–Blood Flow*) It follows from Poiseuille's Law that blood flowing through certain arteries will encounter a resistance of $R(x) = .25(1 + x)^4$, where x is the distance (in meters) from the heart. Find the instantaneous rate of change of the resistance at

(a) 0 meters   (b) 1 meter

71 (*General–Pollution*) The carbon monoxide level in a city is predicted to be $.02x^{3/2} + 1$ ppm (parts per million), where x is the population in thousands. In t years the population of the city is predicted to be $x(t) = 12 + 2t$ thousand people. Therefore, in t years the carbon monoxide level will be

$$P(t) = .02(12 + 2t)^{3/2} + 1 \text{ ppm.}$$

Find $P'(2)$, the rate at which carbon monoxide pollution will be increasing in 2 years.

72 (*Psychology–Learning*) After p practice sessions, a subject could perform a task in $T(p) = 36(p + 1)^{-1/3}$ minutes. Find $T'(7)$ and interpret your answer.

2.7 NONDIFFERENTIABLE FUNCTIONS AND A REVIEW OF CHAPTER TWO

Introduction

Using the rules of differentiation (see the inside back cover) we can differentiate any function composed of the operations of addition, subtraction, multiplication, division, roots, and powers. It is important to master these

---

* Robert L. Hamblin, "Mathematical Experimentation and Sociological Theory: A Critical Analysis," *Sociometry,* **34** 1971), 423–452.

rules, since everything that follows will be based on them. We have four interpretations for the derivative: instantaneous rates of change, slopes, marginals, and velocities. The problems at the end of this section review all these developments.

Despite these powerful differentiation rules, there exist functions that cannot be differentiated. In this section we exhibit such a function (the absolute value function), show that it is not differentiable at $x = 0$, and explain the situation geometrically.

The Absolute Value Function Is Not Differentiable at the Origin

In Chapter 1 we defined the absolute value function,

$$f(x) = |x| = \begin{cases} x & \text{if } x \geq 0 \\ -x & \text{if } x < 0 \end{cases}$$

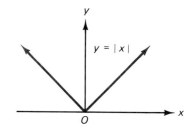

The graph of the absolute value function $f(x) = |x|$ has a "corner" at the origin.

Although the absolute value function is *defined* for *all* values of x, we will show that it is not differentiable at $x = 0$.

We have no "rules" for differentiating the absolute value function, so we must use the definition of the derivative,

$$f'(x) = \lim_{h \to 0} \frac{f(x + h) - f(x)}{h}$$

provided that this limit exists. It is this provision, which until now we have steadfastly ignored, that will be important in this example. That is, we will show that this limit, and hence the derivative, does not exist at $x = 0$.

For $x = 0$ the definition becomes

$$\lim_{h \to 0} \frac{f(0 + h) - f(0)}{h} = \lim_{h \to 0} \frac{f(h) - f(0)}{h} = \lim_{h \to 0} \underbrace{\frac{|h| - |0|}{h}}_{\substack{\text{using} \\ f(x) = |x|}} = \lim_{h \to 0} \frac{|h|}{h}$$

Now h can approach zero through *positive* numbers (like .01, .001, .0001, ...) or through *negative* numbers (like $-.01$, $-.001$, $-.0001$, ...). The limit as $h \to 0$ must be a *single* number, the same regardless of whether h is positive or negative. We will, however, see that $|h|/h$ approaches *two different* numbers, depending upon whether h is positive or negative. This will show that the limit, and hence the derivative, does not exist.

We will use the notation $\lim_{h \to 0^+}$ to mean the limit as h approaches zero through *positive* numbers, and $\lim_{h \to 0^-}$ to mean the limit as h approaches zero through *negative* numbers.

For *positive h*: $\lim\limits_{h\to 0^+} \dfrac{|h|}{h} = \lim\limits_{h\to 0^+} \dfrac{h}{h} = \lim\limits_{h\to 0^+} 1 = 1$

since $|h| = h$ $\dfrac{h}{h} = 1$
for $h > 0$

For *negative h*: $\lim\limits_{h\to 0^-} \dfrac{|h|}{h} = \lim\limits_{h\to 0^-} \dfrac{-h}{h} = \lim\limits_{h\to 0^-} (-1) = -1$

for $h < 0$, $|h| = -h$ $\dfrac{-h}{h} = -1$
(the negative sign
makes the negative
h positive)

These two different results,

$$\lim\limits_{h\to 0^+} \frac{|h|}{h} = +1$$

for positive h, $\dfrac{|h|}{h}$ *approaches + 1*

and

$$\lim\limits_{h\to 0^-} \frac{|h|}{h} = -1$$

for negative h, $\dfrac{|h|}{h}$ *approaches −1*

show that as $h \to 0$, $|h|/h$ approaches two different numbers, $+1$ and -1, depending upon whether h is positive or negative. Therefore, the limit $\lim\limits_{h\to 0} \dfrac{|h|}{h}$ does not exist, so *the derivative does not exist*. This is what we wanted to show—that the absolute value function is not differentiable at $x = 0$.

Geometric Explanation of Nondifferentiability

Looking at the graph of the absolute value function, we can give a geometric and intuitive reason why it is not differentiable at $x = 0$. The graph consists of two straight lines with slopes $+1$ and -1 which meet in a corner at the origin. To the right of the origin the slope is $+1$ and to the left of the origin the slope is -1, but *at* the origin the two conflicting slopes make it impossible to define a *single* slope. Therefore, the slope (and hence the derivative) is undefined at $x = 0$.

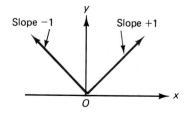

Slope −1   Slope +1

Other Nondifferentiable Functions

For the same reason, at any "corner point" of a graph, where two different slope conflict, the function will not be differentiable.

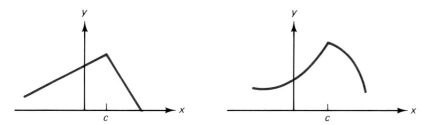

Each of the functions graphed here has a "corner point" at x = c, and so is not differentiable at x = c.

There are other reasons, besides a corner point, why a function may not be differentiable. If a curve has a vertical tangent line at a point, the slope will not be defined at that x-value, since the slope of a vertical line is undefined.

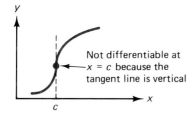

Not differentiable at $x = c$ because the tangent line is vertical

We showed in Section 2.4 that if a function is differentiable, then it is continuous. Therefore, if a function is discontinuous (has a "jump") at some point, then it will not be differentiable at that x-value.

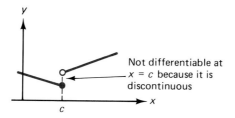

Not differentiable at $x = c$ because it is discontinuous

The fact that all differentiable functions are continuous is illustrated in the diagram on the right, showing the set of differentiable functions contained within the set of continuous functions. Although every differentiable function is continuous, not every continuous function is differentiable, as we know from the absolute value function $f(x) = |x|$.

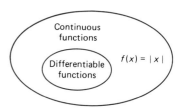

Continuous functions

Differentiable functions

$f(x) = |x|$

Summary

A function $f(x)$ will not be differentiable at a number c in its domain if its graph has either a "*corner point*," a *vertical tangent*, or is *not continuous* at $x = c$. In Chapter 3, when we use calculus for graphing, it will be important to remember these three conditions that make the derivative fail to exist.

Practice Test

Exercises 1, 3, 7, 11, 41, 47, 55, 63, 71, 73, 89

EXERCISES 2.7

Find the indicated limits or state that the limit does not exist.

1 $\lim\limits_{x \to 5} (3x^2 - 10x + 1)$

2 $\lim\limits_{x \to 16} \left(\frac{1}{2} s - s^{1/2}\right)$

3 $\lim\limits_{h \to 0} \dfrac{2x^2h - xh^2}{h}$

4 $\lim\limits_{h \to 0} \dfrac{6xh^2 - x^2h}{h}$

5 $\lim\limits_{x \to 3} \dfrac{1}{x - 3}$

6 $\lim\limits_{x \to 10} \dfrac{1}{x - 10}$

Find the derivative of each function using the definition of the derivative (that is, as you did in Section 2.2). Then check your answer using the "rules of differentiation."

7 $f(x) = 2x^2 + 3x - 1$

8 $f(x) = 3x^2 + 2x - 3$

9 $f(x) = \dfrac{3}{x}$

10 $f(x) = 4\sqrt{x}$

Find the derivative of each function.

11 $f(x) = 6\sqrt[3]{x^5} - \dfrac{4}{\sqrt{x}} + 1$

12 $f(x) = 4\sqrt{x^5} - \dfrac{6}{\sqrt[3]{x}} + 1$

13 $f(x) = (x^2 + 5)(x^2 - 5)$

14 $f(x) = (x^2 + 3)(x^2 - 3)$

15 $y = (x^4 + x^2 + 1)(x^5 - x^3 + x)$

16 $y = (x^5 + x^3 + x)(x^4 - x^2 + 1)$

17 $y = \dfrac{x - 1}{x + 1}$

18 $y = \dfrac{x + 1}{x - 1}$

19 $y = \dfrac{x^5 + 1}{x^5 - 1}$

20 $y = \dfrac{x^6 - 1}{x^6 + 1}$

21 $h(z) = (4z^2 - 3z + 1)^3$

22 $h(z) = (3z^2 - 5z - 1)^4$

23 $g(x) = (100 - x)^5$

24 $g(x) = (1000 - x)^4$

25 $f(x) = \sqrt{x^2 - x + 2}$

26 $f(x) = \sqrt{x^2 - 5x - 1}$

27 $g(y) = \sqrt[3]{2y^3 - 3y^2}$

28 $g(y) = \sqrt[3]{3y^3 - 6y^2}$

29 $w(z) = \sqrt[3]{6z - 1}$

30 $w(z) = \sqrt[3]{3z + 1}$

31 $h(x) = \dfrac{1}{\sqrt[5]{(5x + 1)^2}}$

32 $h(x) = \dfrac{1}{\sqrt[5]{(10x + 1)^3}}$

33 $f(x) = \left(\dfrac{1}{1 - x}\right)^{10}$

34 $f(x) = \left(\dfrac{1}{2 - x}\right)^{6}$

35 $y = x^3 \sqrt[3]{x^3 + 1}$

36 $y = x^4 \sqrt{x^2 + 1}$

37 $f(x) = [(2x^2 + 1)^4 + x^4]^3$

38 $f(x) = [(3x^2 - 1)^3 + x^3]^2$

39 $f(x) = \sqrt{(x^2 + 1)^4 - x^4}$

40 $f(x) = \sqrt{(x^3 + 1)^2 + x^2}$

41 $f(x) = (3x + 1)^4(4x + 1)^3$

42 $f(x) = (4x - 1)^5(5x - 1)^4$

43 $f(x) = (x^2 + 1)^4(x^2 - 1)^3$

44 $f(x) = (x^2 + 1)^3(x^2 - 1)^4$

45 $f(x) = \dfrac{(x + 4)^3}{x^2}$

46 $f(x) = \dfrac{(x - 6)^3}{x^2}$

47 $f(x) = \left(\dfrac{x + 5}{x}\right)^{4}$

48 $f(x) = \left(\dfrac{x + 4}{x}\right)^{5}$

*Find the **second** derivative of each function.*

49 $f(x) = 12\sqrt{x^3} - 9\sqrt[3]{x}$

50 $f(x) = 18\sqrt[3]{x^2} - 4\sqrt{x^3}$

51 $f(x) = \dfrac{1}{3x^2}$

52 $f(x) = \dfrac{1}{2x^3}$

53 $h(w) = (2w^2 - 4)^5$

54 $h(w) = (3w^2 + 1)^4$

55 $g(z) = z^3(z + 1)^3$

56 $g(z) = z^4(z + 1)^4$

Evaluate each expression.

57 If $f(x) = \dfrac{1}{x^2}$, find $f'(\tfrac{1}{2})$.

58 If $f(x) = \dfrac{1}{x}$, find $f'(\tfrac{1}{3})$.

59 If $f(x) = 12\sqrt[3]{x}$, find $f'(8)$.

60 If $f(x) = 6\sqrt[3]{x}$, find $f'(-8)$.

61 If $f(x) = \dfrac{2}{x^3}$, find $f''(-1)$.

62 If $f(x) = \dfrac{3}{x^4}$, find $f''(-1)$.

63 $\dfrac{d^2}{dx^2} x^6 \bigg|_{x=-2}$

64 $\dfrac{d^2}{dx^2} x^{-2} \bigg|_{x=-2}$

65 $\dfrac{d^2}{dx^2} \sqrt{x^5} \bigg|_{x=16}$

66 $\dfrac{d^2}{dx^2} \sqrt{x^7} \bigg|_{x=4}$

67 Find the derivative of $(x^3 - 1)^2$ in two ways:

(a) by the generalized power rule

(b) by "squaring out" the original expression and then differentiating

Your answers should agree.

68 Find the derivative of $g(x) = \dfrac{1}{x^3 + 1}$ in two ways:

(a) by the quotient rule

(b) by the generalized power rule

Your answers should agree.

For each function graphed below, state the values of x for which the derivative does not exist.

69

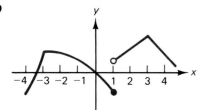

70

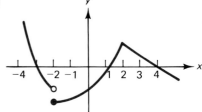

71

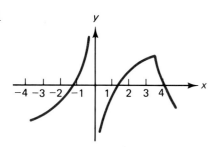

72

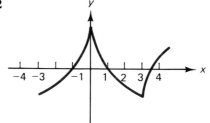

■ **APPLIED EXERCISES**

73 (*General–Temperature*) If the average temperature in North America x years from now is expected to be $T(x) = 65 - 10x^{-1}$ degrees, find $T'(10)$ and $T''(10)$ and interpret your answers.

74 (*General–Velocity*) A rocket rises $s(t) = 8t^{5/2}$ feet in t seconds. Find its velocity and acceleration after 25 seconds.

75 (*General–Velocity*) An object is thrown upward so that its height (in feet) above the ground t seconds after it is thrown is $s(t) = -16t^2 + 48t + 24$. What is the maximum height that it will reach?

76 (*General–Velocity*) The fastest baseball pitch on record (thrown by Lynn Nolan Ryan of the California Angels on August 20, 1974) was clocked at 100.9 miles per hour (148 feet per second). If this pitch had been thrown straight up, its height after t seconds would have been $s(t) = -16t^2 + 148t + 5$. Find the maximum height that the ball would have reached.

77 (*General–Velocity*) The muzzle velocity of a modern recoilless rifle is 1200 miles per hour (1760 feet per second). If the bullet is fired directly upward, its height after t seconds will be $s(t) = -16t^2 + 1760t + 5$. Find the maximum height that the bullet will reach.

Nolan Ryan

78 (*General–Compound Interest*) If $500 is deposited in an account earning interest at r percent annually, after 3 years its value will be $V(r) = 500(1 + .01r)^3$ dollars. Find $V'(8)$ and interpret your answer.

79 (*Business–Marginal Cost*) If a company's cost function is

$$C(x) = 500 + 3x + \frac{10}{x^2}$$

find the marginal cost function.

80 (*Business–Marginal Profit*) If a company's profit function is $P(x) = \sqrt{x^2 + 4x}$, find the marginal profit function.

81 (*Business–Marginal Average Cost*) A company's cost function is $C(x) = 5x + 100$.

(a) Find the average cost function.

(b) Find the marginal average cost function.

82 (*Business–Marginal Average Profit*) A company's profit function is $P(x) = 6x - 200$.

(a) Find the average profit function.

(b) Find the marginal average profit function.

83 (*Sociology–Rumors*) Sociologists have found that under certain circumstances, rumors spread through crowds so that after t minutes, $N(t) = 2t^2$ people have heard the rumor. Find the instantaneous rate of change of this number after 5 minutes and interpret your answers.

84 (*General–Geometry*) The formula for the area of a circle is $A = \pi r^2$, where r is the radius of the circle and π is a constant.

(a) Show that the derivative of the area formula is $2\pi r$, the formula for the circumference of a circle.

(b) Given an explanation for this in terms of rates of change.

85 (*General–Geometry*) The formula for the volume of a sphere is $V = (4/3)\pi r^3$, where r is the radius of the sphere and π is a constant.

(a) Show that the derivative of the volume formula is $4\pi r^2$, the formula for the surface area of a sphere.

(b) Give an explanation for this in terms of rates of change.

86 (*Learning Curves in Industry*) From Section 1.1, the learning curve for building Boeing 707 airplanes is $f(x) = 150x^{-.322}$, where $f(x)$ is the time (in thousands of hours) that it took to build the xth Boeing 707. Find the instantaneous rate of change of this production time for the tenth plane, and interpret your answer.

87 (*Biomedical–Blood Flow*) Blood flowing through an artery encounters a resistance of

$$R(x) = .25\,(.01x + 1)^4$$

where x is the distance (in centimeters) from the heart. Find the instantaneous rate of change of the resistance 100 centimeters from the heart.

88 (*Business–Sales*) The manager of an electronics store estimates that the number of cassette tapes that a store will sell at a price of x dollars is

$$S(x) = \frac{2250}{x + 9}$$

Find the rate of change of this quantity when the price is \$6 per tape, and interpret your answer.

89 (*General–Survival Rate*) Suppose that for a group of 10,000 people, the number who survive to age x is

$$N(x) = 1000\sqrt{100 - x}$$

Find $N'(96)$ and interpret your answer.

3 FURTHER APPLICATIONS OF THE DERIVATIVE

The St. Louis Arch. Note that at the highest point, the slope (steepness) of the curve is zero.

3.1 GRAPHING POLYNOMIALS

Introduction

In this chapter we develop two major applications of differentiation, curve sketching and optimization. Graphs are usually far more informative than algebraic formulas, and curve sketching means making a quick sketch of a graph based on a few well-chosen points. Optimization means finding the largest or smallest values of a function (for example, maximizing profit or minimizing risk). We will begin with curve sketching, since an understanding of graphs will form the basis for optimization in the second half of the chapter.

The Derivative as Slope

The derivative gives the slope of the graph of a function. If the derivative is *positive* at $x = a$, then the curve is *rising* at $x = a$, and if the derivative is *negative* at $x = a$, then the curve is *falling* at $x = a$. (Remember that "rising" and "falling" mean as you move *from left to right* on the graph, the same direction in which you read a book.*)

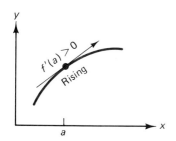

If $f'(a) > 0$, then the graph of $f(x)$ is rising (as you move to the right).

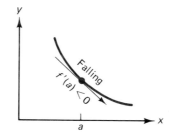

If $f'(a) < 0$, then the graph of $f(x)$ is falling (as you move to the right).

Relative Extreme Points

A *relative maximum point* on a curve is a point on the curve that is at least as high as the points immediately on either side. Similarly, a *relative minimum point* is a point on the curve that is at least as low as the points immediately on either side.

---

* More precisely, "rising at $x = a$" means rising in some small interval surrounding $x = a$.

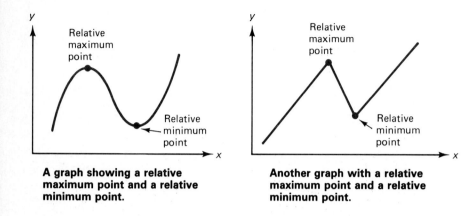

A graph showing a relative maximum point and a relative minimum point.

Another graph with a relative maximum point and a relative minimum point.

Such points are called "relative" maximum and minimum points because, although they may not be the highest and lowest points on the *entire* graph, they are at least the maximum and minimum *relative* to the points immediately on either side. (Later we will use the terms "absolute maximum" and "absolute minimum" to mean the highest and lowest on the entire graph.)

A curve may have several relative maximum points and several relative minimum points. The term *relative extreme point* means a point that is either a relative maximum or a relative minimum point. An endpoint of a curve may be a relative extreme point.

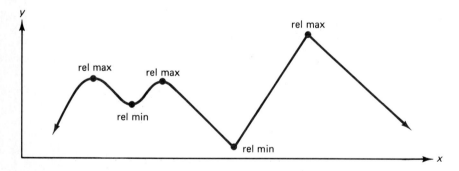

A curve with five relative extreme points. Three are relative maximum points and two are relative minimum points.

Critical Values

Notice that the relative extreme points in the preceding graph occur at points where the slope is zero or undefined (corner points).

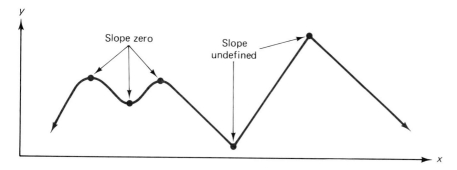

At each relative extreme point the slope is either zero or undefined.

In general:

> At a relative extreme point, the slope must be zero or undefined.

This is because if the slope is positive or negative, the curve is still rising or falling and so cannot have reached its highest or lowest point. The x-values of points where the slope is zero or undefined are called *critical values*.

> A critical value of a function $f(x)$ is an x-value in the domain of the function at which either $f'(x) = 0$ or $f'(x)$ is undefined.

The requirement that x be in the domain is simply to ensure that there is a point on the graph corresponding to that x-value. Points whose x-values are critical values are called *critical points*.

Example 1   Find the critical values of $f(x) = x^3 - 3x^2 - 9x + 15$.

Solution   We want the x-values at which the derivative is zero or undefined. The derivative is

$$f'(x) = 3x^2 - 6x - 9$$

$$= 3(x^2 - 2x - 3) \qquad \textit{factoring out a 3}$$

$$= 3(x - 3)(x + 1) \qquad \textit{factoring further}$$

Since the derivative is a polynomial, there are no x-values at which it is undefined, and the only possible critical values are those at which the derivative is zero,

$$3(x - 3)(x + 1) = 0$$

A product is zero when any one of the factors is zero, so the solutions are

$$x = 3 \qquad \text{\textit{from setting }} (x - 3) = 0$$

$$x = -1 \qquad \text{\textit{from setting }} (x + 1) = 0$$

Since both of these x-values are in the domain of the original function (the domain of a polynomial is $\mathbb{R}$, the set of all real numbers), both are critical values (CVs),

$$\text{CV} \quad \begin{cases} x = 3 \\ x = -1 \end{cases} \qquad \blacksquare$$

This function had two critical values. A function may have any number of critical values, *even none*.

Practice Exercise 1

Find the critical values of $f(x) = x^3 - 12x + 8$. *(solution on page 133)*

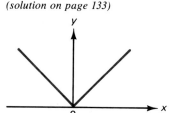

0 is a critical value of
$f(x) = |x|$ **since the derivative**
is undefined at $x = 0$.

In Example 1 the derivative was a polynomial, and polynomials are defined for all values of x, so that there were no critical values resulting from the derivative being undefined. For a function with a critical value at which the derivative is *undefined*, think of the absolute value function, $f(x) = |x|$. The derivative is undefined at $x = 0$ (because the graph has a "corner" at the origin), so $x = 0$ is a critical value of the absolute value function. In Section 3.2 we will see more examples of critical values where the derivative is undefined.

The First Derivative Test

Critical values are the *only* x-values at which relative extreme points (relative maximum and minimum points) can occur. Now we need a way of deciding whether a function has a relative maximum, minimum, or neither at a particular critical value. The whole idea is contained in a diagram.

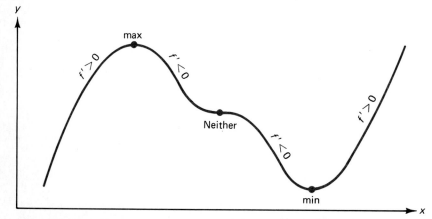

The slopes just before and just after a critical value determine whether there is a relative maximum, relative minimum, or neither at that point.

Notice the relative maximum point. Just before it the slope f' is positive, and just after it the slope is negative (since the curve must rise up to the point and then fall away from it). Similarly, looking at the relative minimum point, just before it the slope is negative and just after it the slope is positive (since the curve must drop down to the point and then rise away from it). Looking at the point in the middle, if the slope is the same on both sides (positive on both sides or negative on both sides), then the curve has *neither* a relative maximum nor a relative minimum at that point. These observations comprise what is known as the *first derivative test*.

First Derivative Test

If a function $f(x)$ has a critical value $x = c$, then at $x = c$ the function has

1. a relative *maximum* if $f' > 0$ just before c and $f' < 0$ just after c;

2. a relative *minimum* if $f' < 0$ just before c and $f' > 0$ just after c;

3. *neither* a relative maximum nor a relative minimum if f' has the same sign just before and just after c.

The first derivative test can be shown in pictures. The important thing to notice is how the slopes just before and just after a critical value determine whether the function has a relative maximum, a relative minimum, or neither at that point.

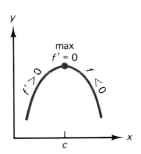 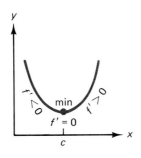 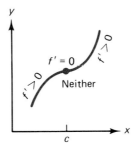 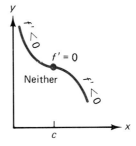

| f' is positive then negative, so $f(x)$ has a relative maximum at $x = c$. | f' is negative then positive, so $f(x)$ has a relative minimum at $x = c$. | f' is positive on both sides, so $f(x)$ has neither at $x = c$. | f' is negative on both sides, so $f(x)$ has neither at $x = c$. |

The diagrams above illustrate the first derivative test for a critical value at which the derivative is zero. The slopes on either side determine the behavior in exactly the same way even if the derivative is *undefined* at the critical value, as the following diagrams show.

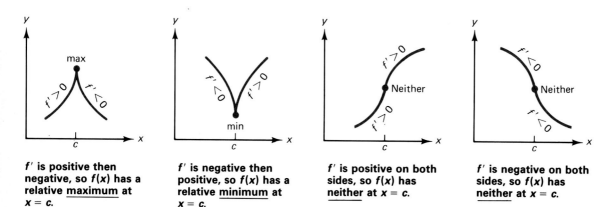

| f' is positive then negative, so $f(x)$ has a relative <u>maximum</u> at $x = c$. | f' is negative then positive, so $f(x)$ has a relative <u>minimum</u> at $x = c$. | f' is positive on both sides, so $f(x)$ has <u>neither</u> at $x = c$. | f' is negative on both sides, so $f(x)$ has <u>neither</u> at $x = c$. |

All of this can be summarized very simply:

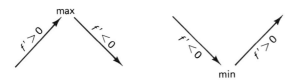

To show where the derivative is positive and where it is negative, we will make a "sign diagram." An example will make the method clear.

Example 2 Make a sign diagram for the derivative of

$$f(x) = x^3 - 3x^2 - 9x + 15$$

Solution We first need the critical values. However, in Example 1 we found that the critical values for this function are

$$\text{CV} \begin{cases} x = 3 \\ x = -1 \end{cases}$$

We draw a copy of the x-axis, writing these x-values below the line and describing the behavior of $f'(x)$ above the line.

$$\underline{\begin{array}{ccc} f' = 0 & & f' = 0 \end{array}} \quad \longleftarrow \text{behavior of } f'$$
$$\begin{array}{ccc} x = -1 & & x = 3 \end{array} \quad \longleftarrow \text{where } f' \text{ is zero or undefined}$$

The derivative can change sign only at points where $f'(x)$ is zero or undefined. Therefore, between each pair of these values the derivative must keep the same sign, always positive or always negative throughout the interval. To find the sign in each interval, we need only "test" the sign of f' at one point in each interval.

Between x = −1 and x = 3. We may use the "test value" $x = 2$ (or any other value between -1 and 3). Substituting $x = 2$ into the derivative

$$f'(x) = 3(x - 3)(x + 1)$$

gives

$$f'(2) = 3 \underbrace{(2 - 3)}_{\text{neg}} \underbrace{(2 + 1)}_{\text{pos}} = \text{negative}$$

$$\underbrace{}_{\text{pos}}$$

(We need only the *sign* of the derivative, not the actual value, so we need only consider the sign of each factor.) Since $f'(2)$ is negative, $f'(x)$ is negative *throughout* the middle interval. We record this fact on the sign diagram by writing $f' < 0$ above the middle interval.

$$
\begin{array}{ccc}
f' = 0 & f' < 0 & f' = 0 \\
\hline
x = -1 & & x = 3
\end{array}
$$

To the right of x = 3. The test value $x = 4$ (or any other value to the right of $x = 3$) substituted into the derivative $f'(x) = 3(x - 3)(x + 1)$ gives

$$f'(4) = 3 \underbrace{(4 - 3)}_{\text{pos}} \underbrace{(4 + 1)}_{\text{pos}} = \textit{positive}$$

$$\underbrace{}_{\text{pos}}$$

which we enter on our sign diagram by writing $f' > 0$:

$$
\begin{array}{cccc}
f' = 0 & f' < 0 & f' = 0 & f' > 0 \\
\hline
x = -1 & & x = 3 &
\end{array}
$$

To the left of x = −1. The test value $x = -2$ (or any other number to the left of $x = -1$) substituted into the derivative $f'(x) = 3(x - 3)(x + 1)$ gives

$$f'(-2) = 3 \underbrace{(-2 - 3)}_{\text{neg}} \underbrace{(-2 + 1)}_{\text{neg}} = \text{positive}$$

$$\underbrace{}_{\text{pos}}$$

which we enter on our sign diagram as $f' > 0$:

$$
\begin{array}{ccccc}
f' > 0 & f' = 0 & f' < 0 & f' = 0 & f' > 0 \\
\hline
& x = -1 & & x = 3 &
\end{array}
$$

The sign diagram shows the sign of $f'(x)$ for all values of x. We then add arrows underneath the sign diagram (rising arrows where $f' > 0$ and falling arrows where $f' < 0$) to show the behavior of $f(x)$.

$f' > 0$	$f' = 0$	$f' < 0$	$f' = 0$	$f' > 0$
	$x = -1$		$x = 3$	
↗	→	↘	→	↗

This shows that the graph is rising, then level, then falling, then level, and finally rising again.

The first part of the sign diagram (the ↗→↘ arrows) shows that the graph has a *relative maximum point* at $x = -1$. The second part of the sign diagram (the ↘→↗ arrows) shows that the graph has a *relative minimum point* at $x = 3$. The y-coordinates of these points are found by substituting the x-coordinates into the original function $f(x) = x^3 - 3x^2 - 9x + 14$:

At $x = -1$: $y = (-1)^3 - 3(-1)^2 - 9(-1) + 15 = 20$ point $(-1, 20)$

At $x = 3$: $y = (3)^3 - 3(3)^2 - 9(3) + 15 = -20$ point $(3, -12)$

Adding these relative extreme points completes the sign diagram

$f' > 0$	$f' = 0$	$f' < 0$	$f' = 0$	$f' > 0$
	$x = -1$		$x = 3$	
↗	→	↘	→	↗
	rel max $(-1, 20)$		rel min $(3, -12)$	

■

Graphing Polynomials

It is a simple matter to graph a polynomial from its sign diagram.

Example 3   Graph the function $f(x) = x^3 - 3x^2 - 9x + 15$.

Solution   The sign diagram (obtained above) for this function is

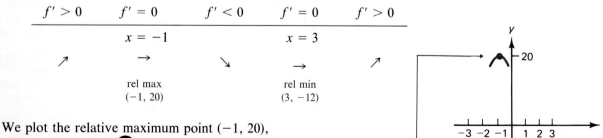

$f' > 0$	$f' = 0$	$f' < 0$	$f' = 0$	$f' > 0$
	$x = -1$		$x = 3$	
↗	→	↘	→	↗
	rel max $(-1, 20)$		rel min $(3, -12)$	

We plot the relative maximum point $(-1, 20)$, with a small "cap" ⌢ (to indicate a maximum) and the relative minimum point $(3, -12)$, with a small "cup" ⌣ (to indicate a minimum)

■

We then sketch the graph by joining these pieces of the curve to give a smooth curve, as in the diagram below.

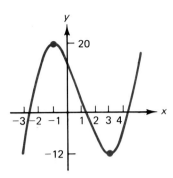

The graph of
$f(x) = x^3 - 3x^2 - 9x + 15$.

Notice that we sketched the function by plotting only two points, the *critical points*. Curve sketching using sign diagrams is very efficient. For greater accuracy we could plot a few more points, but the idea of this section is to get a reasonably accurate sketch quickly by plotting only a few "important" points.

Practice Exercise 2

Make a sign diagram for the derivative and sketch the graph of

$$f(x) = x^3 - 12x + 8$$

(Use the critical values that you found in Practice Exercise 1.)   *(solution on page 133)*

Summary

A sign diagram for $f'(x)$ consists of

$f' > 0$	$f' = 0$	$f' < 0$	$f' = 0$	$f' > 0$	← behavior of f'
	$x = -1$		$x = 3$		← where f' is zero or undefined
↗	→	↘	→	↗	← arrows showing the slope of $f(x)$
	rel max $(-1, 20)$		rel min $(3, -12)$		← critical points

A piece of a sign diagram like

$f' > 0$	$f' = 0$	$f' < 0$
↗	→	↘

means a graph like

rel max

A piece of a sign diagram like

$f' < 0$	$f' = 0$	$f' > 0$
$\searrow$	$\rightarrow$	$\nearrow$

means a graph like

rel
min

A piece of a sign diagram like

$f' > 0$	$f' = 0$	$f' > 0$
$\nearrow$	$\rightarrow$	$\nearrow$

means a graph like

neither

A piece of a sign diagram like

$f' < 0$	$f' = 0$	$f' < 0$
$\searrow$	$\rightarrow$	$\searrow$

means a graph like

neither

("Neither" means that the point is neither a relative maximum point nor a relative minimum point.)

Example 4 Sketch the graph of $f(x) = -x^4 + 4x^3 - 12$.

Solution

$$f'(x) = -4x^3 + 12x^2 \qquad \text{\textit{the derivative}}$$

$$= -4x^2(x - 3) \qquad \text{\textit{factoring}}$$

Critical values:

$$\text{CV} \begin{cases} x = 3 & \text{\textit{from setting } } (x - 3) = 0 \\ x = 0 & \text{\textit{from setting } } 4x^2 = 0 \end{cases}$$

We put these two values on a sign diagram.

$f' = 0$	$f' = 0$
$x = 0$	$x = 3$

Then we determine the sign of $f'(x) = -4x^2(x - 3)$ in each interval (using test points) and add arrows.

$f' > 0$	$f' = 0$	$f' > 0$	$f' = 0$	$f' < 0$
	$x = 0$		$x = 3$	
$\nearrow$	$\rightarrow$	$\nearrow$	$\rightarrow$	$\searrow$

Finally, we interpret the arrows to describe the critical points.

$f' > 0$	$f' = 0$	$f' > 0$	$f' = 0$	$f' < 0$
	$x = 0$		$x = 3$	
↗	→	↗	→	↘
	neither		rel max	
	$(0, -12)$		$(3, 15)$	

the y-values comes from evaluating
$f(x) = -x^4 + 4x^3 - 12$ *at the x-values*

From the sign diagram we draw the curve "pieces" shown on the left below, and joining them gives the completed graph shown on the right.

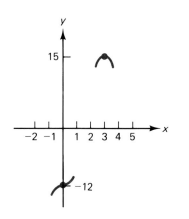
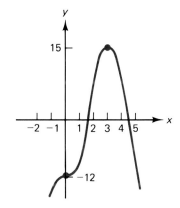

The graph of
$f(x) = -x^4 + 4x^3 - 12$.

■

SOLUTIONS TO PRACTICE EXERCISES

1. $f(x) = x^3 - 12x + 8$

   $f'(x) = 3x^2 - 12 = 3(x^2 - 4) = 3(x + 2)(x - 2)$

   Critical values: $x = -2$, $x = 2$

2.
$f' > 0$	$f' = 0$	$f' < 0$	$f' = 0$	$f' > 0$
	$x = -2$		$x = 2$	
↗	→	↘	→	↗
	rel max		rel min	
	$(-2, 24)$		$(2, -8)$	

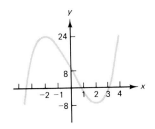

EXERCISES 3.1

For each function find all critical values.

1 $f(x) = x^3 - 48x$

2 $f(x) = x^3 - 27x$

3 $f(x) = x^4 + 4x^3 - 8x^2 + 1$

4 $f(x) = x^4 + 4x^3 - 20x^2 - 12$

5 $f(x) = (2x - 6)^4$

6 $f(x) = (3x - 12)^5$

7 $g(x) = (x^2 + 2x - 3)^2$

8 $g(x) = (x^2 + 6x - 7)^2$

9 $f(x) = 3x + 5$

10 $f(x) = 4x - 12$

Sketch the graphs of each function by making a sign diagram for the derivative and plotting all critical points.

11 $f(x) = x^4 + 4x^3 - 8x^2 + 64$

12 $f(x) = x^4 - 4x^3 - 8x^2 + 64$

13 $f(x) = -x^4 + 4x^3 - 4x^2 + 1$

14 $f(x) = -x^4 - 4x^3 - 4x^2 + 1$

15 $f(x) = 3x^4 - 8x^3 + 6x^2$

16 $f(x) = 3x^4 + 8x^3 + 6x^2$

17 $f(x) = (x - 1)^6$

18 $f(x) = (x + 1)^4$

19 $f(x) = (x + 1)^3$

20 $f(x) = (x - 1)^5$

21 $f(x) = (x^2 - 4)^2$

22 $f(x) = (x^2 - 9)^2$

23 $f(x) = (x^2 + 2x - 3)^2$

24 $f(x) = (x^2 - 2x - 8)^2$

25 $f(x) = -x^2(x - 3)$

26 $f(x) = -x^3(x - 4)$

27 $f(x) = x^2(x - 4)^2$

28 $f(x) = x(x - 4)^3$

29 $f(x) = x^2(x - 5)^3$

30 $f(x) = x^3(x - 5)^2$

31 Derive the formula $x = -\dfrac{b}{2a}$ for the x-coordinate of the vertex of a quadratic function $f(x) = ax^2 + bx + c$ (for constants a, b, c, with $a \neq 0$). (*Hint:* The slope is zero at the vertex, so finding the vertex means finding the critical value.)

32 Derive the formula $x = -b$ for the x-coordinate of the vertex of a quadratic function $f(x) = a(x + b)^2 + c$ (for constants a, b, c, with $a \neq 0$). (*Hint:* The slope is zero at the vertex, so finding the vertex means finding the critical value.)

APPLIED EXERCISES

33 (*Business–Marginal and Average Cost*) A company's cost function is $C(x) = x^3 - 6x^2 + 14x$, and therefore its average cost function, $C(x)/x$, is $AC(x) = x^2 - 6x + 14$ (for $x \geq 0$).

(a) Graph the average cost function $AC(x)$.

(b) Differentiate the cost function $C(x)$ to find the marginal cost function $MC(x)$.

(c) Graph the marginal cost function on the same set of axes that you used for the average cost function. Notice that the two curves intersect at the point where the average cost is minimized. We shall return to this important observation later.

34 (*Business–Marginal and Average Cost*) A company's cost function is $C(x) = x^3 - 12x^2 + 50x$, and therefore its average cost function, $C(x)/x$, is $AC(x) = x^2 - 12x + 50$ (for $x \geq 0$).

(a) Graph the average cost function $AC(x)$.

(b) Differentiate the cost function $C(x)$ to find the marginal cost function $MC(x)$.

(c) Graph the marginal cost function on the same set of axes that you used for the average cost function. Notice that the two curves intersect at the point where the average cost is minimized. We shall return to this important observation later.

35 (*Biomedical–Bacterial Growth*) A population of bacteria grows to size $p(x) = x^3 - 9x^2 + 24x + 10$ after x hours (for $x \geq 0$). Sketch the graph of this growth curve.

36 (*Behavioral Science–Learning Curve*) A learning curve is a function $L(x)$ that gives the amount of time that a person requires to learn x pieces of informa-

tion. Many learning curves take the form

$$L(x) = (x - a)^n + b$$

(for $x \geq 0$), where a, b, and n are positive constants. Graph the learning curve

$$L(x) = (x - 2)^3 + 8$$

3.2 GRAPHING, CONTINUED

Introduction

In Section 3.1, we showed how the first derivative is used for curve sketching. In this section we see how to make even better sketches using the *second* derivative as well. We will define the *concavity* of a curve, and this will lead to another test for relative maximum and minimum points.

Concavity

A curve that "curls upward" (like the curves on the left below) is called *concave up*, and a curve that "curls downward" (like the ones on the right) is called *concave down*.

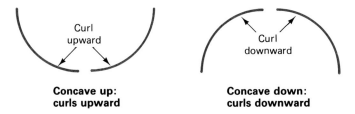

Concave up: Concave down:
curls upward   curls downward

For example, the graphs below show the weight and height of an average child from age 2 to age 15. Notice that the weight curve is concave up, while the height curve is concave down.

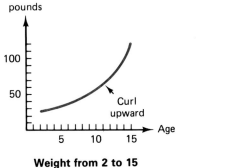

Weight from 2 to 15

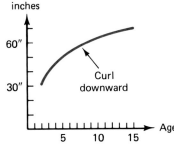

Height from 2 to 15

Distinguish carefully between concavity and slope. A straight line with slope *m* is neither concave up nor concave down (we say that it has zero concavity).

Straight
Zero concavity

However, if we were to bend the two ends *upward*, it would be concave *up*:

Concave up

and if we were to bend the two ends *downward*, it would be concave *down*:

Concave down

That is, *slope* measures the steepness, while *concavity* measures the "curl" away from straightness. These pictures also show that a curve that is concave *up* lies *above* its tangent line, and a curve that is concave *down* lies *below* its tangent line (except of course at the point of tangency).

We speak of a *function* as being concave up or down, meaning that its graph is concave up or down. A function may be concave up for certain values of *x* and concave down for others. A point where the concavity changes (from up to down or down to up) is called an *inflection point*.

> An inflection point is a point on the graph where the concavity changes.

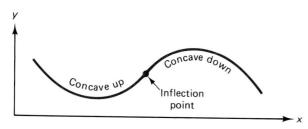

A curve with an inflection point

A function may have any number of inflection points, even none.

Practice Exercise 1

For each curve label the parts that are concave up and the parts that are concave down, and find all inflection points.

(a)

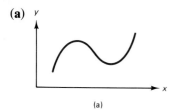

(a)

(b)

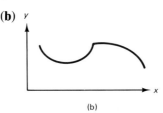

(b)

(solution on page 147)

The Second Derivative Measures Concavity

How can we measure the concavity of a curve? The second derivative is the derivative of the derivative, and so gives the rate of change of the slope. That is, the second derivative tells whether the slope is increasing or decreasing.

   Therefore, if the second derivative is positive, then the slope is *increasing* (as you move to the right), which means that the curve is concave up ("curling" upward).

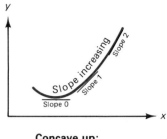

**Concave up:
slope increasing**

   Similarly, if the second derivative is negative, then the slope is *decreasing* (as you move to the right), and so the curve is concave down ("curling" downward).

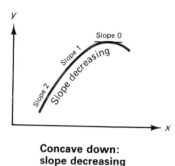

**Concave down:
slope decreasing**

Therefore, the sign of the second derivative determines the concavity.

If $f''(a) > 0$, the graph of $f(x)$ is concave *up* at $x = a$.
If $f''(a) < 0$, the graph of $f(x)$ is concave *down* at $x = a$.

   At an inflection point the second derivative must be either zero or undefined. This is because if the second derivative is positive or negative, the curve is still increasing or decreasing in steepness, rather than changing concavity. Therefore, to find inflection points we first find where the second derivative is either zero or undefined. Then we determine whether the concavity actually changes by making a sign diagram for the *second* derivative. An example will make the method clear.

Example 1   A company's annual profit after x years is $f(x) = x^3 - 9x^2 + 24x$ million dollars (for $x \geq 0$). Graph this function, showing all relative extreme points and inflection points.

Solution

$$f'(x) = 3x^2 - 18x + 24 \qquad\qquad \textit{the derivative}$$

$$= 3(x^2 - 6x + 8) = 3(x - 2)(x - 4) \qquad \textit{factoring}$$

The critical values are $x = 2$ and $x = 4$, and the sign diagram for f' (found in the usual way) is

$f' > 0$	$f' = 0$	$f' < 0$	$f' = 0$	$f' > 0$
	$x = 2$		$x = 4$	
↗	→	↘	→	↗
	rel max (2, 20)		rel min (4, 16)	

To find the inflection points, we calculate the second derivative,

$$f''(x) = 6x - 18 = 6(x - 3)$$

differentiating
$f'(x) = 3x^2 - 18x + 24$

This is zero at $x = 3$, which we enter on a sign diagram for the *second* derivative.

$f'' = 0$
$x = 3$

We use test points to determine the sign of $f''(x) = 6(x - 3)$ on either side of $x = 3$, just as we did for the first derivative.

$f''(2) = 6(2 - 3) < 0$ $f''(4) = 6(4 - 3) > 0$

$f'' < 0$	$f'' = 0$	$f'' > 0$
	$x = 3$	
con dn		con up

concave down, concave up

This shows that the concavity *does* change (from down to up) at $x = 3$, so there *is* an inflection point at $x = 3$.

$f'' < 0$	$f'' = 0$	$f'' > 0$
	$x = 3$	
con dn		con up
	IP (3, 18)	

IP means inflection point. The 18 comes from substituting $x = 3$ into $f(x) = x^3 - 9x^2 + 24x$

This completes the sign diagram for the second derivative. To graph the function, we plot the relative extreme points (from the first derivative sign diagram) using appropriate "caps" and "cups", and then the inflection point (from the second derivative sign diagram). This gives the "pieces" of the curve on the left below, and joining them (according to the sign diagrams) gives the completed graph shown on the right.

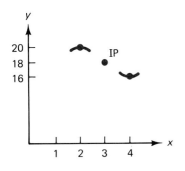

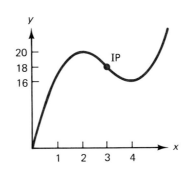

$$f(x) = x^3 - 9x^2 + 24x \text{ for } x \geqslant 0 \quad ■$$

Notice how calculus helped us in this problem. We graphed the function using just three points, the relative extreme points and the inflection point. If we had plotted these three points before we knew calculus, we might have joined them with a straight line, completely mistaking the shape of the graph. Calculus helped by choosing not just *any* three points, but the three most important points on the entire graph.

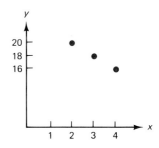

Practice Exercise 2

For the function $f(x) = x^3 - 12x + 8$:

(a) Find the *second* derivative.

(b) Make a sign diagram for the second derivative and find all inflection points.

(c) Look at the graph of this function on page 133 and check that the result you found in part (b) agrees with the graph.

(solutions on page 148)

Locating inflection points requires showing that the concavity (that is, the sign of f'') *actually changes* at the point. For example, if the sign diagram for the second derivative in the previous problem had looked like this,

$f'' < 0$	$f'' = 0$	$f'' < 0$
	$x = 3$	
con dn		con dn

concavity does not change

there would *not* have been an inflection point, since the concavity is the same on both sides. For there to be an inflection point, the sign of f'' must actually change.

Practice Exercise 3

Which of the following sign diagrams indicates an inflection point?

(a)

$$f'' > 0 \qquad f'' = 0 \qquad f'' > 0$$

$$x = a$$

con up   con up

(b)

$$f'' < 0 \qquad f'' = 0 \qquad f'' > 0$$

$$x = a$$

con dn   con up

(solutions on page 148)

Interpretations of Inflection Points

Inflection points have important interpretations. The function in the previous example gave the annual profit for a company after x years. The graph shows that the company's revenue increased for the first 2 years, then decreased until year 4, and then began to increase again. The inflection point at $x = 3$ is the point where the profit *first began to show signs of improvement*. It marked the end of the period of increasingly steep decline and the first sign of an "upturn."

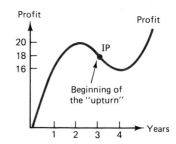

The graph on the right shows a person's temperature during an illness, and the inflection point is where the temperature ends its unchecked rise and begins to moderate, that is, the point at which the illness is *first brought under control*.

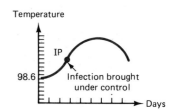

The graph on the right gives a company's total sales after x days of an advertising campaign. The inflection point gives the *point of diminishing returns*: advertising beyond this point will still bring additional sales, but at a slower rate.

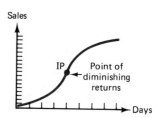

Concavity Is Independent of Slope

Note that all combinations of signs for the first and second derivatives are possible. A graph may be *increasing* and concave *up* ($f' > 0, f'' > 0$), such as

increasing and concave *down* ($f' > 0, f'' < 0$), such as

decreasing and concave *up* ($f' < 0, f'' > 0$), such as

decreasing and concave *down* ($f' < 0, f'' < 0$), such as

The four quarters of a circle illustrate all four possibilities.

$$
\begin{array}{cc}
f' > 0 & f' < 0 \\
f'' < 0 & f'' < 0 \\
f' < 0 & f' > 0 \\
f'' > 0 & f'' > 0
\end{array}
$$

On sign diagrams, we will use the abbreviation "und" for "undefined."

$$\frac{f' \text{ und}}{x = a}$$ means that the first derivative is undefined at $x = a$

$$\frac{f'' \text{ und}}{x = a}$$ means that the second derivative is undefined at $x = a$

The following examples show how to graph critical points at which the derivative is undefined.

Application: Stevens' Law of Psychophysics

Suppose that you are given two weights and asked to judge how much heavier one is than the other. If one weight is actually *twice* as heavy as the other, most people will judge the heavier weight as being *less* than twice as heavy. This is one of the oldest problems of experimental psychology—how sensation (perceived weight) varies with stimulus (actual weight). Similar experiments can be performed for perceived brightness of a light compared to actual brightness, perceived effort compared to actual work, and so on. The results will vary somewhat from person to person, but the diagram on the right shows some typical response curves.

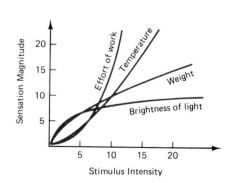

Notice, for example, that perceived effort increases more rapidly than actual work, which suggests that a 10% increase in an employee's work should be rewarded with a *greater* than 10% increase in pay. Such stimulus–

response curves were studied by the psychologist S. S. Stevens* at Harvard, who expressed them in the form

$$\text{response} = a(\text{stimulus})^b \quad \text{or} \quad f(x) = ax^b$$

for constants a and b

Example 2 Sketch the response curve for brightness of light, $f(x) = 9x^{1/3}$.

Solution Although this response curve is meaningful only for $x \geq 0$, we will graph it for *all* values of x to illustrate to our graphing technique. The derivative is

$$f'(x) = 3x^{-2/3} = \frac{3}{\sqrt[3]{x^2}}$$

differentiating $f(x) = 9x^{1/3}$

At $x = 0$ the derivative is *undefined* (because of the zero in the denominator), which we enter on the sign diagram as

$$\underline{\qquad\qquad f' \text{ und} \qquad\qquad}$$
$$x = 0$$

f' is undefined at $x = 0$

The test points $x = -1$ and $x = +1$ show that $f'(x)$ is *positive* on both sides.

$$\underline{f' > 0 \qquad f' \text{ und} \qquad f' > 0}$$
$$x = 0$$
$$\nearrow \qquad\qquad\qquad \nearrow$$

*Do **not** draw a horizontal arrow under $x = 0$ since the derivative there is **undefined** rather than zero.*

The second derivative is

$$f''(x) = -2x^{-5/3} = \frac{-2}{\sqrt[3]{x^5}}$$

differentiating $f'(x) = 3x^{-2/3}$

which is *undefined* at $x = 0$.

$$\underline{\qquad\qquad f'' \text{ und} \qquad\qquad}$$
$$x = 0$$

Using test points on either side we obtain the sign diagram for f''.

$$\underline{f'' > 0 \qquad f'' \text{ und} \qquad f'' < 0}$$
$$x = 0$$
$$\text{con up} \qquad\qquad\qquad \text{con dn}$$
$$\text{IP } (0, 0)$$

the y-coordinate 0 comes from substituting $x = 0$ into $f(x) = 9x^{1/3}$

* S.S. Stevens, "On the Psychophysical Law," *Psychol. Rev.,* **64** (1957), 153–181.

There *is* an inflection point because the concavity changes. The two sign diagrams,

$$f' > 0 \qquad f' \text{ und} \qquad f' > 0$$

$$\overline{\hspace{1cm} x = 0 \hspace{1cm}}$$

↗   ↗

$$f'' > 0 \qquad f'' \text{ und} \qquad f'' < 0$$

$$\overline{\hspace{1cm} x = 0 \hspace{1cm}}$$

con up   con dn

IP (0, 0)

show that to the *left* of the inflection point (0, 0) the curve is increasing and concave up ($f' > 0$, $f'' > 0$), which means a curve like ╱ .

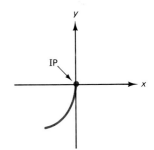

We therefore ''attach'' this curve to the left of the inflection point (0, 0).

To the *right* of $x = 0$ the sign diagrams show that the curve is increasing and concave down ($f' > 0$, $f'' < 0$), which means a curve like ╱ .
We therefore ''attach'' this curve to the right of the inflection point (0, 0).

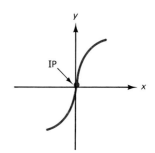

By plotting another point on either side, such as $(-1, -9)$ and $(1, 9)$, we get our final sketch of the curve.

$$f(x) = 9x^{1/3}$$

■

Notice that this curve has a *vertical tangent* at the origin, so that the slope is undefined at $x = 0$, agreeing with the "f' und" on the sign diagram for the first derivative:

$$f' > 0 \qquad f' \text{ und} \qquad f' > 0$$
$$\overline{\hspace{6cm}}$$
$$x = 0$$
$$\nearrow \qquad\qquad\qquad \nearrow$$

Other Fractional Powers

Example 3 Sketch the graph of $f(x) = 18\sqrt[3]{(x-1)^2}$.

Solution

$$f(x) = 18(x-1)^{2/3} \qquad\qquad \textit{f(x) in exponential form}$$

$$f'(x) = 12(x-1)^{-1/3} = \frac{12}{\sqrt[3]{x-1}} \qquad\qquad \textit{the derivative}$$

From this we get the sign diagram for the first derivative.

$$f' < 0 \qquad f' \text{ und} \qquad f' > 0$$
$$\overline{\hspace{6cm}}$$
$$x = 1$$
$$\searrow \qquad\qquad\qquad \nearrow$$
$$\text{rel min } (1, 0)$$

The second derivative is

$$f''(x) = -4(x-1)^{-4/3} = \frac{-4}{\sqrt[3]{(x-1)^4}}$$

giving the sign diagram for f''.

$$f'' < 0 \qquad f'' \text{ und} \qquad f'' < 0$$
$$\overline{\hspace{6cm}}$$
$$x = 1$$
$$\text{con dn} \qquad\qquad\qquad \text{con dn}$$

concavity does not change, so no inflection points

We then plot the relative minimum point (1, 0) (but not with a 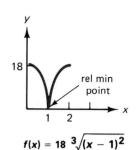 since the derivative is *undefined* rather than zero at the critical value). According to the sign diagrams, to the *left* of $x = 1$ the curve satisfies $f' < 0$ and $f'' < 0$, and to the right it satisfies $f' > 0$ and $f'' < 0$, so we attach the appropriate curves to the left and right of the relative minimum point (1, 0). Plotting two other points, (0, 18) and (2, 18), shows how steeply the curve rises.

$$f(x) = 18 \sqrt[3]{(x - 1)^2}$$

■

The sharp point or "cusp" at $x = 1$ agrees with the "f' und" on the sign diagram, since the derivative is undefined at a corner point. Fractional power curves like this are used in many applications, from economics to medicine.

Graphing Even Roots

For functions containing *even* roots, like $f(x) = \sqrt[4]{x}$, the domain must be restricted to avoid even roots of negative numbers, which are undefined. For example, to graph $f(x) = \sqrt[4]{x}$, we would proceed just as before, looking for relative extreme points and inflection points, but remembering that the domain is $\{x \mid x \geq 0\}$, so that the graph lies entirely to the right of the y-axis.

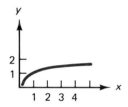

The function
$f(x) = \sqrt[4]{x}$ **is defined only for $x \geq 0$.**

Second Derivative Test

The pictures below show that if a curve is concave *up* at a critical value, then it has a relative *minimum,* while if the curve is concave *down* at the critical value, it has a relative *maximum.*

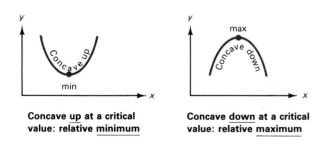

Concave up at a critical value: relative minimum

Concave down at a critical value: relative maximum

Since the second derivative gives concavity, the second derivative may sometimes be used to test whether a function has a relative maximum or a relative minimum at a critical value. This is called the second derivative test.

Second Derivative Test

If $x = c$ is a critical value at which f'' exists, then

    (i) $f''(c) > 0$ means that $f(x)$ has a relative *minimum* at $x = c$;

    (ii) $f''(c) < 0$ means that $f(x)$ has a relative *maximum* at $x = c$;

    (iii) $f''(c) = 0$ means that no conclusion can be drawn.

To use the second derivative test we first find all critical values, then substitute each into the second derivative and determine the sign of the result. A positive result means a *minimum* at the critical value, negative means a *maximum,* and zero means that anything is possible: the function may still have a relative maximum, a relative minimum, or neither.

Example 4   Find all relative extreme points of $f(x) = x^3 - 9x^2 + 24x$ using the second derivative test.

Solution

$$f'(x) = 3x^2 - 18x + 24 \qquad\qquad \textit{the derivative}$$

$$= 3(x^2 - 6x + 8) = 3(x - 2)(x - 4) \qquad\qquad \textit{factoring}$$

$$\text{CV} \begin{cases} x = 2 \\ x = 4 \end{cases}$$

We test the critical values using the second derivative $f''(x) = 6x - 18$.

At $x = 2$:

  $f''(2) = 6 \cdot 2 - 18 = \textit{negative}, \quad$ so $f(x)$ has a relative *maximum* at $x = 2$

At $x = 4$:

  $f''(4) = 6 \cdot 4 - 18 = \textit{positive}, \quad$ so $f(x)$ has a relative *minimum* at $x = 4$

Using this information, we could plot relative maximum and minimum points and sketch the graph of the function. We will not continue further with this example since we graphed this function on page 139. ▪

Example 5   Find all relative extreme points of $f(x) = x^4$ using the second derivative test.

Solution

$$f'(x) = 4x^3$$ *the derivative*

$$\text{CV} \quad x = 0$$ *one critical value*

Testing the critical values using the second derivative,

$$f''(0) = 12(0)^2 = 0$$ $f''(x) = 12x^2$ *at* $x = 0$

The second derivative is zero, so part (iii) of the test says that *no conclusion can be drawn*. That is, the second derivative test provides no information about this critical value. The graph of $f(x) = x^4$ (which can be found using sign diagrams) is shown on the right.

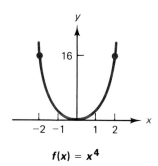

$f(x) = x^4$

Notice that despite the second derivative being zero, the function still has a relative minimum. Because of this possibility that the second derivative test may fail to give any useful information, we prefer sign diagrams to the second derivative test for curve sketching. The second derivative test will be very useful in the second half of this chapter when we study optimization. ▪

Summary

The *second* derivative gives the *concavity* or "curl" of a curve, as shown in the diagram below. An inflection point is a point on the graph where the concavity changes. To locate inflection points, we find where the second derivative is zero or undefined, and then use a sign diagram to see whether f'' actually changes sign. To graph a function, we make sign diagrams for the *first* derivative (to find slopes and relative extreme points) and for the *second* derivative (to find concavity and inflection points).

The second derivative test can sometimes be used to determine whether a curve has a relative maximum or a relative minimum at a critical value. The second derivative test should be thought of as a simple application of concavity.

SOLUTIONS TO PRACTICE PROBLEMS

1. (a)

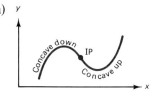

(b)

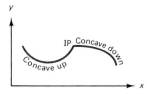

2. **(a)** $f''(x) = 6x$ **(b)**

$f'' < 0$	$f'' = 0$	$f'' > 0$
	$x = 0$	
con dn		con up

IP $(0, 8)$

3. Sign diagram (b)

EXERCISES 3.2

1 The graph of a child's weight on page 135 is concave up. Does this mean that weight gain speeds up or slows down as children become older?

2 The graph of a child's height on page 135 is concave down. Does this mean that height gain speeds up or slows down as children become older?

For each graph, which of the numbered points are inflection points?

3
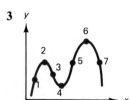

4
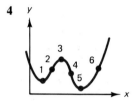

5
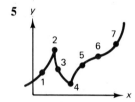

6
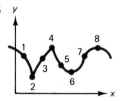

For each function:

(a) Make a sign diagram for the first derivative.

(b) Make a sign diagram for the second derivative.

(c) Sketch the graph, showing all relative extreme points and inflection points.

7 $f(x) = x^3 + 3x^2 - 9x + 5$
8 $f(x) = x^3 - 3x^2 - 9x + 7$
9 $f(x) = x^3 - 6x^2 + 9x + 24$

10 $f(x) = x^3 + 6x^2 + 9x + 12$
11 $f(x) = x^3 - 3x^2 + 3x + 4$
12 $f(x) = x^3 + 3x^2 + 3x + 6$

13 $f(x) = x^4 - 8x^3 + 18x^2 + 2$
14 $f(x) = x^4 + 8x^3 + 18x^2 + 8$
15 $f(x) = x^4 + 4x^3 + 15$

16 $f(x) = x^4 - 4x^3 + 12$
17 $f(x) = 5x^4 - x^5$
18 $f(x) = 3x^2 - x^3$

19 $f(x) = (x - 1)^3 + 1$
20 $f(x) = (x - 2)^3 + 2$
21 $f(x) = (2x + 4)^5$

22 $f(x) = (3x + 6)^5$
23 $f(x) = (2x - 4)^6 + 2$
24 $f(x) = (3x - 6)^6 + 1$

25 $f(x) = x(x - 3)^2$
26 $f(x) = x(x - 6)^2$
27 $f(x) = x^2(x - 3)$

28 $f(x) = x^3(x - 4)$
29 $f(x) = x^{3/5}$
30 $f(x) = x^{1/5}$

31 $f(x) = \sqrt[5]{x^4} + 2$
32 $f(x) = \sqrt[5]{x^3} + 3$
33 $f(x) = \sqrt[5]{x} - 1$

34 $f(x) = \sqrt[5]{x^2} - 1$
35 $f(x) = \sqrt[4]{x^3}$
36 $f(x) = \sqrt[4]{x}$

37 $f(x) = \sqrt{x^5}$
38 $f(x) = \sqrt{x^3}$
39 $f(x) = \sqrt[3]{(x - 1)^2}$

40 $f(x) = \sqrt[5]{(x - 2)^4} + 3$
41 $f(x) = \sqrt[3]{x + 1} + 1$
42 $f(x) = \sqrt[5]{x + 2} + 3$

43 $f(x) = \sqrt[5]{(x - 1)^4}$
44 $f(x) = \sqrt[3]{(x - 1)^5}$
45 $f(x) = \sqrt{(x - 3)^3} + 4$

46 $f(x) = \sqrt{x - 1} + 5$

Use the techniques of this section to graph each function for x ≥ 0. (▦ will be helpful.)

47 $f(x) = x^{.15}$ **48** $f(x) = x^{.35}$ **49** $f(x) = x^{1.15}$ **50** $f(x) = x^{1.35}$

51 (*Concavity of a Parabola*) Show that the quadratic function $f(x) = ax^2 + bx + c$ is concave up if $a > 0$ and is concave down if $a < 0$. (Therefore, the rule that a parabola opens up if $a > 0$ and down if $a < 0$ is merely an application of concavity.)

52 (*Inflection Point of a Cubic*) Show that the general "cubic" function $f(x) = ax^3 + bx^2 + cx + d$ (with $a \neq 0$) has an inflection point at $x = -b/(3a)$.

▨ **APPLIED EXERCISES**

53 (*Business–Revenue*) A company's annual revenue after x years is $f(x) = x^3 - 9x^2 + 15x + 25$ thousand dollars (for $x \geq 0$).

(a) Make sign diagrams for the first and second derivatives.

(b) Sketch the graph of the revenue function, showing all relative extreme points and inflection points.

54 (*Business–Sales*) A company's weekly sales (in thousands) after x weeks is $f(x) = -x^4 + 4x + 70$ (for $0 \leq x \leq 3$).

(a) Make sign diagrams for the first and second derivatives.

(b) Sketch the graph of the sales function, showing all relative extreme points and inflection points.

55 (*General–Temperature*) The temperature in a refining tower after x hours is $f(x) = x^4 - 4x + 112$ degrees Fahrenheit (for $x \geq 0$).

(a) Make sign diagrams for the first and second derivatives.

(b) Sketch the graph of the temperature function, showing all relative extreme points and inflection points.

56 (*Biomedical–Dosage Curve*)   The dose–response curve for x grams of a drug is $f(x) = 8(x - 1)^3 + 8$ (for $x \geq 0$).

(a) Make sign diagrams for the first and second derivatives.

(b) Sketch the graph of the response function, showing all relative extreme points and inflection points.

57 (*Psychology–Stimulus and Response*) Sketch the graph of the brightness response curve $f(x) = x^{2/5}$ for $x \geq 0$, showing all relative extreme points and inflection points.

58 (*Psychology–Stimulus and Response*) Sketch the graph of the loudness response curve $f(x) = x^{4/5}$ for $x \geq 0$, showing all relative extreme points and inflection points.

59–60 (*Sociology–Status*) Sociologists have estimated how a person's "status" in society (as perceived by others) depends upon the person's income and education level. One estimate is that status, S, depends upon income, i, according to the formula $S(i) = 16\sqrt{i}$ (for $i \geq 0$), and that status depends upon education level, e, according to the formula $S(e) = \frac{1}{4}e^2$ (for $e \geq 0$).

59 (a) Sketch the graph of the function $S(i)$.

(b) Is the curve concave up or down? What does this signify about the rate at which status increases at higher income levels?

60 (a) Sketch the graph of the function $S(e)$.

(b) Is the curve concave up or down? What does this signify about the rate at which status increases at higher education levels?

Sketch the graphs of the following functions, showing all relative extreme points and inflection points.

61 $f(x) = 3x^{2/3} - 2x$ **62** $f(x) = 3x^{1/3} - x$ **63** $f(x) = 4x - 5x^{4/5}$

64 $f(x) = 3x - 5x^{3/5}$ **65** $f(x) = 3x^{2/3} - x^2$ **66** $f(x) = 6x^{1/3} - x^2$

3.3 GRAPHING, CONCLUDED

Introduction

In this section we graph rational functions (quotients of polynomials) such as

$$f(x) = \frac{x^2 + 1}{x^2 - 1}, \qquad g(x) = \frac{x^3 - 3x^2 + 1}{5x}, \qquad h(x) = \frac{1}{x^3}$$

We assume that all rational functions have been simplified by canceling common factors from their numerators and denominators. For example, the rational function

$$\frac{(x - 2)(x + 3)}{(x + 5)(x + 3)} \quad \text{should first be simplified to} \quad \frac{x - 2}{x + 5}$$

cancel

Reciprocals of Large and Small Numbers

There is an important reciprocal relationship between large and small numbers. For positive numbers,

the reciprocal of a large number is a small number.

$$\frac{1}{1,000,000} = .000001$$

one over a million   *one millionth*

and

the reciprocal of a small number is a large number.

$$\frac{1}{.000001} = 1,000,000$$

one over a millionth   *one million*

If the *denominator* of a fraction becomes arbitrarily large (with the numerator staying the same), the *value* of the fraction approaches zero. On the other hand, if the denominator of the fraction approaches *zero* (but stays positive), the fraction becomes arbitrarily large ("approaches infinity").

For negative numbers the results are similar except that the signs will be negative.

$$\frac{1}{-1,000,000} = -.000001$$

$$\frac{1}{-.000001} = -1,000,000$$

These ideas will be very useful in curve sketching.

Asymptotes

When a curve straightens out and approaches a line, we call that line an *asymptote* of the curve, and we say that the curve approaches the line *asymptotically*. For example, the dashed lines in the diagrams below are asymptotes of the curves.

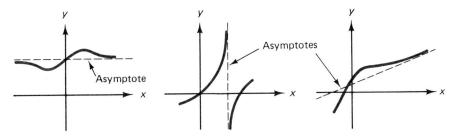

Curves with horizontal, vertical, and diagonal asymptotes

Limits at Positive and Negative Infinity

Earlier we used the notation $\lim_{x \to a} f(x)$ to mean the number (if it exists) that $f(x)$ approaches as x approaches the number a. We will now use the notation $x \to \infty$ (read "x approaches infinity") to mean that x takes on arbitrarily large values, farther and farther to the right on the x-axis. Similarly, the notation $x \to -\infty$ ("x approaches negative infinity") means that x takes on values farther and farther to the *left* on the x-axis.

Therefore,

$\lim_{x \to \infty} f(x)$	means the limit of $f(x)$ as x takes on values arbitrarily far to the right on the x-axis

and

$\lim_{x \to -\infty} f(x)$	means the limit of $f(x)$ as x takes on values arbitrarily far to the left on the x-axis.

We will sometimes refer to these limits as the limits *at* positive and negative infinity. Of course, for a given function these limits may fail to exist. Note that $x \to \infty$ does *not* mean "x approaches the *number* infinity" since there is no number infinity.

Limits of Rational Functions at Positive and Negative Infinity

Recall that the *degree* of a polynomial is the highest power of the variable. There are two useful rules for finding limits of rational functions. The first is that if the degree of the numerator is *less* than the degree of the denominator, then the limits at both positive and negative infinity exist and are zero.

Rational Limit Rule 1

$$\lim_{x \to \pm\infty} \frac{\text{(lower-degree polynomial)}}{\text{(higher-degree polynomial)}} = 0$$

(We use the notation $x \to \pm\infty$ when the limits at positive and negative infinity are the same.) A justification for rule 1 will follow an example.

Example 1   Find the limits as $x \to \pm\infty$ of $f(x) = \dfrac{x + 1}{x^2 + 2x + 2}$.

Solution   Since the degree of the numerator is less than the degree of the denominator,

$$f(x) = \frac{x + 1}{x^2 + 2x + 2} \quad \begin{matrix} \longleftarrow \text{ degree 1} \\ \longleftarrow \text{ degree 2} \end{matrix}$$

we may answer immediately that the limits at $\pm\infty$ are both zero.

$$\lim_{x \to \pm\infty} \frac{x + 1}{x^2 + 2x + 2} = 0$$

■

Justification for Rule 1

To see why rule 1 holds, divide numerator and denominator of this rational function by x^2.

$$\lim_{x \to \pm\infty} \frac{x + 1}{x^2 + 2x + 2} = \lim_{x \to \pm\infty} \frac{x/x^2 + 1/x^2}{x^2/x^2 + 2x/x^2 + 2/x^2} = \lim_{x \to \pm\infty} \frac{1/x + 1/x^2}{1 + 2/x + 2/x^2} = \frac{0}{1} = 0$$

This shows that the limits at positive and negative infinity are indeed both zero. In general, dividing by the highest power of the variable shows that these limits will be zero whenever the degree of the numerator is less than the degree of the demoninator.

Horizontal Asymptotes

If the limits of a rational function at positive and negative infinity exist, they will be the same, and they give the *horizontal asymptote* of the function. For example, in Example 1 the limits at $\pm\infty$ were both 0. Therefore, as x approaches positive or negative infinity, the curve approaches height 0, so the x-axis is a horizontal asymptote.

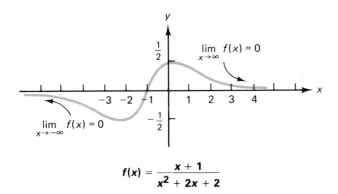

$$f(x) = \frac{x + 1}{x^2 + 2x + 2}$$

If $\lim\limits_{x \to \pm\infty} f(x) = c$, then $y = c$ is the horizontal asymptote of $f(x)$.

Practice Exercise 1

Find the limits as $x \to \pm\infty$ of $f(x) = \dfrac{-2x}{x^2 + 1}$.   *(solution on page 160)*

Rule 2 for calculating limits of rational function says that to calculate the limit we may ignore all but the highest-power term in the top and the highest-power term in the bottom.

Rational Limit Rule 2

The limit of a rational function at positive and negative infinity, if it exists, will be equal to the limit of the fraction

$$\frac{\text{(highest-power term in numerator)}}{\text{(highest-power term in denominator)}}$$

Example 2   Find the limits at positive and negative infinity of

$$f(x) = \frac{2x^2 - 4x + 2}{x^2 - 2x + 3}$$

Solution   Taking only the highest-power term in the numerator (the $2x^2$) and the highest-power term in the denominator (the x^2) gives

$$\lim_{x \to \pm\infty} \frac{2x^2 - 4x + 2}{x^2 - 2x + 3} = \lim_{x \to \pm\infty} \frac{2x^2}{x^2} = \lim_{x \to \pm\infty} \frac{2\cancel{x^2}}{\cancel{x^2}} = \lim_{x \to \pm\infty} 2 = 2$$

highest-power term in the top

highest-power term in bottom

Therefore, the limits at positive and negative infinity exist, and both are equal to 2.   ■

Omitting all but the highest-power terms, of course, does not mean that the original function is *equal* to $2x^2/x^2$, only that to evaluate the *limit* as $x \to \pm\infty$ we may use the "simplified" function $2x^2/x^2$. Since this function approaches 2 as $x \to \pm\infty$, the horizontal line $y = 2$ is an asymptote.

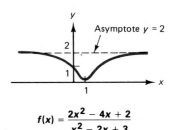

$$f(x) = \frac{2x^2 - 4x + 2}{x^2 - 2x + 3}$$

Justification for Rule 2

Intuitively, the reason that we may consider only the highest-degree terms in the numerator and denominator is that for large values of x, these terms will be so large as to make the lower powers insignificant. A more rigorous justification follows from dividing numerator and denominator by the highest power of the variable in the function, just as we did for rule 1.

Practice Exercise 2

Find the limits as $x \to \pm\infty$ of $f(x) = \dfrac{3x^2 - 2}{x^2 + 1}$.

(solution on page 160)

Example 3   Find the limits as $x \to \pm\infty$ of $f(x) = \dfrac{x^5}{x^2 + 15}$.

Solution   Here the degree of the denominator is *less* than the degree of the numerator. We use rule 2, keeping only the highest-order terms.

$$\lim_{x \to \pm\infty} \frac{x^5}{x^2 + 15} = \lim_{x \to \pm\infty} \frac{x^5}{x^2} = \lim_{x \to \pm\infty} x^3$$

x^5/x^2

In this case, since x^3 becomes arbitrarily large (positively or negatively) as $x \to \pm\infty$, *neither* limit exists. To show that the function becomes arbitrarily large as $x \to \infty$, we write

$$\lim_{x \to \infty} \frac{x^5}{x^2 + 15} = \infty$$

as x approaches infinity, the limit of f(x) is infinite

As x approaches *negative* infinity, x^3 becomes *negatively* large (since an odd power of a negative number is negative). Therefore, as $x \to -\infty$ this function takes on arbitrarily large negative values, which we write as

$$\lim_{x \to -\infty} \frac{x^5}{x^2 + 15} = \lim_{x \to -\infty} x^3 = -\infty$$

as x approaches negative infinity the limit of f(x) is negative infinity   ■

These two results are shown in the graph below.

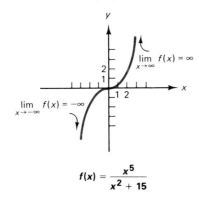

$$f(x) = \frac{x^5}{x^2 + 15}$$

Practice Exercise 3

Find the limits as $x \to \pm\infty$ of $f(x) = \dfrac{3x^4 + x^2 + 1}{x^2 + 2}$.

(solution on page 161)

Summary

We have two rules for finding the limits of a rational function at positive or negative infinity.

> **Rational Limit Rule 1**
>
> If the degree of the denominator exceeds the degree of the numerator, then both limits exist and are zero.
>
> **Rational Limit Rule 2**
>
> The limits, if they exist, will be the same as the limits of the fraction
>
> $$\frac{\text{(highest-power term in numerator)}}{\text{(highest-power term in denominator)}}$$

If the limits at positive and negative infinity exist, they will be the same and the give the horizontal asymptote of the curve.

If $\lim\limits_{x \to \pm\infty} f(x) = c$, then $y = c$ is the horizontal asymptote of $f(x)$.

Graphing Rational Functions

These results will help us to graph rational functions. If the denominator of a rational function is zero at an x-value, the function is *undefined* at that x-value (since division by zero is undefined). As x *approaches* such an x-value, the denominator will approach zero, making the function take on arbitrarily large positive or negative values, giving its graph a *vertical asymptote* at that x-value.*

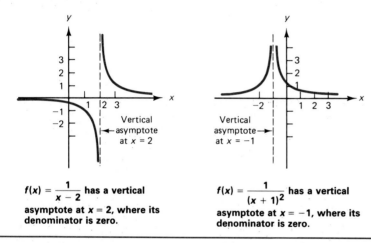

$f(x) = \dfrac{1}{x - 2}$ **has a vertical asymptote at $x = 2$, where its denominator is zero.**

$f(x) = \dfrac{1}{(x + 1)^2}$ **has a vertical asymptote at $x = -1$, where its denominator is zero.**

A rational function has a vertical asymptote wherever its denominator is zero.

We graph rational functions by drawing asymptotes and using sign diagrams.

Example 4   Sketch the graph of $f(x) = \dfrac{1}{x^2 - 1}$.

Solution   Factoring gives

$$f(x) = \frac{1}{x^2 - 1} = \frac{1}{(x + 1)(x - 1)}$$

---

* Since we assume that all common factors have been canceled, the numerator and denominator cannot both be zero at the same value of x. If the numerator and denominator are both zero, the rational function has not been simplified.

which shows that the function is *undefined* at $x = -1$ and $x = 1$. Therefore, the function has vertical asymptotes at these x-values, which we indicate by vertical dashed lines.

The derivative of

$$f(x) = \frac{1}{x^2 - 1} = (x^2 - 1)^{-1}$$

is

$$f'(x) = -(x^2 - 1)^{-2}(2x) = \frac{-2x}{(x^2 - 1)^2}$$

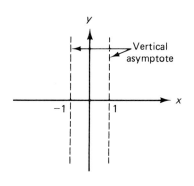

(using the generalized power rule). The derivative is zero at $x = 0$ and *undefined* at $x = 1$ and $x = -1$ (since $x = \pm 1$ make the denominator zero). The sign diagram for f', made in the usual way, is

$f' > 0$	f' und	$f' > 0$	$f' = 0$	$f' < 0$	f' und	$f' < 0$
	$x = -1$		$x = 0$		$x = 1$	
↗		↗	→	↘		↘

rel max $(0, -1)$

There is a relative maximum at $x = 0$, which we plot with a "cap," ⌒ . The sign diagram also shows that between the dashed lines the curve slopes upward to the "cap" and then downward. Since the curve must approach the dashed lines asymptotically, this part must look like

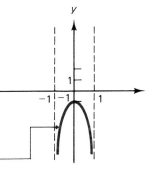

As $x \to \pm\infty$ both limits are zero, since the degree of the denominator in $f(x) = 1/(x^2 - 1)$ exceeds the degree of the numerator. Therefore, the curve asymptotically approaches the x-axis.

Approaching each dashed vertical line the graph must curve steeply upward or downward. To the left of $x = -1$ the curve slopes *upward* (according to the sign diagram), and so must rise from the x-axis toward the vertical asymptote

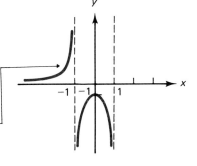

To the right of $x = 1$, the sign diagram shows that curve slopes *downward*, and so it curves downward from the vertical asymptote to the x-axis.

The curve on the right completes the graph of the function.

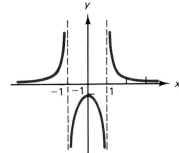

$$f(x) = \frac{1}{x^2 - 1}$$

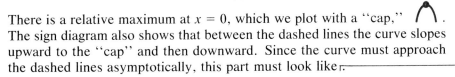

■

We graphed the function without making a sign diagram for the second derivative. For this function, as for many rational functions, the second derivative is difficult to calculate, and so we try to draw the graph without it. If you are asked to show inflection points, you *must* calculate the second derivative and construct its sign diagram, but otherwise you may omit the second derivative sign diagram if you can graph the curve without it.

Example 5   Sketch the graph of $f(x) = \dfrac{x^2 + 2}{x^2 + 1}$.

Solution   The denominator is never zero (since $x^2 + 1$ is always positive), so there are no vertical asymptotes. The derivative (using the quotient rule) is

$$f'(x) = \frac{(x^2 + 1)2x - 2x(x^2 + 2)}{(x^2 + 1)^2} = \frac{2x^3 + 2x - 2x^3 - 4x}{(x^2 + 1)^2} = \frac{-2x}{(x^2 + 1)^2}$$

The sign diagram for f' is

$$f' > 0 \qquad f' = 0 \qquad f' < 0$$

$$x = 0$$

$$\nearrow \qquad \rightarrow \qquad \searrow$$

rel max (0, 2)

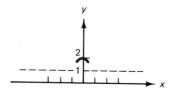

Partial graph

The behavior as $x \to \pm\infty$ is

$$\lim_{x \to \pm\infty} \frac{x^2 + 2}{x^2 + 1} = \lim_{x \to \pm\infty} \frac{x^2}{x^2} = 1$$

└ highest-
power terms

This shows that as $x \to \pm\infty$ the curve asymptotically approaches height 1, so the curve has a *horizontal asymptote* at height 1 (shown in dashes).

Given the relative maximum point, the sign diagram, and the asymptotic behavior as $x \to \pm\infty$, we complete the graph as shown below.

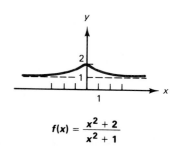

$$f(x) = \frac{x^2 + 2}{x^2 + 1}$$

■

Note that this curve never dips below the horizontal asymptote, for if it did, it would then have to rise back up to the asymptote, resulting in relative minimum points shown on the right. Since the sign diagram shows that there are no such relative minimum points, the curve must remain above its horizontal asymptote.

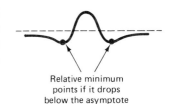

Relative minimum
points if it drops
below the asymptote

As a final example we graph a function of the form $f(x) = \dfrac{a}{x} + b + cx$. Such functions arise in many applications, from drug dosage to inventory control.

Example 6 Sketch the graph of $f(x) = \dfrac{50}{x} + 10 + 2x$.

Solution The denominator is zero at $x = 0$, so the curve has a horizontal asymptote at $x = 0$, which we indicate on the graph at the right by dashes along the y-axis. The derivative of $f(x) = 50x^{-1} + 10 + 2x$ is

$$f'(x) = -50x^{-2} + 2 = -\frac{50}{x^2} + 2$$

We set this equal to zero and solve.

$$-\frac{50}{x^2} + 2 = 0$$

$$2 = \frac{50}{x^2} \qquad \textit{adding } 50/x^2 \textit{ to each side}$$

$$2x^2 = 50 \qquad \textit{multiplying by } x^2$$

$$x^2 = 25 \qquad \textit{dividing by 2}$$

$$\text{CV} \quad x = \pm 5 \qquad \textit{taking square roots}$$

The sign diagram for f' is

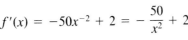

$f' > 0$	$f' = 0$	$f' < 0$	f'und	$f' < 0$	$f' = 0$	$f' > 0$

$\qquad\quad x = -5 \qquad\qquad x = 0 \qquad\qquad x = 5$

$\nearrow \qquad \rightarrow \qquad\quad \searrow \qquad\qquad\qquad \searrow \qquad \rightarrow \qquad \nearrow$

$\quad$ rel max $(-5, -10) \qquad\qquad\qquad$ rel min $(5, 30)$

The behavior as $x \to \pm\infty$ is

$$\lim_{x\to\infty}\left(\underbrace{\frac{50}{x}}_{\substack{\text{approaches}\\0}} + \underbrace{10 + 2x}_{\substack{\text{approaches}\\\infty}} \right) = \infty \quad \text{and} \quad \lim_{x\to-\infty}\left(\underbrace{\frac{50}{x}}_{\substack{\text{approaches}\\0}} + \underbrace{10 + 2x}_{\substack{\text{approaches}\\-\infty}} \right) = -\infty$$

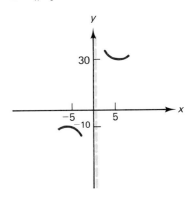

These show that as $x \to \infty$ the curve rises arbitrarily high and as $x \to -\infty$ the curve falls arbitrarily low. (Even more precisely, as $x \to \pm\infty$ the term $50/x$ approaches zero, leaving $10 + 2x$, so the function asymptotically approaches the line $y = 10 + 2x$.) Using all of this information (the vertical asymptotes, the relative extreme points, the sign diagram, and the behavior as $x \to \pm\infty$), we obtain the graph on the right.

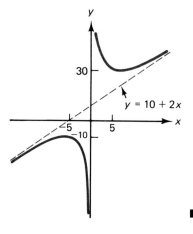

Summary

To sketch a graph:

1. Draw vertical asymptotes (with dashes) at x-values where the denominator is zero. These lines separate the curve into distinct, unconnected parts.

2. Make a sign diagram for the first derivative and plot all critical points, using appropriate "caps" and "cups."

3. If it is not too difficult, make a sign diagram for the second derivative and plot all inflection points.

4. Determine the behavior of the function as $x \to \pm\infty$, showing any horizontal asymptotes by dashed lines.

5. Using all of this information, sketch the curve. The sign diagram for $f'(x)$ will show the slope approaching the vertical asymptotes.

SOLUTIONS TO PRACTICE EXERCISES

1. Since the degree of the numerator is less than the degree of the denominator,

$$\lim_{x \to \pm\infty} \frac{-2x}{x^2 + 1} = 0$$

Therefore, the curve has a horizontal asymptote $y = 0$ (the x-axis), as shown on the graph shown on the right.

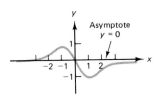

2. $\lim\limits_{x \to \pm\infty} \dfrac{3x^2 - 2}{x^2 + 1} = \lim\limits_{x \to \pm\infty} \dfrac{3x^2}{x^2} = \lim\limits_{x \to \pm\infty} \dfrac{3x^2}{x^2} = 3$

Therefore the graph has a horizontal asymptote $y = 3$, as shown on the graph on the right.

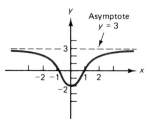

3. $\lim\limits_{x \to \pm\infty} \dfrac{3x^4 + x^2 + 1}{x^2 + 2} = \lim\limits_{x \to \pm\infty} \dfrac{3x^4}{x^2} = \lim\limits_{x \to \pm\infty} 3x^2 = \infty$

Therefore, the curve becomes arbitrarily high as $x \to \infty$ and as $x \to -\infty$, as shown on the graph on the right.

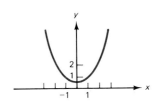

EXERCISES 3.3

Determine the behavior of each function as $x \to \infty$ and as $x \to -\infty$.

1 $f(x) = \dfrac{1}{x^2 - 4}$ **2** $f(x) = \dfrac{1}{x^2 - 9}$ **3** $f(x) = \dfrac{4x}{x^2 + 1}$ **4** $f(x) = \dfrac{8x}{x^2 + 4}$

5 $f(x) = \dfrac{x^2}{x^2 + 1}$ **6** $f(x) = \dfrac{x^2}{x^2 + 2}$ **7** $f(x) = \dfrac{x^2 + 5}{x^2 + 1}$ **8** $f(x) = \dfrac{x^2 + 9}{x^2 + 1}$

9 $f(x) = \dfrac{x^2 - 4x + 7}{x - 3}$ **10** $f(x) = \dfrac{x^2 - 6x + 12}{x - 4}$ **11** $f(x) = \dfrac{x^3}{x - 2}$ **12** $f(x) = \dfrac{x^3}{x - 4}$

Graph each function, showing all relative extreme points.

13 $f(x) = \dfrac{1}{x^2 - 4}$ **14** $f(x) = \dfrac{1}{x^2 - 9}$ **15** $f(x) = \dfrac{1}{x^2 - 4x}$ **16** $f(x) = \dfrac{1}{x^2 - 2x}$

17 $f(x) = \dfrac{4x}{x^2 + 1}$ **18** $f(x) = \dfrac{8x}{x^2 + 4}$ **19** $f(x) = \dfrac{x^2}{x^2 + 1}$ **20** $f(x) = \dfrac{x^2}{x^2 + 2}$

21 $f(x) = \dfrac{1}{x^2 + 1}$ **22** $f(x) = \dfrac{1}{x^2 + 2}$ **23** $f(x) = \dfrac{x^2}{x - 3}$ **24** $f(x) = \dfrac{x^2}{x - 1}$

25 $f(x) = \dfrac{x^2 - 4x + 7}{x - 3}$ **26** $f(x) - \dfrac{x^2 - 6x + 12}{x - 4}$ **27** $f(x) = \dfrac{x^2}{x^2 - 9}$ **28** $f(x) = \dfrac{x^2}{x^2 - 4}$

29 $f(x) = \dfrac{x^2 + 4}{x}$ **30** $f(x) = \dfrac{x^2 + 2x + 1}{x}$ **31** $f(x) = \dfrac{x^3}{x - 2}$ **32** $f(x) = \dfrac{x^3}{x - 4}$

33 $f(x) = \dfrac{4x^3}{x^2 - 2x + 1}$ **34** $f(x) = \dfrac{-4x^3}{x^2 + 2x + 1}$ **35** $f(x) = \dfrac{-2x^3}{x^2 + 4x + 4}$ **36** $f(x) = \dfrac{2x^3}{x^2 - 4x + 4}$

37 $f(x) = \dfrac{1}{x - 1}$ **38** $f(x) = \dfrac{1}{x + 1}$ **39** $f(x) = \dfrac{1}{(x - 1)^2}$ **40** $f(x) = \dfrac{1}{(x + 1)^2}$

41 $f(x) = \dfrac{2x^2}{x^4 + 1}$ **42** $f(x) = \dfrac{3x^2}{x^6 + 2}$ **43** $f(x) = \dfrac{x^4 + x^2 + 1}{x^4 + 1}$ **44** $f(x) = \dfrac{x^4 - 2x^2 + 1}{x^4 + 2}$

45 $f(x) = x + \dfrac{1}{x}$ **46** $f(x) = x + \dfrac{4}{x}$ **47** $f(x) = x + \dfrac{32}{x^2}$ **48** $f(x) = x + \dfrac{4}{x^2}$

49 $f(x) = \dfrac{12}{x} + 2 + 3x$ **50** $f(x) = \dfrac{18}{x} + 4 + 2x$ **51** $f(x) = \dfrac{100}{x} + 10 + 4x$ **52** $f(x) = \dfrac{300}{x} + 10 + 3x$

APPLIED EXERCISES

53 (*Business–Marginal and Average Cost*) If a company's cost function is $C(x) = x^2 + 2x + 4$ where x is the number of units), then its average cost function, $C(x)/x$, is

$$AC(x) = \frac{x^2 + 2x + 4}{x}$$

(a) Differentiate $C(x)$ to find the marginal cost function $MC(x)$.

(b) Graph the average cost function $AC(x)$ and the marginal cost function $MC(x)$ on the same axes (for $x \geq 0$). Notice that the marginal cost function and the average cost function meet at the point where the average cost is minimized. We will return to this observation later.

54 (*Business–Marginal and Average Cost*) If a company's cost function is $C(x) = x^2 + 4x + 9$ (where x is the number of units), then its average cost function, $C(x)/x$, is

$$AC(x) = \frac{x^2 + 4x + 9}{x}$$

(a) Differentiate $C(x)$ to find the marginal cost function $MC(x)$.

(b) Graph the average cost function $AC(x)$ and the marginal cost function $MC(x)$ on the same axes (for $x \geq 0$). Notice that the marginal cost function and the average cost function meet at the point where the average cost is minimized. We will return to this observation later.

55 (*General–Water Purification*) The cost of building a water treatment plant to purify water to x percent purity ($0 \leq x < 100$) is

$$C(x) = \frac{300}{20,000 - 2x^2}$$

million dollars. Graph this function for $0 \leq x < 100$.

56 (*General–Drug Interception*) The cost of a border patrol that intercepts x percent of the illegal drugs crossing a state border is

$$C(x) = \frac{600}{100 - x}$$

million dollars. Graph this function for $0 \leq x < 100$.

57 (*Biomedical–Analgesia*) Clinical studies have shown that the analgesic (pain relieving) effect of aspirin is approximately

$$f(x) = \frac{100x^2}{x^2 + .02}$$

where $f(x)$ is the percentage of relief from x grams of aspirin.

(a) Graph this dose-response curve for $x \geq 0$.

(b) Calculate the height of the curve when $x = .65$ gram. Notice that there is little added effect beyond this level, which is the amount of aspirin in two standard tablets. This is why it is seldom worthwhile to take more than two ordinary tablets, notwithstanding the aspirin companies' promotion of "extra-strength" pills.

3.4 OPTIMIZATION

Introduction

Many problems consist of "optimizing" a function, that is, finding its maximum or minimum values. For example, you may want to maximize your profit, or to minimize the time required to do a task. If you could express your happiness as a function, you would want to maximize it.* One of the

---

* Expressing happiness in numbers has an honorable past: Plato (*Republic* IX, 587), calculated that a king is exactly 729 (= 3^6) times as happy as a tyrant.

principal uses of calculus is that it provides a very general technique for optimization.

We will concentrate on applications of optimization. Accordingly, we will optimize continuous functions that are defined on closed intervals (that is, intervals containing both endpoints, $a \leq x \leq b$), or that have only one critical value in an interval. Most applications fall into these two categories, and the wide range of examples and exercises in these sections will demonstrate the power of these techniques.

Interval Notation

The set of numbers satisfying $2 \leq x \leq 5$ can be written [2, 5], with the square brackets indicating that the endpoints 2 and 5 are *included*. The set of numbers satisfying $2 < x < 5$ can be written (2, 5), with the parentheses indicating that the endpoints 2 and 5 are *excluded*. Such sets are called *intervals*. The interval is *closed* if it includes *both* endpoints, and *open* if it includes *neither* endpoint. For numbers a and b with $a < b$, we can write the four types of intervals shown below. A "solid" dot • on the graph indicates that the point is included in the interval, and an "empty" dot ∘ indicates that the point is excluded.

Interval Notation	Graph	Set Notation	Type
$[a, b]$	•————• a b	$\{x \mid a \leq x \leq b\}$	Closed
(a, b)	∘————∘ a b	$\{x \mid a < x < b\}$	Open
$[a, b)$	•————∘ a b	$\{x \mid a \leq x < b\}$	Half-open or Half-closed
$(a, b]$	∘————• a b	$\{x \mid a < x \leq b\}$	Half-open or Half-closed

Absolute Extreme Values

The *absolute maximum value* of a function is the *largest* value of the function on its domain. Similarly, the *absolute minimum value* of a function is the *smallest* value of the function on its domain. On a graph we set $y = f(x)$, so the maximum and minimum values of $f(x)$ are the y-coordinates of the highest and lowest points. The term *absolute extreme values* will mean the absolute maximum *and* absolute minimum values.

For a given function, one or both of the absolute extreme values may fail to exist.

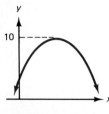
Absolute maximum value = 10; no absolute minimum value.

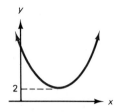
Absolute minimum value = 2; no absolute maximum value.

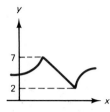
Absolute maximum value = 7; absolute minimum value = 2.

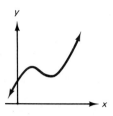
No absolute maximum or minimum value

Notice in these diagrams that the absolute extreme values, if they exist, occur at critical values (where the slope is zero or undefined). For a function defined on a closed interval $[a, b]$, the absolute extreme values may also occur at the end points of the interval, $x = a$ or $x = b$.

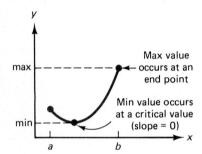

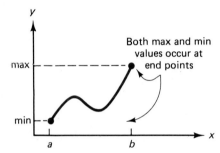

These pictures show that absolute extreme values can occur either at critical values or at endpoints. For a *continuous* function (one whose graph is a single unbroken curve) defined on a *closed* interval, the absolute maximum and minimum values are *guaranteed* to exist. This leads to the following procedure for optimizing a continuous function on a closed interval.

To find the absolute extreme values of a continuous function $f(x)$ on $a \leq x \leq b$:

(i)   Find all critical values of $f(x)$ in the domain $[a, b]$.

(ii) Evaluate the function at these critical values and at the endpoints $x = a$ and $x = b$.

The largest and smallest of the values found in step (ii) are the absolute maximum and absolute minimum values of $f(x)$ on $a \leq x \leq b$.

Simply stated, to find absolute extreme values, we need look only at critical values and endpoints. The reason for this is quite simple. If the derivative $f'(x)$ is positive or negative at an interior point (that is, a point other than an endpoint) of the domain, then the function is increasing or decreasing at that x-value, and so cannot have reached its maximum or minimum value. Therefore, the extreme values can occur only where $f'(x)$ is zero or undefined (critical values), or at the endpoints of the interval.

Since polynomials are continuous everywhere, a polynomial defined on a closed interval will have absolute maximum and minimum values. Similarly, a rational function defined on a closed interval will have absolute maximum and minimum values, provided that its denominator is not zero anywhere in the interval.

Remember to use only critical values in the (stated) domain of the function. The following example will clarify this.

Example 1   Find the absolute extreme values of $f(x) = x^4 + 4x^3 - 20x^2$ on $[-1, 4]$.

Solution   The function is continuous (since it is a polynomial), and the interval is closed, so both extreme values exist.

$$f'(x) = 4x^3 + 12x^2 - 40x \qquad \qquad \textit{differentiating to find critical values}$$

$$= 4x(x^2 + 3x - 10) = 4x(x + 5)(x - 2) \qquad \textit{factoring}$$

$$\text{CV} \begin{cases} x = 0 & \textit{from the } 4x \\ x = -5 & \textit{from } (x + 5) \\ x = 2 & \textit{from } (x - 2) \end{cases}$$

The problem states the domain as $[-1, 4]$, so we eliminate the critical value $x = -5$ since it is not in the domain. We evaluate $f(x) = x^4 + 4x^3 - 20x^2$ at the remaining critical values (CV) and at the endpoints (EP).

$$\text{CV} \begin{cases} x = 0 & f(0) = 0 \\ x = 2 & f(2) = 16 + 4 \cdot 8 - 20 \cdot 4 = -32 \longleftarrow \text{smallest} \end{cases}$$

$$\text{EP} \begin{cases} x = -1 & f(-1) = 1 - 4 - 20 = -23 \\ x = 4 & f(4) = 256 + 4 \cdot 64 - 20 \cdot 16 = 192 \longleftarrow \text{largest} \end{cases}$$

The largest (192) and the smallest (-32) of these resulting values are the maximum and minimum of the function.

Answer: The maximum value of $f(x)$ is 192 (which occurs at $x = 4$). The minimum value of $f(x)$ is -32 (which occurs at $x = 2$).   ■

The graph of this function verifies that the absolute maximum is 192 (at $x = 4$) and the absolute minimum is -32 (at $x = 2$). Notice that one extreme value occurred at an endpoint ($x = 4$) and one occurred at a critical value ($x = 2$). In other problems both might occur at critical values or both might occur at endpoints.

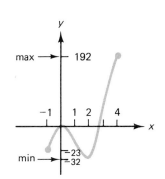

$f(x) = x^4 + 4x^3 - 20x^2$ on $[-1, 4]$

Notice how calculus helped in this example. The absolute extreme values might have occurred at *any* x-value between −1 and 4. Calculus reduced this infinite list of possibilities (*all* numbers from −1 to 4, not just integers) to a mere four possibilities (0, 2, −1, and 4). We then had only to test these four numbers "by hand," finding the ones that made $f(x)$ largest and smallest.

Remember that the critical values and endpoints are only the x-coordinates at which the maximum and minimum values occur. The actual maximum and minimum values of the function are found by substituting these x-values into the function.

Second Derivative Test

Many applications involve finding only one extreme value. Frequently, a function has only one critical value in its domain, and in such cases the second derivative test may be used to show whether the function is maximized or minimized at that critical value.

Second Derivative Test for Absolute Extreme Values

If $f(x)$ has only one critical value $x = c$ in its domain, and

   (i) if $f''(c) > 0$, then $f(x)$ has an absolute *minimum* value at $x = c$;

   (ii) if $f''(c) < 0$, then $f(x)$ has an absolute *maximum* value at $x = c$.

Earlier we used the second derivative test to find *relative* maxima and minima. This result says that when there is only *one* critical value in the interval, the second derivative test also gives *absolute* extreme values. (Some students refer to this as "the only critical point in town" test.) The justification for this test is as follows. A function with a single critical value $x = c$ at which $f''(x) < 0$ must have a relative maximum at $x = c$, so the sign diagram must look like this:

$$f' > 0 \qquad f' = 0 \qquad f' < 0$$
$$\nearrow \qquad\quad \rightarrow \qquad\quad \searrow$$
$$x = c$$

The arrows on the two sides of the sign diagram show that the curve must be lower for all other values of x in the interval, so $f(x)$ must have an *absolute* maximum at $x = c$. Similar reasoning applies to the case where $f''(c) > 0$.

Application: Optimizing the Value of a Timber Forest

If a timber forest is allowed to grow for t years, the value of the timber increases in proportion to the square root of t, while maintenance costs are proportional to t. Therefore, the value of the forest after t years is

$$V(t) = a\sqrt{t} - bt$$

for constants a and b.

Example 2   The value of a timber forest after t years is $V(t) = 96\sqrt{t} - 6t$ thousand dollars (for $t > 0$). Find when its value is maximized.

Solution

$V(t) = 96t^{1/2} - 6t$	*V(t) in exponential form*
$V'(t) = 48t^{-1/2} - 6$	*the derivative*
$48t^{-1/2} - 6 = 0$	*setting V' = 0*
$48t^{-1/2} = 6$	*adding 6 to each side*
$t^{-1/2} = \dfrac{6}{48} = \dfrac{1}{8}$	*dividing by 48*
$\dfrac{1}{\sqrt{t}} = \dfrac{1}{8}$	*expressing $t^{-1/2}$ in radical form*
$\sqrt{t} = 8$	*inverting both sides*
$t = 64$	*squaring both sides*

Since there is a *single* critical value ($t = 0$, which makes the derivative undefined, is not in the domain), we may use the second derivative test. The second derivative is

$$V''(t) = -24t^{-3/2} = -\frac{24}{\sqrt{t^3}} \qquad\qquad \textit{differentiating } V'(t) = 48t^{-1/2} - 6$$

At $t = 64$ this is clearly negative, so $V(t)$ is indeed *maximized* at $t = 64$.

Answer: The value of the forest is maximized after 64 years. The maximum value is $V(64) = 96\sqrt{64} - 6\cdot64 = 384$ thousand dollars, or \$384,000. ■

Until now the function to be optimized has been stated in the problem. In some applications, however, we must first *construct* the function to be optimized.

Maximizing Profit

Many management problems involve maximizing profit. Such problems have only three economic ingredients. The first is that profit equals revenue minus cost*:

$$\boxed{\text{Profit} = \text{Revenue} - \text{Cost}}$$

The second ingredient is that revenue equals price times quantity. For example, if a company sells 100 toasters for $25 each, the revenue will obviously be $25 \cdot 100 = \$2500$.

$$\boxed{\text{Revenue} = (\text{unit price}) \cdot (\text{quantity})}$$

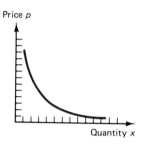

Price p and quantity x are inversely related.

The third economic ingredient reflects the fact that, in general, price and quantity are inversely related: increasing the price decreases sales, while decreasing the price increases sales. To put this another way, "flooding the market" with a product drives the price down, while creating a shortage drives the price up. This relationship between price p and quantity x is illustrated in the graph on the right. The graph may be straight (linear) or curved, depending on the particulars of the situation, but as quantity x increases, price p should decrease, and vice versa.

The equation relating the price p and the quantity x that consumers will demand at that price is called the *demand equation*. If this equation is solved for price, the resulting function $p(x)$ is called the *price function*, since it gives the price at which consumers will buy exactly x units of the product.

Price Functions

The price function $p(x)$ gives the price at which consumers will buy exactly x units of a product.

In actual practice, price functions are very difficult to determine, requiring extensive (and expensive) market research. In this section we will be given the price function. In Section 3.5 we will see how to construct them, at least in simple cases.

Example 3   It costs the American Automobile Company $5000 to produce each automobile, and fixed costs (rent and other costs that do not depend on the amount of production) are $20,000 per week. The company's price function is $p = 19,000 - 70x$, where p is the price at which exactly x cars will be sold.

---

* The Profit, Revenue, and Cost functions mean the *total* profit, revenue, and cost from x items. *Capital P* will stand for Profit and *small p* for selling price.

(a) How many cars should be produced each week to maximize profit?

(b) For what price should they be sold?

(c) What is the company's maximum profit?

Solution ***Revenue*** is price times quantity, $R = p \cdot x$. To express the revenue function as a function of just *one* variable, we use the price function $p = 19{,}000 - 70x$ to replace p by $19{,}000 - 70x$, giving

$$R = p \cdot x = \underbrace{(19{,}000 - 70x)}_{p(x)}x = \underbrace{19{,}000x - 70x^2}_{R(x)}$$ *revenue function*

Cost is the cost per car ($5000) times the number of cars (x) plus the fixed cost ($20{,}000).

$$C(x) = 5000x + 20{,}000$$ *(unit cost)·(quantity) + (fixed costs)*

Profit is revenue minus cost.

$$P(x) = \underbrace{(19{,}000x - 70x^2)}_{R(x)} - \underbrace{(5000x + 20{,}000)}_{C(x)}$$

$$= -70x^2 + 14{,}000x - 20{,}000$$ *profit function*

(a) We maximize the profit by setting its derivative equal to zero.

$$P'(x) = -140x + 14{,}000 = 0$$ *differentiating*
$P = -70x^2 + 14{,}000x - 20{,}000$

$$-140x = -14{,}000$$ *solving*

$$x = \frac{-14{,}000}{-140} = 100$$ *only one critical value*

$$P''(x) = -140$$ *the second derivative is negative, so profit is maximized at the critical value*

(If the second derivative had involved x, we would have substituted the critical value $x = 100$.) Since x is the number of cars, the company should produce 100 cars per week (the time period stated in the problem).

(b) The selling price p is found from the price function.

$$p = 19{,}000 - 70 \cdot 100 = \$12{,}000$$ *$p = 19{,}000 - 70x$ evaluated at $x = 100$*

(c) The maximum profit is found from the profit function.

$$P(100) = -70(100)^2 + 14{,}000(100) - 20{,}000$$ *$P(x) = -70x^2 + 14{,}000x - 20{,}000$ evaluated at $x = 100$*

$$= \$680{,}000$$

Finally, state the answer clearly in words.

Answer: The company should make 100 cars per week and sell them for $12,000 each. The maximum profit will be $680,000.   ■

Graphs of the Revenue, Cost, and Profit Functions

The graphs of the revenue and cost functions are shown on the right. At *x*-values where revenue is above cost there is a profit, and where the cost is above the revenue there is a loss.

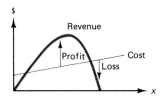

   The height of the profit function at any *x* is the amount by which the revenue is above the cost in the graph above.

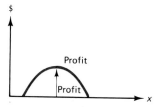

At Maximum Profit, Marginal Revenue Equals Marginal Cost

Since profit equals revenue minus cost, we may differentiate each side of $P(x) = R(x) - C(x)$, obtaining

$$P'(x) = R'(x) - C'(x)$$

This shows that setting $P'(x) = 0$ (which we do to maximize profit) is equivalent to setting $R'(x) - C'(x) = 0$, which is equivalent to $R'(x) = C'(x)$. This last equation may be expressed in marginals, MR = MC, which is a classic economic criterion for maximum profit.

At maximum profit,

   (marginal revenue) = (marginal cost)

However, to maximize profit it is not enough to set marginal revenue equal to marginal cost. The mistake is that MR = MC is equivalent to $P' = 0$, and this is true when profit is *minimized* as well as maximized. To avoid minimizing profit (a rather serious mistake in business) we will continue to maximize profit by solving $P'(x) = 0$ and then using the second derivative test to verify that we have maximized.

Area and Volume Formulas

The following area and volume formulas will be useful.

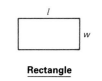

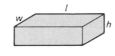

Rectangle	**Circle**	**Rectangular solid**	**Cylinder**	**Sphere**
Area = $l \cdot w$	Area = πr^2	Volume = $l \cdot w \cdot h$	Volume = $\pi r^2 l$	Volume = $\frac{4}{3}\pi r^3$
Perimeter = $2l + 2w$	Circumference = $2\pi r$			Surface area = $4\pi r^2$

Example 4   A farmer has 1000 feet of fence and wants to build a rectangular enclosure along a straight wall. If the side along the wall needs no fence, find the dimensions that make the enclosure as large as possible. Also find the maximum area.

Solution   The largest enclosure means, of course, the largest area. It might seem that the dimensions do not matter as long as all 1000 feet are used. However, the two pictures below show that using the 1000 feet of fence in different ways does give different areas.

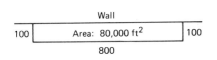
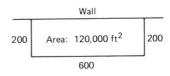

To maximize the area of the rectangle, we let variables stand for the length and width. Let

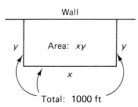

$$x = \text{length (parallel to wall)}$$

$$y = \text{width (perpendicular to wall)}$$

The problem becomes:

$$\text{maximize} \quad A = xy \qquad\qquad \textit{area is length times width}$$

$$\text{subject to} \quad x + 2y = 1000 \qquad \textit{one x side and two y sides from 1000 feet of fence}$$

We must express the area $A = xy$ in terms of one variable.

$$x = 1000 - 2y \qquad\qquad \textit{solving x + 2y = 1000 for x}$$

$$A = xy = \underbrace{(1000 - 2y)}_{x}y = 1000y - 2y^2 \qquad \textit{substituting x = 1000 − 2y into A = xy}$$

$$A' = 1000 - 4y = 0 \qquad\qquad \textit{maximizing A = 1000y − 2y² by setting the derivative equal to zero}$$

$$y = 250 \qquad\qquad \textit{solving 1000 − 4y = 0 for y}$$

Since $A'' = -4$, the second derivative test shows that the area is indeed *maximized* when $y = 250$. The length x is

$$x = 1000 - 2 \cdot 250 = 500 \qquad\qquad \textit{evaluating x = 1000 − 2y at x = 250}$$

Answer: Length (parallel to the wall) is 500 feet, width (perpendicular to the wall) is 250 feet, and area (length times width) is 12,500 square feet.

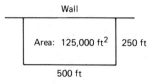

■

Example 5   An open-top box is to be made from a square sheet of metal 12 inches on each side, by cutting a square from each corner and folding up the sides, as in the diagram below. Find the volume of the largest box that can be made in this way.

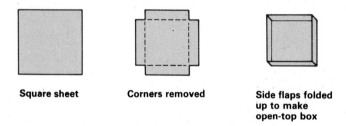

Square sheet   **Corners removed**   **Side flaps folded up to make open-top box**

Solution   Let x = the length of the side of the square cut from each corner.

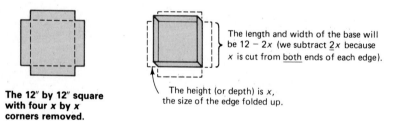

The 12″ by 12″ square with four x by x corners removed.

The length and width of the base will be $12 - 2x$ (we subtract $2x$ because x is cut from <u>both</u> ends of each edge).

The height (or depth) is x, the size of the edge folded up.

Therefore, the volume is

$$V(x) = (12 - 2x)(12 - 2x)x \qquad\qquad \textit{(length)·(width)·(height)}$$

Since x is a length, $x > 0$, and since x inches are cut from *both* sides of each 12-inch edge, we must have $2x < 12$, so $x < 6$. The problem becomes:

maximize $V(x) = (12 - 2x)(12 - 2x)x$ on $0 < x < 6$

$$V(x) = (144 - 48x + 4x^2)x = 4x^3 - 48x^2 + 144x \qquad\qquad \textit{multiplying out}$$

$$V'(x) = 12x^2 - 96x + 144 \qquad\qquad \textit{differentiating}$$

$$= 12(x^2 - 8x + 12) = 12(x - 2)(x - 6) \qquad\qquad \textit{factoring}$$

CV $\begin{cases} x = 2 \\ x = \cancel{6} \quad \text{(not in the domain, so we eliminate it)} \end{cases}$

The second derivative is $V''(x) = 24x - 96$, which at $x = 2$ is

$$V''(2) = 48 - 96 < 0$$

Therefore, the volume is *maximized* at $x = 2$.

   Answer: Maximum volume is 128 cubic inches.   *from V(x) evaluated at x = 2*   ▪

Alternative Solution   We could also have maximized

$$V(x) = 4x^3 - 48x + 144x$$

on the *closed* interval $0 \leq x \leq 6$ (allowing zero lengths). We would then evaluate $V(x)$ at all critical values and endpoints in $[0, 6]$.

$$x = 2: \quad V(2) = 128$$

$$x = 0: \quad V(0) = 0$$

$$x = 6: \quad V(6) = 0$$

again finding that the maximum volume is 128 when $x = 2$.   ▪

Summary

There is no single, all-purpose procedure for solving word problems. You must think about the problem, draw a picture if possible, and express the quantity to be maximized or minimized in terms of some appropriate variable. With practice you can become good at it!

   We have two procedures for optimizing continuous functions on intervals.

   1. If the function has only one critical value in the interval, we may use the second derivative test.

   2. If the interval is closed, we evaluate the function at all critical values and endpoints.

   For functions satisfying neither of these two conditions, graph the function. The maximum (or minimum) values will be the y-coordinate of the highest (or lowest) point on the graph.

▨   **EXERCISES 3.4**

1 $f(x) = x^3 - 6x^2 + 9x + 2$ on $[-2, 5]$ **2** $f(x) = x^3 - 6x^2 + 30$ on $[-1, 5]$

3 $f(x) = x^3 - 12x$ on $[-3, 3]$ **4** $f(x) = x^3 - 27x$ on $[-4, 4]$

5 $f(x) = x^4 + 4x^3 + 4x^2 - 100$ on $[-3, 3]$ **6** $f(x) = x^4 - 4x^3 + 4x^2$ on $[-3, 3]$

7 $f(x) = x^4 + 4x^3 - 8x^2$ on $[-3, 3]$ **8** $f(x) = x^4 - 4x^3 - 8x^2$ on $[-3, 3]$

9 $f(x) = x^3 + 3x^2 - 9x - 11$ on $[-2, 2]$ **10** $f(x) = x^3 - 3x^2 - 9x + 10$ on $[-2, 2]$

11 $f(x) = 5 - x$ on $[0, 5]$

12 $f(x) = 10 - x$ on $[0, 10]$

13 $f(x) = x(20 - x)$ on $[0, 20]$

14 $f(x) = x(100 - x)$ on $[0, 100]$

15 $f(x) = (x^2 - 1)^2$ on $[-1, 1]$

16 $f(x) = (x^2 - 4)^2$ on $[-2, 2]$

17 $f(x) = \sqrt[3]{x^2}$ on $[-1, 8]$

18 $f(x) = \sqrt[3]{x}$ on $[-1, 8]$

19 $f(x) = \sqrt[5]{x^4}$ on $[-1, 1]$

20 $f(x) = \sqrt[5]{x^3}$ on $[-1, 1]$

21 $f(x) = \dfrac{x}{x^2 + 1}$ on $[-3, 3]$

22 $f(x) = \dfrac{1}{x^2 + 1}$ on $[-3, 3]$

▨ APPLIED EXERCISES

23 (*Biomedical–Pollen Count*) The average pollen count in New York City on day x of the pollen season is $P(x) = 8x - .2x^2$ (for $x < 40$). On which day is the pollen count highest?

24 (*General–Fuel Economy*) The fuel economy (in miles per gallon) of an average American compact car is $E(x) = -.015x^2 + 1.14x + 8.3$, where x is the driving speed (in miles per hour). At what speed is fuel economy greatest?

25 (*General–Fuel Economy*) The fuel economy (in miles per gallon) of an average American midsized car is $E(x) = -.01x^2 + .62x + 10.4$, where x is the driving speed (in miles per hour). At what speed is fuel economy greatest?

26 (*Ecology–Water Power*) The proportion of the river's energy that can be obtained from an undershot waterwheel is $E(x) = 2x^3 - 4x^2 + 2x$, where x is the speed of the waterwheel relative to the speed of the river. Find the maximum value of this function on the interval $[0, 1]$, thereby showing that only about 30% of a river's energy can be captured. Your answer should agree with the old millwright's rule that the speed of the wheel should be about one-third of the speed of the river.

27 (*Sociology–Riots*) In the Detroit riot of 1967, the number of reported riot incidents x hours after the riot began was $f(x) = -.25x^2 + 12x + 16$ (for $x \leq 48$). After how many hours was the riot activity the greatest?

28 (*Sociology–Riots*) In the Washington, D.C. riot of 1968, the number of reported riot incidents x hours after the riot began was $f(x) = -.10x^2 + 6x$ (for $x \leq 48$). After how many hours was the riot activity the greatest?

29 (*General–Timber Value*) The value of a timber forest after t years is $V(t) = 480\sqrt{t} - 40t$ (for $0 \leq t \leq 50$). Find when its value is maximized.

30 (*General–Longevity and Exercise*) A recent study of the exercise habits of 17,000 Harvard alumni found that the death rate (deaths per 10,000 man-years) was approximately $R(x) = 5x^2 - 35x + 104$, where x is the weekly amount of exercise in thousands of Calories ($x \leq 4$). Find the exercise level that minimizes the death rate.

31 (*General–Pollution*) Two chemical factories are polluting a large lake, and the pollution level at a point x miles from a factory A toward factory B is

$$P(x) = 3x^2 - 72x + 576 \text{ parts per million}$$

(for $0 \leq x \leq 50$). Find where the pollution is the least.

32 (*Business–Maximum Profit*) City Cycles Incorporated finds that it costs $70 to manufacture each bicycle, and fixed costs are $100 per day. Their price function is $p(x) = 270 - 10x$, where p is the price (in dollars) at which exactly x bicycles will be sold. Find the quantity they should produce and the price they should charge to maximize profit. Also find the maximum profit.

33 (*Business–Maximum Profit*) Country Motorbikes Incorporated finds that it costs $200 to produce each motorbike, and that fixed costs are $1500 per day. Their price function is $p(x) = 600 - 5x$, where p is the price (in dollars) at which exactly x motorbikes will be sold. Find the quantity they should produce and the price they should charge to maximize profit. Also find the maximum profit.

34 (*Business–Maximum Profit*) A retired potter can produce china pitchers at a cost of $5 each. She estimates her price function to be $p = 17 - .5x$, where p is the price at which exactly x pitchers will be sold per week. Find the number of pitchers that she should produce and the price that she should charge in order to maximize profit. Also find the maximum profit.

35 (*General–Parking Lot Design*) A company wants to build a parking lot along the side of one of its buildings using 800 feet of fence. If the side along the building needs no fence, what are the dimensions of the largest possible parking lot?

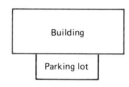

36 (*General–Area*) A farmer wants to make two identical rectangular enclosures along a straight river, as in the diagram shown below. If he has 600 yards of fence, and if the sides along the river need no fence, what should be the dimensions of each enclosure if the total area is to be maximized?

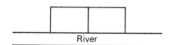

37 (*General–Area*) A farmer wants to make three identical rectangular enclosures along a straight river, as in the diagram shown below. If he has 1200 yards of fence, and if the sides along the river need no fence, what should be the dimensions of each enclosure if the total area is to be maximized?

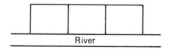

38 (*General–Area*) What is the area of the largest rectangle whose perimeter is 100 feet?

39 (*General–Package Design*) An open-top box is to be made from a square piece of cardboard that measures 18 inches by 18 inches by removing a square from each corner and folding up the sides. What are the dimensions and volume of the largest box that can be made in this way?

40 (*General*) A long gutter is to be made from a 12-inch-wide strip of metal by folding up the two edges. How much of each edge should be folded up in order to maximize the capacity of the gutter? (*Hint:* Maximizing the capacity means maximizing the cross-sectional area, shown below.)

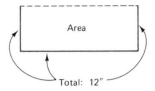

41 (*General*) Find two numbers whose sum is 50 and such that their product is a maximum.

42 (*General*) Show that the largest rectangle with fixed perimeter is a square.

43 (*Biomedical–Coughing*) When you cough you are using a high-speed stream of air to clear your trachea (windpipe). During a cough your trachea contracts, forcing the air to move faster, but also increasing the friction. If a trachea contracts from a normal radius of 3 centimeters to a radius of r centimeters, the velocity of the airstream is $V(r) = c(3 - r)r^2$, where c is a constant depending on the length and the elasticity of the trachea. Find the radius r that maximizes this velocity. (X-ray pictures verify that the trachea does indeed contract to this radius.)

44 (*General–"Efishency"*) At what speed should a fish swim upstream so as to reach its destination with the least expenditure of energy? The energy depends on the friction of the fish through the water and on the duration of the trip. A fish swimming with velocity v encounters friction equal to av^k, where a is a constant and k is a constant (greater than 2) that depends upon the shape of the fish. If the velocity of the current is c, the time (distance divided by speed) required to swim s miles upstream is $s/(v - c)$. Therefore, the energy required is $av^k s/(v - c)$, where a, c, and s are constants. Minimizing this energy expenditure is equivalent to minimizing the function

$$E(v) = \frac{v^k}{v - c}$$

where c is the speed of the current. For a fish whose friction constant is $k = 3$, this energy expenditure becomes

$$E(v) = \frac{v^3}{v - c}$$

Find the speed v at which the fish should swim in order to minimize its energy expenditure. (Your answer will depend upon c, the speed of the current.)

45 (*General–Athletic Fields*) A running track consists of a rectangle with a semicircle at each end, as in the diagram. If the perimeter is to be exactly 440 yards, find the dimensions (x and r) that maximize the area of the rectangle. (*Hint:* The perimeter is $2x + 2\pi r$.)

46 (*General–Window Design*) A Norman window consists of a rectangle topped by a semicircle, as in the diagram. If the perimeter is to be 18 feet, find the dimensions (x and r) that maximize the area of the window. (*Hint:* The perimeter is $2x + 2r + \pi r$.)

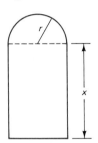

3.5 FURTHER APPLICATIONS OF OPTIMIZATION

Constructing Revenue Functions

In Section 3.4, when we maximized profit, we were given the price function, from which we calculated revenue as price times quantity. In this section we will not be given price functions, but only certain information describing how changes in price will affect the quantity sold. From this information we will construct revenue and cost functions. We will also see that sometimes x should be chosen as something other than quantity sold.

Example 1 A store can sell 20 bicycles per week at a price of $400 each. The manager estimates that for each $10 price reduction he can sell two more bicycles per week. The bicycles cost the store $200 each. If x stands for *the number of $10 price reductions,* express the price p and the quantity q as functions of x.

Solution Let

$$x = \text{the number of \$10 price reductions}$$

For example, $x = 4$ means that the price is reduced by $40 (four $10 price reductions). Therefore, in general, if there are x $10 price reductions from the original $400 price, then the price $p(x)$ is

$$p(x) = 400 - 10x$$

price ⏜ original price ↑ └— less x $10 price reductions

The quantity sold $q(x)$ will be

$$q(x) = \underbrace{20}_{\substack{\text{original} \\ \text{quantity}}} + \overbrace{2x}^{\text{plus two for each price reduction}}$$

▪

We will return to this example and maximize the store's profit after a practice exercise.

Practice Exercise 1

A computer manufacturer can sell 1500 personal computers per month at a price of $3000 each. The manager estimates that for each $200 price reduction he will sell 300 more each month. If x equals *the number of $200 price reductions*, express the price p and the quantity q as functions of x.

(solution on page 182)

Example 2   (continuation of Example 1) Using the information in Example 1, find the price of the bicycles and the quantity that maximize profit. Also find the maximum profit.

Solution   In Example 1 we found

$$p(x) = 400 - 10x \qquad\qquad \textit{price}$$

$$q(x) = 20 + 2x \qquad\qquad \textit{quantity sold at that price}$$

Revenue is price times quantity, $p(x)q(x)$.

$$R(x) = (400 - 10x)(20 + 2x) \qquad\qquad \textit{p(x)q(x)}$$

$$= 8000 + 600x - 20x^2 \qquad\qquad \textit{multiplying out and simplifying}$$

The cost function is unit cost times quantity.

$$C(x) - \underbrace{200}_{\substack{\text{unit} \\ \text{cost}}}\underbrace{(20 + 2x)}_{\substack{\text{quantity} \\ q(x)}} = 4000 + 400x \qquad\qquad \textit{if there were a fixed cost, we would add it}$$

Profit is revenue minus cost.

$$P(x) = \underbrace{(8000 + 600x - 20x^2)}_{R(x)} - \underbrace{(4000 + 400x)}_{C(x)} = 4000 + 200x - 20x^2 \qquad \textit{simplifying}$$

We maximize profit by setting the derivative equal to zero.

$$P'(x) = 200 - 40x = 0 \qquad\qquad \begin{array}{l}\textit{differentiating}\\ \textit{P = 4000 + 200x - 20x}^2\end{array}$$

The critical value is $x = 5$. The second derivative, $P''(x) = -40$, shows that the profit is *maximized* at $x = 5$. Since x is the number of $10 price reductions, the original price of $400 should be lowered by $50 ($10 five times), from $400 to $350. The quantity sold is found from the quantity function.

$$q(5) = 20 + 2 \cdot 5 = 30$$
q(x) = 20 + 2x at x = 5

Answer: Sell the bicycles for $350 each; quantity sold, 30 per week, maximum profit, $4500.
from P(x) = 4000 + 200x − 20x²
at x = 5   ■

Choosing Variables

Notice that in Example 2 we did not choose x to be the quantity sold, but instead, the number of $10 price reductions. (Therefore, a negative x would have meant a price *increase*.) We chose this x because from it we could easily calculate both the new price and the new quantity. Other choices for x are also possible, but in situations where a certain price change will make one quantity rise and another fall, it is often easiest to choose x to be the *number of such changes*.

Example 3   An orange grower finds that if he plants 80 orange trees per acre, each tree will yield 60 bushels of oranges. He estimates that for each addditional tree that he plants per acre, the yield of each tree will decrease by 2 bushels. How many trees should he plant per acre to maximize his harvest?

Solution   We take x equal to the number of "changes," that is, let

$$x = \text{the number of added trees per acre}$$

With x extra trees per acre,

$$\text{trees per acre} = 80 + x$$
original 80 plus x more

$$\text{yield per tree} = 60 - 2x$$
original yield less 2 per extra tree

Therefore, the total yield per acre will be

$$Y(x) = \underbrace{(60 - 2x)}_{\substack{\text{yield} \\ \text{per tree}}}\underbrace{(80 + x)}_{\substack{\text{trees} \\ \text{per acre}}} = 4800 - 100x - 2x^2$$

We maximize this by setting the derivative equal to zero.

$$-100 - 4x = 0$$
differentiating Y = 4800 − 100x − 2x²

$$x = -25$$
negative!

The number of *added* trees is negative, meaning that he should plant 25 *fewer* trees per acre. The second derivative, $Y''(x) = -4$, shows that the yield is indeed maximized at $x = -25$.

   Answer: Plant 55 trees per acre.   *80 − 25 = 55*

                                                                         ▪

Minimizing the Amount of Materials

Earlier problems involved maximizing areas and volumes using only a fixed amount of material (such as a fixed length of fence). Instead, we could minimize the amount of materials for a fixed area or volume.

Example 4 A moving company wishes to design an open-top box with a square base whose volume is exactly 32 cubic feet. Find the dimensions of the box requiring the least amount of materials.

Solution The base is square, so we define

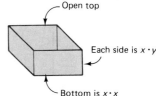

$$x = \text{length of side of base}$$

$$y = \text{height}$$

The volume (length·width·height) is $x·x·y$ or x^2y, which (according to the problem) must equal 32 cubic feet.

$$x^2y = 32$$

The box consists of a bottom (area x^2) and four sides (each of area xy). Minimizing the amount of materials means minimizing the *surface area* of the bottom and four sides,

$$A = x^2 + 4xy$$

$\left(\begin{matrix}area\ of\\bottom\end{matrix}\right) + \left(\begin{matrix}area\ of\\four\ sides\end{matrix}\right)$

As usual, we must express this area in terms of just *one* variable. We solve the volume equation, $x^2y = 32$, for y.

$$y = \frac{32}{x^2}$$

(We could also have solved for x, but it is easier to solve $x^2y = 32$ for y.) The area function becomes

$$A = x^2 + 4x\,\frac{32}{x^2}$$ $A = x^2 + 4xy$ *with y replaced by* $32/x^2$

$$= x^2 + \frac{128}{x}$$ *simplifying*

$$= x^2 + 128x^{-1}$$ *writing 1/x as* x^{-1}

We mimimize this by differentiating.

$$A'(x) = 2x - 128x^{-2}$$

$$2x - \frac{128}{x^2} = 0 \qquad \textit{setting the derivative equal to zero}$$

$$2x^3 - 128 = 0 \qquad \textit{multiplying by } x^2 \textit{ (since } x > 0\text{)}$$

$$x^3 = 64 \qquad \textit{adding 128 and then dividing by 2}$$

$$x = 4 \qquad \textit{taking cube roots}$$

The second derivative

$$A''(x) = 2 + 256x^{-3} = 2 + \frac{256}{x^3} \qquad \textit{from } A'(x) = 2x - 128x^{-2}$$

is positive at $x = 4$, so the area is minimized.

   Answer:

base: 4 feet on each side
height: 2 feet

$\textit{from } y = \dfrac{32}{x^2} \textit{ at } x = 4$

■

Minimizing the Cost of Materials

How would this problem have changed if the material for the bottom of the box had been more costly than the material for the sides? If, for example, the material for the sides cost $2 per square foot, and if the materials for the base, needing greater strength, cost $4 per square foot, then instead of simply minimizing the surface area, we would minimize *total cost*.

$$\text{Cost} = \left(\begin{smallmatrix} \text{area of} \\ \text{bottom} \end{smallmatrix} \right)\left(\begin{smallmatrix} \text{cost of bottom} \\ \text{per square foot} \end{smallmatrix} \right) + \left(\begin{smallmatrix} \text{area of} \\ \text{sides} \end{smallmatrix} \right)\left(\begin{smallmatrix} \text{cost of sides} \\ \text{per square foot} \end{smallmatrix} \right)$$

Since the areas would be just as before, this cost would be

$$\text{cost} = (x^2)(4) + (4xy)(2) = 4x^2 + 8xy$$

From here on we would proceed just as before, eliminating the y (using the volume relationship $x^2y = 32$) and then setting the derivative equal to zero.

Maximizing Tax Revenue

The next example concerns maximizing tax revenues for the government. Governments raise money by collecting taxes. If a sales tax or an import tax is too high, trade will be discouraged and tax revenues will fall. If, on the other hand, the tax rate is too low, trade may flourish but tax revenues will again fall. Government economists try to predict the relationship between the tax rate on an item and the total sales of the item so that they can find the tax rate that maximizes revenues.

Example 5   Suppose that economists predict that the relationship between the tax rate t on an item and the total amount of sales, S, of that item is

$$S(t) = 9 - 20\sqrt{t} \qquad \text{(for } 0 \le t \le .20)$$

where total sales, S, is measured in millions of dollars. Find the tax rate that maximizes revenue to the government.

Solution   To understand the problem, consider various tax rates. For example, if the tax rate is $t = 0$ (0%), then total sales will be

$$S(0) = 9 - 20\sqrt{0} = 9 \qquad\qquad \textit{\$9 million}$$

while if the tax rate is raised to $t = .16$ (16%), sales would be

$$S(.16) = 9 - 20\sqrt{.16} = 9 - (20)(.4)$$

$$= 9 - 8 = 1 \qquad\qquad \textit{\$1 million}$$

That is, raising the tax rate from 0% to 16% has discouraged $8 million worth of sales. The graph of $S(t)$ in the diagram below shows that total sales decrease as the tax rate increases.

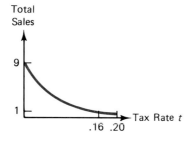

The government's revenue R is the tax rate t times the total sales $S(t) = 9 - 20\sqrt{t}$.

$$R(t) = \underbrace{t(9 - 20t^{1/2})}_{S(t)} = 9t - 20t^{3/2}$$

The graph of this function is shown on the right.* To maximize it, we set its derivative equal to zero.

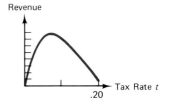

---

   * Such rate-revenue curves, although they been around for a long time, have recently been called Laffer curves (after the economist Arthur Laffer). See Bruce Bender, "An Analysis of the Laffer Curve," *Economic Inquiry*, July 1984, 414–420.

$$9 - 30t^{1/2} = 0 \qquad \text{\textit{derivative of} } 9t - 20t^{3/2}$$

$$9 = 30t^{1/2} \qquad \text{\textit{adding} } 30t^{1/2} \text{ \textit{to each side}}$$

$$t^{1/2} = 9/30 = .3 \qquad \text{\textit{dividing by} } 30$$

$$t = .09 \qquad \text{\textit{squaring both sides}}$$

This gives a tax rate of $t = 9\%$. The second derivative,

$$R''(t) = -30\tfrac{1}{2}t^{-1/2} = -\frac{15}{\sqrt{t}} \qquad \text{\textit{from} } R' = 9 - 30t^{1/2}$$

is negative at $t = .09$, showing that the revenue is maximized.

Answer: A tax rate of 9% maximizes revenue for the government.   ■

SOLUTION TO PRACTICE EXERCISE

1. Quantity, $q(x) = 1500 + 300x$; price, $p(x) = 3000 - 200x$.

EXERCISES 3.5

1 (*Business–Maximum Profit*) An automobile dealer can sell 12 cars per day at a price of $15,000. He estimates that for each $300 price reduction he can sell two more cars per day. If each car costs him $12,000, and fixed costs are $1000, what price should he charge to maximize his profit? How many cars will he sell at this price? (*Hint*: Let $x =$ the number of $300 price reductions.)

2 (*Business–Maximum Profit*) An automobile dealer can sell four cars per day at a price of $12,000. He estimates that for each $200 price reduction he can sell two more cars per day. If each car costs him $10,000, and his fixed costs are $1000, what price should he charge to maximize his profit? How many cars will he sell at this price? (*Hint*: Let $x =$ the number of $200 price reductions.)

3 (*Business–Maximum Revenue*) An airline finds that if it prices a cross-country ticket at $200, it will sell 300 tickets per day. It estimates that each $10 price reduction will result in 30 more tickets sold per day. Find the ticket price (and the number of tickets sold) that will maximize the airline's revenue.

4 (*Economics–Oil Prices*) An oil-producing country can sell 1 million barrels of oil a day at a price of $25 per barrel. If each $1 price increase will result in a sales decrease of 50,000 barrels per day, what price will maximize the country's revenue? How many barrels will they sell at that price?

5 (*Business–Maximum Revenue*) Rent-A-Reck Incorporated finds that it can rent 60 cars if it charges $80 for a weekend. It estimates that for each $5 price increase it will rent three fewer cars. What price should it charge to maximize its revenue? How many cars will it rent at this price?

6 (*General–Maximum Yield*) A peach grower finds that if he plants 40 trees per acre each tree will yield 60 bushels of peaches. He also estimates that for each additional tree that he plants per acre, the yield of each tree will decrease by 2 bushels. How many trees should he plant per acre to maximize his harvest?

7 (*General–Maximum Yield*) An apple grower finds that if he plants 20 trees per acre each tree will yield 90 bushels of apples. He also estimates that for each additional tree that he plants per acre, the yield of each tree will decrease by 3 bushels. How many trees should he plant per acre to maximize his harvest?

8 (*General–Fencing*) A farmer has 1200 feet of fence and wishes to build two identical rectangular enclosures, as in the diagram. What should be the dimensions of each enclosure if the total area is to be a maximum?

9 (*General–Minimum Materials*) An open-top box with a square base is to have a volume of 4 cubic feet. Find the dimensions of the box that can be made with the smallest amount of materials.

10 (*General–Minimum Materials*) An open-top box with a square base is to have a volume of 108 cubic inches. Find the dimensions of the box that can be made with the smallest amount of materials.

11 (*General–Largest Postal Package*) The U.S. Postal Service will accept a package if its length plus its girth (the distance all the way around) does not exceed 84 inches. Find the dimensions and volume of the largest package with a square base that can be mailed.

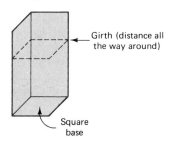

Girth (distance all the way around)

Square base

12 (*General–Fencing*) A homeowner wants to build, along his driveway, a garden surrounded by a fence. If the garden is to be 800 square feet, and the fence along the driveway costs $6 per foot while on the other three sides it costs only $2 per foot, find the dimensions that will minimize the cost. Also find the minimum cost.

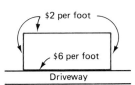

$2 per foot

$6 per foot

Driveway

13 (*General–Fencing*) A homeowner wants to build, along his driveway, a garden surrounded by a fence. If the garden is to be 5000 square feet, and the fence along the driveway costs $6 per foot while on the other three sides it costs only $2 per foot, find the dimensions that will minimize the cost. Also find the minimum cost. (See the diagram at the bottom of the left-hand column.)

14–15 (*Economics–Tax Revenue*) Suppose that the relationship between the tax rate t on imported shoes and the total sales S (in millions of dollars) is given by the function below. Find the tax rate t that maximizes revenue for the government.

14 $S = 4 - 6\sqrt[3]{t}$ **15** $S = 8 - 15\sqrt[3]{t}$

16 (*Biomedical–Drug Concentration*) If the amount of a drug in a person's blood after t hours is

$$f(t) = t/(t^2 + 9)$$

when will the drug concentration be the greatest?

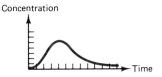

Concentration

Time

17 (*General–Wine Storage*) A case of wine appreciates in value each year, but there is also an annual storage charge. The net value of the wine after t years is $V(t) = 2000 + 96\sqrt{t} - 12t$ dollars (for $t < 25$). Find the storage time that will maximize the value of the wine.

18 (*General–Bus Shelter Design*) A bus stop shelter, consisting of two square sides, a back, and a roof, as shown below, is to have volume 1024 cubic feet. What are the dimensions that require the least amount of materials?

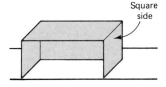

Square side

19 (*General–Area*) Show that the rectangle of fixed area whose perimeter is a minimum is a square.

20 (*Political Science–Campaign Expenses*) A politician estimates that by campaigning in a county for x days, she will gain $2x$ (thousand) votes, but her campaign expenses will be $5x^2 + 500$ dollars. She wants to campaign for the number of days that maximizes the number of votes per dollar,

$$f(x) = \frac{2x}{5x^2 + 500}$$

For how many days should she campaign?

21 (*General–Page Design*) A page of 96 square inches is to have margins of 1 inch on either side and $1\frac{1}{2}$ inches at the top and bottom, as in the diagram. Find the dimensions of the page that maximize the (inner) print area.

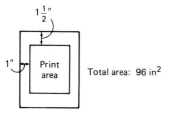

22 (*Biomedical–Contagion*) If an epidemic spreads through a town at a rate that is proportional to the number of infected people and to the number of uninfected people, then the rate is $R(x) = cx(p - x)$ where x is the number of infected people and c and p (the population) are constants. Show that the rate $R(x)$ is greatest when half of the population is infected.

23 (*Biomedical–Contagion*) If an epidemic spreads through a town at a rate that is proportional to the number of uninfected people and to the square of the number of infected people, then the rate is

$$R(x) = cx^2(p - x)$$

where x is the number of infected people and c and p (the population) are constants. Show that the rate $R(x)$ is greatest when two-thirds of the population is infected.

3.6 OPTIMIZING LOT SIZE AND HARVEST SIZE

Introduction

In this section we discuss two important applications of optimization, one economic and one ecological. The first concerns the most efficient way for a business to order merchandise (or for a manufacturer to produce merchandise), and the second concerns the preservation of animal populations that are harvested by man. Either of these applications can be read without the other.

Minimizing Inventory Costs

A business encounters two kinds of costs in maintaining inventory: storage costs (warehouse and insurance costs for merchandise not yet sold) and reorder costs (delivery and bookkeeping costs for each order). For example, if a furniture store expects to sell 250 sofas in a year, it could order all 250 at once (incurring high storage costs), or it could order them in many small lots, say 50 orders of five each, spaced throughout the year (incurring high reorder costs). Obviously, the best order size (or "lot" size) is the one that minimizes the total of storage plus reorder costs.

Example 1   A furniture showroom expects to sell 250 sofas a year. Each sofa costs the store $300, and there is a fixed charge of $500 per order. If it costs $100 to store a sofa for a year, how large should each order be and how often should orders be placed to minimize inventory costs?

Solution   Let

$$x = \text{lot size.}$$

x is the number of sofas in each order

Storage Costs: If the sofas sell steadily throughout the year, and if the store reorders x more whenever the stock runs out, then its inventory during the year looks like the following graph.

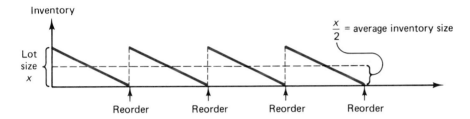

Notice that the inventory level varies from the lot size x down to zero, with an average inventory of $x/2$ sofas throughout the year. Since it costs $100 to store a sofa for a year, the total (annual) storage costs are

$$\begin{pmatrix} \text{storage} \\ \text{costs} \end{pmatrix} = \begin{pmatrix} \text{storage} \\ \text{per item} \end{pmatrix} \cdot \begin{pmatrix} \text{average num-} \\ \text{ber of items} \end{pmatrix}$$

$$= \quad 100 \cdot \frac{x}{2} \quad = 50x$$

Reorder Costs: Each sofa costs $300, so an order of lot size x costs $300x$, plus the fixed order charge of $500.

$$\begin{pmatrix} \text{cost} \\ \text{per order} \end{pmatrix} = 300x + 500$$

The yearly supply of 250 sofas, with x sofas in each order, require $250/x$ orders. (For example, 250 sofas at 5 per order require $250/5 = 50$ orders.) Therefore, the yearly reorder costs are

$$\begin{pmatrix} \text{reorder} \\ \text{costs} \end{pmatrix} = \begin{pmatrix} \text{cost} \\ \text{per order} \end{pmatrix} \cdot \begin{pmatrix} \text{numbers} \\ \text{of orders} \end{pmatrix}$$

$$= (300x + 500) \cdot \left(\frac{250}{x}\right)$$

Total Cost: $C(x)$ is storage costs plus reorder costs.

$$C(x) = \left(\begin{array}{c}\text{storage}\\\text{costs}\end{array}\right) + \left(\begin{array}{c}\text{reorder}\\\text{costs}\end{array}\right)$$

$$= 100\,\frac{x}{2} + (300x + 500)\left(\frac{250}{x}\right) \qquad \textit{using storage and reorder costs found earlier}$$

$$= 50x + 75{,}000 + 125{,}000x^{-1} \qquad \textit{simplifying}$$

To minimize $C(x)$, we differentiate.

$$C'(x) = 50 - 125{,}000x^{-2} = 50 - \frac{125{,}000}{x^2} \qquad \textit{differentiating } C = 50x + 75{,}000 + 125{,}000x^{-1}$$

$$50 - \frac{125{,}000}{x^2} = 0 \qquad \textit{setting the derivative equal to zero}$$

$$50x^2 = 125{,}000 \qquad \textit{multiplying by } x^2 \textit{ and adding } 125{,}000 \textit{ to each side}$$

$$x^2 = \frac{125{,}000}{50} = 2500 \qquad \textit{dividing each side by 50}$$

$$x = 50 \qquad \textit{taking square roots } (x > 0) \textit{ gives lot size 50}$$

$$C''(x) = 250{,}000x^{-3} = 250{,}000\,\frac{1}{x^3} \qquad C'' \textit{ is positive, so } C \textit{ is minimized at } x = 50$$

At 50 sofas per order, the yearly 250 will require $250/50 = 5$ orders.

Answer: Lot size is 50 sofas, with orders placed five times a year.   ■

Modifications and Assumptions

If the number of orders per year is not a whole number, say 7.5 orders per year, we just interpret it as 15 orders in two years, and handle it accordingly.
   We made two major assumptions in Example 1. We assumed a steady demand, and that orders were equally spaced throughout the year. These are reasonable assumptions for many products. Even for seasonal products like bathing suits or fur coats, separate calculation can be done for the "on" and "off" season.

Production Runs

Similar analysis applies to manufacturing. For example, if a book publisher can estimate the yearly demand for a book, she may print the yearly total all at once, incurring high storage costs, or she may print them in several smaller runs throughout the year, incurring setup costs for each run. Here the setup costs for each printing run play the role of the reorder costs for a store.

Example 2 A publisher estimates the annual demand for a book to be 4000 copies. Each book costs $5 to print, and setup costs are $1000 for each printing. If storage costs are $2 per book per year, find how many books should be printed per run and how many printings will be needed if costs are to be minimized.

Solution   Let

$$x = \text{the number of books in each run.}$$

Storage costs: As before, an average of $x/2$ books are stored throughout the year, at a cost of $2 each, so annual storage costs are

$$\left(\begin{matrix}\text{storage}\\\text{costs}\end{matrix}\right) = \left(\frac{x}{2}\right)\cdot 2 = x$$

Production costs: The cost per run is

$$\left(\begin{matrix}\text{costs}\\\text{per run}\end{matrix}\right) = 5x + 1000 \qquad \textit{x books at \$5 each, plus \$1000 setup costs}$$

The 4000 books at x books per run will require $4000/x$ runs. Therefore, production costs are

$$\left(\begin{matrix}\text{production}\\\text{costs}\end{matrix}\right) = (5x + 1000)\left(\frac{4000}{x}\right) \qquad \textit{cost per run times number of runs}$$

Total cost: The total cost is storage costs plus production costs.

$$C(x) = x + (5x + 1000)\left(\frac{4000}{x}\right) \qquad \textit{storage + production}$$

$$= x + 20{,}000 + 4{,}000{,}000x^{-1} \qquad \textit{multiplying out}$$

We differentiate, set the derivative equal to zero, and solve, just as before (omitting the details), obtaining $x = 2000$. The second derivative test will show that costs are minimized at 2000 books per run. The 4000 books require $4000/2000 = 2$ printings.

Answer: Print 2000 books per run, with two printings. ■

Maximum Sustainable Yield

The next application involves industries like fishing, in which a naturally occurring animal population is "harvested." Taking too large a harvest will kill off the animal population (like the bowhead whale, hunted almost to extinction in the nineteenth century). We want the "maximum sustainable yield" that may be taken, year after year, while preserving a stable animal population.

For some animals one can determine a *reproduction function*, $f(p)$, which gives the expected population a year from now if the present population is p.

Reproduction Function

A reproduction function $f(p)$ gives the population a year from now if the current population is p.

For example, the reproduction function $f(p) = -(1/4)p^2 + 3p$ (where p and $f(p)$ are measured in thousands) means that if the population is now $p = 6$ (thousand), then a year from now the population will be

$$f(6) = -\frac{1}{4} \cdot 6^2 + 3 \cdot 6 = -\frac{1}{4} \cdot 36 + 18 = -9 + 18 = 9 \quad \text{(thousand)}$$

Therefore, during the year the population will increase from 6000 to 9000.

If, on the other hand, the present population is $p = 10$ (thousand), a year later the population will be

$$f(10) = -\frac{1}{4} \cdot 10^2 + 3 \cdot 10 = -25 + 30 = 5 \quad \text{(thousand)}$$

That is, during the year the population will decline from 10,000 to 5000 (perhaps due to inadequate food to support such a large population). In actual practice, reproduction functions are very difficult to calculate, but can sometimes be estimated by analyzing previous population and harvest data.*

Suppose that we have a reproduction function f and a current population of size p, which will therefore grow to size $f(p)$ next year. The *amount of growth* in the population during that year is

$$\left(\begin{array}{c} \text{amount} \\ \text{of growth} \end{array} \right) = f(p) - p$$

current population
next year's population

Harvesting this amount removes only the *growth*, returning the population to its former size p. The population will then repeat this growth, and taking the same harvest $f(p) - p$ will cause this situation to repeat itself year after year. The quantity $f(p) - p$ is called the *sustainable yield*.

Sustainable Yield

For reproduction function $f(p)$, the sustainable yield is

$$Y(p) = f(p) - p$$

---

* For more information, see J. Blower, L. Cook, and J. Bishop, *Estimating the Size of Animal Populations* (George Allen and Unwin (Publisher) Ltd., London, 1981).

We want the population size p that maximizes the sustainable yield $Y(p)$. We set the derivative equal to zero.

$$Y'(p) = f'(p) - 1 = 0$$ *derivative of $Y = f(p) - p$*

$$f'(p) = 1$$ *solving for $f'(p)$*

For a given reproduction function $f(p)$ we find the maximum sustainable yield by solving this equation [provided that the second derivative test gives $Y''(p) = f''(p) < 0$].

Maximum Sustainable Yield

For reproduction function $f(p)$, the population p that results in the maximum sustainable yield is the solution to

$$f'(p) = 1$$

[provided that $f''(p) < 0$]. The maximum sustainable yield is then

$$Y(p) = f(p) - p$$

Once we calculate the population p that gives the maximum sustainable yield, we wait until the population reaches this size and then harvest, year after year, an amount $Y(p)$.

Note that to find the maximum sustainable yield, we set the derivative $f'(p)$ equal to 1, not 0. This is because we are maximizing not the reproduction function $f(p)$ but the yield function $Y(p) = f(p) - p$.

Example 3   The reproduction function for the American lobster in an east coast fishing area is $f(p) = -.02p^2 + 2p$ [where p and $f(p)$ are in thousands]. Find the population p that gives the maximum sustainable yield and the size of the yield.

Solution   We set the derivative of the reproduction function equal to 1.

$$f'(p) = -.04p + 2 = 1$$ *differentiating $f(p) = -.02p^2 + 2p$*

$$-.04p = -1$$ *subtracting 2 from each side*

$$p = \frac{-1}{-.04} = 25$$ *dividing by $-.04$*

The second derivative $f''(p) = -.04$ is negative, showing that $p = 25$ (thousand) is the population that gives the maximum sustainable yield. The actual yield is found from the yield function $Y(p) = f(p) - p$.

$$Y(p) = \underbrace{-.02p^2 + 2p}_{f(p)} - p = -.02p^2 + p$$

$$Y(25) = -.02(25)^2 + 25 \qquad \textit{evaluating at } p = 25$$

$$= -12.5 + 25 = 12.5 \quad \text{(thousand)}$$

Answer: The population size for the maximum sustainable yield is 25,000, and the yield is 12,500 lobsters.

25,000 from p = 25

EXERCISES 3.6

LOT SIZE

1 A supermarket expects to sell 4000 boxes of sugar in a year. Each box costs the store $2, and there is a fixed delivery charge of $20 per order. If it costs $1 to store a box for a year, what is the lot size and how many times a year should the orders be placed to minimize inventory costs?

2 A supermarket expects to sell 5000 boxes of rice in a year. Each box costs the store $2, and there is a fixed delivery charge of $50 per order. If it costs $2 to store a box for a year, what is the order size and how many times a year should the orders be placed to minimize inventory costs?

3 A liquor warehouse expects to sell 10,000 bottles of scotch whiskey in a year. Each bottle costs the store $12, plus a fixed charge of $125 per order. If it costs $10 to store a bottle for a year, how many bottles should be ordered at a time and how many orders should it place in a year to minimize inventory costs?

4 A wine warehouse expects to sell 30,000 bottles of wine in a year. Each bottle costs the store $5, plus a fixed charge of $200 per order. If it costs $3 to store a bottle for a year, how many bottles should be ordered at a time and how many orders should it place in a year to minimize inventory costs?

5 An automobile dealer expects to sell 800 cars a year. The cars cost $5000 each plus a fixed charge of $1000 per delivery. If it costs $1000 to store a car for a year, find the order size and the number of orders that minimize inventory costs.

6 An automobile dealer expects to sell 400 cars a year. The cars cost $6000 each plus a fixed charge of $500 per order. If it costs $1000 to store a car for a year, find the order size and the number of orders that minimize inventory costs.

PRODUCTION RUNS

7 A toy manufacturer estimates the demand for a game to be 2000 per year. Each game costs $3 to manufacture, plus setup costs of $500 for each production run. If a game can be stored for a year for a cost of $2, how many should be manufactured at a time and how many production runs should there be to minimize costs?

8 A toy manufacturer estimates the demand for a doll to be 10,000 per year. Each doll costs $5 to manufacture, plus setup costs of $800 for each production run. If it costs $4 to store a doll for a year, how many should be manufactured at a time and how many production runs should there be to minimize costs?

9 A record manufacturer estimates the yearly demand for a record to be 1,000,000 records. It costs $800 to set the presses for the record, plus $10 for each record produced. If it costs the company $1 to store a record for a year, how many should be pressed at a time and how many production runs will be needed to minimize costs?

10 A record manufacturer estimates the yearly demand for a record to be 10,000. It costs $400 to set the presses for the record, plus $3 for each record produced. If it costs the company $2 to store a record for a year, how many should be pressed at a time and how many production runs will be needed to minimize costs?

MAXIMUM SUSTAINABLE YIELD

11 Fishermen estimate the reproduction curve for swordfish in the Georges Bank fishing grounds to be

$$f(p) = -.01p^2 + 5p$$

where p and $f(p)$ are in hundreds. Find the population that gives the maximum sustainable yield, and the size of the yield.

12 The reproduction function for the Hudson's Bay lynx is estimated to be $f(p) = -.02p^2 + 5p$, where p and $f(p)$ are in thousands. Find the population that gives the maximum sustainable yield, and the size of the yield.

13 The reproduction function for the Antarctic blue whale is estimated to be $f(p) = -.0004p^2 + 1.06p$, where p and $f(p)$ are in thousands. Find the population that gives the maximum sustainable yield, and the size of the yield.

14 The reproduction function for the Canadian snowshoe hare is estimated to be $f(p) = -.025p^2 + 4p$, where p and $f(p)$ are in thousands. Find the population that gives the maximum sustainable yield, and the size of the yield.

15 A fishermen estimates the reproduction curve for rainbow trout in a large lake to be $f(p) = 50\sqrt{p}$, where p and $f(p)$ are in thousands. Find the population that gives the maximum sustainable yield, and the size of the yield.

16 The reproduction curve for oysters in a large bay is $f(p) = 30\sqrt[3]{p^2}$, where p and $f(p)$ are in pounds. Find the size of the population that gives the maximum sustainable yield, and the size of the yield.

3.7   IMPLICIT DIFFERENTIATION AND RELATED RATES

Introduction

A function $y = f(x)$ is defined *explicitly* if it is defined by a formula, such as $y = \sqrt{x^3 - 1}$. A function $y = f(x)$ is defined *implicitly* if it is implied by an equation in x and y, such as

$$x^3 + y^5 - y = 8$$

If we were to solve this equation for y, we would obtain an explicitly defined function, which could then be differentiated to find dy/dx. Solving equations, however, can be difficult, and sometimes even impossible. The technique of *implicit differentiation* enables us to calculate the derivative dy/dx of an implicitly defined function without having to solve for y.

Implicit Differentiation

Think of the equation above, $x^3 + y^5 - y = 8$, as defining one (or more) functions y of x, which we write as $y(x)$. To find the derivative y', we differentiate both sides of $x^3 + y^5 - y = 8$ with respect to x, and then solve for y'. Remember, however, that y is a *function* of x, so differentiating y^n means differentiating a *function to a power*, which requires the generalized power rule,

$$\frac{d}{dx}\, y^n = n \cdot y^{n-1} \cdot y'$$

Example 1   For $x^3 + y^5 - y = 8$, find dy/dx.

Solution   We differentiate both sides with respect to x.

$$x^3 + y^5 - y = 8$$ *original equation*

$$3x^2 + 5y^4y' - y' = 0$$ *taking $\frac{d}{dx}$ of both sides*

We now solve for y'.

$$5y^4y' - y' = -3x^2$$ *subtracting $3x^2$, so only y' terms are on the left*

$$y'(5y^4 - 1) = -3x^2$$ *factoring out y'*

$$y' = \frac{-3x^2}{5y^4 - 1}$$ *dividing by $(5y^4 - 1)$ to solve for y'*

This is the answer, the derivative dy/dx when x and y are related by the equation $x^3 + y^5 - y = 8$.   ■

Notice that the formula for y' involves both x and y. Implicit differentiation enables us to find derivatives that would otherwise be difficult or impossible to calculate, but at a "cost"—the result depends on both x and y.

The x and y play very different roles. We regard y as a *function of x*, and the prime means *derivative with respect to x*. Therefore, when differentiating y^n we must include a y', but when differentiating x^n we do not (since $dx/dx = 1$).

Example 2 (continuation of Example 1)   Evaluate the derivative in Example 1 at $x = 2$ and $y = 1$.

Solution

$$y' = \frac{-3x^2}{5y^4 - 1}$$ *from Example 1*

$$= \frac{-3(2)^2}{5(1)^4 - 1} = \frac{-12}{4} = -3$$ *evaluating at $x = 2$, $y = 1$*   ■

A derivative should be evaluated only at a point on the curve, so we evaluate y' only at x- and y-values *satisfying the original equation*. It is easy to check that the values $x = 2$ and $y = 1$ in Example 2 do satisfy the original equation $x^3 + y^5 - y = 8$.

The following problems are typical "pieces" that appear in implicit differentiation problems.

Example 3

(a) $\dfrac{d}{dx} y^3 = 3y^2y'$ *differentiating y^3 so include y'*

(b) $\dfrac{d}{dx} x^3 = 3x^2$ *differentiating x^3 so no primed variable*

(c) $\dfrac{d}{dx}(x^3 y^5) = 3x^2 y^5 + x^3 5y^4 y'$ *using the product rule*

$$\underbrace{\dfrac{d}{dx} x^3} \qquad \underbrace{\dfrac{d}{dx} y^5}$$

$$= 3x^2 y^5 + 5x^3 y^4 y'$$ *putting the 5 first*

Suggestion: Try to do problems like Example 3c in one step, putting the constants in front from the start.

(d) $\dfrac{d}{dx}(xy) = y + xy'$ *using the product rule, and $\dfrac{d}{dx} x = 1$*

(e) $\dfrac{d}{dx}(1 + y^2)^4 = 4(1 + y^2)^3 \cdot 2yy'$ *using the generalized power rule*   ▪

Practice Exercise 1

Find

(a) $\dfrac{d}{dx} x^4$ **(b)** $\dfrac{d}{dx} y^2$

(c) $\dfrac{d}{dx}(x^2 y^3)$ **(d)** $\dfrac{d}{dx}(1 + y^3)^2$ *(solutions on page 197)*

Example 4 For $y^4 + x^4 - 2x^2 y^2 = 32$, find dy/dx and evaluate it at $x = 2$, $y = 1$.

Solution

$$4y^3 y' + 4x^3 - 4xy^2 - 4x^2 yy' = 0$$ *differentiating with respect to x, putting constants first*

$$4y^3 y' - 4x^2 yy' = -4x^3 + 4xy^2$$ *collecting y' terms on the left, others on the right*

$$y'(4y^3 - 4x^2 y) = -4x^3 + 4xy^2$$ *factoring out y'*

$$y' = \dfrac{-4x^3 + 4xy^2}{4y^3 - 4x^2 y} = \dfrac{-x^3 + xy^2}{y^3 - x^2 y}$$ *dividing by $4y^3 - 4x^2 y$ to solve for y', and simplifying*

At $x = 2$, $y = 1$,

$$y' = \dfrac{-(2)^3 + (2)(1)^2}{(1)^3 - (2)^2(1)} = \dfrac{-8 + 2}{1 - 4} = \dfrac{-6}{-3} = 2$$ ▪

Implicit Differentiation in General

To find dy/dx from an equation in x and y:

1. Differentiate both sides of the equation *with respect to x.*

2. Collect all terms involving y' on one side, and all others on the other side.

3. Factor out the y' and solve for y'.

Practice Exercise 2

For $x^3 + y^3 - xy = 7$, find dy/dx and evaluate it at $x = 2$, $y = 1$. *(solution on page 197)*

Application to Economics

Recall that the *demand equation* is the relationship between the price p of an item and the quantity x that consumers will demand at that price. (All prices are in dollars, unless otherwise stated.)

Example 5   For the demand equation $x = \sqrt{1900 - p^3}$, use implicit differentiation to find dp/dx. Then evaluate it at $x = 30$, $p = 10$ and interpret your answer.

Solution

$$x = (1900 - p^3)^{1/2} \qquad\qquad \textit{in power form}$$

$$1 = \frac{1}{2}(1900 - p^3)^{-1/2}(-3p^2 p') \qquad\qquad \textit{differentiating with respect to x, using the generalized power rule}$$

$$1 = \frac{-3p^2}{2\sqrt{1900 - p^3}}\, p' \qquad\qquad \textit{simplifying and factoring out p'}$$

$$p' = -\frac{2\sqrt{1900 - p^3}}{3p^2} \qquad\qquad \textit{solving for p'}$$

$$p' = -\frac{2\sqrt{1900 - (10)^3}}{3(10)^2} \qquad\qquad \textit{evaluating at p = 10, x = 30 (except that there is no x)}$$

$$= -\frac{2\sqrt{900}}{300} = -\frac{60}{300} = -.2 \qquad\qquad \textit{simplifying}$$

Interpretation: $dp/dx = -.2$ says that the rate of change of price with respect to quantity is $-.2$, so increasing quantity by 1 means decreasing price by .20 (or 20 cents). Therefore, each 20-cent price decrease results in one more sale (at the given values of x and p). ■

Notice that this demand function $x = \sqrt{1900 - p^3}$ can be solved *explicitly* for p.

$$x^2 = 1900 - p^3 \qquad\qquad \textit{squaring}$$

$$p^3 = 1900 - x^2 \qquad\qquad \textit{adding } p^3 \textit{ and subtracting } x^2$$

$$p = (1900 - x^2)^{1/3} \qquad\qquad \textit{taking cube roots}$$

We can differentiate this *explicitly* with respect to x.

$$p' = \frac{1}{3}(1900 - x^2)^{-2/3}(-2x) \qquad\qquad \textit{using the generalized power rule}$$

$$= -\frac{2}{3}x(1900 - x^2)^{-2/3} \qquad\qquad \textit{simplifying}$$

Evaluating at the given values, $x = 30$ and $p = 10$, gives

$$p' = -\frac{2}{3} \cdot 30(1900 - 30^2)^{-2/3} \qquad \text{\textit{substituting }} x = 30$$

$$= -20(1000)^{-2/3} = -\frac{20}{100} = -.2$$

This agrees with the answer by implicit differentiation.

Related Rates

Sometimes in an equation, *both* variables will be functions of a *third* variable, usually t for time. For example, with a seasonal product like fur coats, the price p and weekly sales x will be related by an equation (the demand equation), and both p and x depend on the time of year. Differentiating both sides of the demand equation with respect to t will give an equation relating the derivatives dp/dt and dx/dt. Such "related rate" equations show how fast one quantity is changing relative to another. First, an "everyday" example.

Example 6   A pebble thrown into a pond causes circular ripples to radiate outward. If the radius of the outer ripple is growing by 2 feet per second, how fast is the area of its circle growing at the moment when the radius is 10 feet?

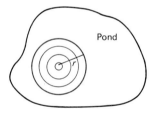

Solution   The formula for the area of a circle is $A = \pi r^2$. Both the area and the radius of the circle increase with time, so both A and r are functions of t. Finding how fast the area is growing means finding dA/dt, so we differentiate both sides of $A = \pi r^2$ with respect to t.

$$A' = 2\pi r \cdot r' \qquad \text{\textit{prime means derivative with respect to }} t$$

The radius r is growing by 2 feet per second ($dr/dt = 2$), so we substitute $r' = 2$ and the given radius $r = 10$.

$$A' = 2\pi \cdot 10 \cdot 2 = 40\pi \approx 125.6 \qquad \text{\textit{using }} \pi \approx 3.14$$

$$\underset{r \quad r'}{}$$

Therefore at the moment when the radius is 10 feet, the area of the circle is growing at the rate of about 126 square feet per second.   ▪

We should be ready to interpret any *rate* as a derivative, just as we interpreted the radius growing by 2 feet per second as $dr/dt = 2$.

Example 7   A boat yard's total profit from selling x outboard motors is $P = -x^2 + 1000x - 2000$. If the outboards are selling at the rate of 20 per week, how fast is the profit changing when 400 motors have been sold?

Solution   Profit P and quantity x both change with time, so both are functions of t. We differentiate both sides of $P = -x^2 + 1000x - 2000$ with respect to t.

$$\frac{dP}{dt} = -2x \frac{dx}{dt} + 1000 \frac{dx}{dt}$$

written in Leibniz's d/dt notation, just for variety

The quantity sold, x, is growing by 20 per week, so we substitute $dx/dt = 20$ and the given sales level $x = 400$.

$$\frac{dP}{dt} = -2\underbrace{(400)}_{x}\underbrace{20}_{dx/dt} + 1000 \cdot \underbrace{20}_{dx/dt}$$

$$= -16{,}000 + 20{,}000 = 4000$$

Therefore, the company's profits are growing at the rate of $4000 per week. ■

Example 8   An environmental study predicts that sulfur oxide pollution in a city will be $S = 2 + 20x + .1x^2$ units, where x is the population (in thousands). The population of the city t years from now is expected to be $x = 800 + 20\sqrt{t}$ thousand people. Find how rapidly pollution will be increasing (units per year) 4 years from now.

Solution   To find dS/dt we differentiate $S = 2 + 20x + .1x^2$ with respect to t.

$$\frac{dS}{dt} = 20 \frac{dx}{dt} + .2x \frac{dx}{dt}$$

since x is a function of t

We then find dx/dt by differentiating $x = 800 + 20t^{1/2}$ with respect to t.

$$\frac{dx}{dt} = 10t^{-1/2}$$

$$= 10 \cdot 4^{-1/2} = 10 \cdot \frac{1}{2} = 5$$

substituting the given t = 4

We substitute this relation, $dx/dt = 5$, and $x = 800 + 20t^{1/2}$ (evaluated at $t = 4$) into the equation above for dS/dt.

$$\frac{dS}{dt} = 20 \cdot \underbrace{5}_{dx/dt} + .2\underbrace{(800 + 20\sqrt{4})}_{x}\underbrace{5}_{dx/dt}$$

substituting x = 800 + 20√t and t = 4

$$= 100 + .2(840)5 = 100 + 840 = 940$$

Therefore, in 4 years, pollution will be increasing at the rate of 940 units per year. ■

SOLUTIONS TO PRACTICE EXERCISES

1. (a) $\dfrac{d}{dx}\, x^4 = 4x^3$ **(b)** $\dfrac{d}{dx}\, y^2 = 2yy'$

(c) $\dfrac{d}{dx}\,(x^2y^3) = 2xy^3 + 3x^2y^2y'$ **(d)** $\dfrac{d}{dx}\,(1 + y^3)^2 = 2(1 + y^3)\cdot 3y^2y'$

2. $x^3 + y^3 - xy = 7$

$3x^2 + 3y^2y' - y - xy' = 0$

$3y^2y' - xy' = -3x^2 + y$

$y'(3y^2 - x) = -3x^2 + y$

$y' = \dfrac{-3x^2 + y}{3y^2 - x}$ and at $x = 2$, $y = 1$, $y' = \dfrac{-12 + 1}{3 - 2} = -11$

EXERCISES 3.7

For each equation use implicit differentiation to find dy/dx.

1 $y^3 - x^2 = 4$ **2** $y^2 = x^4$

3 $x^3 = y^2 - 2$ **4** $x^2 + y^2 = 1$

5 $y^4 - x^3 = 2x$ **6** $y^2 = 4x + 1$

7 $(x + 1)^2 + (y + 1)^2 = 18$ **8** $xy = 12$

9 $x^2y = 8$ **10** $x^2y + xy^2 = 4$

11 $xy - x = 9$ **12** $x^3 + 2xy^2 + y^3 = 1$

13 $x(y - 1)^2 = 6$ **14** $(x - 1)(y - 1) = 25$

15 $y^3 - y^2 + y - 1 = x$ **16** $x^2 + y^2 = xy + 4$

17 $\dfrac{1}{x} + \dfrac{1}{y} = 2$ **18** $\sqrt[3]{x} + \sqrt[3]{y} = 2$

19 $x^3 = (y - 2)^2 + 1$ **20** $\sqrt{xy} = x + 1$

For each equation find dy/dx evaluated at the given values.

21 $y^2 - x^3 = 1$ at $x = 2$, $y = 3$ **22** $x^2 + y^2 = 25$ at $x = 3$, $y = 4$

23 $y^2 = 6x - 5$ at $x = 1$, $y = -1$ **24** $xy = 12$ at $x = 6$, $y = 2$

25 $x^2y + y^2x = 0$ at $x = -2$, $y = 2$ **26** $y^2 + y + 1 = x$ at $x = 1$, $y = -1$

27 $x^2 + y^2 = xy + 7$ at $x = 3$, $y = 2$ **28** $\sqrt[3]{x} + \sqrt[3]{y} = 3$ at $x = 1$, $y = 8$

For each demand equation use implicit differentiation to find dp/dx.

29 $p^2 + p + 2x = 100$ **30** $p^3 + p + 6x = 50$

31 $12p^2 + 4p + 1 = x$ **32** $8p^2 + 2p + 100 = x$

33 $xp^3 = 36$

35 $(p + 5)(x + 2) = 120$

34 $xp^2 = 96$

36 $(p - 1)(x + 5) = 24$

APPLIED EXERCISES ON IMPLICIT DIFFERENTIATION

37 (*Business–Demand Equation*) A company's demand equation is $x = \sqrt{68 - p^2}$. Find dp/dx when $p = 2$ and interpret your answer.

38 (*Business–Demand Equation*) A company's demand equation is $x = \sqrt{650 - p^2}$. Find dp/dx when $p = 5$ and interpret your answer.

39 (*Business–Sales*) If a company spends r million dollars on research, its sales will be s million dollars, where r and s are related by $s^2 = r^3 - 55$.

(a) Find ds/dr by implicit differentiation and evaluate it at $r = 4$, $s = 3$. (*Hint:* Differentiate the equation with respect to r.)

(b) Find dr/ds by implicit differentiation and evaluate it at $r = 4$, $s = 3$. (*Hint:* Differentiate the original equation with respect to s.)

(c) Interpret your answers to parts (a) and (b) as rates of change.

40 (*Biomedical–Muscle Contraction*) When a muscle lifts a load, it does so according to the "fundamental equation of muscle contraction"

$$(L + m)(V + n) = k$$

where L is the load that the muscle is lifting, V is the velocity of contraction of the muscle, and m, n, and k are constants. Use implicit differentiation to find dV/dL.

In each equation x and y are functions of t. Differentiate with respect to t to find a relation between dx/dt and dy/dt.

41 $x^3 + y^2 = 1$

45 $3x^2 - 7xy = 12$

42 $x^5 - y^3 = 1$

46 $2x^3 - 5xy = 14$

43 $x^2y = 80$

47 $x^2 + xy = y^2$

44 $xy^2 = 96$

48 $x^3 - xy = y^3$

APPLIED EXERCISES ON RELATED RATES

49 (*General–Snowballs*) A large snowball is melting so that its radius is decreasing at the rate of 2 inches per hour. How fast is the volume decreasing at the moment when the radius is 3 inches? [*Hint:* The volume of a sphere of radius r is $V = (4/3)\pi r^3$.]

50 (*General–Hailstones*) A hailstone (a small sphere of ice) is forming in the clouds so that its radius is growing at the rate of 1 mm (millimeter) per minute. How fast is its volume is growing at the moment when the radius is 2 mm? [*Hint:* the volume of a sphere of radius r is $V = (4/3)\pi r^3$.]

51 (*Biomedical–Tumors*) The radius of a spherical tumor is growing by ½ cm (centimeter) per week. Find how rapidly the volume is increasing at the moment when the radius is 4 cm. [*Hint:* The volume of a sphere of radius r is $V = (4/3)\pi r^3$.]

52 (*Business–Profit*) A company's profit from selling x units of an item is $P = 1000x - \frac{1}{2}x^2$ dollars. If sales are growing at the rate of 20 per day, find how rapidly profit is growing (in dollars per day) when 600 units have been sold.

53 (*Business–Revenue*) A company's revenue from selling x units of an item is $R = 1000x - x^2$ dollars. If sales are increasing at the rate of 80 per day, find how rapidly revenue is growing (in dollars per day) when 400 units have been sold.

54 (*Social Science–Accidents*) The number of traffic accidents per year in a city of population p is predicted to be $T = .002p^{3/2}$. If the population is growing by 500 people a year, find the rate at which traffic accidents will be rising when the population is $p = 40,000$.

55 (*Social Science–Welfare*) The number of welfare cases in a city having population p is expected to be $W = .003p^{4/3}$. If the population is growing by 1000 people per year, find the rate at which welfare cases will be rising when the population is $p = 1,000,000$.

56 (*General–Rockets*) A rocket fired straight up is being tracked by a radar station 3 miles from the launching pad. If the rocket is traveling at 2 miles per second, how fast is the distance between the rocket and the tracking station changing at the moment when the rocket is 4 miles up? (*Hint:* The distance D in the illustration in the next column satisfies $D^2 = 9 + y^2$. To find D, solve $D^2 = 9 + 4^2$.)

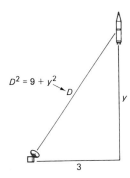

$D^2 = 9 + y^2$

57 (*Biomedical–Poiseuille's Law*) Blood flowing through an artery flows faster in the center of the artery and more slowly near the sides (due to friction). The speed of the blood is $V = c(R^2 - r^2)$, where R is the radius of the artery, r is the distance of the blood from the center of the artery, and c is a constant. Suppose that arteriosclerosis is narrowing the artery at the rate of $dR/dt = -.01$ mm (millimeters) per year. Find the rate at which blood flow is being reduced in an artery whose radius is $R = .05$ *mm*. (*Hint:* Using $c = 500$, find dV/dt, considering r to be a constant. The units of dV/dt will be mm/sec per year.)

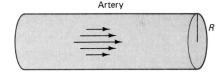

Artery

3.8 REVIEW OF CHAPTER THREE

Introduction

In this chapter we have put calculus to several uses: graphing, optimization, and calculating rates of change. The exercises in this section review all these developments.

Relative Versus Absolute Extremes

What is the difference between relative and absolute extreme values? A function may have several *relative* maximum points (high points compared to their neighbors), but it can have at most one *absolute* maximum (the largest value of the function on its entire domain). Similarly, a function may have several relative minimum points but can have at most one absolute minimum point. *Relative* extremes are used in graphing, and *absolute* extremes are used in optimization (when we want only the highest or lowest y-value).

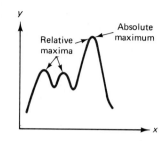

Graphing

To graph a function, make a sign diagram for the first derivative (to show slope and relative extreme points) and a sign diagram for the second derivative (to show concavity and inflection points). For some rational functions we omit the second derivative sign diagram but draw the vertical asymptotes (where the denominator is zero), and horizontal asymptotes (for $x \to \pm\infty$).

Optimization

In optimization we do not want the entire graph, but only a little information from it, the highest and lowest y-values. To optimize a continuous function on an interval we have two methods. Each begins by first finding all critical values.

Second Derivative Test Method

If there is only one critical value $x = c$ in the domain, and

   (i)   if $f''(c) > 0$, then $f(x)$ is *minimized* at $x = c$;

   (ii) If $f''(c) < 0$, then $f(x)$ is *maximized* at $x = c$.

Endpoint and Critical Value Method

For a continuous function defined on a *closed* interval, evaluate the function at all critical values and endpoints. The largest and smallest of these function values are the maximum and minimum values.

In applications, one good strategy is this. First find all critical values in the domain. If there is only one critical value, use the second derivative test. Otherwise, try to define the function on a closed interval and then evaluate it at the critical values and endpoints. If all else fails, sketching the graph will show its highest and lowest y-values (if they exist).

Graphs of Average and Marginal Cost

Several exercises in this chapter involved graphing average cost and marginal cost functions on the same axes, and observing that the marginal cost curve intersects the average cost curve where the average cost is minimized. To see that this is true in general, reason as follows.

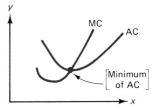

The marginal cost function pierces the average cost function at its minimum.

The average cost function, $AC(x)$, is the total cost $C(x)$ divided by quantity x.

$$AC(x) = \frac{C(x)}{x} \qquad \textit{average cost}$$

To minimize the average cost, we differentiate.

$$AC'(x) = \frac{xC'(x) - C(x)}{x^2} \qquad \textit{using the quotient rule}$$

$$= \frac{1}{x}\left[\frac{xC'(x) - C(x)}{x}\right] \qquad \textit{taking out a 1/x}$$

$$= \frac{1}{x}\left[C'(x) - \frac{C(x)}{x}\right] \qquad \textit{simplifying}$$

this is the average cost

this is the marginal cost

$$= \frac{1}{x}[MC(x) - AC(x)]$$

At the point where the average cost is minimized, its derivative AC' must be zero, which means that the expression in brackets must be zero,

$$MC(x) - AC(x) = 0$$

Therefore,

$$MC(x) = AC(x) \qquad \textit{marginal cost = average cost}$$

This is what we wanted to show.

> At the point where average cost is minimized, marginal cost equals average cost.

This relationship is important in economics. The intuitive reason for this fact is as follows. If marginal cost is lower than average cost, then additional ("marginal") units are cheaper than "average" units, so producing more will bring down the average cost curve until it meets the marginal cost curve. If marginal cost is higher than average cost, then additional units are more expensive than "average" units, so producing more will *raise* the average cost curve until it meets the marginal cost curve.

Practice Test

Exercises 3, 11, 21, 25, 33, 35, 39, 43, 45, 47, 53, 55.

▨ EXERCISES 3.8

Graph each function, showing all relative extreme points and inflection points.

1 $f(x) = x^3 - 3x^2 - 9x + 12$

2 $f(x) = x^3 + 3x^2 - 9x - 7$

3 $f(x) = x^4 - 4x^3 + 15$

4 $f(x) = x^4 + 4x^3 + 17$

5 $f(x) = x(x + 3)^2$

6 $f(x) = x(x - 6)^2$

7 $f(x) = x(x - 4)^3$

8 $f(x) = x(x + 4)^3$

9 $f(x) = \sqrt[7]{x^5} + 1$

10 $f(x) = \sqrt[7]{x^6} + 1$

11 $f(x) = \sqrt[7]{x^4} + 1$

12 $f(x) = \sqrt[7]{x^3} + 1$

Graph each function, showing all relative extreme points.

13 $f(x) = \dfrac{1}{x^2 - 6x}$

14 $f(x) = \dfrac{1}{x^2 + 4x}$

15 $f(x) = \dfrac{x^2}{x^4 + 1}$

16 $f(x) = \dfrac{2x^2}{1 + x^2}$

17 $f(x) = \dfrac{1 - 2x}{x^2}$

18 $f(x) = \dfrac{1 - x}{x^2}$

Find the absolute extreme values of each function on the given interval.

19 $f(x) = 2x^3 - 6x$ on $[0, 5]$

20 $f(x) = 2x^3 - 24x$ on $[0, 5]$

21 $f(x) = x^4 - 4x^3 - 8x^2 + 64$ on $[-1, 5]$

22 $f(x) = x^4 - 4x^3 + 4x^2 + 1$ on $[0, 10]$

23 $h(x) = (x - 1)^{2/3}$ on $[0, 9]$

24 $f(x) = \sqrt{100 - x^2}$ on $[-10, 10]$

25 $g(w) = (w^2 - 4)^2$ on $[-3, 3]$

26 $g(x) = x(8 - x)$ on $[0, 8]$

27 $f(x) = \dfrac{x}{x^2 + 1}$ on $[-3, 3]$

28 $f(x) = \dfrac{x}{x^2 + 4}$ on $[-4, 4]$

▨ APPLIED EXERCISES

29 (*Business–Average and Marginal Cost*) For the cost function $C(x) = 10,000 + x^2$, graph the marginal cost function $MC(x)$ and also the average cost function $AC(x) = C(x)/x$ on the same graph showing that they intersect at the point where the average cost is minimized.

30 (*Social Science–Riots*) Analysis of the 1965 riot in Los Angeles, the 1967 riot in Detroit, and the 1967 riot in Washington shows that the level of riot activity varied with the time of day. For these riots, the average level of activity x hours after midnight was given by $R(x) = x^3 - 6x^2 - 36x + 300$ for $-12 < x < 12$. (Positive values of x correspond to hours after midnight, and negative values of x mean hours before midnight. For example $x = 3$ means 3 A.M., while $x = -3$ means 9 P.M.) At what time was riot activity

(a) the greatest?   **(b)** the least?

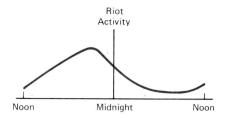

Riot Activity

Noon   Midnight   Noon

31 (*General–Fuel Efficiency*) At what speed should a tugboat travel upstream so as to use the least amount of fuel to reach its destination? If the tugboat's speed through the water is v, and if the speed of the current (relative to the land) is c, then the energy used is proportional to

$$E(v) = \frac{v^4}{v - c}$$

Find the velocity v that minimizes the energy $E(v)$. Your answer will depend upon c, the speed of the current.

32 (*General–Fencing*) A homeowner wants to enclose three adjacent rectangular pens of equal size along a straight wall, as in the diagram on the right. If the side along the wall needs no fence, what is the largest total area that can be enclosed using only 240 feet of fence?

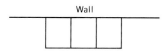

Wall

33 (*General*) A homeowner wants to enclose three adjacent rectangular pens of equal size, as in the diagram on the right. What is the largest total area that can be enclosed using only 240 feet of fence?

34 (*General–Unicorns*) To celebrate the acquisition of Styria in 1261, Ottokar II sent hunters into the Bohemian woods to capture a unicorn. To display the unicorn at court, the king built a rectangular cage. The material for three sides of the cage cost three ducats per running cubit, while the fourth was to be gilded and cost 51 ducats per running cubit. In 1261 it was well known that a happy unicorn requires 2025 square cubits. Find the dimensions that will keep the unicorn happy at the lowest cost.

35 (*General–Box Design*) An open-top box with a square base is to have a volume of exactly 500 cubic inches. Find the dimensions of the box that can be made with the smallest amount of materials.

36 (*General–Packaging*) Find the dimensions of the cylindrical tin can with volume 16π cubic inches that can be made from the least amount of tin. (*Note*: $16\pi \approx 50$ cubic inches.)

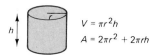

$V = \pi r^2 h$
$A = 2\pi r^2 + 2\pi rh$

37 (*General–Packaging*) Find the dimensions of the open-top cylindrical tin can with volume 8π cubic inches that can be made from the least amount of tin. (*Note*: $8\pi \approx 25$ cubic inches.)

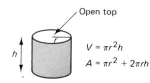

Open top

$V = \pi r^2 h$
$A = \pi r^2 + 2\pi rh$

38 (*General–Bird Flight*) Let v be the flying speed of a bird, and let w be its weight. The power P that the bird must maintain during flight is

$$P = \frac{aw^2}{v} + bv^3$$

where a and b are constants depending on the shape of the bird and the density of the air. Find the speed v that minimizes the power P.

39 (*Business–Maximum Profit*) A computer dealer can sell 12 personal computers per week at a price of $2000 each. He estimates that each $400 price decrease will result in three more sales per week. If the computers cost him $1200 each, what price should he charge to maximize his profit? How many will he sell at that price?

40 (*General–Farming*) A peach tree will yield 100 pounds of peaches now, which will sell for 40 cents a pound. Each week that the farmer waits will increase the yield by 10 pounds, but his selling price will decrease by 2 cents per pound. How long should the farmer wait to pick the fruit in order to maximize his revenue?

41 (*General*) A cable is to connect a power plant to an island, which is 1 mile offshore and 3 miles downshore from the power plant. It costs $5000 per mile to lay a cable underwater and $3000 per mile to lay it underground. If the cost of laying the cable is to be minimized, find the distance x downshore from the island where the cable should meet the land.

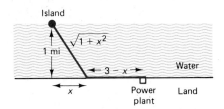

42 (*Business–Production Runs*) A wallpaper company estimates the demand for a certain pattern to be 900 rolls per year. It costs $800 to set up the presses to print the pattern, plus $200 to print each roll. If the company can store a roll of wallpaper for a year at a cost of $4, how many rolls should they print at a time and how many printing runs will they need in a year to minimize production costs?

43 (*Business–Lot Size*) A motorcycle shop estimates that it will sell 500 motorbikes in a year. Each bike costs the store $300, plus a fixed charge of $500 per order. If it costs the store $200 to store a motorbike for a year, what is the order size and how many orders will be needed in a year to minimize inventory costs?

44 (*Maximum Sustainable Yield*) The reproduction function for the North American Duck is estimated to be $f(p) = -.02p^2 + 7p$, where p and $f(p)$ are measured in thousands. Find the size of the population that allows the maximum sustainable yield, and also find the size of the yield.

45 (*Maximum Sustainable Yield*) Fishermen estimate the reproduction function for striped bass in an Each Coast fishing ground to be $f(p) = 60\sqrt{p}$, where p and $f(p)$ are measured in thousands. Find the size of the population that allows the maximum sustainable yield, and also the size of the yield.

For each equation use implicit differentiation to find dy/dx

46 $6x^2 + 8xy + y^2 = 100$ **47** $8xy^2 - 8y = 1$

48 $2xy^2 - 3x^2y = 0$ **49** $\sqrt{x} - \sqrt{y} = 10$

For each equation find dy/dx evaluated at the given values.

50 $x + y = xy$ at $x = 2$, $y = 2$

51 $y^3 - y^2 - y = x$ at $x = 2$, $y = 2$

52 $xy^2 = 81$ at $x = 9$, $y = 3$

53 $x^2y^2 - xy = 2$ at $x = -1$, $y = 1$

54 (*General–Melting Ice*) A cube of ice is melting so that each edge is decreasing at the rate of 2 inches per hour. Find how fast the volume of the ice is decreasing at the moment when each edge is 10 inches long.

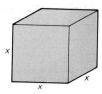

55 (*Business–Profit*) A company's profit from selling x units of an item is $P = 2x^2 - 20x$ dollars. If sales are growing at the rate of 30 per day, find the rate of change of profit when 40 units have been sold.

56 (*Business–Revenue*) A company's revenue from selling x units of a item is $R = x^2 + 500x$ dollars. If sales are increasing at the rate of 50 per month, find the rate of change of revenue when 200 units have been sold.

4

EXPONENTIAL AND LOGARITHMIC FUNCTIONS

Populations are often modeled by exponential functions.

4.1 EXPONENTIAL FUNCTIONS

Introduction

In this chapter we introduce exponential and logarithmic functions, and apply them to a wide variety of problems. We begin with exponential functions, showing how they are used to model the processes of growth and decay.

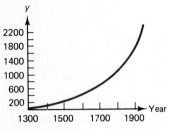

The number of important discoveries and inventions

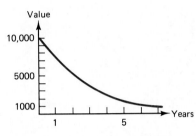

Depreciation of an automobile

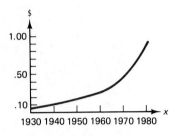

Price of an ice cream cone

We will also define e, a very important mathematical constant.

Calculators

For this chapter you should have a pocket calculator, one with an $\boxed{e^x}$ key for exponentials, an $\boxed{\ln x}$ key for natural logarithms, and a key for powers ($\boxed{x^y}$, $\boxed{y^x}$, or $\boxed{a^x}$ on different calculators).

Exponential Function $y = 2^x$

A function that has a variable in an exponent, such as $f(x) = 2^x$, is called an *exponential function*.

$$f(x) = 2^{\overset{\text{exponent}}{x}}$$
$$\text{base}$$

The table below shows values of the exponential function $f(x) = 2^x$ for various values of x. These points are plotted on the right, showing the graph of $f(x) = 2^x$.

x	$y = 2^x$
-2	$2^{-2} = 1/4$
-1	$2^{-1} = 1/2$
0	$2^0 = 1$
1	$2^1 = 2$
2	$2^2 = 4$
3	$2^3 = 8$

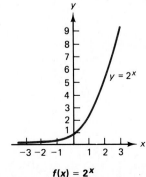

$f(x) = 2^x$

The curve rises very steeply on the right, and approaches the x-axis on the left. Clearly, the exponential function $y = 2^x$ (a constant to a variable power) is quite different from the parabola $y = x^2$ (a variable to a constant power).

Exponential Function $y = (1/2)^x$

The exponential function $f(x) = (1/2)^x$ has base 1/2. The table below shows its values for certain values of x, and its graph is shown on the right.

x	$y = (1/2)^x$
-3	$(1/2)^{-3} = 8$
-2	$(1/2)^{-2} = 4$
-1	$(1/2)^{-1} = 2$
0	$(1/2)^{0} \ \ = 1$
1	$(1/2)^{1} \ \ = 1/2$
2	$(1/2)^{2} \ \ = 1/4$

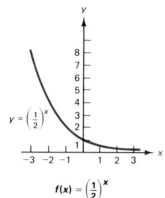

$$f(x) = \left(\frac{1}{2}\right)^{x}$$

Notice that the graph of $y = (1/2)^x$ approaches the x-axis on the right and rises very steeply on the left. It is the mirror image of the curve $y = 2^x$.

Exponential Function $y = a^x$

We can define an exponential function $f(x) = a^x$ for any positive base a. We always take the base to be positive, so for the rest of this section *the letter a will stand for a positive constant*.

Exponential functions with bases $a > 1$ are used to model *growth* (as in populations or savings accounts), and exponential functions with bases $a < 1$ are used to model *decay* (as in depreciation).

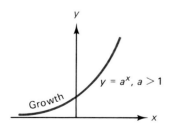

$y = a^x$ slopes upward for base $a > 1$.

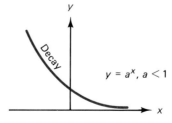

$y = a^x$ slopes downward for $a < 1$.

Population Growth

Many populations grow exponentially.

Example 1 Suppose that in a population of rabbits we count only the females, and that each female rabbit gives birth to four more female rabbits in each generation. How many female rabbits in the tenth generation will be descended from the original female rabbit?

Solution Let $p(n)$ be the size of the nth generation ($n = 1, 2, \ldots, 10$) descended from the original rabbit. The first generation will consist of four rabbits.

$$p(1) = 4 \qquad \qquad \textit{first generation of size 4}$$

Each of these four will give birth to four more in the second generation.

$$p(2) = 4 \cdot 4 = 4^2 = 16 \qquad \qquad \textit{second generation of size } 4^2$$

Each of these 16 will give birth to four more in the third generation.

$$p(3) = 16 \cdot 4 = 4^3 = 64 \qquad \qquad \textit{third generation of size } 4^3$$

Clearly, the nth generation will be of size $p(n) = 4^n$, so for the tenth generation

$$p(10) = 4^{10} = 1{,}048{,}576$$

That is, the original rabbit will have more than one million descendants in the tenth generation. ■

Of course, no actual population could grow at this rate for very long. Limitations of food and space would soon restrict the growth.

Exponential growth governs human populations as well, but of course at slower growth rates. It was exactly this picture of exponential population growth, together with food supplies that increase much more slowly, that caused the great nineteenth-century essayist Thomas Carlyle to dub economics the "dismal science." He was commenting not on how interesting economics is, but on the grim conclusions that follow from populations outstripping their food supplies.

Compound Interest

Money invested at compound interest grows exponentially. (The word "compound" means that the interest is added to the account, earning more interest.) For example, if you invest P dollars (the "principal") a 8% interest compounded annually, then after 1 year you will have P dollars plus 8% of P dollars.

$$\begin{pmatrix} \text{Value after} \\ \text{1 year} \end{pmatrix} = P + .08P = P(1 + .08)$$

Notice that increasing a quantity by 8% is the same as multiplying it by

(1 + .08). Therefore, to find the value after n years of compounding, we simply multiply by (1 + .08) n times.

$$\left(\begin{matrix}\text{Value after}\\ n \text{ years}\end{matrix}\right) = \overbrace{P(1 + .08)\cdot(1 + .08)\cdot \ \ldots \ \cdot(1 + .08)}^{n \text{ times}} = P(1 + .08)^n$$

The .08 can be replaced by any interest rate r.

Compound Interest

For P dollars invested at interest rate r per period,

$$\left(\begin{matrix}\text{value after}\\ n \text{ periods}\end{matrix}\right) = P(1 + r)^n$$

Remember, however, that banks always state *annual* interest rates, but the compounding may be done more frequently, such as quarterly or daily. The r in the formula above is the interest rate *per compounding period,* and the n is the number of *periods.* For example, if a bank offers 8% compounded *quarterly,* the interest must be calculated on a *quarterly* basis: the annual 8% is equivalent to a *quarterly* rate of 2% (one quarter of 8%), so $r = .02$, and n is the number of *quarters.*

Example 2 Find the value of $1000 invested for 2 years at 8% compounded quarterly.

Solution The principal is $P = 1000$. Since the compounding is done quarterly, the interest rate *per period* is one quarter of the annual rate.

$$r = \frac{1}{4}\cdot 8\% = 2\% = .02$$

The number of quarters in 2 years is $n = 8$, so the compound interest formula gives

$$P(1 + r)^n = 1000\cdot(1 + .02)^8 = 1000\cdot(1.02)^8 \approx \$1171.66$$

We may interpret the formula $P\cdot(1 + .02)^8$ intuitively as follows. Multiplying the principal by (1 + .02) means that you keep the original amount (the "1") plus some interest (the .02), and the exponent 8 means that this is done a total of 8 times.

Practice Exercise 1

Find the value of $2000 invested for 2 years at 36% compounded monthly. (*Hint*: Convert the 36% to a monthly rate and let n be the number of months in 2 years. You will need a calculator.)

(solution on page 216)

Present Value

The value to which a sum will grow under compound interest could be called its *future value*. That is, Example 2 showed that the future value of $1000 (at 8% compounded quarterly for 2 years) is $1171.66.

Reversing the order, we can speak of the *present value* of a future payment. Example 2 shows that if a payment of $1171.66 is to be made in 2 years, its *present value* is $1000 (at this interest rate). That is, a promise to pay you $1171.66 in two years is worth exactly $1000 now, since $1000 deposited in a bank now would be worth that much in two years (at the stated interest rate). To find the *future* value we *multiply* P by $(1 + r)^n$, and therefore to find the *present* value, we simply *divide* P by $(1 + r)^n$.

Present Value

At interest rate r per period for n interest periods,

$$\binom{\text{present value}}{\text{of } P \text{ dollars}} = \frac{P}{(1 + r)^n}$$

Example 3 Find the present value of $5000 to be paid 8 years from now, at 10% interest compounded semiannually.

Solution The principal is $P = 5000$, semiannual (half-yearly) compounding means that $r = .05$ (half of 10%), and the number of interest periods is $n = 16$ (the number of half-years in 8 years). The present value formula gives

$$\frac{P}{(1 + r)^n} = \frac{5000}{(1 + .05)^{16}} = \frac{5000}{1.05^{16}} = \$2290.56 \qquad \textit{using a calculator}$$

Answer

Depreciation by a Fixed Percentage

Depreciation by a fixed percentage means that a piece of equipment loses a fixed percentage (say 30%) of its value each year. Losing a percentage of value is like compound interest but with a *negative* interest rate. Therefore, we use the compound interest formula, $P(1 + r)^n$, but for depreciation r is *negative*.

Example 4 A car worth $10,000 depreciates in value by 40% each year. How much is it worth after 3 years?

SEC. 4.1 EXPONENTIAL FUNCTIONS ■ **211**

Solution The car loses 40% of its value each year, which is equivalent to an interest rate of *negative* 40%. The compound interest formula with $P = 10,000$, $r = -.40$, and $n = 3$ gives

$$P(1 + r)^n = 10,000(1 - .40)^3$$

$$= 10,000(.60)^3 = \$2160$$

using a
calculator

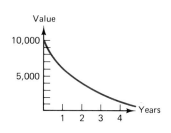

Practice Exercise 2

A printing press, originally worth $50,000, loses 20% of its value each year. What is its value after 4 years?

(solution on page 216)

Fixed Percentage Versus Straight-Line Depreciation

The graph in Example 4 shows that depreciation by a fixed percentage is quite different from "straight-line" depreciation. Under straight-line depreciation the same *dollar* value is lost each year, while under fixed percentage depreciation the same *percentage* of value is lost each year, resulting in larger dollar losses in the early years and smaller dollar losses in later years. Depreciation by a fixed percentage is one type of "accelerated" depreciation. The method of depreciation that one uses depends on how one chooses to estimate value, and in practice is often determined by the tax laws.

The Number e

Imagine that a bank offers 100% interest, and that you deposit $1 for 1 year. Let us see how the value changes under different types of compounding.

For *annual* compounding, your $1 would in a year grow to $2 (the original dollar plus a dollar interest).

For *quarterly* compounding, we would use the compound interest formula with $P = 1$, $n = 4$, and $r = \frac{1}{4} \cdot 100\% = 25\% = .25$.

$$P(1 + r)^n = 1(1 + .25)^4 = (1.25)^4 \approx 2.44$$

using a calculator

or 2 dollars and 44 cents, an improvement of 44 cents over annual compounding.

For *daily* compounding the value after a year would be

$$\left(1 + \frac{1}{365}\right)^{365} \approx 2.71$$

$n = 365$ *periods*
$r = \dfrac{100\%}{365} = \dfrac{1}{365}$

an increase of 27 cents over quarterly compounding. Clearly, if the interest rate, the principal, and the amount of time stays the same, the value increases as the compounding is done more frequently.

In general, if the compounding were done n times a year, the value of the dollar after a year will be

$$\left(\begin{matrix} \text{value of \$1 after 1 year at 100\%} \\ \text{interest compounded } n \text{ times a year} \end{matrix} \right) = \left(1 + \frac{1}{n} \right)^n$$

The following table shows the value of $(1 + 1/n)^n$ for various values of n.

Value of \$1 at 100% Interest Compounded n Times a Year for 1 Year

Compounding	n	$(1 + 1/n)^n$	Answer (to five decimal places)
Annually	1	$\left(1 + \frac{1}{1} \right)^1 =$	2.00000
Semiannually	2	$\left(1 + \frac{1}{2} \right)^2 =$	2.25000
Quarterly	4	$\left(1 + \frac{1}{4} \right)^4 \approx$	2.44141
Daily	365	$\left(1 + \frac{1}{365} \right)^{365} \approx$	2.71457
Hourly	8760	$\left(1 + \frac{1}{8760} \right)^{8760} \approx$	2.71812
	10,000	$\left(1 + \frac{1}{10,000} \right)^{10,000} \approx$	2.71815
	100,000	$\left(1 + \frac{1}{100,000} \right)^{100,000} \approx$	2.71825
	1,000,000	$\left(1 + \frac{1}{1,000,000} \right)^{1,000,000} \approx$	2.71828
	10,000,000	$\left(1 + \frac{1}{10,000,000} \right)^{10,000,000} \approx$	2.71828

Notice that the numbers in the right-hand column change less and less as n becomes larger and larger. For example, the last five numbers agree to three decimal places, and the last two numbers agree to all five decimal places. That is, as n becomes larger and larger the value of $(1 + 1/n)^n$ approaches a definite limit, which, to five decimal places, is 2.71828. This particular number is very important in mathematics, and is given the name e (just as the number $3.14159\ldots$ is given the name π).

$$e = \lim_{n \to \infty} \left(1 + \frac{1}{n}\right)^n \approx 2.71828 \ldots$$

n → ∞ is read "n approaches infinity." The dots mean that the decimal expansion goes on forever.

The same e appears in statistics, in the formula for the "bell-shaped" or "normal" curve. Its value has been calculated to several hundred decimal places. Its value to 15 decimal places is $e \approx 2.718281828459045$.

Continuous Compounding of Interest

This kind of compound interest, with the compounding frequency approaching infinity, is called *continuous* compounding. We have shown that $1 at 100% interest compounded continuously for 1 year would be worth precisely e dollars (about $2.71). The formula for continuous compound interest at other rates is as follows.

Continuous Compounding

For P dollars invested at interest rate r compounded continuously for n years,

$$\left(\begin{array}{c}\text{value after}\\ n \text{ years}\end{array}\right) = Pe^{rn}$$

A justification for this formula is given at the end of the section. For calculating e to a power we use the $\boxed{e^x}$ key on a calculator, or tables.

Example 5   Find the value of $1000 at 8% interest compounded continuously for 20 years.

Solution   We use the formula Pe^{rn} with $P = 1000$, $r = .08$, and $n = 20$.

$$Pe^{rn} = 1000 \cdot e^{(.08)(20)} = 1000 \cdot e^{1.6} \approx \$4953.00 \leftarrow$$

$$\underbrace{\phantom{1000 \cdot e^{1.6}}}_{4.95303} \qquad \text{answer}$$

(by pressing 1.6, then $\boxed{e^x}$ on a calculator)   ■

Present Value with Continuous Compounding

As before, the value that a sum will attain in n years could be called its *future* value, and the current value of a future sum is its *present* value. Under continuous compounding, to find future value we multiply P by e^{rn} so to find *present* value we *divide* by e^{rn}.

Present Value under Continuous Compounding

At interest rate r compounded continuously for n years,

$$\left(\begin{array}{c}\text{present value}\\ \text{of } P \text{ dollars}\end{array}\right) = \frac{P}{e^{rn}} = Pe^{-rn}$$

Example 6 The present value of $5000 to be paid in 10 years, at 7% interest compounded continuously, is

$$\frac{5000}{e^{(.07)(10)}} = \frac{5000}{e^{.7}} \approx \$2482.93$$

■ *using a calculator*

Intuitive Meaning of Continuous Compounding

Under quarterly compounding, your money is, in a sense, earning interest throughout the quarter, but the interest is not added to your account until the end of the quarter. Under continuous compounding, the interest is added to your account *as it is earned,* with no delay. The extra earnings in continuous compounding come from this "instant crediting" of interest, since then your interest starts earning more interest immediately.

How to Compare Interest Rates

Notice that in the compound interest formula $P(1 + r)^n$ the interest rate r always has a 1 added to it. For example, an interest rate of 5% ($r = .05$) appears as $P(1.05)^n$. Therefore, to recover an interest rate from the compound interest formula we must subtract the 1. For example, an interest rate of 6% compounded quarterly will, over a year's time, increase a deposit of P dollars to

$$P\left(1 + \frac{.06}{4}\right)^4 = P(1.015)^4 = P(1.0614)$$

using a calculator

The principal P is multiplied by 1.0614, and to find the actual gain we subtract the 1 (giving .0614) and express the result as a percent (6.14%). Therefore, the original deposit is really being increased by 6.14%. That is, 6% compounded quarterly is equivalent to an *annual yield* of 6.14%. This is the way to compare two interest rates that are compounded differently: convert them both to annual yields.

In banking, the annual yield, 6.14%, is called the *effective* rate of interest, while the stated rate, 6%, is called the *nominal* rate of interest. The Truth in

Lending Act passed by Congress in 1968 requires that all contracts involving interest state the effective rate of interest.

Example 7   A bank offers 5% interest compound continuously. Calculate the effective rate of interest.

Solution   We use the continuous compounding formula Pe^{rn} with $r = .05$ and $n = 1$.

$$Pe^{rn} = Pe^{.05} \approx P(1.0513) \qquad \qquad \text{using a calculator}$$

P is multiplied by 1.0513, and subtracting 1 gives .0513, which expressed as a percent is 5.13%.

Answer: The effective rate of interest is 5.13%.   ■

The Function $y = e^x$

The number e gives us a new exponential function $f(x) = e^x$. This function is used extensively in business, economics, and all areas of science. The table below shows the values of e^x for various values of x. These values lead to the graph of $f(x) = e^x$ shown on the right.

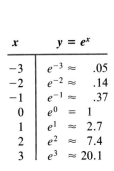

x	$y = e^x$
-3	$e^{-3} \approx$ .05
-2	$e^{-2} \approx$ .14
-1	$e^{-1} \approx$ .37
0	$e^0 =$ 1
1	$e^1 \approx$ 2.7
2	$e^2 \approx$ 7.4
3	$e^3 \approx$ 20.1

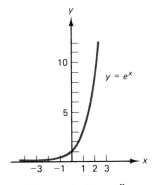

The graph of $y = e^x$

Note that e^x is *positive* for all values of x, even when x is negative.

The Function $y = e^{kx}$

On the following page the function $f(x) = e^{kx}$ is graphed for various values of the constant k. For positive values of k the curve rises, and for negative values of k the function falls (as you move to the right). For higher values of k the curve rises more steeply.

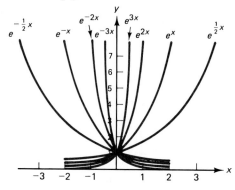

$f(x) = e^{kx}$ for various values of k.

Exponential Growth

All exponential growth, whether continuous or discrete, has one common characteristic: the amount of growth is proportional to the size of the quantity. For example, the interest that a bank account earns is proportional to size of the account, and the growth of a population is proportional to the size of the population. This is in contrast, for example, to a person's height, which does not increase exponentially. That is, exponential growth occurs in those situations when a quantity grows *in proportion to its size*.

Justification of the Formula for Continuous Compounding

The compound interest formula Pe^{rn} is derived as follows. P dollars invested for n years at interest rate r compounded k times a year yields

$$P\left(1 + \frac{r}{k}\right)^{kn}$$

Define m by $m = k/r$, so that $k = rm$. Replacing k by rm and letting k (and therefore m) approach ∞, this becomes

$$P\left(1 + \frac{r}{rm}\right)^{rmn} = P\left[\underbrace{\left(1 + \frac{1}{m}\right)^{m}}_{\substack{\text{approaches} \\ e \text{ as } m \to \infty}}\right]^{rn}$$

As $m \to \infty$ this quantity approaches Pe^{rn}, which is the continuous compounding formula.

SOLUTIONS TO PRACTICE EXERCISES

1   $2000(1 + .03)^{24} = 2000(1.03)^{24} = 2000(2.032794) = \$4065.59$

2   $50{,}000(1 - .20)^4 = 50{,}000(.8)^4 = 50{,}000(.4096) = \$20{,}480$

■ **EXERCISES 4.1** (Most need 🖩 .)

Graph each function.

1 $y = 3^x$

2 $y = 5^x$

3 $y = \left(\dfrac{1}{3}\right)^x$

4 $y = \left(\dfrac{1}{5}\right)^x$

Calculate each value of e^x using a calculator.

5 $e^{1.74}$

6 $e^{2.15}$

7 $e^{-.05}$

8 $e^{-.09}$

■ **APPLIED EXERCISES**

9 (*Business–Interest*) Find the value of $1000 deposited in a bank at 10% interest for 8 years compounded

(a) annually

(b) quarterly

(c) continuously

10 (*Business–Interest*) Find the value of $1000 deposited in a bank at 12% interest for 8 years compounded

(a) annually

(b) quarterly

(c) continuously

11 (*General–Interest*) A loan shark lends you $100 at 2% compound interest per week (this is a *weekly*, not an annual rate).

(a) How much will you owe after 3 years?

(b) In "street" language, the profit on such a loan is known as the "vigorish" or the "vig." Find the shark's vig.

12 (*General–Compound Interest*) In 1626, Peter Minuit purchased Manhattan Island from the Indians for $24 worth of trinkets and beads. Find what the $24 would be worth in the year 1990 if it had been deposited in a bank paying 5% interest compounded quarterly.

13 (*General–Interest*) Find the error in the advertisement shown below, which appeared in a New York newspaper. (*Hint*: Check that the nominal rate is equivalent to the effective rate. For daily compounding some banks use 365 days and some use 360 days in the year. Try it both ways.)

> **If you're really serious about saving, open a Savings Account at T&M Bank**
>
> At T&M Bank, flexibility is the key word. You can choose the length of time and the amount of your deposit, which will earn an annual yield of 9.825% based on a rate of 9.25% compounded daily.
>
> **T&M Bank**   Member FDIC

14 (*General–Interest*) Find the error in the advertisement shown below, which appeared in a Washington, D.C. newspaper. Assume that the compounding is done daily. (*Hint*: Check that the nominal rate is equivalent to the effective rate. For daily compounding some banks use 365 days and some use 360 days in the year. Try it both ways.)

> **Your Money's in 7 Heaven**
>
> Get Hans Johnson's "sky high" return. No time restrictions or withdrawal penalties! Funds available when you want them.
>
> Current Annual Yield: 7.19%   7% Regular Passbooks
>
> **Hans Johnson**
> SAVINGS & LOAN INC.

15 (*General–Present Value*) A rich uncle wants to make you a millionaire. How much money must he deposit in a trust fund paying 8% compounded quarterly at the time of your birth to yield $1,000,000 when you retire at age 60?

16 (*General–Present Value*) If a college education costs $50,000, how large a trust fund must be established at a child's birth (paying 8% compounded quarterly) to ensure sufficient funds at age 18?

17 (*General–Compound Interest*) Which is better: 10% interest compounded quarterly or 9.8% compounded continuously?

18 (*General–Compound Interest*) Which is better: 8% interest compounded quarterly or 7.8% compounded continuously?

19 (*Business–Depreciation*) A $15,000 automobile depreciates by 35% per year. Find its value after

(a) 4 years   (b) 6 months

20 (*Business–Appreciation*) A $10,000 painting appreciates in value by 12% each year. Find its value after 16 years.

21 (*General–Population*) In 1985 the population of the world was approximately 5 billion people and growing at 1% annually. Assuming that this growth rate continues, find the world population in the year 2055.

22 (*General–Population*) The most populous country is China, with a (1980) population of 1 billion, and growing at the rate of 1.8% a year. The 1980 population of India was 710 million, growing at 2.1% a year. Assuming that these growth rates continue, which will be larger in the year 2100?

23 (*General–Nuclear Meltdown*) According to the Nuclear Regulatory Commission,* the probability of a "severe core meltdown accident" at a nuclear reactor in the United States within n years of 1985 is

$$P(n) = 1 - (.9997)^{100n}$$

Find the probability of a meltdown

(a) by the year 2010

(b) by the year 2025

(The 1986 core meltdown in the Chernobyl reactor in the Soviet Union spread radiation over much of Eastern Europe, leading to an undetermined number of fatalities.)

---

* *New York Times*, April 17, 1985, p. A16.

24 (*General–Mosquitoes*) Female mosquitoes (*Culex pipiens*) feed on blood (only the females drink blood) and then lay several hundred eggs. In this way each mosquito can, on the average, breed another 300 mosquitoes in about 9 days. Find the number of great-grandchildren mosquitoes that will be descended from one female mosquito, assuming that all eggs hatch and mature.

Female mosquito as seen with a scanning electron microscope

25 (*Biomedical–Light*) According to the Bouguer–Lambert law, the proportion of light that penetrates ordinary seawater to a depth of x feet is $e^{-.44x}$. Find the proportion of light that penetrates to a depth of

(a) 3 feet   (b) 10 feet

26 (*Biomedical–Drug Dosage*) If a dosage d of a drug is administered to a patient, the amount of the drug remaining in the bloodstream t hours later will be

$$f(t) = d \cdot e^{-kt}$$

where k (the "absorption constant") depends upon the drug and the age of the patient. If the dose is $d = 1.2$ cc and the absorption constant is $k = .05$, find the amount of the drug remaining in the patient's bloodstream after

(a) 3 hours   (b) 6 hours

27 (*Business–Advertising*) A company finds that x days after the conclusion of an advertising campaign the daily sales of a new product are $S(x) = 100 + 800e^{-.2x}$. Find the daily sales 10 days after the advertising campaign.

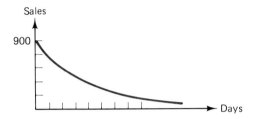

28 (*Business–Quality Control*) A company finds that the proportion of its light bulbs that will burn continuously for longer than t weeks is $p(t) = e^{-.01t}$. Find the proportion of bulbs that burn for longer than 10 weeks.

29 (*General–Temperature*) A mug of coffee originally at 200 degrees, if left for t hours in a room whose temperature is 70 degrees, will cool to a temperature of $T(t) = 70 + 130e^{-1.8t}$ degrees. Find the temperature of the coffee after

(a) 15 minutes   (b) half an hour

30 (*Behavioral Science–Learning*) In certain experiments the percentage of items that are remembered after t times units is

$$p(t) = 100 \frac{1 + e}{1 + e^{t+1}}$$

Such curves are called "forgetting" curves. Find the percentage remembered after

(a) 0 time units   (b) 2 time units

31 (*Biomedical–Epidemics*) The Reed–Frost model for the spread of an epidemic predicts that the number I of newly infected people is $I = S(1 - e^{-rx})$, where S is the number of susceptible people, r is the effective contact rate, and x is the number of infectious people. Suppose that a school reports an outbreak of measles with $x = 10$ cases, and that the effective contact rate is $r = .01$. If the number of susceptibles is $S = 400$, use the Reed–Frost model to estimate how many students will be newly infected during this stage of the epidemic.

32 (*Social Science–Election Cost*) The cost of winning a seat in the House of Representatives in recent years has been approximately $C(x) = 87e^{.16t}$ thousand dollars, where x is the number of years since 1976. Estimate the cost of winning a House seat in the year 1996.

4.2   LOGARITHMIC FUNCTIONS

Introduction

In this section we introduce logarithmic functions, emphasizing the *natural* logarithm function. We apply natural logarithms to a wide variety of problems, from doubling of money under compound interest to drug dosage.

Common Logarithms

The word *logarithm* (abbreviated "log") means *power* or *exponent*. The expression

$$\log_{10} 1000 \qquad\qquad \text{\textit{log (base 10) of 1000}}$$

means the exponent of 10 that gives 1000. Since $10^3 = 1000$, the exponent is 3, so the log is 3.

$$\log_{10} 1000 = 3 \qquad\qquad \text{\textit{since } } 10^3 = 1000$$

Example 1 Find $\log_{10} 100$.

Solution

$$\log_{10} 100 = y \qquad \text{is equivalent to} \qquad 10^y = 100$$

the logarithm y is the exponent that solves

Since the exponent is 2, the answer is 2.

$$\log_{10} 100 = 2 \qquad \text{because} \qquad 10^2 = 100$$

answer

■

The subscript 10 in $\log_{10} 100 = 2$ is called the *base* of the logarithm (since it is the number that is raised to the power).

Example 2 Find $\log_{10} \dfrac{1}{10}$.

Solution

$$\log_{10} \frac{1}{10} = y \qquad \text{is equivalent to} \qquad 10^y = \frac{1}{10}$$

the logarithm y is the exponent that solves

Since the exponent is -1, the answer is -1.

$$\log_{10} \frac{1}{10} = -1 \qquad \text{because} \qquad 10^{-1} = \frac{1}{10}$$

answer

■

Practice Exercise 1

Find

(a) $\log_{10} 10,000$ (b) $\log_{10} \dfrac{1}{100}$

(*Hint*: Find the correct exponent of 10.)   *(solutions on page 230)*

Logarithms to the base 10 are called *common* logarithms. We may calculate logarithms to other bases as well.

Logarithms to Other Bases

We may use *any* positive number (except 1) as a base for logarithms.

Example 3   Find $\log_2 32$.

Solution

$$\log_2 32 = y \qquad \text{is equivalent to} \qquad 2^y = 32$$

Since $2^5 = 32$, the exponent is 5, so the log is 5.

$$\log_2 32 = 5 \qquad\qquad \textit{since } 2^5 = 32 \qquad ▪$$

Example 4   Find $\log_9 3$.

Solution

$$\log_9 3 = y \qquad \text{is equivalent to} \qquad 9^y = 3$$

Since $9^{1/2} = \sqrt{9} = 3$, the exponent is 1/2, so the log is 1/2.

$$\log_9 3 = \frac{1}{2} \qquad\qquad \textit{since } 9^{1/2} = 3 \qquad ▪$$

In general, for any positive base a other than 1,

$$\log_a x = y \quad \text{is equivalent to} \quad a^y = x$$

In words,

$$\log_a x \text{ means the exponent of } a \text{ that gives } x.$$

Practice Exercise 2

Find

(**a**) $\log_5 125$ (**b**) $\log_8 2$ (**c**) $\log_2 \dfrac{1}{4}$ *(solutions on page 230)*

Natural Logarithms

The most widely used of all bases is e, the number (approximately 2.718) that we defined in Section 4.1. Logarithms to the base e are called *natural* or *Napierian logarithms*.* The natural logarithm of x is written $\ln x$ ("n" for "natural") instead of $\log_e x$.

$$\ln x = \text{the logarithm of } x \text{ to the base } e$$

---

* After John Napier, a seventeenth-century Scottish mathematician and, incidentally, the inventor of the decimal point.

Natural logarithms are most easily found using the $\boxed{\ln \text{ x}}$ (or $\boxed{\ln}$) key on a calculator. (The $\boxed{\log}$ key gives base 10 logarithms. Natural logarithms can also be found in tables.)

Practice Exercise 3

Use a calculator to find ln 8.34.   *(solution on page 231)*

The Graph of the Natural Logarithm Function

The table below shows some values of the natural logarithm function

$$f(x) = \ln x$$

These values are plotted on the right, showing the graph of $y = \ln x$.

x	$y = \ln x$
.1	−2.3
.5	− .69
1	0
2	.69
5	1.6
10	2.3
15	2.7

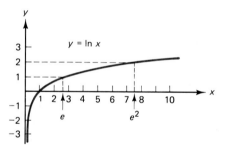

The graph of the natural logarithm function

The natural logarithm function increases very gradually for higher values of x. It is used for modeling growth that continually slows. Notice that ln x is defined only for $x > 0$. This is because e to any power is positive.

The graph below shows logarithm functions of several different bases. Notice that each is defined only for $x > 0$, and that they all pass through the point $(1, 0)$, since $a^0 = 1$.

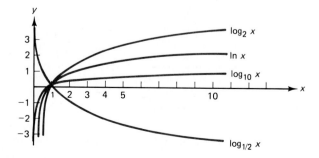

Properties of Natural Logarithms

Since logs are exponents, each of the laws of exponents can be restated as a law of logarithms. The first three properties show that some natural logarithms can be found without using a calculator.

	In Words	**Exponential Form**
1. $\ln 1 = 0$	the natural log of 1 is 0 because	$e^0 = 1$
2. $\ln e = 1$	the natural log of e is 1 because	$e^1 = e$
3. $\ln e^x = x$	the natural log of e to a because power is just the power	$e^x = e^x$

Note that the first two are special cases of the third, with $x = 0$ and $x = 1$.

Example 5

(a) $\ln e^7 = 7$

(b) $\ln e^{3x} = 3x$

(c) $\ln \sqrt[3]{e^2} = \ln e^{2/3} = \frac{2}{3}$

the natural log of e to a power is just the power

▪

The next three properties enable us to simplify logs of products, quotients, and powers. For any positive numbers M and N,

4. $\ln (MN) = \ln M + \ln N$	*the log of a product is the sum of the logs*
5. $\ln \left(\dfrac{M}{N}\right) = \ln M - \ln N$	*the log of a quotient is the difference of the logs*
6. $\ln (M^N) = N \cdot \ln M$	*the log of a number to a power is the power times the log*

A justification for these three properties is given at the end of this section.

Example 6

(a) $\ln (2 \cdot 3) = \ln 2 + \ln 3$ *property 4*

(b) $\ln \dfrac{2}{3} = \ln 2 - \ln 3$ *property 5*

(c) $\ln (2^3) = 3 \ln 2$ *$\ln (2^3) = 3 \ln 2$*

▪

Property 6, $\ln(M^N) = N \cdot \ln M$, which is illustrated in Example 6c, can be stated simply in words: *logarithms bring down exponents.*

These properties are very useful for simplifying functions, as the following examples show.

Example 7

$$f(x) = \ln(2x) - \ln 2$$

$$= \ln 2 + \ln x - \ln 2 \qquad \text{\textit{since} } ln(2x) = ln 2 + ln x$$

by property 4

$$= \ln x \qquad \text{\textit{canceling}} \qquad ■$$

Example 8

$$f(x) = \ln\left(\frac{x}{e}\right) + 1$$

$$= \ln x - \ln e + 1 \qquad \text{\textit{since} } ln(x/e) = ln x - ln e$$

by property 5

$$= \ln x - 1 + 1 \qquad \text{\textit{since} } ln e = 1 \text{ \textit{by property 2}}$$

$$= \ln x \qquad \text{\textit{canceling}} \qquad ■$$

Example 9

$$f(x) = \ln(x^5) - \ln(x^3)$$

$$= 5\ln x - 3\ln x \qquad \text{\textit{bringing down exponents by property 6}}$$

$$= 2\ln x \qquad \text{\textit{simplifying}} \qquad ■$$

Practice Exercise 4

Simplify

(a) $f(x) = \ln(4x) - \ln 4$

(b) $f(x) = \ln(x/2) - \ln x$

(c) $f(x) = 12\ln(\sqrt{x^3}) - \ln(x^{10})$ *(solutions on page 231)*

Ln x and eˣ Are Inverse Functions

Property 3, $\ln e^x = x$,   shows that raising e to the power x and then taking the natural logarithm of the result gives back the original number x. That is, the natural logarithm function "undoes" or "inverts" the function e^x. This inverse relationship also holds in the opposite order.

$$e^{\ln x} = x$$

e to the natural log of a number is just the number

Example 10

$$e^{\ln 7} = 7 \qquad ■$$

The natural log function ln x and the exponential function e^x are *inverse functions:* either "undoes" the other. This inverse relationship is equivalent to the geometric fact that the graphs of $y = e^x$ and $y = \ln x$ are reflections of each other in the diagonal line $y = x$.

Intuitively, the reason that these functions are inverses is that e^x raises x up to the exponent, and the natural log function brings down exponents, so either function "undoes" the other.

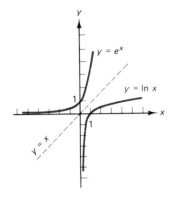

$y = e^x$ and $y = \ln x$
are inverse functions.

Doubling Under Compound Interest

How soon will money invested at compound interest double in value? The solution to this question makes important use of the property
$\ln (M^N) = N \ln M$ ("logs bring down exponents").

Example 11   A sum is invested at 12% interest compounded quarterly. How soon will it double in value?

Solution   We use the formula $P(1 + r)^n$ with (quarterly) interest rate $r = \frac{1}{4} \cdot 12\% = 3\% = .03$. Since double P dollars is $2P$ dollars, we want to solve

$$P(1 + .03)^n = 2P \qquad\qquad\qquad\qquad \text{\itshape n is the number of quarters}$$

$$\underset{\text{\footnotesize P dollars doubled}}{\underline{\hspace{1.5cm}}}$$

$$1.03^n = 2 \qquad\qquad\qquad\qquad \text{\itshape canceling the Ps and simplifying}$$

The variable is in the *exponent,* so we take logarithms to bring it down.

$$\ln (1.03^n) = \ln 2 \qquad\qquad \text{\itshape taking natural logs of both sides}$$

$$n \ln 1.03 = \ln 2 \qquad\qquad \text{\itshape bringing down the exponent (property 6 of logarithms)}$$

$$n = \frac{\ln 2}{\ln 1.03} \qquad\qquad \text{\itshape dividing each side by ln 1.03 to solve for n}$$

$$\approx \frac{.6931}{.0296} \approx 23.4 \qquad\qquad \text{\itshape evaluating the logs using a calculator (answer means 23.4 quarters)}$$

Since n is in *quarters* we divide by 4 to convert to years, $\dfrac{23.4}{4} \approx 5.9$.

Answer: A sum at 12% compounded quarterly doubles in about 5.9 years.

■

Of course, for quarterly compounding you will not get the $2P$ dollars until the end of the quarter, by which time it will be slightly more.

Notice that the principal P canceled out after the first step. This shows that *any* sum will double in the same amount of time: one dollar will double into two dollars in exactly the same time that a million dollars will double into two million dollars.

To find how soon the value would *triple,* we would solve $P(1 + r)^n = 3P$, and similarly for any other multiple. To find how soon the value would *increase by 50%* (that is, to become 1.5 times its original value) we would solve

$$P(1 + r)^n = 1.5P \qquad\qquad \textit{multiplying by 1.5 increases P by 50\%}$$

For the value to increase by 25% we would solve

$$P(1 + r)^n = 1.25P \qquad\qquad \textit{multiplying by 1.25 increases P by 25\%}$$

Practice Exercise 5

A sum is invested at 10% interest compounded semiannually (twice a year). How soon will it increase by 60%? *(solution on page 231)*

Example 12 A sum is invested at 15% compounded continuously. How soon will it triple?

Solution The idea is the same as before, but for continuous compounding we use the formula Pe^{rn}.

$$Pe^{.15n} = 3P \qquad\qquad \textit{3P since the value triples}$$

$$e^{.15n} = 3 \qquad\qquad \textit{canceling the Ps}$$

$$\underbrace{\ln e^{.15n}}_{.15n} = \ln 3 \qquad\qquad \textit{taking the natural log of both sides}$$

$$.15n = \ln 3 \qquad\qquad \textit{since the natural log of e to a power is just the power}$$

$$n = \frac{\ln 3}{.15} \qquad\qquad \textit{dividing by .15 to solve for n}$$

$$\approx \frac{1.0986}{.15} \approx 7.3 \qquad\qquad \textit{using a calculator}$$

Answer: A sum at 15% compounded continuously triples in about 7.3 years.

■

Logarithms in Biology and Medicine

The amount of a drug that remain in a person's bloodstream decreases exponentially with time. If the initial concentration is c (milligrams per milliliter of blood), the concentration t hours later will be

$$C(t) = ce^{-kt}$$

where the "absorption constant" k measures how rapidly the drug is absorbed.

Every medicine has a minimum concentration below which it is not effective. When the concentration falls to this level, another dose should be administered. If doses are administered regularly over a period of time, the concentration will look like this.

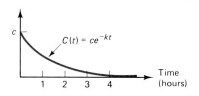

Drug concentration in the blood

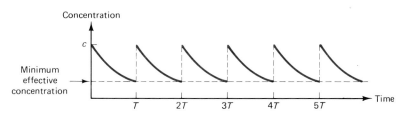

Drug concentration with repeated doses

The problem is to determine the time T at which the dose should be repeated so as to maintain an effective concentration.

Example 13   The absorption constant for penicillin is $k = 0.11$, and the minimum effective concentration is 2. If the original concentration is $c = 5$, find when another dose should be administered in order to maintain an effective concentration.

Solution   The concentration formula $C(t) = ce^{-kt}$ with $c = 5$ and $k = .11$ is

$$C(t) = 5e^{-.11t}$$

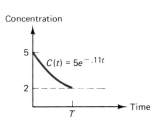

To find the time when this concentration reaches the minimum effective level of 2, we solve

$$5e^{-.11t} = 2$$

$$e^{.11t} = .4 \qquad \text{\textit{dividing by 5}}$$

$$\underbrace{\ln e^{-.11t}}_{-.11t} = \ln .4 \qquad \text{\textit{taking natural logs to bring down the exponent}}$$

$$-.11t = \ln .4$$

$$t = \frac{\ln .4}{-.11} \approx \frac{-.9163}{-.11} \approx 8.3 \qquad \text{\textit{solving for t using a calculator}}$$

The concentration will reach the minimum effective level in 8.3 hours, so the dose should be repeated approximately every 8 hours.   ■

Examples 12 and 13 have led to equations of the form $e^{at} = b$. Such equations occur frequently in applications, and are solved by taking the natural log of each side to bring down the exponent.

Carbon 14 Dating

All living things absorb small amounts of radioactive carbon 14 from the atmosphere. When they die, the carbon 14 stops being absorbed and decays exponentially into ordinary carbon. Therefore, the proportion of carbon 14 still present in a fossil or other ancient remain can be use to estimate how old it is.* The proportion of the original carbon 14 that will be present after t years is

$$\left(\begin{array}{c}\text{proportion of carbon 14} \\ \text{remaining after } t \text{ years}\end{array}\right) = e^{-.00012t}$$

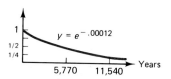

Example 14   The Dead Sea Scrolls, discovered in a cave near the Dead Sea in Jordan, are among the earliest documents of Western civilization. Estimate the age of the Dead Sea Scrolls if the animal skins on which some were written contained 78% of their original carbon 14.

Solution   The proportion of carbon 14 remaining after t years is $e^{-.00012t}$. We equate this formula to the actual proportion (expressed as a decimal).

$$e^{-.00012t} = .78 \qquad\qquad\qquad \textit{equating the proportions}$$

$$\ln e^{-.00012t} = \ln .78 \qquad\qquad \textit{taking natural logs}$$

$$-.00012t = \ln .78 \qquad\qquad \textit{ln } e^{-.00012t} = -.00012t \textit{ by property 6}$$

$$t = \frac{\ln .78}{-.00012} \approx \frac{-.24846}{-.00012} \approx 2071 \qquad \textit{solving for t using a calculator}$$

Therefore, the Dead Sea Scrolls are approximately 2070 years old.   ■

Behavioral Science: Learning Theory

One's ability to do a task generally improves with practice. Frequently, one's skill after t units of practice is given by a function of the form

$$S(t) = c(1 - e^{-kt})$$

where c and k are positive constants.

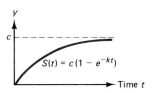

Example 15   After t weeks of practice, a typist can type

$$S(t) = 100(1 - e^{-.25t})$$

words per minute. How soon will the typist type 80 words per minute?

Solution   We solve for t in the following equation.

---

* Carbon 14 dating was devised by Willard Libby, for which he received a Nobel prize in 1960. For extremely old remains such as dinosaur fossils, archeologists use longer-lasting radioactive elements, like potassium 40.

$$100(1 - e^{-.25t}) = 80 \qquad \text{\textit{setting } S(t) \text{ \textit{equal to} 80}}$$

$$1 - e^{-.25t} = .80 \qquad \text{\textit{dividing by} 100}$$

$$-e^{-.25t} = -.20 \qquad \text{\textit{subtracting} 1}$$

$$e^{-.25t} = .20 \qquad \text{\textit{multiplying by} -1}$$

$$-.25t = \ln .20 \qquad \text{\textit{taking natural logs}}$$

$$t = \frac{\ln .20}{-.25} \approx \frac{-1.6094}{-.25} \approx 6.4 \qquad \text{\textit{solving for t using a calculator}}$$

The typist will reach 80 words per minute in about $6\frac{1}{2}$ weeks. ■

Social Science: Diffusion of Information by Mass Media

When a news bulletin is repeatedly broadcast over radio and television, the proportion of people who hear the bulletin within t hours is

$$p(t) = 1 - e^{-kt}$$

for some constant k.

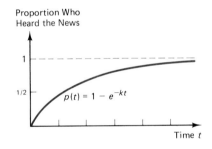

Example 16 A storm warning is broadcast, and the proportion of people who hear the bulletin within t hours of its first broadcast is $p(t) = 1 - e^{-.30t}$. When will 75% of the people have heard the bulletin?

Solution We want to solve

$$1 - e^{-.30t} = .75 \qquad \text{\textit{equating the proportions}}$$

$$-e^{-.30t} = -.25 \qquad \text{\textit{subtracting} 1}$$

$$e^{-.30t} = .25 \qquad \text{\textit{multiplying by} -1}$$

$$-.30t = \ln .25 \qquad \text{\textit{taking natural logs}}$$

$$t = \frac{\ln .25}{-.30} \approx \frac{-1.386}{-.30} \approx 4.6 \qquad \text{\textit{solving for t using a calculator}}$$

Therefore, it takes about $4\frac{1}{2}$ hours for 75% of the people to hear the news.

■

Summary

We defined the logarithm (base a) of x as follows.

$$\log_a x = y \quad \text{is equivalent to} \quad a^y = x$$

The most widely used logarithm is the natural (base e) logarithm, $\ln x$.

$$\ln x = y \quad \text{is equivalent to} \quad e^y = x$$

The properties of natural logarithms are as follows.

1. $\ln 1 = 0$

2. $\ln e = 1$

3. $\ln e^x = x$

4. $\ln (MN) = \ln M + \ln N$

5. $\ln \left(\dfrac{M}{N}\right) = \ln M - \ln N$

6. $\ln (M^N) = N \ln M$

7. $e^{\ln x} = x$

Properties 1–3 and 7 were justified earlier.

Property 4 follows from the addition law of exponents, $e^x \cdot e^y = e^{x+y}$ ("the exponent of a product is the sum of the exponents"). Since logs are exponents, this can be restated "the log of a product is the sum of the logs," which is just property 4.

Property 5 follows from the subtraction law of exponents, $e^x/e^y = e^{x-y}$ ("the exponent of a quotient is the difference of the exponents"). This translates into "the log of a quotient is the difference of the logs," which is just property 5.

Property 6 comes from the law of exponents $(e^x)^y = e^{x \cdot y}$, which says that the exponent y can be "brought down" and multiplied by the x. Since logs are exponents, this says that in $\log_a (M^N)$ the exponent N can be brought down and multiplied by the logarithm, $N \log_a M$, giving property 6.

SOLUTIONS TO PRACTICE EXERCISES

1. **(a)** $\log_{10} 10,000 = 4$ *since $10^4 = 10,000$*

(b) $\log_{10} \dfrac{1}{100} = -2$ *since $10^{-2} = \dfrac{1}{100}$*

2. **(a)** $\log_5 125 = 3$ *since $5^3 = 125$*

(b) $\log_8 2 = \dfrac{1}{3}$ *since $8^{1/3} = \sqrt[3]{8} = 2$*

(c) $\log_2 \dfrac{1}{4} = -2$ *since $2^{-2} = \dfrac{1}{4}$*

3. $\ln 8.34 \approx 2.121$ (to three decimal places)

4. **(a)** $f(x) = \ln (4x) - \ln 4 = \ln 4 + \ln x - \ln 4 = \ln x$

　 (b) $f(x) = \ln \left(\dfrac{x}{2}\right) - \ln x = \ln x - \ln 2 - \ln x = -\ln 2$

　 (c) $f(x) = 12 \ln (x^{3/2}) - \ln (x^{10}) = \dfrac{3}{2} \cdot 12 \ln x - 10 \ln x$

　　　 $= 18 \ln x - 10 \ln x = 8 \ln x$

5. $P(1 + .05)^n = 1.60 \cdot P$ 　　　　　　　　　　　　　 *multiplying by 1.60 increases P by 60%*

　　　　　 $(1.05)^n = 1.6$

　　　　　 $\ln (1.05)^n = \ln 1.6$

　　　　　 $n \ln 1.05 = \ln 1.6$

　　　　　 $n = \dfrac{\ln 1.6}{\ln 1.05} \approx 9.6$ half-years $= 4.8$ years

EXERCISES 4.2

*Find each logarithm **without** using a calculator or tables.*

1 (a) $\log_5 25$　　　(b) $\log_3 81$　　　(c) $\log_3 \dfrac{1}{3}$　　　(d) $\log_3 \dfrac{1}{9}$　　　(e) $\log_4 2$　　　(f) $\log_4 \dfrac{1}{2}$

2 (a) $\log_3 27$　　　(b) $\log_2 16$　　　(c) $\log_{16} 4$　　　(d) $\log_4 \dfrac{1}{4}$　　　(e) $\log_2 \dfrac{1}{8}$　　　(f) $\log_9 \dfrac{1}{3}$

3 (a) $\ln (e^{10})$　　　(b) $\ln \sqrt{e}$　　　(c) $\ln \sqrt[3]{e^4}$　　　(d) $\ln 1$　　　(e) $\ln (\ln (e^e))$　　　(f) $\ln \left(\dfrac{1}{e^3}\right)$

4 (a) $\ln (e^{-5})$　　　(b) $\ln e$　　　(c) $\ln \sqrt[3]{e}$　　　(d) $\ln \sqrt{e^5}$　　　(e) $\ln \left(\dfrac{1}{e}\right)$　　　(f) $\ln (\ln e)$

Use the properties of natural logarithms to simplify each function.

5 $f(x) = \ln (9x) - \ln 9$ 　　　　　　　　　**6** $f(x) = \ln \left(\dfrac{x}{2}\right) + \ln 2$

7 $f(x) = \ln (x^3) - \ln x$ 　　　　　　　　　**8** $f(x) = \ln (4x) - \ln 4$

9 $f(x) = \ln \left(\dfrac{x}{4}\right) + \ln 4$ 　　　　　　　　**10** $f(x) = \ln(x^5) - 3 \ln x$

11 $f(x) = \ln (e^{5x}) - 2x - \ln 1$ 　　　　　　**12** $f(x) = \ln (e^{-x}) + x$

13 $f(x) = \ln (x^9) - \ln (x^6)$ 　　　　　　　　**14** $f(x) = \ln (e^{-2x}) + 3x + \ln 1$

15 $f(x) = 8x - e^{\ln x}$ 　　　　　　　　　　**16** $f(x) = e^{\ln x} + \ln (e^{-x})$

▨ **APPLIED EXERCISES** (Most Require ▦ .)

17 (*Business–Interest*) An investment grows at 24% compounded monthly. How many years will it take to

(a) double?   (b) increase by 50%?

18 (*Business–Interest*) An investment grows at 36% compounded monthly. How many years will it take to

(a) double?   (b) increase by 50%?

19 (*Business–Interest*) A bank offers 7% compounded continuously. How soon will a deposit

(a) triple?   (b) increase by 25%?

20 (*Business–Interest*) A bank offers 6% compounded continuously. How soon will a deposit
(a) quadruple?   (b) increase by 75%?

21 (*General–Depreciation*) An automobile depreciates by 30% per year. How soon will it be worth only half its original value? (*Hint:* Depreciation is like interest but at a negative rate r.)

22 (*General–Population Decline*) The population of Detroit is decreasing by 2% a year. Assuming that this continues, when will the population be half its present size? (*Hint:* Think of growth but with a negative growth rate r.)

23 (*Economics–Electricity Demand*) The demand for electricity in the United States is increasing at the rate of 2.6% per year. Assuming that this continues, how soon will the demand double?

24 (*Business–Appreciation*) An art collection appreciates in value by 8% per year. How soon will its value double?

25–26 (*Business–Advertising*) After t days of advertising a new laundry detergent, the proportion of shoppers in a town who have seen the ads is

$$p(t) = 1 - e^{-.03t}$$

How long must the ad run to reach

25 90% of the shoppers?

26 99% of the shoppers?

27 (*Behavioral Science–Forgetting*) The proportion of students in a psychology experiment who could remember an eight-digit number correctly for t minutes was $p(t) = 0.9 - 0.2 \ln t$ (for $t > 1$). Find the proportion that remembered the number for 5 minutes.

28 (*Social Science–Diffusion of Information by Mass Media*) Election returns are broadcast in a town of 1 million people, and the number of people who have heard the news within t hours is

$$P(t) = 1,000,000 \, (1 - e^{-.4t})$$

How long will it take for 900,000 people to hear the news?

29 (*Behavioral Science–Learning*) After t weeks of practice a typing student can type $s(t) = 100(1 - e^{-.4t})$ words per minute (wpm). How soon will the student type 80 wpm?

30 (*Biomedical–Drug Dose*) If the concentration of a drug in a patient's bloodstream is c (milligrams per milliliter), t hours later the concentration will be $C(t) = c \cdot e^{-kt}$, where k is the absorption constant. If the original concentration is $c = 4$ and the absorption constant is $k = .25$, when should the drug be re-administered so that the concentration does not fall below the minimum effective concentration of 1.4?

31–32 (*General–Carbon 14 Dating*) The proportion of carbon 14 still present in a sample after t years is $e^{-.00012t}$.

31 Estimate the age of the Lascaux cave paintings in France from the pieces of charcoal found nearby that contained only 20% of their original carbon 14.

32 Estimate the age of the last ice age from some tree trunks found in Wisconsin that are believed to have been crushed by the glacier if they contained only 27% of their original carbon 14.

33–34 (*General–Potassium 40 Dating*) The radioactive isotope potassium 40 is used to date very old remains. The proportion of potassium 40 that remains after t million years is $e^{-.00054t}$. Use this function to estimate the age of the following fossils.

33 The most complete skeleton of an early human ancestor ever found was discovered in Kenya in 1984. Estimate the age of the remains if they contained 99.91% of their original potassium 40.

34 (*Dating Older Women*) Estimate the age of the partial skeleton of *Australopithecus afarensis* (known as "Lucy") that was found in Ethiopia in 1977 if it had 99.82% of its original potassium 40.

35 (*General–Radioactive Waste*) Hospitals use radioactive tracers in many medical tests. After the tracer is used, it must be stored as radioactive waste until its radioactivity has decreased enough to be disposed of as ordinary chemical waste. For the radioactive iso- tope potassium, the proportion of radioactivity remaining after t days is $e^{-.05t}$. How soon will this radioactivity decrease to .001, so that it can be disposed of as ordinary chemical waste?

4.3 DIFFERENTIATION OF LOGARITHMIC AND EXPONENTIAL FUNCTIONS

Introduction

You may have noticed that Sections 4.1 and 4.2 contained no calculus. Their purpose was to introduce logarithmic and exponential functions. In this section we differentiate these new functions and use their derivatives for graphing, optimization, and calculating rates of change. We emphasize *natural* (base e) logs and exponentials, since most applications use these exclusively.

Derivative of Ln x

The rule for differentiating the natural logarithm function is as follows.

$$\frac{d}{dx} \ln x = \frac{1}{x}$$

the derivative of ln x is 1 over x

A verification of this rule (and all other differentiation rules of this section) is given at the end of the section.

Example 1   Differentiate $f(x) = x^3 \ln x$.

Solution   The function is a product, x^3 times $\ln x$, so we use the product rule.

$$\frac{d}{dx}(x^3 \ln x) = 3x^2 \ln x + x^3 \frac{1}{x} = 3x^2 \ln x + x^2$$

from $x^3 \frac{1}{x} = x^2$

derivative of the first / second left alone / first left alone / derivative of ln x   ■

Practice Exercise 1

Differentiate $f(x) = x \ln x$.

(solution on page 244)

Example 2   If $f(x) = \dfrac{\ln x}{x}$, find

(a) $f'(x)$ **(b)** $f'(e)$

Solution

(a) This function is a quotient, so we use the quotient rule.

$$f'(x) = \frac{d}{dx}\left(\frac{\ln x}{x}\right) = \frac{x(1/x) - 1\,\ln x}{x^2} = \frac{1 - \ln x}{x^2}$$

(with annotations: derivative of $\ln x$; derivative of x; $x(1/x) = 1$)

(b) To find $f'(e)$ we evaluate this at $x = e$.

$$f'(e) = \frac{1 - \ln e}{e^2} = \frac{1 - 1}{e^2} = \frac{0}{e^2} = 0$$

(with annotations: substituting $x = e$; using $\ln e = 1$; answer) ■

Practice Exercise 2

If $f(x) = \dfrac{\ln x}{x^2}$, find

(a) $f'(x)$ **(b)** $f'(1)$ *(solutions on page 244)*

Derivative of the Natural Logarithm of a Function

The preceding rule, together with the chain rule, shows how to differentiate the natural logarithm of a *function*. For any differentiable function $f(x)$,

$$\frac{d}{dx}\ln f(x) = \frac{f'(x)}{f(x)}$$

the derivative of the natural log of a function is the derivative of the function over the function

Notice that the right-hand side does not involve logarithms at all.

Example 3

$$\frac{d}{dx}\ln(x^2 + 1) = \frac{2x}{x^2 + 1}$$

← derivative of $x^2 + 1$
← original function (without the ln) ■

As we observed, the answer does not involve logarithms.

Practice Exercise 3

Find $\dfrac{d}{dx}\ln(x^3 - 5x + 1)$.   *(solution on page 244)*

Example 4   Find the derivative of $f(x) = \ln(x^4 - 1)^3$.

Solution   For this problem we need the rule for differentiating the natural logarithm of a function, together with the generalized power rule [for differentiating $(x^4 - 1)^3$].

$$\frac{d}{dx} \ln (x^4 - 1)^3 = \frac{\frac{d}{dx} (x^4 - 1)^3}{(x^4 - 1)^3}$$ *using $\frac{d}{dx} \ln f = \frac{f'}{f}$*

$$= \frac{3(x^4 - 1)^2 4x^3}{(x^4 - 1)^3}$$ *using the generalized power rule*

$$= \frac{12x^3}{x^4 - 1}$$ *dividing top and bottom by $(x^4 - 1)^2$*   ■

Alternative Solution   It is easier if we simplify first, using property 6 of logarithms (see the inside back cover) to bring down the exponent 3.

$$\ln (x^4 - 1)^3 = 3 \ln (x^4 - 1)$$ *using $\ln (M^N) = N \ln M$*

Now we differentiate the simplified expression $3 \ln (x^4 - 1)$.

$$\frac{d}{dx} 3 \ln (x^4 - 1) = 3 \frac{4x^3}{x^4 - 1} = \frac{12x^3}{x^4 - 1}$$ *giving the same answer as above*   ■

Moral: Simplifying $\ln (\cdots)^n$ into $n \ln (\cdots)$ makes differentiation easier.

Derivative of e^x

The rule for differentiating the exponential function e^x is as follows.

$$\frac{d}{dx} e^x = e^x$$ *the derivative of e^x is simply e^x*

This shows the rather surprising fact that e^x is its own derivative. Stated another way, the function e^x is unchanged by the operation of differentiation.

Graphical Interpretation

This rule shows that if $y = e^x$, then $y' = e^x$, so that $y = y'$. This means that on the graph of $y = e^x$, the slope y' always equals the y-coordinate, as shown on the right.

For the function $y = e^x$, $y' = y$.

Example 5   Find $\dfrac{d}{dx} \left(\dfrac{e^x}{x} \right)$.

Solution   Since the function is a quotient, we use the quotient rule.

derivative of e^x

derivative of x

$$\frac{d}{dx}\left(\frac{e^x}{x}\right) = \frac{x \cdot e^x - 1 \cdot e^x}{x^2} = \frac{xe^x - e^x}{x^2}$$ ▪

Example 6   If $f(x) = x^2 e^x$, find $f'(1)$.

Solution

$$f'(x) = 2xe^x + x^2 e^x \qquad\qquad \textit{using the product rule on } x^2 e^x$$

so

$$f'(1) = 2(1)e^1 + (1)^2 e^1 \qquad\qquad \textit{substituting } x = 1$$

$$= 2e + e = 3e \qquad\qquad \textit{simplifying} \qquad\qquad ▪$$

In these problems we leave our answers in their "exact" form, leaving e as e. Later, in applied problems, we will approximate our answers using $e \approx 2.718$.

Practice Exercise 4

If $f(x) = xe^x$, find $f'(1)$.   *(solution on page 244)*

Derivative of $e^{f(x)}$

For any differentiable function $f(x)$, the derivative of $e^{f(x)}$ is given by the following rule.

$$\frac{d}{dx}\,e^{f(x)} = e^{f(x)}f'(x)$$

the derivative of e to a function is e to the function times the derivative of the function

That is, to differentiate $e^{f(x)}$ we simply "copy" the original $e^{f(x)}$ and then multiply by the derivative of the exponent.

Example 7

$$\frac{d}{dx}\,e^{x^4+1} = e^{x^4+1}(4x^3) = 4x^3 e^{x^4+1} \qquad\qquad \textit{reversing the order}$$

copied   └ derivative of the exponent   ▪

Example 8   Find $\dfrac{d}{dx}\,e^{x^2/2}$.

Solution   It is better to write the exponent $x^2/2$ as $\frac{1}{2}x^2$, a constant times x to a power, since then its derivative is easily seen to be x.

$$\frac{d}{dx}\, e^{x^2/2} = \frac{d}{dx}\, e^{\frac{1}{2}x^2} = e^{\frac{1}{2}x^2}(x) = xe^{\frac{1}{2}x^2}$$

└─ derivative of the exponent   ▪

Practice Exercise 5

Find $\dfrac{d}{dx}\, e^{1-4x^3}$.   *(solution on page 244)*

Summary of the Differentiation Formulas for Logarithmic and Exponential Functions

The formulas for differentiating natural logarithmic and exponential functions are summarized below, with $f(x)$ written simply as f.

Logarithmic Formulas	Exponential Formulas
$\dfrac{d}{dx}\ln x = \dfrac{1}{x}$	$\dfrac{d}{dx}\, e^x = e^x$
$\dfrac{d}{dx}\ln f = \dfrac{f'}{f}$	$\dfrac{d}{dx}\, e^f = e^f f'$

The formulas in the top row apply only to the particular functions $\ln x$ and e^x, while the lower formulas are more general, applying to $\ln f$ and e^f for *any* differentiable function f.

e^x Versus x^n

Notice that we do *not* take the derivative of e^x by the power rule,

$$\frac{d}{dx}\, x^n = nx^{n-1}$$

This is because the power rule applies to x^n, *a variable to a constant power*, while e^x is *a constant to a variable power*. The two types of function are quite different, as their graphs show.

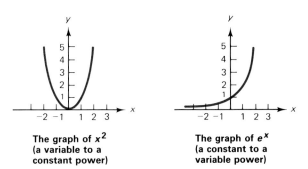

The graph of x^2
(a variable to a
constant power)

The graph of e^x
(a constant to a
variable power)

Each type of function has its own differentiation formula.

$$\frac{d}{dx} x^n = nx^{n-1} \qquad \frac{d}{dx} e^x = e^x$$

for a variable x to   for a constant e to
a constant power n a variable power x

Further Examples

Example 9   Find the derivative of ln $(1 + e^x)$.

Solution

$$\frac{d}{dx} \ln (1 + e^x) = \frac{\dfrac{d}{dx}(1 + e^x)}{1 + e^x} = \frac{e^x}{1 + e^x}$$

using $\dfrac{d}{dx} \ln f = \dfrac{f'}{f}$ working out
the numerator   ■

Practice Exercise 6

Find the derivative of $f(x) = \ln (e^x + 1)^2$. (*Hint*: Simplify before differentiating.)

(solution on page 244)

Functions of the form e^{kx} (for constant k) arise in many applications. The derivative of e^{kx} is as follows.

$$\frac{d}{dx} e^{kx} = e^{kx}k = ke^{kx}$$

using $\dfrac{d}{dx} e^f = e^f \cdot f'$

└ derivative of the exponent

This result is so useful that we record it as a separate formula.

$$\boxed{\frac{d}{dx} e^{kx} = ke^{kx}}$$

for any constant k

This formula says that the rate of change (the derivative) of e^{kx} is proportional to itself. We noted this earlier when we observed that in exponential growth a quantity *grows in proportion to itself* (as in populations and savings accounts.)

Application to Sports

These differential formulas enable us to find instantaneous rates of change of logarithmic and exponential functions. In many applications the variable stands for time, so we use t instead of x.

Example 10 After t weeks of practice a pole vaulter can vault

$$H(t) = 14(1 - e^{-.10t}) \quad \text{feet}$$

Find the rate of change of the athlete's jumps after

(a) 0 weeks (at the beginning of training)   **(b)** 12 weeks

Solution

$$H(t) = 14(1 - e^{-.10t}) = 14 - 14e^{-.10t} \qquad\qquad \textit{H(t) multiplied out}$$

We differentiate to find the rate of change.

$$H'(t) = \underbrace{-14(-.10)e^{-.10t}}_{\text{using } \frac{d}{dt} e^{kt} = ke^{kt}} = \underbrace{1.4e^{-.10t}}_{\text{simplifying}} \qquad\qquad \textit{differentiating } 14 - 14e^{-.10t}$$

(a) For the rate of change after 0 weeks we evaluate at $t = 0$.

$$H'(0) = 1.4e^{-.10(0)} = 1.4e^0 = 1.4$$

(b) After 12 weeks:

$$H'(12) = 1.4e^{-.10(12)} = 1.4e^{-1.2} \approx 1.4(.30) = .42$$

⌞———————using a calculator

Answer: At first the vaults increased by 1.4 feet per week. After 12 weeks, the gain was only .42 foot (about 5 inches) per week. ▪

This is typical of learning a new skill: early improvement is rapid, later improvement is slower.

Recall from Section 4.1 that *e to any power is positive*. For example,

$$e^{-1} \approx \frac{1}{2.718} \approx .368 > 0 \qquad\qquad \textit{e}^{-1} \textit{ is positive}$$

This fact will be very useful in the following examples.

Applications to Economics: Consumer Expenditure

The amount of a commodity that consumers will buy depends on the price of the commodity. For a commodity whose price is p, let the consumer demand be given by a function $D(p)$. Multiplying the number of units $D(p)$ by the price p gives the total *consumer expenditure* for the commodity.

> Let $D(p)$ be the consumer demand at price p. Then consumer expenditure is
>
> $$E(p) = p \cdot D(p)$$

Example 11 If consumer demand for a commodity is $D(p) = 10,000e^{-.02p}$ units per week, where p is the selling price, find the price that maximizes consumer expenditure.

Solution Using the formula above for consumer expenditure,

$$E(p) = p \cdot 10,000e^{-.02p} = 10,000pe^{-.02p}$$

$E(p) = p \cdot D(p)$

To maximize $E(p)$ we differentiate.

$$E'(p) = \underbrace{10,000e^{-.02p}}_{\substack{\text{derivative} \\ \text{of } 10,000p}} + \underbrace{10,000p(-.02)e^{-.02p}}_{\substack{\text{derivative} \\ \text{of } e^{-.02p}}}$$

from $10,000pe^{-.02p}$, using the product rule

$$= 10,000e^{-.02p} - 200pe^{-.02p}$$

simplifying

$$= 200e^{-.02p}(50 - p)$$

factoring

Since e to any power is positive, $e^{-.02p}$ can never be zero. However, the other factor, $(50 - p)$, is zero at $p = 50$, which is the only critical value.

$$\text{CV:} \quad p = 50$$

We calculate E'' for the second derivative test.

$$E''(p) = 200(-.02)e^{-.02p}(50 - p) + 200e^{-.02p}(-1)$$

from $E'(p) = 200e^{-.02p}(50 - p)$ using the product rule

$$= -4e^{-.02p}(50 - p) - 200e^{-.02p}$$

simplifying

At the critical value $p = 50$,

$$E''(50) = -4e^{-.02(50)}(50 - 50) - 200e^{-.02(50)}$$

substituting $p = 50$

$$= -200e^{-1} = \frac{-200}{e}$$

simplifying

E'' is negative, so the expenditure $E(p)$ is maximized at $p = 50$.
   Answer: Consumer expenditure is maximized at price $p = \$50$.   ■

Graphing Logarithmic and Exponential Functions

To graph logarithmic and exponential functions we make sign diagrams for the first and second derivatives.

Example 12 Graph the function $f(x) = e^{-x^2/2}$.

Solution As before, we write the function as $f(x) = e^{-\frac{1}{2}x^2}$. The derivative is

$$f'(x) = e^{-\frac{1}{2}x^2}(-x) = -xe^{-\frac{1}{2}x^2}$$

↑———derivative of the exponent

using $\frac{d}{dx}e^f = e^f f'$

Since e to any power is positive, $e^{-\frac{1}{2}x^2}$ can never be zero. However, the other factor in the derivative, $-x$, *is* zero at $x = 0$. We enter this critical value on the sign diagram below. Evaluating the derivative

$$f'(x) = -xe^{-\frac{1}{2}x^2}$$

at test points on either side gives the signs of f', again using the fact that e to any power is positive.

$f' > 0$	$f' = 0$	$f' < 0$
	$x = 0$	
↗	→	↘
	rel max: (0, 1)	

The second derivative is

$$f''(x) = (-1)e^{-\frac{1}{2}x^2} - xe^{-\frac{1}{2}x^2}(-x)$$ *from $f'(x) = -x \cdot e^{-\frac{1}{2}x^2}$ using the product rule*

$$= -e^{-\frac{1}{2}x^2} + x^2 e^{-\frac{1}{2}x^2}$$ *simplifying*

$$= e^{-\frac{1}{2}x^2}(-1 + x^2)$$ *factoring*

$$= (x^2 - 1)e^{-\frac{1}{2}x^2}$$ *rearranging*

$$= (x + 1)(x - 1)e^{-\frac{1}{2}x^2}$$ *factoring*

Since f'' is zero at $x = \pm 1$, we obtain the following sign diagram for f''. As before, test points show the concavity in each interval.

$f'' > 0$	$f'' = 0$	$f'' < 0$	$f'' = 0$	$f'' < 0$
	$x = -1$		$x = 1$	
con up		con dn		con up
	IP $(-1, e^{-1/2})$		IP $(-1, e^{-1/2})$	

$\approx .6$ using a calculator

Plotting the relative maximum point and the inflection points gives the graph on the right.

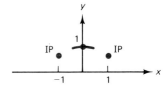

For very large (positive or negative) values of x the exponent in $e^{-x^2/2}$ is negatively large, and e to such a power is close to 0, making the curve approaches the x-axis. Using this information together with the partial graph above, we draw the completed graph, shown on the right.

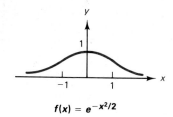

$$f(x) = e^{-x^2/2}$$ ■

This function, multiplied by the constant $1/\sqrt{2\pi}$, is the famous "bell-shaped curve" of statistics. It is used for predicting everything from the IQs of newborn babies to stock prices.

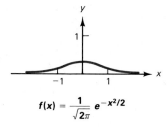

$$f(x) = \frac{1}{\sqrt{2\pi}} e^{-x^2/2}$$

Verification of the Power Rule for Arbitrary Powers

In Chapter 2 we proved the power rule, $\dfrac{d}{dx} x^n = nx^{n-1}$, for positive integer exponents. We can now show that it holds for *all* exponents.

$$f(x) = x^n$$

x to a constant power

$$\ln f(x) = n \ln x$$

taking logarithms of each side, using property 6 of natural logarithms on the right side

$$\frac{f'(x)}{f(x)} = n \frac{1}{x}$$

differentiating both sides

$$f'(x) = n \frac{1}{x} f(x)$$

multiplying each side by $f(x)$

$$= n \frac{1}{x} x^n = nx^{n-1}$$

replacing $f(x)$ by x^n and then simplifying

This is what we wanted to show—that the derivative of $f(x) = x^n$ is

$$f'(x) = nx^{n-1}$$

Verification of the Differentiation Formulas

The formula for the derivative of the natural logarithm function comes from applying the definition of the derivative to $f(x) = \ln x$.

$$f'(x) = \lim_{h \to 0} \frac{f(x + h) - f(x)}{h} = \lim_{h \to 0} \frac{\ln (x + h) - \ln x}{h}$$

definition of $f'(x)$

$$= \lim_{h \to 0} \frac{1}{h} [\ln (x + h) - \ln x]$$

dividing by h is equivalent to multiplying by $1/h$

$$= \lim_{h \to 0} \frac{1}{h} \ln \left(\frac{x + h}{x}\right) = \lim_{h \to 0} \frac{1}{h} \ln \left(1 + \frac{h}{x}\right)$$

using property 5 of logarithms (see the inside back cover), and then simplifying

$$= \lim_{h \to 0} \frac{1}{x} \frac{x}{h} \ln \left(1 + \frac{h}{x}\right) = \lim_{h \to 0} \frac{1}{x} \ln \left(1 + \frac{h}{x}\right)^{x/h}$$

dividing and multiplying by x, and then using property 6 of logarithms

$$= \lim_{n \to \infty} \frac{1}{x} \ln \underbrace{\left(1 + \frac{1}{n}\right)^{n}}$$

replacing x/h by n, so $h \to 0$ implies $n \to \infty$

approaches e as $n \to \infty$

$$= \frac{1}{x} \ln e = \frac{1}{x}$$

since $\ln e = 1$

This is the result that we wanted to show—that the derivative of $f(x) = \ln x$ is $f'(x) = 1/x$. For the rule to differentiate the natural logarithm of a *function*, we begin with

$$\frac{d}{dx} f(g(x)) = f'(g(x))g'(x)$$

chain rule (from page 106)

$$\frac{d}{dx} \ln (g(x)) = \frac{1}{g(x)} g'(x) = \frac{g'(x)}{g(x)}$$

taking $f(x) = \ln x$ so $f'(x) = 1/x$

Replacing g by f, this is exactly the formula we wanted to show,

$$\frac{d}{dx} \ln f(x) - \frac{f'(x)}{f(x)}$$

To derive the rule for differentiating e^x we begin with

$$\ln e^x = x$$

property 3 of natural logarithms

and differentiate both sides

$$\frac{\frac{d}{dx} e^x}{e^x} = 1$$

using $\frac{d}{dx} \ln f = \frac{f'}{f}$ on the left side

Multiplying each side by e^x gives

$$\frac{d}{dx} e^x = e^x$$

which is the rule for differentiating e^x. The rule together with the chain rule gives the rule for differentiating e to a function, just as before.

SOLUTIONS TO PRACTICE EXERCISES

1. $f'(x) = \ln x + x \dfrac{1}{x} = \ln x + 1$

2. **(a)** $f'(x) = \dfrac{x^2(1/x) - 2x \ln x}{x^4} = \dfrac{x - 2x \ln x}{x^4} = \dfrac{x(1 - 2x \ln x)}{x^4} = \dfrac{1 - 2x \ln x}{x^3}$

   (b) $f'(1) = \dfrac{1 - 2 \ln 1}{1} = 1 - 2 \cdot 0 = 1$

3. $\dfrac{d}{dx} \ln (x^3 - 5x + 1) = \dfrac{3x^2 - 5}{x^3 - 5x + 1}$

4. $f'(x) = e^x + xe^x$

   $f'(1) = e^1 + 1e^1 = 2e$ (which is approximately $2 \cdot 2.718 = 5.436$)

5. $\dfrac{d}{dx} e^{1-4x^3} = e^{1-4x^3}(-12x^2) = -12x^2 e^{1-4x^3}$

6. $\ln (e^x + 1)^2 = 2 \ln (e^x + 1)$

   $\dfrac{d}{dx} 2 \ln (e^x + 1) = 2 \dfrac{e^x}{e^x + 1}$

EXERCISES 4.3

Find the derivative of each function.

1 $x^2 \ln x$	**2** $\dfrac{\ln x}{x^4}$	**3** $\dfrac{\ln x}{x^3}$	**4** $x^3 \ln x$
5 $\ln x^2$	**6** $\ln x^5$	**7** $\ln (x^3 + 1)$	**8** $\ln (x^4 - 1)$
9 $\ln \sqrt{x}$	**10** $\sqrt{\ln x}$	**11** $\ln (x^2 + 1)^3$	**12** $\ln (x^4 + 1)^2$
13 $\ln (-x)$	**14** $\ln (5x)$	**15** $x^2 e^x$	**16** $\dfrac{e^x}{x^3}$
17 $\dfrac{e^x}{x^2}$	**18** $x^3 e^x$	**19** e^{x^3+2x}	**20** $2e^{7x}$
21 $e^{x^3/3}$	**22** $4e^{\sqrt{x}}$	**23** $10e^{x/2}$	**24** $\ln (e^x - 2x)$
25 $x - e^{-x}$	**26** $x \ln x - x$	**27** $\ln e^{2x}$	**28** $\ln e^x$
29 e^{1+e^x}	**30** $\ln (e^x + e^{-x})$	**31** x^e	**32** ex
33 e^3	**34** $\sqrt{e}$	**35** $\ln (x^4 + 1) - 4e^{x/2} - x$	
36 $x^2 e^x - 2 \ln x + (x^2 + 1)^3$	**37** $x^2 \ln x - \frac{1}{2}x^2 + e^{x^2} + 5$	**38** $e^{-2x} - x \ln x + x - 7$	

For each function, evaluate the indicated expressions.

39 $f(x) = \dfrac{\ln x}{x^5}$,   find   (a) $f'(x)$ (b) $f'(1)$

40 $f(x) = x^4 \ln x$,   find   (a) $f'(x)$ (b) $f'(1)$

41 $f(x) = \ln(x^4 + 48)$,   find   (a) $f'(x)$ (b) $f'(2)$

42 $f(x) = x^2 \ln x - x^2$,   find   (a) $f'(x)$ (b) $f'(e)$

43 $f(x) = x^3 e^x$,   find   (a) $f'(x)$ (b) $f'(1)$

44 $f(x) = \dfrac{e^x}{x^2}$,   find   (a) $f'(x)$ (b) $f'(2)$

45 $f(x) = \ln(e^x - 3x)$,   find   (a) $f'(x)$ (b) $f'(0)$

46 $f(x) = \ln(e^x + e^{-x})$,   find   (a) $f'(x)$ (b) $f'(0)$

▦ *For each function:*
(a) Evaluate the given expression.
(b) Use a calculator to approximate your answer to three decimal places.

47 $f(x) = 5x \ln x$, find and approximate $f'(2)$.

48 $f(x) = e^{x^2/2}$, find and approximate $f'(2)$.

49 $f(x) = \dfrac{e^x}{x}$, find and approximate $f'(3)$.

50 $f(x) = \ln(e^x - 1)$, find and approximate $f'(3)$.

*Find the **second** derivative of each function.*

51 $f(x) = e^{-x^5/5}$ **52** $f(x) = e^{-x^6/6}$

By calculating the first few derivatives, find a formula for the nth derivative of each function (k is a constant).

53 $f(x) = e^{kx}$ **54** $f(x) = e^{-kx}$

Sketch the graph of each function, showing all relative extreme points and inflection points. (▦ *will be useful.)*

55 $f(x) = e^{-2x^2}$ **56** $f(x) = e^{2x^2}$ **57** $f(x) = \ln(x^2 + 1)$ **58** $f(x) = -\ln(1 + x^2)$

▦ **APPLIED EXERCISES** (Most require ▦)

Approximate answers using a calculator.

59 (*General–Compound Interest*) A sum of $1000 at 5% interest compounded continuously will grow to $V(t) = 1000e^{.05t}$ dollars in t years. Find the rate of growth after

(a) 0 years (the time of the original deposit)

(b) 10 years

(*Hint:* The rate of growth means the derivative.)

60 (*General–Depreciation*) A $10,000 automobile depreciates so that its value after t years is

$$V(t) = 10,000e^{-.35t} \text{ dollars}$$

Find the rate of change of its value

(a) when it is new ($t = 0$)   (b) after 2 years

(*Hint:* The rate of change means the derivative.)

61 (*General–Population*) The world population (in billions) is predicted to be $P(t) = 4.3e^{.01t}$, where t is the number of years after 1980. Find the rate of change of the population in the year 2000. (*Hint:* The rate of change means the derivative.)

62 (*Behavioral Science–Ebbinghaus Memory Model*) According to the Ebbinghaus model of memory, if one is shown a list of items, the percentage of items that one will remember t time units later is

$$P(t) = (100 - a)e^{-bt} + a$$

where a and b are constants. For $a = 25$ and $b = .2$ this function becomes

$$P(t) = 75e^{-.2t} + 25$$

Find the rate of change of this percentage

(a) at the beginning of the test ($t = 0$)

(b) after 3 time units

(*Hint*: The rate of change means the derivative.)

63 (*Biomedical–Drug Dosage*) A patient receives an injection of 1.2 cc of a drug, and the amount remaining in the bloodstream t hours later is $A(t) = 1.2e^{-.05t}$. Find the rate of change of this amount

(a) just after the injection (at time $t = 0$)

(b) after 2 hours

(*Hint*: The rate of change means the derivative.)

64 (*General–Temperature*) A cup of coffee at 200 degrees, if left in a 70-degree room, will cool to

$$T(t) = 70 + 130e^{-2.5t} \text{ degrees in } t \text{ hours}$$

Find the rate of change of the temperature

(a) at time $t = 0$ (b) after 1 hour

65 (*Business–Sales*) The weekly sales (in thousands) of a new product after x weeks of advertising are predicted to be $S(x) = 1000 - 900e^{-.1x}$. Find the rate of change of sales after

(a) 1 week (b) 10 weeks

66 (*Social Science–Diffusion of Information by Mass Media*) The number of people in a town of 50,000 who have heard an important news bulletin within t hours of its first broadcast is $N(t) = 50,000(1 - e^{-.4t})$. Find the rate of change of the number of informed people

(a) at time $t = 0$ (b) after 8 hours

67–68 (*Economics–Consumer Expenditure*) If consumer demand for a commodity is given by the function below (where p is the selling price in dollars), find the price that maximizes consumer expenditure.

67 $D(p) = 5000e^{-.01p}$ **68** $D(p) = 8000e^{-.05p}$

69–70 (*Business–Maximizing Revenue*) The function below is a company's price function, where p is the price (in dollars) at which quantity x (in thousands) will be sold.

(a) Find the revenue function $R(x)$. (*Hint*: Revenue is price times quantity, $p \cdot x$.)

(b) Find the quantity and price that will maximize revenue.

69 $p = 400e^{-.20x}$ **70** $p = 4 - \ln x$

71 (*Biomedical–Drug Concentration*) If a drug is injected intramuscularly, the concentration of the drug in the bloodstream after t hours will be

$$A(t) = \frac{c}{b - a}(e^{-at} - e^{-bt})$$

If the constants are $a = .4$, $b = .6$, and $c = .1$, find the time of maximum concentration.

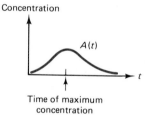

Concentration

$A(t)$

t

Time of maximum concentration

72 (*Biomedical–Reynolds Number*) An important characteristic of blood flow is the "Reynolds number." As the Reynolds number increases, blood flows less smoothly. For blood flowing through certain arteries, the Reynolds number is

$$R(r) = a \ln r - br$$

where a and b are positive constants and r is the radius of the artery. Find the radius r that maximizes the Reynolds number R. (Your answer will involve the constants a and b.)

Exponential and Logarithmic Functions to Other Bases

The rules for differentiating exponential functions with (positive) base a are as follows.

$$\frac{d}{dx}\,a^x = (\ln a)a^x \qquad \frac{d}{dx}\,a^{f(x)} = (\ln a)a^{f(x)}f'(x)$$

For example,

$$\frac{d}{dx}\,2^x = (\ln 2)2^x \qquad \frac{d}{dx}\,5^{3x^2+1} = (\ln 5)5^{3x^2+1}(6x) = 6(\ln 5)x\,5^{3x^2+1}$$

These formulas are more complicated than the corresponding base e formulas (page 237), which is why e is called the "natural" base: it makes the derivative formulas simplest. These formulas reduce to the natural (base e) formulas if $a = e$.

Use the above formulas to find the derivatives of each function.

73 (a) $f(x) = 10^x$ (b) $f(x) = 3^{x^2+1}$ (c) $f(x) = 2^{3x}$ (d) $f(x) = 5^{3x^2}$ (e) $f(x) = 2^{4-x}$

74 (a) $f(x) = 5^x$ (b) $f(x) = 2^{x^2-1}$ (c) $f(x) = 3^{4x}$ (d) $f(x) = 9^{5x^2}$ (e) $f(x) = 10^{1-x}$

The rules for differentiating logarithmic functions with (positive) base a are as follows.

$$\frac{d}{dx}\,\log_a x = \frac{1}{(\ln a)x} \qquad \frac{d}{dx}\,\log_a f(x) = \frac{f'(x)}{(\ln a)f(x)}$$

For example,

$$\frac{d}{dx}\,\log_5 x = \frac{1}{(\ln 5)x} \qquad \frac{d}{dx}\,\log_2 (x^3 + 1) = \frac{3x^2}{(\ln 2)(x^3 + 1)}$$

These formulas are more complicated than the corresponding base e formulas (page 237), and again the simplicity of the base e formulas is why e is called the "natural" base. As before, these formulas reduce to the natural (base e) formulas if $a = e$.

Use the formulas above to find the derivatives of each function.

75 (a) $\log_2 x$ (b) $\log_{10} (x^2 - 1)$ (c) $\log_3 (x^4 - 2x)$

76 (a) $\log_3 x$ (b) $\log_2 (x^2 + 1)$ (c) $\log_{10} (x^3 - 4x)$

4.4 TWO APPLICATIONS TO ECONOMICS: RELATIVE RATES AND ELASTICITY OF DEMAND

Introduction

In this section we define *relative* rates of change and see how they are used in economics. (Relative rates are not the same as related rates, discussed in Section 3.7.) We then define the very important economic concept of elasticity of demand.

Relative Versus Absolute Rates of Change

The derivative of a function gives its rate of change. For example, if $f(t)$ is the cost of a pair of shoes at time t years, then $f'(t)$ is the rate of change of cost (in dollars per year). That is, $f' = 2$ would mean that the price of shoes is increasing at the rate of $2 per year. Similarly, if $g(t)$ is the price of a new automobile at time t years, then $g' = 200$ would mean that automobile prices are increasing at the rate of $200 per year.

Does this mean that car prices are rising 100 times as fast as shoe prices? In absolute terms, yes. However, this does not take into account the enormous price difference between automobiles and shoes.

Relative Rates of Change

If shoe prices are increasing at the rate of $2 per year, and if the current price of a pair of shoes is $40, the *relative* rate of increase is $2/40 = 1/20 = .05$, which means that shoe prices are increasing at the *relative* rate of 5% per year. Similarly, if the price of an average automobile is $10,000, then an increase of $200 relative to this price is $200/10,000 = 1/50 = .02$, for a relative rate of 2% per year. Therefore, in a *relative* sense (that is, as a fraction of the current price), car prices are increasing *less* rapidly than shoe prices.

In general, if $f(t)$ is the price of an item at time t, then the rate of change is $f'(t)$, and the *relative* rate of change is $f'(t)/f(t)$, the derivative divided by the function. We will sometimes call the derivative $f'(x)$ the "absolute" rate of change to distinguish it from the relative rate of change $f'(x)/f(x)$.

Relative rates are often more meaningful than absolute rates. For example, it is easier to grasp the fact that the gross national product is growing at the relative rate of 10% a year than that it is growing at the absolute rate of $200,000,000,000 per year.

Calculating Relative Rates

The expression $f'(x)/f(x)$ is the derivative of the natural logarithm of $f(x)$.

$$\frac{d}{dx} \ln f(x) = \frac{f'(x)}{f(x)}$$

This provides an alternative expression for the relative rate of change, in terms of logarithms. In general, for any differentiable function f,

$$\binom{\text{relative rate of}}{\text{change of } f(t)} = \frac{d}{dt} \ln f(t) = \frac{f'(t)}{f(t)}$$

We use the variable t since it often stands for time. The middle expression is sometimes called the *logarithmic derivative*, since it is found by first taking the logarithm and then differentiating.

   The relative rate of change, being a ratio or a percent, does not depend on the units of the function (whether dollars or rubles, pounds or kilos). Therefore, relative rates can be compared between different products, and even between different nations. This is in contrast to absolute rates of change (that is, derivatives), which *do* depend upon the units (for example, dollars per year).

Example 1   If the gross national product t years from now is predicted to be $G(t) = 1.2 e^{\sqrt{t}}$ trillion dollars, find the relative rate of change 25 years from now. We give two solutions, showing the use of both formulas.

Solution   [Using the derivative-of-the-logarithm formula, $\frac{d}{dt} \ln G(t)$] First we take the logarithm of $G(t)$.

$$\ln (1.2 e^{\sqrt{t}}) = \ln 1.2 + \ln e^{\sqrt{t}} \qquad \text{\small\textit{since the log of a product is the sum of the logs}}$$

$$= \ln 1.2 + \sqrt{t} \qquad \text{\small\textit{since } \ln e^{\sqrt{t}} = \sqrt{t} \textit{ by property 6 of logs (see the inside back cover)}}$$

$$= \ln 1.2 + t^{1/2} \qquad \text{\small\textit{in exponent form}}$$

Then we differentiate:

$$\frac{d}{dx} (\ln 1.2 + t^{1/2}) = 0 + \frac{1}{2} t^{-1/2} = \frac{1}{2} t^{-1/2} \qquad \text{\small\textit{ln 1.2 is a constant so its derivative is zero}}$$

Finally, we evaluate at the given time $t = 25$.

$$\frac{1}{2} (25)^{-1/2} = \frac{1}{2} \frac{1}{\sqrt{25}} = \frac{1}{2} \frac{1}{5} = \frac{1}{10} = .10 \qquad \text{\small\textit{$\frac{1}{2} t^{-1/2}$ evaluated at $t = 25$}}$$

Therefore, in 25 years the gross national product will be increasing at the relative rate of 10% per year.   ∎

Alternative Solution   [using the $f'(t)/f(t)$ formula]

$$G(t) = 1.2e^{\sqrt{t}} = 1.2e^{t^{1/2}}$$

writing $G(t)$ with fractional exponents

$$G'(t) = 1.2e^{t^{1/2}} \left(\frac{1}{2} t^{-1/2} \right)$$

differentiating

└──── derivative of the exponent

Therefore, the relative rate of change $G'(t)/G(t)$ is

$$\frac{G'(t)}{G(t)} = \frac{1.2e^{t^{1/2}}[(\frac{1}{2}) t^{-1/2}]}{1.2e^{t^{1/2}}} = \frac{1}{2} t^{-1/2}$$

same as with the first formula

At $t = 25$,

$$\frac{1}{2} (25)^{-1/2} = \frac{1}{2} \frac{1}{\sqrt{25}} = \frac{1}{2} \frac{1}{5} = \frac{1}{10} = .10$$

again the same

Therefore, the relative growth rate is 10%, just as we found above.   ■

Both formulas give the same answer, so you should use the one that is easier to apply in any particular problem. The $\frac{d}{dx} \ln f(t)$ formula sometimes allows simplification *before* the differentiation, while the $f'(t)/f(t)$ formula often involves simplification afterward.

Practice Exercise 1

An investor estimates that if a piece of land is held for t years, it will be worth $f(t) = 300 + t^2$ thousand dollars. Find the relative rate of change at time $t = 10$ years. [*Hint:* Use the $f'(t)/f(t)$ formula.]

(solution on page 254)

Elasticity of Demand

Farmers are aware of the paradox that an abundant harvest usually produces *lower* total revenue than a poor harvest. The reason is simply that the larger quantities in an abundant harvests result in lower prices, which in turn cause increased demand, but the demand does *not* increase enough to compensate for the lower prices.

Revenue is price times quantity, and when one of these quantities increases, the other generally decreases. The question is whether the increase in one is enough to compensate for the decrease in the other. The concept of *elasticity of demand* was invented to answer this question.

Intuitively, elasticity of demand is a measure of how *responsive* demand is to price changes. Think of "elastic" as meaning "very responsive." If demand is elastic, a small price cut will bring a large increase in demand, so total revenue will rise. On the other hand, if demand is *in*elastic, a price cut will bring only a slight increase in demand, so total revenue will fall.

In general, *elastic* demand means that consumers will purchase significantly more or less in response to price changes. *Inelastic* demand means that consumers will buy only slightly more or less in response to price changes. (This is the cause of the farmers' difficulties: demand for farm products is inelastic.)

Demand Functions

In general, if the price of an item rises, the demand will fall, and vice versa. If the relationship between price p of an item and the quantity x that will be sold at that price can be expressed with x as a function of p, the function is called the *demand function*.

Demand Function

$$x = D(p)$$

gives the quantity of an item that will be demanded by consumers at price p.

Since, in general, demand falls as prices rise, the slope of the demand function will be negative, as shown on the right. This is known as the *law of downward-sloping demand*.

The law of downward sloping demand

Calculating Elasticity of Demand

For a demand function $D(p)$, let us calculate the relative rate of change of demand divided by the relative rate of change of price. Using the derivative-of-the-logarithm formula,

$$\frac{\left(\begin{array}{c}\text{relative rate of} \\ \text{change of demand}\end{array}\right)}{\left(\begin{array}{c}\text{relative rate of} \\ \text{change of price}\end{array}\right)} = \frac{\dfrac{d}{dp}\ln D(p)}{\dfrac{d}{dp}\ln p} = \frac{\dfrac{D'(p)}{D(p)}}{\dfrac{1}{p}} = \underbrace{\frac{pD'(p)}{D(p)}}_{\text{simplified}}$$

Because most demand functions are downward-sloping, the derivative $D'(p)$ is generally negative. Economists prefer to work with positive numbers, so the *elasticity of demand* is taken to be the negative of this quantity (in order to make it positive).

Elasticity of Demand

For a demand function $D(p)$ the elasticity of demand is

$$E(p) = \frac{-p \cdot D'(p)}{D(p)}$$

Demand is *elastic* if $E(p) > 1$ and *inelastic* if $E(p) < 1$.

Elasticity, being composed of *relative* rates of change, does not depend on the units of the demand function. Therefore, elasticities can be compared between different products, and even between different countries.

Example 2 A bus line estimates the demand function for its daily commuter tickets to be $D(p) = 81 - p^2$ (in thousands of tickets), where p is the price (in dollars) Find the elasticity of demand when the price is

(a) $3   (b) $6

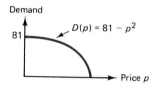

Demand

$D(p) = 81 - p^2$

81

Price p

Solution

$$E(p) = \frac{-p \, D'(p)}{D(p)}$$

definition of elasticity

$$= \frac{-p(-2p)}{81 - p^2}$$

substituting $D(p) = 81 - p^2$ so $D'(p) = -2p$

$$= \frac{2p^2}{81 - p^2}$$

simplifying

(a) Evaluating at $p = 3$ gives

$$E(3) = \frac{2(3)^2}{81 - (3)^2} = \frac{18}{81 - 9} = \frac{18}{72} = \frac{1}{4}$$

substituting $p = 3$

Interpretation: The elasticity is less than 1, so demand for tickets is *inelastic* at a price of $3. This means that small price changes (up or down from this level) will cause only *slight* changes in demand. More precisely, elasticity of $\frac{1}{4}$ means that a 1% price change will cause only about a $\frac{1}{4}$% change in demand.

(b) At the price of $6, the elasticity of demand is

$$E(6) = \frac{2(6)^2}{81 - (6)^2} = \frac{72}{81 - 36} = \frac{8}{5} = 1.6$$

substituting $p = 6$

Interpretation: The elasticity is greater than 1, so demand is *elastic* at a price of $6. This means that small changes in price (up or down from this level) will cause a relatively large change in demand. In particular, an elasticity of 1.6 means that a price change of 1% will cause about a 1.6% change in demand.   ■

The changes in demand are, of course, in the opposite direction from the changes in price. That is, if prices are *raised* by 1% (from the $6 level), demand will *fall* by 1.6%, while if prices are *lowered* by 1%, demand will *rise* by 1.6%. In the future we will assume that the *direction* of the change is clear, and we say simply that a 1% change in price will cause about a 1.6% change in demand.

Practice Exercise 2

For demand function $D(p) = 100 - 2p$, find the elasticity of demand $E(p)$ and evaluate it at $p = 20$ and $p = 30$.   *(solution on page 255)*

Using Elasticity to Increase Revenues

In Example 2 we found that at a price of $3 demand is inelastic ($E = \frac{1}{4} < 1$), so demand responds only *weakly* to price changes. Therefore, to increase revenue the company should *raise* prices, since the higher prices will drive away only a relatively small number of customers. On the other hand, at a price of $6 demand is elastic ($E = 1.6 > 1$), so demand is very responsive to price changes. In this case to increase revenue the company should *lower* prices, since this will attract more than enough new customers to compensate for the price decrease. In general,

---

To Increase Revenues

> *Raise* prices if demand is *inelastic* ($E < 1$).
> *Lower* prices if demand is *elastic* ($E > 1$).

This statement shows why elasticity of demand is important to any company that cuts prices in an attempt to boost revenues, as it is for any utility that raises prices in order to increase revenues. Elasticity shows whether the strategy will succeed.

The borderline case, where elasticity equals 1 (called *unitary elasticity*), is where revenue is not changed by small price changes in either direction. The statement in the box above shows that elasticity must be unitary when revenue is maximized.

At maximum revenue, elasticity of demand must equal 1.

We could use this fact as a basis for a new method for maximizing revenue, but instead, we will stick with our old (and easier) method of maximizing functions by finding critical values.

Verification of the Relationship between Elasticity and Revenue

We may verify the relationship between elasticity of demand and revenue as follows. Revenue is price p times quantity x.

$$R = px = p \cdot D(p)$$ *using $x = D(p)$*

Differentiating will show how revenue responds to price changes.

$$R'(p) = D(p) + pD'(p)$$ *using the product rule on $pD(p)$*

$$= D(p)\left[1 + \frac{pD'(p)}{D(p)}\right]$$ *factoring out $D(p)$*

$$= D(p)\left[1 - \frac{-pD'(p)}{D(p)}\right]$$ *replacing the plus sign by two minus signs*

this is the definition of elasticity $E(p)$

$$= D(p)\left[1 - E(p)\right]$$ *replacing $\frac{-pD'(p)}{D(p)}$ by $E(p)$*

If demand is *elastic*, $E > 1$, then the quantity in brackets is negative, so the derivative $R'(p)$ is negative, showing that revenue *decreases* as price increases. Therefore, to increase revenue, one should *lower* prices. On the other hand, if the demand is *inelastic*, $E < 1$, then the quantity in brackets is positive, so the derivative $R'(p)$ is positive, showing that revenue *increases* as price increases. In this case to increase revenue, one should *raise* prices. This proves the two statements in the boxes on the previous page.

Elasticity is Not the Same as Slope

Do not confuse elasticity of demand with the slope of the demand curve. The two ideas are quite different. For example, a demand curve with unit slope downward does not have elasticity equal to 1 all along it. A demand function with such a slope means that each dollar price increase decreases demand by 1 unit. A loss of one sale would be insignificant for a supermarket that deals in large volumes, but would be very significant for a manufacturer of giant supercomputers that sells only a few a year. That is, to see how revenue changes, we must look not at absolute rates of change of price and quantity but at their *relative* rates of change, which is what elasticity of demand is all about.

SOLUTIONS TO PRACTICE EXERCISES

1. $\dfrac{f'(t)}{f(t)} = \dfrac{2t}{300 + t^2}$, and at $t = 10$, $\dfrac{2 \cdot 10}{300 + (10)^2} = \dfrac{20}{400} = \dfrac{1}{20} = .05$ or 5%

and at $t = 10$, $\dfrac{2 \cdot 10}{300 + (10)^2} = \dfrac{20}{400} = \dfrac{1}{20} = .05$ or 5%

2. $E(p) = \dfrac{-pD'(p)}{D(p)} = \dfrac{-p(-2)}{100 - 2p} = \dfrac{2p}{100 - 2p}$

At $p = 20$, $E(20) = \dfrac{40}{100 - 40} = \dfrac{40}{60} = \dfrac{2}{3} \approx .67$ (demand is inelastic)

At $p = 30$, $E(30) = \dfrac{60}{100 - 60} = \dfrac{60}{40} = \dfrac{3}{2} = 1.5$ (demand is elastic)

■ EXERCISES 4.4

For each function:
(a) Find the relative rate of change.
(b) Evaluate the relative rate of change at the given values of t.

1 $f(t) = t^2$, $t = 1$ and $t = 10$

2 $f(t) = t^3$, $t = 1$ and $t = 10$

3 $f(t) = 100e^{.2t}$, $t = 5$

4 $f(t) = 100e^{.5t}$, $t = 4$

5 $f(t) = e^{t^2}$, $t = 10$

6 $f(t) = e^{t^3}$, $t = 5$

7 $f(t) = e^{-t^2}$, $t = 10$

8 $f(t) = e^{-t^3}$, $t = 5$

9 $f(t) = 25\sqrt{t - 1}$, $t = 6$

10 $f(t) = 100\sqrt[3]{t + 2}$, $t = 8$

■ APPLIED EXERCISES ON RELATIVE RATES   (▦ needed.)

11 If the national debt t years from now is estimated to be $N(t) = .5 + 1.1e^{.01t}$ trillion dollars, find the relative rate of change of the debt 10 years from now.

12 If the national debt t years from now is estimated to be $N(t) = .4 + 1.2e^{.01t}$ trillion dollars, find the relative rate of change of the debt 10 years from now.

■ EXERCISES ON ELASTICITY OF DEMAND

For each demand function D(p):
(a) Find the elasticity of demand E(p).
(b) Determine whether the demand is elastic, inelastic, or unitary elastic at the given price p.

13 $D(p) = 200 - 5p$, $p = 10$

14 $D(p) = 60 - 8p$, $p = 5$

15 $D(p) = 300 - p^2$, $p = 10$

16 $D(p) = 100 - p^2$, $p = 5$

17 $D(p) = \dfrac{300}{p}$, $p = 4$

18 $D(p) = \dfrac{500}{p}$, $p = 2$

19 $D(p) = \sqrt{175 - 3p}$, $p = 50$

20 $D(p) = \sqrt{100 - 2p}$, $p = 20$

21 $D(p) = \dfrac{100}{p^2}$, $p = 40$

22 $D(p) = \dfrac{600}{p^3}$, $p = 25$

23 $D(p) = 4000e^{-.01p}$, $p = 200$

24 $D(p) = 6000e^{-.05p}$, $p = 100$

APPLIED EXERCISES ON ELASTICITY OF DEMAND

25 An automobile dealer is selling cars at a price of $12,000. The demand function is

$$D(p) = 2(15 - .001p)^2$$

where p is the price of a car. Should the dealer raise or lower price to increase revenue?

26 A liquor distributor wants to increase its revenues by discounting its best-selling liquor. If the demand function for this liquor is $D(p) = 60 - 3p$, where p is the price per bottle, and if the current price is $15, will the discounts succeed?

27 A city bus line estimates its demand function to be $D(p) = 140,000\sqrt{100 - p}$, when the fare is p cents. The bus line currently charges a fare of 75 cents, and it plans to raise the fare to increase its revenues. Will it succeed?

28 The demand function for a newspaper is

$$D(p) = 80,000\sqrt{75 - p}$$

where the price is p cents. The publisher currently charges 50 cents, and it plans to raise the price to increase revenues. Will the publisher succeed?

29 An electrical utility asks the Federal Regulatory Commission for permission to raise rates to increase revenues. The utility's demand function is

$$D(p) = \frac{120}{10 + p}$$

where p is the price (in cents) of a kilowatt-hour of electricity. If the utility currently charges 6 cents per kilowatt-hour, should the commission grant the request?

30 A Middle Eastern oil-producing country estimates that the demand for oil (in millions of barrels) is $D(p) = 28e^{-.03p}$, where p is the price of a barrel of oil. To raise its revenues, should it raise or lower its price from its current level of $30 per barrel?

31 A European oil-producing country estimates that the demand for its oil (in millions of barrels) is

$$D(p) = 41e^{-.04p}$$

where p is the price of a barrel of oil. To raise its revenues, should it raise or lower its price from its current level of $30 per barrel?

---

32 (a) Show that for a demand function of the form $D(p) = \dfrac{c}{p^n}$, where c and n are positive constants, the elasticity is constant.

(b) What type of demand function has elasticity equal to 1 for every value of p?

33 Show that for a demand function of the form

$$D(p) = ae^{-cp}$$

where a and c are positive constants, the elasticity of demand is $E(p) = cp$.

34–35 (*Elasticity of Supply*) A supply function $S(p)$ gives the total amount of a product that producers are willing to supply at a given price p. The elasticity of supply is defined as

$$E_s(p) = \frac{p \cdot S'(p)}{S(p)}$$

Elasticity of supply measures the relative increase in supply resulting from a small relative increase in price. It is less useful than elasticity of demand, however, since it is not related to total revenue.

34 Find the elasticity of supply for a supply function of the form $S(p) = ae^{cp}$, where a and c are positive constants.

35 Find the elasticity of supply for a supply function of the form $S(p) = ap^n$, where a and n are positive constants.

4.5 REVIEW OF CHAPTER FOUR

Exponential Functions

This chapter introduced logarithmic and exponential functions and illustrated some of their many applications, from dating ancient remains to forecasting economic growth. Exponential and logarithmic functions often involve the constant e.

$$e = \lim_{n \to \infty} \left(1 + \frac{1}{n}\right)^n \approx 2.718$$

The formulas for the value of a sum invested at compound interest are exponential functions.

Continuous Compounding	Discrete Compounding
Pe^{rn}	$P(1 + r)^n$

In these formulas P is the principal, r the interest rate per period, and n the number of periods (for the continuous formula the period is always a year). For negative values of r these formulas model exponential decay, as in depreciation.

Logarithmic Functions

The logarithm of x to the base e (the "natural" logarithm) is written $\ln x$, and is defined as the power of e that gives x. For example,

$$\ln 5 \approx 1.609 \quad \text{because} \quad e^{1.609} \approx 5$$

as may be verified using a calculator. Natural logarithms have several useful properties.

1. $\ln 1 = 0$
2. $\ln e = 1$
3. $\ln e^x = x$
4. $\ln (MN) = \ln M + \ln N$
5. $\ln \left(\dfrac{M}{N}\right) = \ln M - \ln N$
6. $\ln (M^N) = N \ln M$
7. $e^{\ln x} = x$

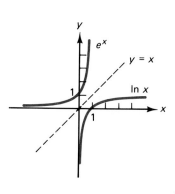

Property 6 ("logarithms bring down exponents") is used for solving equations in which the variable is in the exponent. Properties 3 and 7 show that e^x and $\ln x$ are "inverse functions" in that applying either to a number and then applying the other to the result simply gives back the original number. This is equivalent to the geometric fact that the graphs of e^x and $\ln x$ are reflections of each other in the diagonal line $x = y$.

Derivatives of Logarithmic and Exponential Functions

The formulas for differentiating logarithmic and exponential functions are as follows.

$$\frac{d}{dx} \ln x = \frac{1}{x} \qquad \frac{d}{dx} e^x = e^x$$

$$\frac{d}{dx} \ln f = \frac{f'}{f} \qquad \frac{d}{dx} e^f = e^f f'$$

A useful special case of the last formula occurs when $f(x) = kx$.

$$\frac{d}{dx} e^{kx} = ke^{kx} \qquad \qquad \textit{for any constant } k$$

We used these formulas to optimize, sketch graphs, and calculate marginals and rates of change of exponential and logarithmic functions.

Relative Rates and Elasticity of Demand

Section 4.4 discussed *relative* rates of change (the rate of change as a fraction of the quantity's current size).

$$\left(\begin{array}{c}\text{Relative rate of}\\\text{change of } f(t)\end{array}\right) = \frac{d}{dt} \ln f(t) = \frac{f'(t)}{f(t)}$$

Relative rates led to the important concept of *elasticity of demand*.

$$\left(\begin{array}{c}\text{Elasticity of demand for}\\\text{a demand function } D(p)\end{array}\right) = \frac{-p \cdot D'(p)}{D(p)}$$

Elasticity of demand measures consumer responsiveness to changes in price. All of these developments are reviewed in the following exercises.

Practice Test

Exercises 1, 5, 15, 21, 25, 29, 31, 37, 41, 43

■ **EXERCISES 4.5** (Most require 🖩)

1 (*General–Interest*) Find the value of $10,000 invested for 8 years at 8% interest if the compounding is done

(a) quarterly   (b) continuously

2 (*General–Interest*) One bank offers 6% compounded quarterly and a second offers 5.9% compounded continuously. Where should you take your money?

3 (*Business–Depreciation*) An $800,000 computer loses 20% of its value each year.

(a) Give a formula for its value after t years.

(b) Find its value after 4 years.

4 (*Biomedical–Drug Concentration*) If the concentration of a drug in a patient's bloodstream is c (milligrams per milliliter), then t hours later the concentration will be $C(t) = ce^{-kt}$, where k is a constant (the "elimination constant"). Two drugs are being compared: drug A (with initial concentration $c = 2$ and elimination constant $k = 0.2$) and drug B (with initial concentration $c = 3$ and elimination constant $k = .25$). Which drug will have the greater concentration 4 hours later?

5 (*General–Interest*) Find how soon an investment at 10% interest compounded semiannually will

(a) double in value   (b) increase by 50%

6 (*General–Interest*) Find how soon an investment at 7% interest compounded continuously will

(a) double in value   (b) increase by 50%

7–8 (*General–Fossils*) In the following exercises use the fact that the proportion of potassium 40 remaining after t million years is $e^{-.00054t}$.

7 In 1984 in the Wind River Basin of Wyoming, scientists discovered a fossil of a small, three-toed horse, an ancestor of the modern horse. Estimate the age of this fossil if it contained 97.3% of its original potassium 40.

8 Estimate the age of a skull found in 1959 in Tanzania (dubbed "Nutcracker Man" because of its huge jawbone) which contained 99.9% of its original potassium 40.

9 (*Social Science–Diffusion of Information by Mass Media*) In a city of 1 million people, news of election results broadcast over radio and television will reach $N(t) = 1,000,000 (1 - e^{-.3t})$ people within t hours. Find when the news will have reached 500,000 people.

10 (*Economics–Oil Demand*) The demand for oil in the United States is decreasing by 1% per year, due mainly to conservation and the use of alternative fuels. Assuming that this rate continues, how soon will demand fall to 80% of its present level?

11–12 (*Business–Rule of 72*) If a sum is invested at interest rate r compounded continuously, the doubling time (the time in which it will double in value) is found by solving the equation $Pe^{rt} = 2P$. The solution (by the usual method of cancelling the P and taking logs) is $t = \ln 2/r \approx .69/2$. For *annual* compounding, the doubling time should be somewhat longer, and may be estimated by replacing 69 by 72.

> **Rule of 72**
>
> For $r\%$ interest compounded annually the doubling time is approximately $\dfrac{72}{r}$.

The 72, however, is only a rough "upward adjustment" of 69, and the rule is most accurate for interest rates around 9%. For example, to estimate the doubling time for an investment at 8% compounded annually we would divide 72 by 8, giving $72/8 = 9$ years.
   For each interest rate:

(a) Use the rule of 72 to estimate the doubling time for annual compounding.

(b) Use the compound interest formula $P(1 + r)^n$ to find the actual doubling time for annual compounding.

11 6%

12 1% (This shows that for interest rates very different from 9% the rule of 72 is less accurate.)

Find the derivative of each function.

13 $\ln 2x$

14 $\ln (x^2 - 1)^2$

15 $\ln (1 - x)$

16 $\ln \sqrt{x^2 + 1}$

17 $\ln \sqrt[3]{x}$

18 $\ln e^x$

19 $\ln x^2$

20 $x \ln x - x$

21 e^{-x^2}

22 e^{1-x}

23 $\ln e^{x^2}$

24 $e^{x^2 \ln x - \frac{1}{2}x^2}$

25 $5x^2 + 2x \ln x + 1$

26 $2x^3 + 3x \ln x - 1$

27 $2x^3 - 3xe^{2x}$

28 $4x - 2x^2 e^{2x}$

Graph each function, showing all relative extreme points and inflection points.

29 $f(x) = \ln(x^2 + 4)$ **30** $f(x) = 16e^{-x^2/8}$

31 (*Business–Sales*) The weekly sales (in thousands) of a new product after x weeks of advertising is

$$S(x) = 2000 - 1500e^{-.1x}$$

Find the rate of change of sales after

(a) 1 week (b) 10 weeks

32 (*Biomedical–Drug Dosage*) A patient receives an injection of 1.5 cc of a drug, and the amount remaining in the bloodstream t hours later is $A(t) = 1.5e^{-.08t}$. Find the instantaneous rate of change of this amount

(a) immediately after the injection (time $t = 0$)

(b) after 5 hours

33 (*Behavioral Science–Learning*) In a test of short-term memory, the percent of subjects who remember an eight-digit number for at least t seconds is

$$P(t) = 100 - 20\ln(t + 1)$$

Find the rate of change of this percent after 5 seconds.

34 (*General–Temperature*) A thermos bottle that is chilled to 35 degrees and then left in a 70-degree room will warm to a temperature of $T(t) = 70 - 35e^{-.1t}$ degrees after t hours. Find the rate of change of the temperature

(a) at time $t = 0$ (b) after 5 hours

35 (*Social Science–Diffusion of Information by Mass Media*) The number of people in a town of 30,000 who have heard an important news bulletin within t hours of its first broadcast is $N(t) = 30,000(1 - e^{-.3t})$. Find the instantaneous rate of change of the number of informed people after

(a) 1 hour (b) 8 hours

36 (*Business–Maximizing Present Value*) A new company is growing so that its value t years from now will be $50t^2$ dollars. Therefore, its present value (at the rate of 8% compounded continuously) is

$$V(t) = 50t^2e^{-.08t} \text{ dollars (for } t > 0)$$

Find the number of years that maximizes the present value.

37–38 (*Business–Maximizing Revenue*) The function given below is a company's price function, where x is the quantity (in thousands) that will be sold at price p dollars.

(a) Find the revenue function $R(x)$. (*Hint*: Revenue is price times quantity, $p \cdot x$).

(b) Find the quantity and the price that will maximize revenue.

37 $p = 200e^{-.25x}$ **38** $p = 5 - \ln x$

39 (*Economics–Maximizing Consumer Expenditure*) Consumer demand for a commodity is estimated to be $D(p) = 25,000e^{-.02p}$ units per month, where p is the selling price in dollars. Find the selling price that maximizes consumer expenditure. [*Hint*: Consumer expenditure is price times demand, $p \cdot D(p)$.]

40–41 (*Economics–Relative Rate of Change*) The gross national product of a developing country is forecast to be $G(t) = 5 + 2e^{.01t}$ million dollars t years from now. Find the relative rate of change

40 20 years from now **41** 10 years from now

42 (*Elasticity of Demand*) A South American country exports coffee and estimates the demand function to be $D(p) = 63 - 2p^2$. If the country wants to raise revenues to improve its balance of payments, should it raise or lower the price from the present level of $3 per pound?

43 (*Economics–Elasticity of Demand*) A South African country exports gold and estimates the demand function to be $D(p) = 400\sqrt{600 - p}$. If the country wants to raise revenues to improve its balance of payments, should it raise or lower the price from the present level of $350 per ounce?

5 INTEGRATION AND ITS APPLICATIONS

A roller coaster. The area under the curve can be estimated from the rectangular grid.

5.1 ANTIDERIVATIVES

Introduction

The part of calculus consisting of differentiation and its application is known as *differential calculus*. We now turn our attention to *integral calculus* and the operation of *integration,* which is essentially the inverse operation of differentiation.

Integration has many uses. For example, differentiation turns a cost function into a marginal cost function, and integration does the reverse, turning marginal cost back into cost.

Later we will use integration to calculate areas. For example, the area between the import curve and the export curve shown below gives the cumulative United States trade deficit for 1981–1988. As another example, the graph on the right below shows the annual traffic fatalities from 1975 to 1985, and also the predicted fatalities if everyone used seat belts. The area between the curves gives the total number of lives that could have been saved by seat belts.

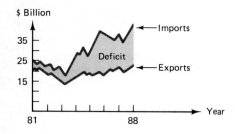 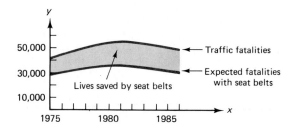

Antidifferentiation: A First Example

Integration, being the reverse of differentiation, is sometimes called *antidifferentiation.* Differentiation begins with a function like x^2 and finds its derivative, $2x$. *Anti*differentiation begins with the derivative $2x$, and recovers the original function x^2. This can be stated:

an antiderivative of $2x$ is x^2

since the derivative of x^2 is $2x$

There are, however, other antiderivatives of $2x$. Each of the following is an antiderivative of $2x$ (since the derivative of each is $2x$).

$$x^2 + 1, \qquad x^2 - 17, \qquad x^2 + e$$

We may add *any* constant C to x^2 and the derivative will still be $2x$, so any function of the form $x^2 + C$ is an antiderivative of $2x$. Furthermore, it can be shown that there are no other antiderivatives of $2x$, so *the most general antiderivative of $2x$ is $x^2 + C$,* where C is any constant.

Indefinite Integrals

The most general antiderivative of a function is called the *indefinite integral* of the function. To state that the indefinite integral of $2x$ is $x^2 + C$ we write the $2x$ between an *integral sign* $\int$ and a "*dx*."

$$\int 2x \, dx = x^2 + C$$

integral sign — integrand — arbitrary constant

the indefinite integral of 2x is x² + C

The function to be integrated (here, $2x$) is called the *integrand*. The *dx* reminds us that the variable of integration is x. The constant C is called an "arbitrary constant" because it may take any value, positive, negative, or zero.

Integration problems can always be checked by differentiation: the derivative of the answer must give the integrand. For example, to check that $\int 2x \, dx = x^2 + C$, we simply differentiate the answer.

$$\int 2x \, dx = x^2 + C$$

the derivative of the answer is the integrand

Since the derivative of $x^2 + C$ is $2x$, the answer is correct. In general:

$$\int f(x) \, dx = g(x) + C$$

if and only if

$$g'(x) = f(x)$$

the integral of f(x) is g(x) + C
if and only if
the derivative of g(x) is f(x)

Indefinite integrals are often referred to simply as integrals.

Power Rule for Integration

There are several "rules" that simplify integration. The first shows how to integrate x to a constant power.

The Power Rule for Integration

$$\int x^n \, dx = \frac{1}{n+1} x^{n+1} + C \qquad (n \neq -1)$$

to integrate x to a power, add 1 to the power and multiply by 1 over the new power

This is one of the most useful rules in all of integral calculus.

Example 1

$$\int x^3 \, dx = \frac{1}{4} x^4 + C$$

using the power rule with n = 3

Differentiating the answer 1/4 x^4 + C immediately gives the integrand x^3, so the answer is correct.

Practice Exercise 1

Solve $\int x^2 \, dx$ and check your answer by differentiation.

(solution on page 271)

The proof of the power rule for integration consists simply of differentiating the right-hand-side.

$$\frac{d}{dx} \left(\frac{1}{n + 1} x^{n+1} + C \right) = \frac{1}{n + 1} (n + 1)x^n = x^n$$

the power $n + 1$ ⟶ ⌐the power ↘simplified
brought down   decreased by 1

Since the derivative is the integrand x^n, the power rule for integration is correct.

To integrate functions like $\sqrt{x}$ and $1/x^2$ we first express them as powers.

Example 2

$$\int \sqrt{x} \, dx = \int x^{1/2} \, dx = \frac{2}{3} x^{3/2} + C$$

using the power rule for integration with n = 1/2

$n = \frac{1}{2}$

$n + 1 = \frac{1}{2} + 1 = \frac{3}{2}$

$\frac{1}{n + 1} = \frac{1}{\frac{3}{2}} = \frac{2}{3}$

Differentiating the answer $(2/3)x^{3/2}$ + C gives $x^{1/2}$, so the answer is correct. Note that the new power is 3/2, and we multiply by its reciprocal, 2/3. In general, if the new power is a/b, then we multiply by b/a.

Example 3

$$\int \frac{1}{x^2} \, dx = \int x^{-2} \, dx = \frac{1}{-1} x^{-1} + C = -x^{-1} + C$$

$n = -2$ $n + 1 = -1$ simplified
answer

If the integrand is a fraction, the *dx* may be written in the numerator instead of at the end of the integral. That is, the integral in Example 3 could have been written

$$\int \frac{dx}{x^2} \quad \text{instead of} \quad \int \frac{1}{x^2}\, dx$$

Practice Exercise 2

Solve $\int \dfrac{dx}{x^3}$. *(solution on page 272)*

Example 4

$$\int 1\, dx = \int x^0\, dx = \frac{1}{1} x^1 + C = x + C$$

$$\underset{n\,=\,0}{\rule{0pt}{0pt}} \qquad \underset{-n+1\,=\,1}{\rule{0pt}{0pt}} \qquad ■$$

This result is so useful that it should be memorized along with the power rule.

$$\boxed{\int 1\, dx = x + C}$$ *the integral of 1 is x + C*

Integrating with Respect to Other Variables

We integrate functions of other variables in the same way, replacing the *x* in the *dx* by the new variable.

Example 5

(a) $\int t^3\, dt = \dfrac{1}{4} t^4 + C$ *using the power rule (n = 3) with dt since the variable is t*

(b) $\int u^{-4}\, du = \dfrac{1}{-3} u^{-3} + C = -\dfrac{1}{3} u^{-3} + C$ *using the power rule (n = −4) with du since the variable is u* ■

In Example 5(a) we are integrating "with respect to *t*," and in Example 5b "with respect to *u*."

Practice Exercise 3

Solve $\int z^{-1/2}\, dz$. *(solution on page 272)*

Notice what happens if we try to integrate x^{-1}.

$$\int x^{-1}\, dx = \frac{1}{0} x^0 + C$$ *undefined because of the $\dfrac{1}{0}$*

$$\underset{n\,=\,-1}{\rule{0pt}{0pt}} \qquad \underset{n+1\,=\,0}{\rule{0pt}{0pt}}$$

The power rule for integration fails for the exponent -1 because it leads to the undefined expression $1/0$. For this reason the power rule for integration includes the restriction "$n \neq -1$." It can integrate any power of x *except* x^{-1}.

Sum and Constant Multiple Rules for Integration

The sum and constant multiple rules for differentiation (pages 69 and 70) lead immediately to analogous rules for integration. For any differentiable functions $f(x)$ and $g(x)$,

---

Sum Rule for Integration

$$\int [f(x) + g(x)] \, dx = \int f(x) \, dx + \int g(x) \, dx$$

---

the integral of a sum is the sum of integrals

For any constant c and differentiable function $f(x)$,

Constant Multiple Rule for Integration

$$\int c \cdot f(x) \, dx = c \int f(x) \, dx$$

---

the integral of a constant times a function is the constant times the integral of the function

The sum rules says that a sum of terms can be integrated one at a time. This rule can be extended to integrate the sum or difference of any number of terms. The constant multiple rule says that constants can be moved across the integral sign. Note that only constants, not variables, can be moved across the integral sign.

Example 6

$$\int (x^2 + x^3) \, dx = \int x^2 \, dx + \int x^3 \, dx$$

using the sum rule to break the integral into two integrals

$$= \frac{1}{3} x^3 + \frac{1}{4} x^4 + C$$

using the power rule on each ■

Example 7

$$\int 6x^2 \, dx = 6 \int x^2 \, dx$$

using the constant multiple rule to move the constant outside

$$= 6 \cdot \frac{1}{3} x^3 + C$$

using the power rule with $n = 2$

$$= 2x^3 + C$$

simplifying ■

Example 8

$$\int 7 \, dx = 7 \int 1 \, dx = 7x + C$$

└─ moving the   └─ the integral of 1 is x
constant   (plus C)
outside

■

This leads to a very useful general rule. For any constant k,

$$\int k\ dx = kx + C$$

the integral of a constant is the constant times x (plus C)

Further Examples

Regardless of how many integrations there are in a problem, we only use one C at the end. This is because any number of constants of integration can be added together to give just one C.

Example 9

$$\int (6x^2 - 3x^{-2} + 5)\ dx = 6\int x^2\ dx\ -\ 3\int x^{-2}\ dx\ +\ 5\int 1\ dx$$

breaking up the integral up and moving constants outside

$$= 6\cdot\frac{1}{3}x^3 - 3\frac{1}{-1}x^{-1} + 5x + C$$

integrating each separately

└─ from integrating the 1

$$= 2x^3 + 3x^{-1} + 5x + C$$

simplifying ■

Practice Exercise 4

Solve $\int (9x^2 - 4x^{-3} + 2)\ dx$.

(solution on page 272)

Remember that the only function that we can integrate so far is x^n, x to a power, along with sums, differences, and constant multiples of such functions. Therefore, any integrand must first be expressed in terms of *powers*.

Example 10

$$\int \left(\frac{3\sqrt{x}}{2} - \frac{2}{\sqrt{x}}\right) dx = \int \left(\frac{3}{2}x^{1/2} - 2x^{-1/2}\right) dx$$

written as powers of x

$$= \frac{3}{2}\frac{2}{3}x^{3/2} - 2\frac{2}{1}x^{1/2} + C$$

integrating each term separately

$$= x^{3/2} - 4x^{1/2} + C$$

simplifying ■

Practice Exercise 5

Solve $\int \left(\sqrt[3]{w} - \frac{4}{w^3}\right) dw$.

(solution on page 272)

Some integrals are so simple that they can be integrated "at sight."

Example 11

(a) $\int 4x^3\ dx = x^4 + C$

by remembering that $4x^3$ is the derivative of x^4

(b) $\int 7x^6\ dx = x^7 + C$

by remembering that $7x^6$ is the derivative of x^7 ■

Each of these integrands is of the form nx^{n-1}, which is the derivative of x^n, x to a power. Try to "spot" integrands of the form nx^{n-1} and solve them instantly as x^n without working through the power rule.

Practice Exercise 6

Solve each integral "at sight" by recognizing the integrand as the derivative of x to a power.

(a) $\int 5x^4 \, dx$ **(b)** $\int 3x^2 \, dx$ *(solutions on page 272)*

Algebraic Simplification of Integrals

Sometimes an integrand needs to be multiplied out or otherwise simplified before it can be integrated.

Example 12
Solve the integral $\int x^2(x + 6)^2 \, dx$.

Solution

$$\int x^2(x + 6)^2 \, dx = \int x^2\underbrace{(x^2 + 12x + 36)}_{(x + 6)^2} \, dx \qquad \text{"squaring out" the } (x + 6)^2$$

$$= \int (x^4 + 12x^3 + 36x^2) \, dx \qquad \text{multiplying out}$$

$$= \frac{1}{5} x^5 + 12 \frac{1}{4} x^4 + 36 \frac{1}{3} x^3 + C \qquad \text{integrating each term separately}$$

$$= \frac{1}{5} x^5 + 3x^4 + 12x^3 + C \qquad \text{simplifying}$$

■

Practice Exercise 7

Solve the integral $\int \frac{6t^2 - t}{t} \, dt$. (*Hint*: First simplify the integrand.) *(solution on page 272)*

Now that we can calculate integrals, let us see what they are used for.

Application: Integrals Recover Cost from Marginal Cost

Since differentiation turns a cost function into a marginal cost function, integration turns a marginal cost function back to a cost function. To determine the entire cost function, however, we need both the marginal and fixed costs.

Example 13 If a company's marginal cost function is $MC(x) = 6\sqrt{x}$ and the fixed cost is $1000, find the cost function.

Solution

$$C(x) = \int MC(x)\, dx = \int 6\sqrt{x}\, dx = 6\int x^{1/2}\, dx$$

integrating marginal cost to get cost

$$= 6 \cdot \frac{2}{3} x^{3/2} + K = 4x^{3/2} + K$$

using the power rule

(We write the arbitrary constant as K to avoid confusion with the cost function C). The cost function evaluated at $x = 0$ always give the fixed cost (because when nothing is produced, only the fixed cost remains):

$$\left(\begin{array}{c}\text{fixed}\\\text{costs}\end{array}\right) = C(0) = 4\underbrace{(0)^{3/2}}_{0} + K = K$$

evaluating $C(x) = 4x^{3/2} + K$ at $x = 0$

Therefore, the constant K is the fixed cost, which is given as $1000. The completed cost function is obtained by replacing K by 1000.

$$C(x) = 4x^{3/2} + 1000$$

$C(x) = 4x^{3/2} + K$ with $K = 1000$ ■

Notice that finding the cost function involved two steps:

(a) Integrating the marginal cost to find the cost function

(b) Using the fixed cost to evaluate the arbitrary constant

Application: Integrals Recover Total Quantities from Rates of Change

Integration, being the reverse of differentiation, can recover *any* quantity from its rate of change (together with one fixed value). For example, given the size of an economy and its rate of growth, we can integrate to find the size of the economy at any time in the future.

Example 14 The gross national product (GNP) of a country is $78 billion and growing at the rate of $r(t) = 4.4t^{-1/3}$ billion dollars per year. Find a formula for the GNP after t years. Then find the GNP after 8 years.

Solution

$$G(t) = \int 4.4t^{-1/3}\, dt = 4.4 \int t^{-1/3}\, dt$$

integrating the rate of change

$$= 4.4 \left(\frac{3}{2}\right) t^{2/3} + C = 6.6t^{2/3} + C$$

using the power rule and then simplifying

As before, evaluating $G(t)$ at $t = 0$ gives $G(0) = C$, which shows that the constant C is the GNP at time $t = 0$, which is given as $78 billion. Therefore, $C = 78$.

$$G(t) = 6.6t^{2/3} + 78$$

$G(t) = 6.6t^{2/3} + C$ with $C = 78$

This is a formula for the GNP at any time t. The GNP in 8 years will be

$$G(8) = 6.6 \cdot 8^{2/3} + 78$$

$G(t) = 6.6t^{2/3} + 78$ *evaluated at* $t = 8$

$$= 6.6(4) + 78 = \$104.4 \text{ billion}$$

▪

Note that C gave the "initial value" (the GNP at time 0) for the function. The applied exercises will give many other examples in which the C is the initial value for a function.

Geometrical Meaning of the Arbitrary Constant

Besides being an "initial value" or "starting point," the constant C has a graphical meaning. For example, the integral of $2x$ is $\int 2x\, dx = x^2 + C$, and the solution $x^2 + C$ is actually a whole collection of functions, a different one for each value of C.

If C is:	Then $x^2 + C$ is:
-1	$x^2 - 1$
0	x^2
$+1$	$x^2 + 1$
	etc.

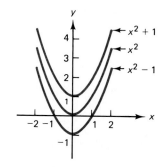

For different values of C, $x^2 + C$ gives parallel curves, all having the same slope at any given x-value.

Geometrically, solving the integral $\int 2x\, dx$ means finding a curve whose derivative (slope) at x is given by $2x$. The arbitrary constant means that there are infinitely many such curves, all of the form x^2 shifted up or down by an arbitrary amount C.

The Integral Is a Continuous Sum

We have seen that differentiation is a kind of "breaking down," changing total cost into marginal (per unit) cost. Integration, the reverse process is therefore a kind of "adding up," recovering the total from its parts. For example, integration recovers the total cost from marginal (per unit) costs, and integration adds up the growth of an economy over several years to give its total size.

To put this another way, ordinary addition is for summing "chunks," like $2 plus $3 equals $5, while integration is for summing *continuous* change, like the slow but steady growth of an economy over time.

> Integration is continuous summation.

The coming sections give further examples of integrals as continuous sums.

Summary

Keep in mind the two different lines of development of this section: one the *techniques* of integration, the other the *meaning* or *uses* of integration.

On the technical side, we defined indefinite integration as the reverse process of differentiation. The power rule enables us to integrate powers.

$$\int x^n \, dx = \frac{1}{n+1} x^{n+1} + C \qquad\qquad n \neq -1$$

Increasing the exponent by 1 should seem quite reasonable: differentiation lowers the exponent by 1, so integration, the reverse process, should raise it by 1. The sum and constant multiple rules extended integration to more complicated functions such as polynomials. Sometimes algebra is necessary to express a function in power form. *Reminder*: Do not forget the C.

As for the *uses* of integration, it recovers cost revenue and profit from their marginals, and in general it recovers any quantity from its rate of change (together with an initial value). The inverse relationship between integration and differentiation can be shown in a diagram.

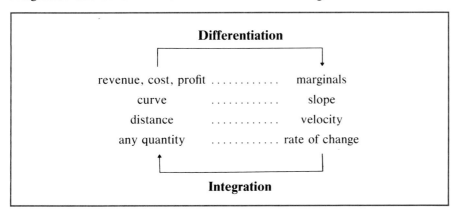

SOLUTIONS TO PRACTICE EXERCISES

1. $\displaystyle\int x^2 \, dx = \frac{1}{3} x^3 + C \qquad Check: \frac{d}{dx}\left(\frac{1}{3} x^3 + C\right) = x^2$

2. $\int x^{-3} \, dx = -\dfrac{1}{2} x^{-2} + C$

3. $\int z^{-1/2} \, dx = 2z^{1/2} + C$

4. $\int (9x^2 - 4x^{-3} + 2) \, dx = 9 \int x^2 \, dx - 4 \int x^{-3} \, dx + 2 \int 1 \, dx$

$= 9 \cdot \dfrac{1}{3} x^3 - 4 \left(-\dfrac{1}{2}\right) x^{-2} + 2x + C = 3x^3 + 2x^{-2} + 2x + C$

5. $\int (w^{1/3} - 4w^{-3}) \, dw = \int w^{1/3} \, dw - 4 \int w^{-3} \, dw$

$= \dfrac{3}{4} w^{4/3} - 4 \left(-\dfrac{1}{2}\right) w^{-2} + C = \dfrac{3}{4} w^{4/3} + 2w^{-2} + C$

6. (a) $\int 5x^4 \, dx = x^5 + C$ (b) $\int 3x^2 \, dx = x^3 + C$

7. $\int \dfrac{6t^2 - t}{t} \, dt = \int \dfrac{t(6t - 1)}{t} \, dt = \int (6t - 1) \, dt = 6 \int t \, dt - \int 1 \, dt$

$= 6 \dfrac{1}{2} t^2 - t + C = 3t^2 - t + C$

EXERCISES 5.1

Solve each integral.

1 $\int x^4 \, dx$
2 $\int x^7 \, dx$
3 $\int x^{2/3} \, dx$
4 $\int x^{3/2} \, dx$

5 $\int \sqrt{u} \, du$
6 $\int \sqrt[3]{u} \, du$
7 $\int \dfrac{dw}{w^4}$
8 $\int \dfrac{dw}{w^2}$

9 $\int \dfrac{dz}{\sqrt{z}}$
10 $\int \dfrac{dz}{\sqrt[3]{z}}$
11 $\int 6x^5 \, dx$
12 $\int 9x^8 \, dx$

13 $\int (8x^3 - 3x^2 + 2) \, dx$
14 $\int (12x^3 + 3x^2 - 5) \, dx$
15 $\int \left(6\sqrt{x} + \dfrac{1}{\sqrt[3]{x}}\right) dx$
16 $\int \left(3\sqrt{x} + \dfrac{1}{\sqrt{x}}\right) dx$

17 $\int \left(16 \sqrt[3]{x^5} - \dfrac{16}{\sqrt[3]{x^5}}\right) dx$
18 $\int \left(14 \sqrt[4]{x^3} - \dfrac{3}{\sqrt[4]{x^3}}\right) dx$
19 $\int \left(10 \sqrt[3]{t^2} + \dfrac{1}{\sqrt[3]{t^2}}\right) dt$
20 $\int \left(21 \sqrt{t^5} + \dfrac{6}{\sqrt{t^5}}\right) dt$

---

21 $\int (x - 1)^2 \, dx$
22 $\int (x + 2)^2 \, dx$
23 $\int (1 + 10w) \sqrt{w} \, dw$
24 $\int (1 - 7w) \sqrt[3]{w} \, dw$

25 $\int \dfrac{6x^3 - 6x^2 + x}{x} \, dx$
26 $\int \dfrac{4x^4 + 4x^2 - x}{x} \, dx$
27 $\int (x - 2)(x + 4) \, dx$
28 $\int (x + 5)(x - 3) \, dx$

29 $\int (r - 1)(r + 1) \, dr$
30 $\int (3s + 1)(3s - 1) \, ds$
31 $\int \dfrac{x^2 - 1}{x + 1} \, dx$
32 $\int \dfrac{x^2 - 1}{x - 1} \, dx$

33 $\int (t + 1)^3 \, dt$
34 $\int (t - 1)^3 \, dt$

35 Evaluate

(a) $\int \frac{1}{x^3}\, dx$ **(b)** $\dfrac{\int 1\, dx}{\int x^3\, dx}$

Notice that the answers are not the same, showing that you do not integrate a fraction by integrating the numerator and denominator separately.

36 Evaluate

(a) $\int x\, dx$ **(b)** $x \int 1\, dx$

Notice that the two answers are not the same, showing that a variable cannot be moved across the integral sign.

■ **APPLIED EXERCISES**

37 (*Business–Cost*) A company's marginal cost function is MC $= 20x^{3/2} - 15x^{2/3} + 1$, where x is the number of units, and fixed costs are $4000. Find the cost function.

38 (*Business–Cost*) A company's marginal cost function is MC $= 21x^{4/3} - 6x^{1/2} + 50$, where x is the number of units, and fixed costs are $3000. Find the cost function.

39 (*Business–Revenue*) A company's marginal revenue function is MR $= 12\sqrt[3]{x} + 3\sqrt{x}$, where x is the number of units. Find the revenue function. (Evaluate C so that revenue is zero when nothing is produced.)

40 (*Business–Revenue*) A company's marginal revenue function is MR $= 15\sqrt{x} + 4\sqrt[3]{x}$, where x is the number of units. Find the revenue function. (Evaluate C so that revenue is zero when nothing is produced.)

41 (*General–Velocity*) A Porsche 928 can accelerate from a standing start to a speed of $v(t) = -.09t^2 + 8t$ feet per second after t seconds (for $t < 35$).

(a) Find a formula for the distance that it will travel from its starting point in the first t seconds. (*Hint*: Integrate velocity to find distance, and then use the fact that distance is 0 at time $t = 0$.)

(b) Use the formula that you found in part (a) to find the distance that the car will travel in the first 10 seconds.

42 (*General–Velocity*) A BMW 733i can accelerate from a standing start to a speed of $v(t) = -.09t^2 + 6t$ feet per second after t seconds (for $t < 40$).

(a) Find a formula for the distance that it will travel from its starting point in the first t seconds. (*Hint*: Integrate velocity to find distance, and then use the fact that distance is 0 at time $t = 0$.)

(b) Use the formula that you found in part (a) to find the distance that the car will travel in the first 10 seconds.

43 (*General–Learning*) A person can memorize words at the rate of $r(t) = 3/\sqrt{t}$ words per minute.

(a) Find a formula for the total number of words that can be memorized in t minutes. (*Hint*: Evaluate C so that 0 words have been memorized at time $t = 0$.)

(b) Use the formula that you found in part (a) to find the total number of words that can be memorized in 25 minutes.

44 (*Biomedical–Temperature*) A patient's temperature is 108 degrees and is changing at the rate of

$$r(t) = t^2 - 4t \text{ degrees per hour}$$

where t is the number of hours since taking fever-reducing medication ($t \leq 3$).

(a) Find a formula for the patient's temperature after t hours. (*Hint*: Evaluate the constant C so that the temperature is 108 at time $t = 0$.)

(b) Use the formula that you found in part (a) to find the patients temperature after 3 hours.

45 (*General–Pollution*) A chemical plant is adding pollution to a lake at the rate of $r(t) = 40\sqrt{t^3}$ tons per year, where t is the number of years that the plant has been in operation.

(a) Find a formula for the total amount of pollution that will enter the lake in the first t years of the plant's operation. (*Hint*: Evaluate C so that no pollution has been added at time $t = 0$.)

(b) Use the formula that you found in part (a) to find how much pollution will enter the lake in the first 4 years of the plant's operation.

(c) If all life in the lake will cease when 400 tons of pollution have entered the lake, will the lake "live" beyond 4 years?

46 (*Business–Appreciation*) A $20,000 art collection is increasing in value at the rate of $r(t) = 300\sqrt{t}$ (dollars per year) after t years.

(a) Find a formula for its value after t years. (*Hint*: Evaluate C so that its value at time $t = 0$ is $20,000.)

(b) Use the formula that you found in part (a) to find its value after 25 years.

■ **REVIEW EXERCISE**

(This exercise will be important in the next section.)

47 Find $\dfrac{d}{dx} \ln(-x)$.

5.2 INTEGRATION USING LOGARITHMIC AND EXPONENTIAL FUNCTIONS

Introduction

In Section 5.1 we defined integration as the reverse of differentiation, and we introduced several integration formulas. In this section we develop integration formulas involving logarithmic and exponential functions. One of these formulas will answer a question that we could not answer earlier, namely, how to integrate x^{-1}, the one power not covered by the power rule.

The Integral $\int e^{ax} \, dx$

For any constant $a \neq 0$, the integral of the exponential function e^{ax} is

$$\int e^{ax} \, dx = \frac{1}{a} e^{ax} + C$$

the integral of e to a constant times x is one over the constant times the original function

The restriction $a \neq 0$ prevents the $1/a$ on the right-hand side from becoming undefined.

Example 1

$$\int e^{2x} \, dx = \frac{1}{2} e^{2x} + C$$

$$\underset{a\,=\,2}{\uparrow} \quad \underset{1/a}{\uparrow} \quad \underset{\text{the original function}}{\uparrow}$$

using the formula with a = 2

■

As always, we may check the answer by differentiation.

$$\frac{d}{dx}\left(\frac{1}{2}e^{2x} + C\right) = \frac{1}{2}\cdot(2)e^{2x} = e^{2x}$$

using the differentiation formula
$$\frac{d}{dx}e^{ax} = ae^{ax}$$

The result is the integrand e^{2x}, so the integration is correct.
   The proof of this rule consists simply of differentiating the right-hand side.

$$\frac{d}{dx}\left(\frac{1}{a}e^{ax} + C\right) = \frac{1}{a}ae^{ax} = e^{ax}$$

using the differentiation formula
$$\frac{d}{dx}e^{ax} = ae^{ax}$$

The result is the integrand, so the integration formula is correct.

Example 2

(a) $\displaystyle\int e^{(1/2)x}\, dx = 2e^{(1/2)x} + C$

$a = \dfrac{1}{2}\qquad so\qquad \dfrac{1}{a} = \dfrac{1}{1/2} = 2$

(b) $\displaystyle\int 6e^{-3x}\, dx = 6\int e^{-3x}\, dx = 6\left(-\frac{1}{3}\right)\!\cdot e^{-3x} + C = -2e^{-3x} + C$

$\underset{a = -3}{\uparrow}\qquad \underset{1/a = -\frac{1}{3}}{\uparrow}\qquad \underbrace{\qquad\qquad}_{\text{simplified}}$

(c) $\displaystyle\int e^{x}\, dx = 1e^{x} + C = e^{x} + C$

$a = 1,\ so\ 1/a = 1$

Each of these answers may be checked by differentiation. Example 2c says that e^{x} is the integral of itself, just as e^{x} is the derivative of itself.

$$\int e^{x}\, dx = e^{x} + C$$

the integral of e^{x} is e^{x} (plus C)

Practice Exercise 1

Solve

(a) $\displaystyle\int 12e^{4x}\, dx$

(b) $\displaystyle\int e^{(1/3)x}\, dx$

(solutions on page 281)

Application

When using these new integration formulas to solve applied problems, be careful to evaluate the constant C correctly. It will not always be equal to the initial value of the function.

Example 3   An influenza epidemic hits a large city and spreads at the rate of $r(t) = 12e^{.2t}$ new cases per day, where t is the number of days since the epidemic began. If the epidemic began with four cases, find a formula for the total number of flu cases in the first t days of the epidemic. Also find the number of cases during the first 30 days.

Solution   To find the total number of cases, we integrate the growth rate $r(t) = 12e^{.2t}$.

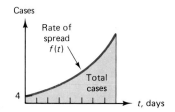

$$\binom{\text{total number}}{\text{of flu cases}} = f(t) = \int 12e^{.2t}\,dt = 12\int e^{.2t}\,dt \qquad \textit{taking out the constant}$$

$$= 12\,\frac{1}{.2}\,e^{.2t} + C = 60e^{.2t} + C \qquad \textit{using the} \int e^{ax}\,dx\ \textit{formula}$$

$$\underset{5}{\underbrace{\phantom{\frac{1}{.2}}}}$$

Evaluating $f(t)$ at $t = 0$ must give the initial number of cases.

$$\binom{\text{initial number}}{\text{of cases}} = f(0) = 60e^{.2(0)} + C = 60 + C \qquad \begin{array}{l}\textit{f(t) evaluated at t = 0 (the beginning}\\ \textit{of the epidemic)}\end{array}$$

$$\underbrace{e^0 = 1}$$

This initial quantity must equal the given initial number, 4.

$$60 + C = 4$$

$$C = 4 - 60 = -56 \qquad \textit{solving for C}$$

Replacing C by -56 gives the total number of flu cases within t days.

$$\binom{\text{flu cases}}{\text{within } t \text{ days}} = f(t) = 60e^{.2t} - 56 \qquad \textit{f(t) = 60e}^{.2t}\textit{ + C with C = −56}$$

To find number within 30 days we evaluate $f(t)$ at $t = 30$.

$$f(30) = 60e^{.2(30)} - 56 \qquad \textit{f(t) = 60e}^{.2t}\textit{ − 56 at t = 30}$$

$$= 60e^6 - 56 \approx 24{,}124 \qquad \textit{using a calculator}$$

Therefore, within 30 days the epidemic will have grown to more than 24,000 cases. ■

Evaluating the Constant C

Notice that we did *not* simply replace the constant C by the initial number of cases, 4. Had we done that, we would have obtained $C(t) = 60e^{.2t} + 4$, which would have meant that at time $t = 0$ there were

$$C(0) = 60e^0 + 4 = 60 + 4 = 64$$

cases of flu, rather than the correct number, 4. Instead, we evaluated the function at the initial time $t = 0$ and set it equal to the initial number of cases.

$$\underbrace{60e^0 + C}_{\substack{C(t)\ evaluated\\ at\ t\ =\ 0}} = \underbrace{4}_{\substack{given\\ initial\ value}}$$

We then solved to find $C = -56$. With this correct value for C the function gives the correct initial value, $C(0) = 60e^0 - 56 = 60 - 56 = 4$.

In general, to evaluate the constant C in $f(t) + C$:

1. Evaluate $f(t) + C$ at the given number (usually $t = 0$) and set the result equal to the stated initial value.

2. Solve for C.

3. Replace the C in $f(t) + C$ by its correct value.

The Integral $\int \frac{1}{x}\, dx$

The differentiation formula $\dfrac{d}{dx} \ln x = \dfrac{1}{x}$ can be read "backward" as an integration formula

$$\int \frac{1}{x}\, dx = \ln x + C$$

the integral of 1 over x is the natural log of x (plus C)

This formula, however, is restricted to $x > 0$, for only then is $\ln x$ defined. For $x < 0$ we can differentiate $\ln(-x)$, giving

$$\frac{d}{dx} \ln(-x) = \frac{-1}{-x} = \frac{1}{x}$$

using $\dfrac{d}{dx} \ln f = \dfrac{f'}{f}$ and simplifying

This says that for $x < 0$ the integral of $1/x$ is $\ln(-x)$. The negative sign in $\ln(-x)$ serves only to make the already negative x positive, and this could be accomplished just as well with absolute value bars.

$$\int \frac{1}{x}\, dx = \ln |x| + C$$

the integral of 1 over x is the natural logarithm of the absolute value of x

This formula holds for negative *and* positive values of x, since in both cases $\ln |x|$ is defined. The integrand can be written in different ways.

$$\int \frac{1}{x}\, dx = \int \frac{dx}{x} = \int x^{-1}\, dx = \ln |x| + C$$

All three integrals are equal, and the solution to each is $\ln |x| + C$.

Example 4

$$\int \frac{5}{2x}\, dx = \int \frac{5}{2}\frac{1}{x}\, dx = \frac{5}{2}\int \frac{1}{x}\, dx = \frac{5}{2} \ln |x| + C$$

taking out the constant   using the ln formula   ▪

Example 5

$$\int (x^{-1} + x^{-2})\, dx = \underbrace{\int x^{-1}\, dx + \int x^{-2}\, dx}_{\substack{\text{separating into} \\ \text{two integrals}}} = \ln |x| - x^{-1} + C$$

separating into   from the ⌐   ⌐ from the
two integrals   natural log   power rule
            formula   with $n = -2$ ■

Practice Exercise 2

Solve $\int \dfrac{3}{4x}\, dx$.

(solution on page 281)

Application: Sales

Example 6 When IBM announced in 1985 that it would discontinue its PCjr personal computer, a computer dealer decided to sell his inventory of 500 PCjr's at a discount. The dealer predicted that the sales rate (sales per month) during month t of the sale would be $r(t) = 200/t$, where $t = 1$ corresponded to the beginning of the sale. Find a formula for the number of computers that will be sold up to month t, and find whether the entire inventory will be sold by time $t = 12$ months.

Solution To find the *total* sales we integrate the sales *rate* $r(t) = 200/t$.

$$\left(\begin{matrix}\text{Total} \\ \text{sales}\end{matrix}\right) = S(t) = \int \frac{200}{t}\, dt = 200 \int \frac{1}{t}\, dt = 200 \ln |t| + C$$

We omit the absolute value bars around t, since the number of months must be positive. Evaluating this sales function $S(t) = 200 \ln t + C$ at $t = 1$, the beginning of the sale, gives the initial number of sales.

$$\left(\begin{matrix}\text{Initial} \\ \text{sales}\end{matrix}\right) = S(1) = 200 \underbrace{\ln 1}_{0} + C = C$$

This must be 0 (since the sale began then), giving $C = 0$. Substituting $C = 0$ into the sales function $S(t) = 200 \cdot \ln t + C$, we obtain the formula for the total number sold up to month t.

$$S(t) = 200 \ln t$$

To find the total sales up to month 12, we evaluate this at $t = 12$.

$$S(12) = 200 \ln 12 \approx 200(2.485) = 497 \qquad \text{\textit{using a calculator}}$$

Therefore, at this rate, all but 3 of the original 500 computers will have been sold by time $t = 12$ months. ■

Total sales is the integral of monthly sales.

In the example the initial time (the beginning of the sale) was given as $t = 1$ rather than the more usual $t = 0$. The initial time will be clear from the problem.

Application: Consumption of Natural Resources

Just as the world population grows exponentially, so does the world's annual consumption of natural resources. We can estimate the total consumption at any time in the future by integrating the rate of consumption, and from this predict when the known reserves will be exhausted.

Example 7   According to estimates by the Bureau of Mines, the annual world consumption of tin will be $r(t) = .27e^{.01t}$ million metric tons per year, where t is the number of years since 1985. Find a formula for the total tin consumption within t years of 1985 and estimate when the known world resources of 35 million metric tons will be exhausted.*

Solution   To find the *total* consumption we integrate the *rate* $r(t) = .27e^{.01t}$.

$$\left(\begin{array}{c}\text{Total tin} \\ \text{consumption}\end{array}\right) = C(t) = \int .27e^{.01t}\, dt = .27 \int e^{.01t}\, dt \qquad \textit{taking out the constant}$$

$$= .27\,\underbrace{\frac{1}{.01}}_{100}\, e^{.01t} + C = 27e^{.01t} + C \qquad \textit{using the } \int e^{ax}dx \textit{ formula}$$

$C(t)$ is the total consumption *since 1985* (year $t = 0$), so $C(t)$ should be zero at $t = 0$.

$$C(0) = 27e^0 + C = 27 + C \qquad \textit{C(t) = 27e}^{.01t}\textit{ + C at t = 0}$$

For this to be zero, C must be -27. Substituting $C = -27$ gives the formula for the total consumption of tin within t years of 1985.

$$C(t) = 27e^{.01t} - 27 \qquad \textit{C(t) = 27e}^{.01t}\textit{ + C with C = −27}$$

To predict when the total world reserves of 35 million tons will be exhausted, we set this function equal to 35 and solve for t.

$$27e^{.01t} - 27 = 35$$

$$27e^{.01t} = 62 \qquad \textit{adding 27}$$

$$e^{.01t} = \frac{62}{27} \approx 2.3 \qquad \textit{dividing by 27}$$

---

* See H. E. Goeller and A. Zucker, "Infinite Resources: The Ultimate Strategy," *Science*, **223** (1984), 456–462.

$$\underbrace{\ln e^{.01t}}_{.01t} = \ln 2.3$$

taking natural logs (as we did in Section 3.4)

$$.01t = \ln 2.3$$

using ln $e^x = x$ (property 3 of natural logarithms)

$$t = \frac{\ln 2.3}{.01} \approx \frac{.83}{.01} = 83$$

dividing by .01 and using a calculator

Therefore, the known world supply of tin will be exhausted in about 83 years of 1985, which means about the year 2068 (assuming that consumption continues at this rate). ■

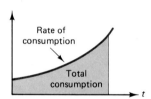

Total tin consumption is the integral of the consumption rate.

Incidentally, tin is used mainly for coating steel. For example, a "tin" can is actually a steel can with a thin protective coating of tin to prevent rust. The predicted unavailability of tin is already causing major changes in the food-packaging industry.

Power Rule for Integration, Revisited

Our new integration formula shows how to integrate x^{-1}, the only power not covered by the power rule. Therefore, we may now write one "combined" formula for the integral $\int x^n \, dx$ for *any n*.

$$\int x^n \, dx = \begin{cases} \dfrac{1}{n+1} x^{n+1} + C & \text{if } n \neq -1 \\[2ex] \ln |x| + C & \text{if } n = -1 \end{cases}$$

← *use the power rule if n is other than* −1

← *use the ln formula if n equals* −1

It is a curious fact that every power of x integrates to another power of x, with the single exception of x^{-1}, which integrates to an entirely different kind of function, the natural logarithm.

Summary

We have three integration formulas.

$$\int x^n \, dx = \frac{1}{n+1} x^{n+1} + C \qquad \text{for } n \neq -1$$

$$\int \frac{1}{x} \, dx = \int \frac{dx}{x} = \int x^{-1} \, dx = \ln|x| + C$$

$$\int e^{ax} \, dx = \frac{1}{a} e^{ax} + C \qquad \text{for } a \neq 0$$

The absolute value bars in the middle formula should be omitted if x is positive. In applications we now evaluate C by setting the function (evaluated at the given number) equal to the stated initial value, and solving for C.

SOLUTIONS TO PRACTICE EXERCISES

1. (a) $\displaystyle \int 12e^{4x} \, dx = 12 \int e^{4x} \, dx = 12 \cdot \frac{1}{4} e^{4x} + C = 3e^{4x} + C$

(b) $\displaystyle \int e^{(1/3)x} \, dx = 3e^{(1/3)x} + C$

2. $\displaystyle \int \frac{3}{4x} \, dx = \frac{3}{4} \int \frac{1}{x} \, dx = \frac{3}{4} \ln|x| + C$

EXERCISES 5.2

Solve each integral.

1 $\displaystyle \int e^{3x} \, dx$

2 $\displaystyle \int e^{4x} \, dx$

3 $\displaystyle \int e^{(1/4)x} \, dx$

4 $\displaystyle \int e^{(1/3)x} \, dx$

5 $\displaystyle \int e^{.05x} \, dx$

6 $\displaystyle \int e^{.02x} \, dx$

7 $\displaystyle \int e^{-2y} \, dy$

8 $\displaystyle \int e^{-3y} \, dy$

9 $\displaystyle \int e^{-.5x} \, dx$

10 $\displaystyle \int e^{-.4x} \, dx$

11 $\displaystyle \int 6e^{(2/3)x} \, dx$

12 $\displaystyle \int e^{(3/4)x} \, dx$

13 $\displaystyle \int 18e^{-(3/4)u} \, du$

14 $\displaystyle \int 24e^{-(2/3)u} \, du$

15 $\displaystyle \int \frac{2}{x} \, dx$

16 $\displaystyle \int \frac{3}{x} \, dx$

17 $\displaystyle \int -5x^{-1} \, dx$

18 $\displaystyle \int -\frac{1}{2} x^{-1} \, dx$

19 $\displaystyle \int \frac{3 \, dx}{x}$

20 $\displaystyle \int \frac{2 \, dx}{x}$

21 $\displaystyle \int \frac{3 \, dx}{x^2}$

22 $\displaystyle \int \frac{2 \, dx}{x^2}$

23 $\displaystyle \int \frac{dv}{3v}$

24 $\displaystyle \int \frac{dv}{4v}$

25 $\displaystyle \int \frac{3}{2x} \, dx$

26 $\displaystyle \int \frac{2}{3x} \, dx$

27 $\displaystyle \int \left(e^{3x} - \frac{3}{x} \right) dx$

28 $\displaystyle \int \left(e^{2x} + \frac{2}{x} \right) dx$

29 $\int (3e^{.5t} - 2t^{-1})\, dt$

30 $\int (5e^{-.5t} - 4t^{-1})\, dt$

31 $\int (x^2 + x + 1 + x^{-1} + x^{-2})\, dx$

32 $\int (x^{-2} - x^{-1} + 1 - x + x^2)\, dx$

33 $\int (5e^{.02t} - 2e^{.01t})\, dt$

34 $\int (3e^{.05t} - 2e^{.04t})\, dt$

APPLIED EXERCISES (Most require 🖩 .)

35–36 (Biomedical–Epidemics) A flu epidemic hits a college community, beginning with five cases on day $t = 0$. The rate of growth of the epidemic (new cases per day) is given by the function $r(t)$ below, where t is the number of days since the epidemic began.

(a) Find a formula for the total number of cases of flu in the first t days.

(b) Use your answer to part (a) to find the total number of cases in the first 20 days.

35 $r(t) = 18e^{.05t}$ **36** $r(t) = 20e^{.04t}$

37 (Business–Sales) In an effort to reduce its inventory, a record store runs a sale on its least popular records. The sales rate (records sold per day) on day t of the sale is predicted to be $r(t) = 50/t$ $(t > 1)$, where $t = 1$ corresponds to the beginning of the sale, at which time none of the inventory of 200 records had been sold.

(a) Find a formula for the total number of records sold up to day t.

(b) Will the store have sold its inventory of 200 records by day $t = 30$?

38 (Business–Sales) In an effort to reduce its inventory, a book store runs a sale on its least popular mathematics books. The sales rate (books sold per day) on day t of the sale is predicted to be $r(t) = 60/t$ $(t > 1)$, where $t = 1$ corresponds to the beginning of the sale, at which time none of the inventory of 350 books had been sold.

(a) Find a formula for the number of books sold up to day t.

(b) Will the store have sold its inventory of 350 books by day $t = 30$?

39 (General–Consumption of Natural Resources) World consumption of silver is running at the rate of $r(t) = 14e^{.02t}$ thousand metric tons per year, where t is measured in years and $t = 0$ corresponds to 1985.

(a) Find a formula for the total amount of silver that will be consumed within t years of 1985.

(b) When will the known world resources of 770 thousand metric tons of silver be exhausted? (Silver is used extensively in photography.)

40 (General–Consumption of Natural Resources) World consumption of copper is running at the rate of $r(t) = 10e^{.04t}$ million metric tons per year, where t is measured in years and $t = 0$ corresponds to 1985.

(a) Find a formula for the total amount of copper that will be used within t years of 1985.

(b) When will the known world resources of 1600 million metric tons of copper be exhausted?

41 (General–Cost of Maintaining a Home) The cost of maintaining a home generally increases as the home becomes older. Suppose that the rate of cost (dollars per year) for a home that is x years old is

$$r(x) = 200e^{.4x}.$$

(a) Find a formula for the total maintenance cost during the first x years. (Total maintenance should be zero at $x = 0$.)

(b) Use your answer to part (a) to find the total maintenance cost during the first 5 years.

42 (Biomedical–Cell Growth) A culture of bacteria is growing at the rate of $r(t) = 20e^{.8t}$ cells per day, where t is the number of days since the culture ways started. Suppose that the culture began with 50 cells.

(a) Find a formula for the total number of cells in the culture after t days.

(b) If the culture is to be stopped when the population reaches 500, when will this occur?

43 (General–Freezing of Ice) An ice cube tray filled with tap water is placed in the freezer, and the temperature is changing at the rate of $r(t) = -12e^{-.2t}$ degrees per hour after t hours. The temperature of the tap water is 70 degrees.

(a) Find a formula for the temperature of water that has been in the freezer for t hours.

(b) When will the ice be ready? (Water freezes at 32 degrees.)

44 (*Social Science–Divorces*) Since 1980 there have been approximately $r(t) = 1.2e^{.01t}$ million divorces per year, where t is measured in years and $t = 0$ corresponds to 1980. Find a formula for the total number of divorces expected within t years of 1980.

45 (*Business–Total Savings*) A factory installs new equipment that is expected to generate savings at the rate of $r(t) = 800e^{-.2t}$ dollars per year, where t is the number of years that the equipment has been in operation.

(a) Find a formula for the total savings that the equipment will generate during its first t years.

(b) If the equipment originally cost $2000, when will it "pay for itself?"

46 (*Business–Total Savings*) A company installs a new computer that is expected to generate savings at the rate of $r(t) = 20,000e^{-.02t}$ dollars per year, where t is the number of years that the computer has been in operation.

(a) Find a formula for the total savings that the computer will generate during its first t years.

(b) If the computer originally cost $250,000, when will it "pay for itself?"

Solve each integral. (*Hint:* Use some algebra first.)

47 $\displaystyle\int \frac{(x + 1)^2}{x}\, dx$

48 $\displaystyle\int \frac{(x - 1)^2}{x}\, dx$

49 $\displaystyle\int \frac{(t - 1)(t + 3)}{t^2}\, dt$

50 $\displaystyle\int \frac{(t + 2)(t - 4)}{t^2}\, dt$

51 $\displaystyle\int \frac{(x - 2)^3}{x}\, dx$

52 $\displaystyle\int \frac{(x + 2)^3}{x}\, dx$

5.3 DEFINITE INTEGRALS AND AREAS

Introduction

In this section we discuss *definite* integrals, which are indefinite integrals with an added step of evaluation. We will use definite integrals for calculating areas and numerous other purposes.

Vertical Bar Evaluation Notation

For a function $f(x)$ and numbers a and b we use a vertical bar to indicate evaluation.

$$f(x)\ \Big|_a^b = \underbrace{f(b)}_{\substack{\text{evaluation} \\ \text{at upper} \\ \text{number}}} - \underbrace{f(a)}_{\substack{\text{evaluation} \\ \text{at lower} \\ \text{number}}}$$

the vertical bar means: evaluate the function at the upper number, and subtract the evaluation at the lower number

Example 1

$$x^2 \Big|_3^5 = \quad 5^2 \quad - \quad 3^2 \quad = 25 - 9 = 16$$

x^2 evaluated at $x = 5$ x^2 evaluated at $x = 3$ ■

Example 2

$$\sqrt{x} \Big|_4^9 = \sqrt{9} - \sqrt{4} = 3 - 2 = 1$$ ■

Example 3

$$(x^2 - 2x) \Big|_{-2}^4 = (16 - 8) - (4 + 4) = 8 - 8 = 0$$

$x^2 - 2x$ evaluated at $x = 4$ $x^2 - 2x$ evaluated $x = -2$ ■

Practice Exercise 1

Evaluate $\dfrac{1}{x} \Big|_1^2$

(solutions on page 293)

Definite Integral

> For a function $f(x)$ continuous on the interval [a, b], the definite integral of $f(x)$ from a to b is defined as
> $$\int_a^b f(x)\, dx = F(b) - F(a)$$
> where $F(x)$ is any antiderivative of $f(x)$.

The numbers a and b are called the *lower* and *upper limits of integration*, and the integral from a to b is sometimes called the integral *over the interval* [a, b]. Solving a definite integral involves two steps:

(a) Finding the *indefinite* integral (omitting the constant C)

(b) *Evaluating* at the upper and lower limits of integration and subtracting

For the evaluation step we will use the vertical bar notation.

Example 4

$$\int_0^3 x^2 \, dx = \frac{1}{3} x^3 \Big|_0^3 \qquad\qquad \textit{indefinite integral of } x^2 \textit{ followed by the evaluation bar}$$

$$= \frac{1}{3} (3)^3 - \frac{1}{3} (0)^3 \qquad\qquad \textit{evaluating at 3 and 0}$$

$$= \frac{1}{3} \cdot 27 - \frac{1}{3} \cdot 0 = 9 - 0 = 9 \qquad\qquad \textit{simplifying}$$ ■

Notice that definite integral equals a *number*, rather than a function plus an arbitrary constant.

Example 5

$$\int_0^1 e^{2x} \, dx = \frac{1}{2} e^{2x} \Big|_0^1 \qquad\qquad \textit{indefinite integral of } e^{2x} \textit{ followed by the evaluation bar (using the } \int e^{ax} \, dx \textit{ formula)}$$

$$= \frac{1}{2} e^{2(1)} - \frac{1}{2} e^{2(0)} \qquad\qquad \textit{substituting 1 and 0}$$

$$= \frac{1}{2} e^2 - \frac{1}{2} e^0 = \frac{1}{2} e^2 - \frac{1}{2} \qquad\qquad \textit{simplifying, using } e^0 = 1$$ ■

Example 6

$$\int_{-1}^{-3} \frac{1}{x} \, dx = \ln |x| \Big|_{-1}^{-3} \qquad\qquad \textit{indefinite integral of } 1/x \textit{ followed by the evaluation bar}$$

$$= \ln |-3| - \ln |-1| \qquad\qquad \textit{evaluating at } -3 \textit{ and } -1$$

$$= \ln 3 - \ln 1 = \ln 3 \qquad\qquad \textit{simplifying (using } \ln 1 - 0). \textit{ Notice that the absolute value bars are important here.}$$ ■

We leave these answers in their *exact* form (including expressions like e^2 and ln 3), rather than approximating them. In applied problems, however, we will approximate answers using a calculator.

Practice Exercise 2

Solve

(a) $\displaystyle\int_0^4 x^3 \, dx$ (b) $\displaystyle\int_0^5 e^{3x} \, dx$ (c) $\displaystyle\int_{-1}^{-5} x^{-1} \, dx$ *(solutions on page 293)*

Example 7

$$\int_1^{e^2} \frac{1}{x} \, dx = \ln |x| \, \Big|_1^{e^2}$$

indefinite integral of $1/x$ followed by the evaluation bar

$$= \ln |e^2| - \ln |1| = \ln e^2 - \ln 1$$

evaluating at e^2 and 1

$$= 2 - 0 = 2$$

using $\ln e^M = M$ and $\ln 1 = 0$ (properties 3 and 1 of logs)

■

Omit the *C* in Definite Integrals

To understand why we omit the *C* in definite integrals, let us recalculate our very first definite integral, $\int_0^3 x^2 \, dx$, but this time including the *C*.

$$\int_0^3 x^2 \, dx = \left(\frac{1}{3} x^3 + C \right) \Big|_0^3$$

indefinite integral, including the C

$$= \left(\frac{1}{3} \cdot 27 + C \right) - \left(\frac{1}{3} \cdot 0 + C \right)$$

evaluating at $x = 3$ and $x = 0$

$$= 9 + \underbrace{C - C}_{Cs \text{ cancel}} = 9$$

simplifying, obtaining the same answer as before

The constant *C* cancels out, leaving the same answer as before. Since this will always happen, we omit the *C* in definite integrals. (However, this does not change what we said earlier: With *indefinite* integrals always include the *C*.)

Properties of Definite Integrals

Definite integrals inherit many of the properties of indefinite integrals.

> Constants may be taken outside the integral sign:
>
> $$\int_a^b c \, f(x) \, dx = c \int_a^b f(x) \, dx$$

for any constant c

> The integral of a sum or difference is the sum or difference of the integrals:
>
> $$\int_a^b [f(x) \pm g(x)] \, dx = \int_a^b f(x) \, dx \pm \int_a^b g(x) \, dx$$

read both plus signs or both minus signs

Example 8

$$\int_0^2 (3x^2 + 6x)\, dx = \left(x^3 + 6\cdot\frac{1}{2}x^2\right)\Big|_0^2 = (x^3 + 3x^2)\Big|_0^2 \qquad \textit{indefinite integral}$$

$$= (8 + 12) - (0) = 20 \qquad \textit{evaluating}$$

■

Now that we can calculate definite integrals, let us see what they are used for.

Areas Under Curves

Definite integrals give areas under curves.

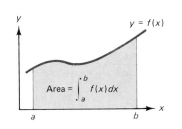

The Fundamental Theorem of Integral Calculus

For a function $f(x)$ that is continuous and nonnegative on an interval $[a, b]$, the area under the curve and above the x-axis from $x = a$ to $x = b$ is given by

$$\left(\begin{array}{l}\text{the area under } f(x) \\ \text{from } x = a \text{ to } x = b\end{array}\right) = \int_a^b f(x)\, dx$$

(We will speak of the area "under a curve," meaning the area under the curve *and above the x-axis*.) Notice that the *smaller* x-value is the *lower* limit of integration and the *larger* x-value is the *upper* limit. A justification of this theorem is given at the end of this section.

Example 9   Find the area under the curve $y = 12 - 3x^2$ from $x = -1$ to $x = 1$.

Solution   The area is shown on the right. The function $f(x) = 12 - 3x^2$ is continuous (since it is a polynomial) and nonnegative between $x = -1$ and $x = 1$, so to find the area we integrate from $x = -1$ to $x = 1$.

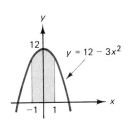

$$\int_{-1}^1 (12 - 3x^2)\, dx = (12x - x^3)\Big|_{-1}^1 = (12 - 1) - (-12 + 1)$$

$$= 11 - (-11) = 22 \text{ square units} \qquad \textit{answer}$$

■

"Square units" means that whatever the units on the graph are (inches, feet, etc.), the area is in *square* units (square inches, square feet, etc.).

Example 10   Find the area under the curve $y = e^{-x}$ from $x = 0$ to $x = 2$.

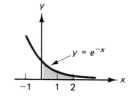

Solution   The area is shown on the right. The function is continuous and nonnegative (since e to any power is positive), so we integrate between the given values.

$$\int_0^2 e^{-x} \, dx = -e^{-x} \Big|_0^2 = -e^{-2} - (-e^0)$$ 　　　*using the $\int e^{ax} \, dx$ formula*

$$= -e^{-2} + 1 = 1 - e^{-2} \text{ square units}$$ 　　　*simplifying (using $e^0 = 1$)*

▪

Application: Total Cost of a Succession of Units

The definite integral of a marginal cost function gives the total cost of a succession of units. For example, given a marginal cost function, to find the total cost of producing units 500 to 600 we would proceed as follows: integrate the marginal cost to find the total cost $C(x)$, evaluate $C(x)$ at 600 to find the cost up to unit 600, and subtract the evaluation at 500 to leave just the cost of units 500 to 600. This process of integrating, evaluating, and subtracting, however, is exactly the process of definite integration. In general,

---

For a marginal cost function $MC(x)$,

$$\begin{pmatrix} \text{total cost of} \\ \text{units } m \text{ to } n \end{pmatrix} = \int_m^n MC(x) \, dx$$

Example 11   A company's marginal cost function is $MC(x) = 10{,}000/x$ (for $x > 1$), where x is the number of units. Find the total cost of units 500 to 600.

Solution   We integrate the marginal cost between the given values.

$$\begin{pmatrix} \text{Total cost of units} \\ \text{500 to 600} \end{pmatrix} = \int_{500}^{600} \frac{10{,}000}{x} \, dx = 10{,}000 \int_{500}^{600} \frac{1}{x} \, dx$$ 　　　*taking out the constant*

$$= 10{,}000 \cdot \ln x \Big|_{500}^{600} = 10{,}000 \cdot \ln 600 - 10{,}000 \cdot \ln 500$$ 　　　*integrating*

$$\approx 10{,}000 \cdot 6.397 - 10{,}000 \cdot 6.215 = 1820$$ 　　　*using a calculator*

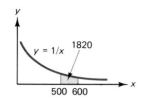

Total cost shown as an area.

Therefore, the cost of manufacturing units 500 to 600 is $1820.   ■

Application: Total Productivity

In repetitive tasks, a worker's productivity usually increases with time until it is slowed by monotony. The definite integral of the worker's rate of productivity gives the total output over an interval.

Example 12   A technician can test computer chips at the rate of $r(t) = -3t^2 + 18t + 15$ chips per hour (for $t \leq 6$), where t is the number of hours after 9:00 A.M. How many chips can be tested between 10:00 A.M. and 1:00 P.M.?

Solution   We integrate the rate from $t = 1$ (10 A.M.) to $t = 4$ (1 P.M.).

$$\int_1^4 \left(-3t^2 + 18t + 15\right) dt = \left(-t^3 + 18\frac{1}{2}t^2 + 15t\right)\Big|_1^4 = \left(-t^3 + 9t^2 + 15t\right)\Big|_1^4$$

$$= -64 + 144 + 60 - (-1 + 9 + 15) = 117$$

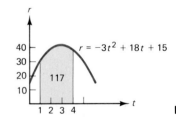

That is, between 10 A.M. and 1 P.M., 117 chips can be tested.   ■

In problems like this where the answer must be a whole number, we round any fractional value.

Application: Cigarette Smoking

The toxic material in cigarette smoke is called "tar." Most cigarettes have filters that absorb some of the tar. The tobacco near the filter acts like an additional filter until it is itself smoked, at which time it releases all of its accumulated tar. Using definite integrals we can find how much tar is inhaled from any part of a cigarette.

Example 13   A typical cigarette has 8 cm (centimeters) of tobacco that is followed by a filter. As the cigarette is smoked, tar is inhaled at the rate of $r(x) = 300e^{.025x} - 240e^{.02x}$ mg (milligrams) of tar per centimeter of tobacco, where x is the distance along the cigarette. Find the amounts of tar inhaled from the first and last centimeters of the cigarette.

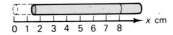

Solution For the tar in first centimeter we integrate the rate from $x = 0$ (the beginning of the cigarette) to $x = 1$.

$$\begin{pmatrix} \text{Tar from} \\ \text{first centimeter} \end{pmatrix} = \int_0^1 (300e^{.025x} - 240e^{.02x}) \, dx$$

From first cm   From last cm

$$= \left[300 \, \frac{1}{.025} \, e^{.025x} - 240 \, \frac{1}{.02} \, e^{.02x} \right] \Big|_0^1 \qquad \text{integrating}$$

$$= \left[12,000 e^{.025x} - 12,000 e^{.02x} \right] \Big|_0^1 \qquad \text{simplifying}$$

$$= 12,000 e^{.025} - 12,000 e^{.02} - (1200 e^0 - 1200 e^0) \qquad \text{evaluating}$$

$$\approx 61 \text{ milligrams of tar} \qquad \text{using a calculator}$$

For the last centimeter we integrate from $x = 7$ to $x = 8$.

$$\begin{pmatrix} \text{Tar from} \\ \text{last centimeter} \end{pmatrix} = \int_7^8 (300e^{.025x} - 240e^{.02x}) \, dx$$

$$\approx 83 \text{ milligrams of tar} \qquad \text{omitting the details of the integration}$$

■

Notice that the last centimeter releases significantly more tar (about 36% more) than the first centimeter.

Moral: The Surgeon General has determined that smoking is hazardous to your health, and the last puffs are 36% more hazardous than the first.

Definite Integrals Give Total Accumulation Over an Interval

We have seen three examples of definite integrals adding up continuous change to give the total accumulation over an interval:

1. Summing marginal cost (cost per unit) to give the total cost for a succession of units

2. Summing the productivity per hour to give the total output over a 3-hour period

3. Summing the tar from smoking different parts of a cigarette

In general, integrating any rate gives the total accumulation over the interval.

$$\begin{pmatrix} \text{Total accumulation at rate} \\ f(x) \text{ from } x = a \text{ to } x = b \end{pmatrix} = \int_a^b f(x) \, dx$$

Since definite integrals give areas, we can show total accumulations geometrically as areas under rate curves.

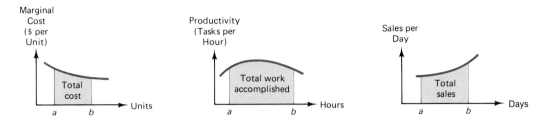

Verification of the Fundamental Theorem of Integral Calculus

Earlier we stated the *fundamental theorem of integral calculus,* which says that definite integrals give areas under curves. The following is a geometric and intuitive justification of the fundamental theorem.

For a function $f(x)$ that is continuous and nonnegative on an interval $[a, b]$, we define a new function $A(x)$ as the *area under the curve from a to x,* as shown on the right. Therefore,

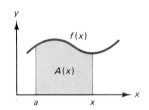

$$A(b) = \begin{pmatrix} \text{area} \\ \text{from } a \text{ to } b \end{pmatrix}$$

and

$$A(a) = 0 \qquad \text{(since the area}$$
$$\text{``from } a \text{ to } a\text{'' is zero)}$$

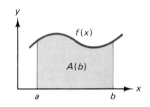

We differentiate the area function $A(x)$ using the definition of the derivative.

$$A'(x) = \lim_{h \to 0} \frac{A(x + h) - A(x)}{h}$$

The expression $A(x + h)$ is the area under the curve up to $x + h$, and $A(x)$ is the area up to x. Subtracting them, $A(x + h) - A(x)$, leaves just the area from x to $x + h$, shown on the right below.

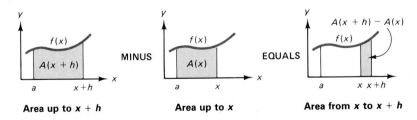

When h is small, this last area can be approximated by a rectangle of base h and height $f(x)$, where the approximation becomes exact as h approaches zero. Therefore, in the limit we may replace $A(x + h) - A(x)$ by the area of the rectangle $h \cdot f(x)$.

$$A'(x) = \lim_{h \to 0} \frac{A(x + h) - A(x)}{h} = \lim_{h \to 0} \frac{h \cdot f(x)}{h} = \lim_{h \to 0} \frac{\not{h} \cdot f(x)}{\not{h}} = f(x)$$

This equation says that $A'(x) = f(x)$, which means that $A(x)$ is an anti-derivative of $f(x)$, so we may use $A(x)$ to evaluate the definite integral $\int_a^b f(x)\,dx$.

$$\int_a^b f(x)\,dx = A(b) - \underbrace{A(a)}_{0} = A(b)$$

⌐— this is just the area from a to b

This is the result that we wanted to show, that definite integrals give areas under curves.

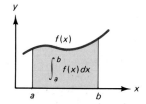

The reason for the grand title "the fundamental theorem of integral calculus" is historical. Before the invention of calculus, two of the most important problems in mathematics were the calculation of rates of change and the calculation of areas. Differentiation (developed by Newton and Leibniz) solved the problem of finding rates of change, and it came as a remarkable surprise that the reverse process, integration, also solved the problem of finding areas. It was this deep and unexpected connection between two seemingly unrelated problems that makes this result so "fundamental."

Summary

The definite integral of a continuous function f on an interval $[a, b]$ is defined as

$$\int_a^b f(x)\,dx = F(b) - F(a)$$

for any antiderivative $F(x)$ of $f(x)$. For a nonnegative function $f(x)$ the definite integral has two interpretations, one geometric and the other analytic (in terms of functions).

$$\int_a^b f(x)\,dx = \binom{\text{area under } f(x)}{\text{from } a \text{ to } b} = \binom{\text{total accumulation at}}{\text{rate } f(x) \text{ from } a \text{ to } b}$$

The exercises give further examples of both of these uses of definite integrals.

SOLUTIONS TO PRACTICE EXERCISES

1. $\dfrac{1}{x}\Big|_1^2 = \dfrac{1}{2} - \dfrac{1}{1} = -\dfrac{1}{2}$

2. **(a)** $\displaystyle\int_0^4 x^3\,dx = \dfrac{1}{4}x^4\Big|_0^4 = \dfrac{1}{4}\cdot 256 - 0 = 64$

(b) $\displaystyle\int_0^5 e^{3x}\,dx = \dfrac{1}{3}e^{3x}\Big|_0^5 = \dfrac{1}{3}\cdot e^{15} - \dfrac{1}{3}\cdot e^0 = \dfrac{1}{3}\cdot e^{15} - \dfrac{1}{3}$

(c) $\displaystyle\int_{-1}^{-5} x^{-1}\,dx = \ln|x|\Big|_{-1}^{-5} = \ln|-5| - \ln|-1| = \ln 5 - \ln 1 = \ln 5$

EXERCISES 5.3

Evaluate each definite integral.

1 $\displaystyle\int_0^2 4x^3\,dx$ **2** $\displaystyle\int_0^2 5x^4\,dx$ **3** $\displaystyle\int_0^1 4e^{2x}\,dx$ **4** $\displaystyle\int_0^1 6e^{3x}\,dx$

5 $\displaystyle\int_1^5 \dfrac{1}{x}\,dx$ **6** $\displaystyle\int_1^7 \dfrac{1}{x}\,dx$ **7** $\displaystyle\int_1^2 (6x^2 + 4x - 1)\,dx$ **8** $\displaystyle\int_1^2 (9x^2 - 6x + 1)\,dx$

9 $\displaystyle\int_2^4 (1 + x^{-2})\,dx$ **10** $\displaystyle\int_1^2 (x^{-2} - 1)\,dx$ **11** $\displaystyle\int_4^9 \dfrac{dx}{\sqrt{x}}$ **12** $\displaystyle\int_8^{27} \dfrac{dx}{\sqrt[3]{x}}$

13 $\displaystyle\int_{-2}^2 (2w + 4)\,dw$ **14** $\displaystyle\int_{-2}^2 (3w^2 + 2w)\,dw$ **15** $\displaystyle\int_1^2 (6x^2 - 2x^{-2})\,dx$ **16** $\displaystyle\int_1^2 (9x^2 - 4x^{-2})\,dx$

17 $\displaystyle\int_{-1}^1 e^{-x}\,dx$ **18** $\displaystyle\int_{-2}^0 e^{-x}\,dx$ **19** $\displaystyle\int_2^4 e^{-(1/2)x}\,dx$ **20** $\displaystyle\int_3^9 e^{-(1/3)x}\,dx$

21 $\displaystyle\int_0^1 (e^x - 2x)\,dx$ **22** $\displaystyle\int_0^1 (6x^2 - e^x)\,dx$ **23** $\displaystyle\int_1^4 \dfrac{9}{x}\,dx$ **24** $\displaystyle\int_1^3 \dfrac{8}{x}\,dx$

25 $\displaystyle\int_1^3 (x^{-2} - x^{-1})\,dx$ **26** $\displaystyle\int_1^2 (x^{-1} - x^{-2})\,dx$ **27** $\displaystyle\int_{-2}^{-1} (x^{-1} - x)\,dx$ **28** $\displaystyle\int_{-2}^{-1} (x - x^{-1})\,dx$

29 $\displaystyle\int_e^{e^2} \dfrac{dx}{x}$ **30** $\displaystyle\int_e^{e^3} \dfrac{dx}{x}$ **31** $\displaystyle\int_0^{\ln 3} e^x\,dx$ **32** $\displaystyle\int_0^{\ln 2} e^x\,dx$

33 $\displaystyle\int_1^2 \dfrac{(x+1)^2}{x}\,dx$ **34** $\displaystyle\int_1^2 \dfrac{(x-1)^2}{x}\,dx$

For each function find the area under the curve and above the x-axis between the given x-values.

35 $f(x) = x^2$ from $x = 1$ to $x = 4$

36 $f(x) = x^3$ from $x = 1$ to $x = 5$

37 $f(x) = \dfrac{1}{x^3}$ from $x = 1$ to $x = 4$

38 $f(x) = \dfrac{1}{x^2}$ from $x = 1$ to $x = 3$

39 $f(x) = 12 - 3x^2$ from $x = -1$ to $x = 2$

40 $f(x) = 27 - 3x^2$ from $x = -1$ to $x = 3$

41 $f(x) = 8 - 4\sqrt[3]{x}$ from $x = 0$ to $x = 8$

42 $f(x) = 9 - 3\sqrt{x}$ from $x = 0$ to $x = 9$

43 $f(x) = 6e^{2x}$ from $x = 1$ to $x = 4$

44 $f(x) = 9e^{3x}$ from $x = 2$ to $x = 3$

45 $f(x) = e^{x/2}$ from $x = 0$ to $x = 2$

46 $f(x) = e^{x/3}$ from $x = 0$ to $x = 3$

47 $f(x) = \dfrac{1}{x}$ from $x = 1$ to $x = 5$

48 $f(x) = \dfrac{1}{x}$ from $x = 1$ to $x = 4$

---

49 (a) Find the area under $f(x) = x$ from $x = 0$ to $x = 2$.

(b) *Graph the straight line $f(x) = x$ and show that the area under it from $x = 0$ to $x = 2$ is a triangle. Then use the formula for the area of a triangle to calculate this area. Your answer should agree with the answer to part (a).*

50 (a) Find the area under $f(x) = 3$ and above the x-axis from $x = 0$ to $x = 2$.

(b) Graph the straight line $f(x) = 3$ and show that the area under it from $x = 0$ to $x = 2$ is a rectangle. Then use the formula for the area of a rectangle to calculate this area. Your answer should agree with the answer to part (a).

51 Find the area under the curve $y = 1/x$ from $x = 1$ to $x = a$ (for any number $a > 1$).

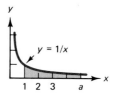

52 (a) Solve the integral $\displaystyle\int_{-1}^{1} x^{-2}\, dx$ (if possible).

(b) Explain why the answer is negative despite the fact that the integrand is nonnegative and so the integral should give the (positive) area. (*Hint:* Is the integrand continuous on the interval of integration? Look back at the definition of definite integrals on page 284.)

![calculator icon] **APPLIED EXERCISES** (Most Require 🖩 .)

53 (*General–Electricity Consumption*) On a hot summer afternoon a city's electricity consumption is $r(t) = -3t^2 + 18t + 10$ units per hour, where t is the number of hours after noon ($t \le 6$). Find the total consumption of electricity between the hours of 1 and 5 pm.

54 (*General–Weight*) An average child of age x years gains weight at the rate of $r(x) = 3.9x^{1/2}$ pounds per year (for $x \le 16$). Find the total weight gain from age 1 to age 9.

55–56 (*General–Repetitive Tasks*) After t hours of work a bank clerk can process checks at the rate of $r(t)$ checks per hour for the function $r(t)$ given below. How many checks will the clerk process during the first 3 hours (time 0 to time 3)?

55 $r(t) = -t^2 + 90t + 5$ ($t \le 6$)

56 $r(t) = -t^2 + 60t + 9$ ($t \le 6$)

57– 58 (*Business–Cost*) A company's marginal cost function is MC(x) (given below), where x is the number of units. Find the total cost of the first hundred units (units $x = 0$ to $x = 100$).

57 MC(x) = $6e^{-.02x}$ **58** MC(x) = $8e^{-.01x}$

59 (*General–Price Increase*) The price of a double-dip ice cream cone is increasing at the rate of $r(t) = 6e^{.08t}$ cents per year, where t is measured in years and $t = 0$ corresponds to 1985. Find the total price increase between the years 1985 and 2000.

60 (*Business–Sales*) An automobile dealer estimates that the newest model car will sell at the rate of $r(t) = 30/t$ cars per month, where t is measured in months and $t = 1$ corresponds to the beginning of January. Find the number of cars that will be sold from the beginning of January to the beginning of May.

61 (*Business–Tin Consumption*) World consumption of tin is running at the rate of $r(t) = .27e^{.01t}$ million tons per year, where t is measured in years and $t = 0$ corresponds to the beginning of 1985. Find the total consumption of tin from the beginning of 1990 to the beginning of the year 2000.

62 (*Sociology–Marriages*) There are approximately $r(t) = 2.4e^{.01t}$ million marriages per year in America, where t is the number of years since 1980. Assuming that this rate continues, find the number of marriages from the year 1990 to the year 2000.

63 (*Behavioral Science–Learning*) A student can memorize words at the rate of $r(t) = 6e^{-t/5}$ words per minute after t minutes. Find the total number of words that the student can memorize in the first 10 minutes.

64 (*Biomedical–Epidemics*) An epidemic is spreading at the rate of $r(t) = 12e^{.2t}$ new cases per day, where t is the number of days since the epidemic began. Find the total number of new cases in the first 10 days of the epidemic.

65 (*Economics–Pareto's Law*) The economist Pareto estimated that the number of people who have an income between A and B dollars ($A < B$) is given by a definite integral of the form

$$N = \int_A^B ax^{-b}\, dx \qquad (b \neq 1)$$

where a and b are constants. Solve this integral.

66 (*Biomedical–Poiseuille's Law*) According to Poiseuille's law, the speed of blood in a blood vessel is given by

$$V = \frac{p}{4Lv}(R^2 - r^2)$$

where R is the radius of the blood vessel, r is the distance of the blood from the center of the blood vessel, and p, L, and v are constants determined by the pressure and viscosity of the blood and the length of the vessel. The total blood flow is then given by

$$\left(\begin{matrix}\text{total blood}\\\text{flow}\end{matrix}\right) = \int_0^R 2\pi \frac{p}{4Lv}(R^2 - r^2)r\, dr$$

Find the total blood flow by solving this integral (p, L, v, and R are constants).

67–68 (*Business–Capital Value of an Asset*) The *capital value* of an asset (such as an oil well) that produces a continuous stream of income is the sum of the present value of all future earnings from the asset. Therefore, the capital value of an asset that produces income at the rate of $r(t)$ dollars per year (at continuous interest rate r) is

$$\left(\begin{matrix}\text{capital}\\\text{value}\end{matrix}\right) = \int_0^T r(t)e^{-it}\, dt$$

where T is the expected life (in years) of the asset.

67 Use the formula on the left-hand column to find the capital value (at interest rate $i = .06$) of an oil well that produces income at the constant rate of $r(t) = 240{,}000$ dollars per year for 10 years.

68 Use the formula on the left-hand column to find the capital value (at interest rate $i = .05$) of a mine that produces income at the (constant) rate of $r(t) = 560{,}000$ dollars per year for 20 years.

5.4 FURTHER APPLICATIONS OF DEFINITE INTEGRALS: AVERAGE VALUE AND AREA BETWEEN CURVES

Introduction

In this section we use definite integrals to find the average value of a function, and also to find the area between two curves.

Average Value of a Function

Averages are used everywhere: birth weights of babies are judged against average weights, and retirement benefits are determined by average income. The average value eliminates the up and down fluctuations in a quantity, reducing a collection of numbers to a single "representative" value.

The average of n numbers is found by adding the numbers and dividing by n. For example,

$$\begin{pmatrix} \text{average of} \\ a, b, \text{ and } c \end{pmatrix} = \frac{a + b + c}{3} = \frac{1}{3}(a + b + c)$$

We want to generalize this idea to find the average value of a *function* on an interval. If the graph on the right shows the temperature over a 24-hour period, how can we calculate the *average temperature* over the period? We could, of course, just take the temperature every hour and then average these 24 numbers, but this would ignore the temperature at all the intermediate times.

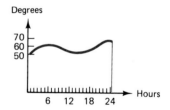

Intuitively, the average value should represent a "leveling off" of the curve to a uniform height, the dashed line shown on the right. This leveling should maintain the same total area, using the high parts to fill in the low parts. Therefore, the area under the dashed line should be the same as the area under the original curve. The area under the line is a rectangle whose area is base $(b - a)$ times height, and this should equal the area under the curve, the definite integral of the function.

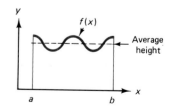

$$\underbrace{(b - a)\begin{pmatrix}\text{average} \\ \text{height}\end{pmatrix}}_{\text{area under line}} = \underbrace{\int_a^b f(x)\, dx}_{\text{area under curve}}$$

Dividing by $(b - a)$ gives

$$\begin{pmatrix}\text{average} \\ \text{height}\end{pmatrix} = \frac{1}{b - a}\int_a^b f(x)\, dx$$

This formula leads to a definition of the average (or "mean") value of a continuous function on an interval.

$$\left(\begin{array}{c}\text{Average value}\\\text{of } f(x) \text{ on } [a, b]\end{array}\right) = \frac{1}{b - a} \int_a^b f(x) \, dx$$

the average value is the definite integral of the function divided by the length of the interval

Integrating and dividing by $b - a$ is analogous to averaging n numbers by adding and dividing by n.

Example 1   Find the average value of $f(x) = \sqrt{x}$ from $x = 0$ to $x = 9$.

Solution

$$\left(\begin{array}{c}\text{Average}\\\text{value}\end{array}\right) = \frac{1}{9 - 0} \int_0^9 \sqrt{x} \, dx = \frac{1}{9} \int_0^9 x^{1/2} \, dx$$

integral divided by the length of the interval

$$= \frac{1}{9} \frac{2}{3} x^{3/2} \bigg|_0^9 = \frac{2}{27} 9^{3/2} - \frac{2}{27} 0^{3/2}$$

integrating and evaluating

$$= \frac{2}{27} (\sqrt{9})^3 - 0 = \frac{2}{27} 27 = 2$$

Therefore, the average value of $f(x) = \sqrt{x}$ over the interval $[0, 9]$ is 2.   ■

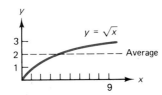

Application: Average Population

The population of the United States is predicted to be $P(t) = 227e^{.01t}$, where t is the number of years since 1985. Find the average population between the year 1990 and 2000.

Solution   We integrate from $t = 5$ (1990) to $t = 15$ (2000).

$$\left(\begin{array}{c}\text{Average}\\\text{value}\end{array}\right) = \frac{1}{15 - 5} \int_5^{15} 227e^{.01t} \, dt = \frac{227}{10} \int_5^{15} e^{.01t} \, dt$$

the integral divided by the length of the interval

$$= 22.7 \frac{1}{.01} e^{.01t} \bigg|_5^{15}$$

using the $\int e^{ax} \, dx$ formula

$$= 2270e^{.15} - 2270e^{.05} \approx 251$$

evaluating, using a calculator

Therefore, the average population from 1990 to 2000 will be about 251 million.

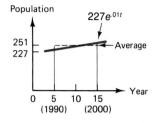

Practice Exercise 1

Find the average value of $f(x) = 3x^2$ from $x = 0$ to $x = 2$. *(solution on page 305)*

Areas Between Curves: the Integral of "Upper Minus Lower"

We know that definite integrals give areas under curves. To calculate the area *between* two curves, we take the area under the *upper* curve and subtract the area under the *lower* curve.

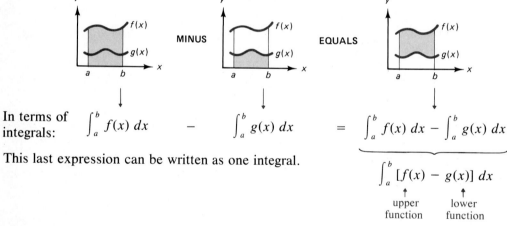

In terms of integrals:

$$\int_a^b f(x)\,dx \quad - \quad \int_a^b g(x)\,dx \quad = \quad \int_a^b f(x)\,dx - \int_a^b g(x)\,dx$$

This last expression can be written as one integral.

$$\int_a^b [f(x) - g(x)]\,dx$$

upper function   lower function

Therefore, the area between two curves is the integral of the upper function minus the lower function between the two extreme x-values.

For continuous functions $f(x)$ and $g(x)$ satisfying $f(x) \geq g(x)$ on the interval $[a, b]$, the area between the curves from $x = a$ to $x = b$ is

$$\int_a^b [f(x) - g(x)]\,dx$$

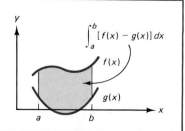

Integrating "upper minus lower" works regardless of whether one or both curves dip below the *x*-axis.

Example 2   Find the area between $y = 3x^2 + 4$ and $y = 2x - 1$ from $x = -1$ to $x = 2$.

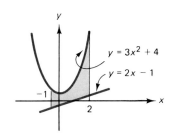

Solution   The area is shown in the diagram on the right. (You may need to make a similar rough sketch for each problem to see which curve is "upper" and which is "lower.") We integrate "upper minus lower" between the given *x*-values.

$$\underbrace{\int_{-1}^{2} [(3x^2 + 4)}_{\text{upper}} - \underbrace{(2x - 1)]}_{\text{lower}} \, dx = \int_{-1}^{2} (3x^2 + 4 - 2x + 1) \, dx \qquad \text{\textit{simplifying}}$$

$$= \int_{-1}^{2} (3x^2 - 2x + 5) \, dx = (x^3 - x^2 + 5x)\Big|_{-1}^{2} \qquad \text{\textit{integrating}}$$

$$= (8 - 4 + 10) - (-1 - 1 - 5) \qquad \text{\textit{evaluating}}$$

$$= 21 \text{ square units} \qquad ▪$$

Practice Exercise 2

Find the area between $y = 2x^2 + 1$ and $y = -x^2 - 1$ from $x = -1$ to $x = 1$.   *(solution on page 305)*

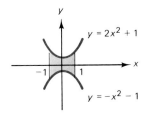

If two curves represent rates,* the area between the curves gives the *total accumulation* at the upper rate minus the lower rate.

Application: Automobile Seatbelts

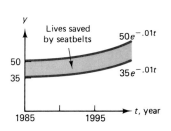

All new cars come equipped with seat belts, and yet only about 1 in 7 people use them. The graph on the right shows the number of automobile fatalities per year, and below it the predicted fatalities if all people wore seat belts. The area between these rates gives the number of lives that would be saved by seat belts over a period of years.

---

* Recall that a rate is one unit per another unit, such as dollars per hour, marriages per year, or people per square mile.

Example 3   The present fatality rate is $50e^{.01t}$ thousand per year, and the predicted rate with increased seat belt use is $35e^{.01t}$, where t is the number of years since 1985. Find how many lives could be saved by seat belts during the decade 1985 to 1995.

Solution   The area between the curves is the integral of "upper minus lower" from $t = 0$ (1985) to $t = 10$ (1995).

$$\int_0^{10} (50e^{.01t} - 35e^{.01t})\, dt = \int_0^{10} 15e^{.01t}\, dt \qquad \text{\textit{simplifying}}$$

$$= 15\,\frac{1}{.01}\, e^{.01t}\, \Big|_0^{10} \qquad \text{\textit{integrating}}$$

$$= 1500e^{.1} - 1500e^{0} \approx 158 \qquad \text{\textit{using a calculator}}$$

These rates are in thousands, so about 158,000 lives would be saved during the 10 years by using seatbelts. ■

Application: World Oil Conservation

The Arab oil embargo of 1973–1974 quadrupled the price of crude oil, and as a result, the rate of oil consumption dropped from the upper curve (in the diagram) to the lower curve. The area between these two rates measures the oil saved by decreasing the rate of consumption.

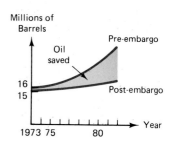

Example 4   The higher (pre-embargo) rate of consumption was $r(t) = 16e^{.08t}$ (million barrels per year), and the lower rate is $r(t) = 15e^{.01t}$, where t is in years and $t = 0$ corresponds to 1973. Find the total amount of oil saved from 1975 to 1980.

Solution   We integrate the higher rate minus the lower rate from $t = 2$ (1975) to $t = 7$ (1980).

$$\int_2^7 (16e^{.08t} - 15e^{.01t})\, dx = \left(16\,\frac{1}{.08}\, e^{.08t} - 15\,\frac{1}{.01}\, e^{.01t}\right)\Big|_2^7$$

$$= (200e^{.08t} - 1500e^{.01t})\,\Big|_2^7$$

$$= (200e^{.56} - 1500e^{.07}) - (200e^{.16} - 1500e^{.02})$$

$$\approx 37$$

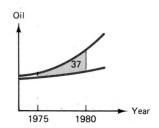

Therefore, the saving during the 5 years was about 37 million barrels of oil. ■

Areas Bounded by Curves

Some problems ask for the *area bounded by two curves*, without giving the starting and ending x-values. In such problems the curves completely enclose an area, and the x-values for the upper and lower limits of integration are found by setting the functions equal to each other and solving.

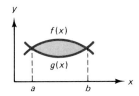

The x-values a and b are where the curves meet.

Example 5   Find the area bounded by the curves

$$y = 3x^2 - 12 \text{ and } y = 12 - 3x^2$$

Solution   The x-values for the upper and lower limits of integration are not given, so we find them by setting the functions equal to each other and solving.

$3x^2 - 12 = 12 - 3x^2$	*setting the functions equal to each other*
$6x^2 - 24 = 0$	*combining everything on one side*
$6(x^2 - 4) = 0$	*factoring*
$6(x + 2)(x - 2) = 0$	*factoring further*
$x = -2 \quad \text{and} \quad x = 2$	*solving*

The smaller of these, $x = -2$, is the lower limit of integration and the larger, $x = 2$, is the upper limit. To determine which function is "upper" and which is "lower," we choose a "test value" between $x = -2$ and $x = 2$ ($x = 0$ will do), which we substitute into each function to see which is larger. Evaluating each of the original functions at the test point $x = 0$ yields

$3x^2 - 12 = 3(0)^2 - 12 = -12$	(smaller)	*$3x^2 - 12$ evaluated at $x = 0$*
$12 - 3x^2 = 12 - 3(0)^2 = 12$	(larger)	*$12 - 3x^2$ evaluated at $x = 0$*

Therefore, $y = 3x^2 - 12$ is the "upper" function (since it gives a higher y-value) and $y = 3x^2 - 12$ is the "lower" function. We then integrate upper minus lower between the x-values found earlier.

$$\int_{-2}^{2} [12 - 3x^2 - \underbrace{(3x^2 - 12)}_{\text{lower}}] \, dx = \int_{-2}^{2} (24 - 6x^2) \, dx \qquad \textit{simplifying}$$

where the first bracket term is the *upper*.

$$= (24x - 2x^3) \Big|_{-2}^{2} \qquad \textit{integrating}$$

$$= (48 - 16) - (-48 + 16) = 64 \qquad \textit{evaluating}$$

Therefore, the area bounded by the two curves is 64 square units. ■

The two curves $y = 12 - 3x^2$ and $y = 3x^2 - 12$ are shown below. Notice that we were able to calculate the area between them without having to graph them.

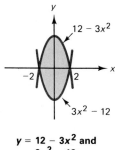

**y = 12 − 3x² and
y = 3x² − 12.**

Practice Exercise 3

Find the area bounded by $y = 2x^2 - 1$ and $y = 2 - x^2$. *(solution on page 305)*

Curves that Intersect at More than Two Points

For curves that intersect at *more* than two points, several integrals may be needed to calculate the area. For example in the diagram below, f is above g from $x = 1$ to $x = 3$, but then the curves cross, and g is above f from $x = 3$ to $x = 5$. Therefore, the total area bounded by the cruves is sum of two integrals, upper minus lower over each interval.

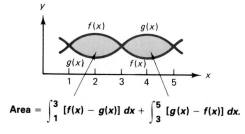

Area $= \int_{1}^{3} [f(x) - g(x)] \, dx + \int_{3}^{5} [g(x) - f(x)] \, dx.$

The numbers 1, 3, and 5 would come from setting the two functions equal to each other and solving. Test values in each interval would show which function is upper and which is lower.

Example 6 Find the area bounded by the curves $y = x^3 - 3x^2$ and $y = 3x^2 - 8x$.

Solution

$$x^3 - 3x^2 = 3x^2 - 8x \qquad \text{\textit{functions set equal to each other}}$$

$$x^3 - 6x^2 + 8x = 0 \qquad \text{\textit{combining all on one side}}$$

$$x(x^2 - 6x + 8) = 0 \qquad \text{\textit{factoring}}$$

$$x(x - 2)(x - 4) = 0 \qquad \text{\textit{factoring further}}$$

$$x = 0, \quad x = 2, \quad x = 4 \qquad \text{\textit{solutions}}$$

These give two intervals, [0, 2] and [2, 4].

Interval [0, 2]. Use test point $x = 1$:

$$x^3 - 3x^2 = (1)^3 - 3(1)^2 = 1 - 3 = -2 \quad \text{(upper)} \qquad \text{\textit{$x^3 - 3x^2$ at $x = 1$}}$$

$$3x^2 - 8x = 3(1)^2 - 8(1) = 3 - 8 = -5 \quad \text{(lower)} \qquad \text{\textit{$3x^2 - 8x$ at $x = 1$}}$$

so on [0, 2] the integral of upper minus lower is

$$\int_0^2 [(x^3 - 3x^2) - (3x^2 - 8x)] \, dx$$

Interval [2, 4]. Use test point $x = 3$:

$$x^3 - 3x^2 = (3)^3 - 3(3)^2 = 27 - 27 = 0 \quad \text{(lower)} \qquad \text{\textit{$x^3 - 3x^2$ at $x = 3$}}$$

$$3x^2 - 8x = 3(3)^2 - 8(3) = 27 - 24 = 3 \quad \text{(upper)} \qquad \text{\textit{$3x^2 - 8x$ at $x = 3$}}$$

so on [2, 4] the integral of upper minus lower is

$$\int_2^4 [(3x^2 - 8x) - (x^3 - 3x^2)] \, dx$$

We now add the two integrals and solve.

$$\int_0^2 [(x^3 - 3x^2) - (3x^2 - 8x)] \, dx + \int_2^4 [(3x^2 - 8x) - (x^3 - 3x^2)] \, dx \qquad \text{\textit{upper-lower on each interval}}$$

$$= \int_0^2 (x^3 - 6x^2 + 8x) \, dx + \int_2^4 (-x^3 + 6x^2 - 8x) \, dx \qquad \text{\textit{simplifying}}$$

$$= \left(\frac{1}{4} x^4 - 2x^3 + 4x^2 \right) \Big|_0^2 + \left(-\frac{1}{4} x^4 + 2x^3 - 4x^2 \right) \Big|_2^4 \qquad \text{\textit{integrating}}$$

$$= (4 - 16 + 16) - (0) + (-64 + 128 - 64) - (-4 + 16 - 16) \qquad \text{\textit{evaluating}}$$

$$= 8 \text{ square units}$$

The area is shown below. Again it was not necessary to graph the functions.

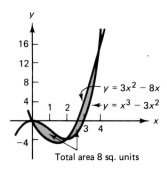

Total area 8 sq. units   ▪

Be sure to integrate upper minus lower over each interval, or else some of the area will be counted as negative. For example, in the diagram below, integrating $f(x) - g(x)$ from $x = 1$ to $x = 5$ would count the first area as positive (since $f - g$ is upper minus lower) but the second area as negative (since $f - g$ is *lower minus upper* on this interval), and this would give not the total area but the *difference* between the areas.

$\int_1^5 [f(x) - g(x)]\, dx$ would give

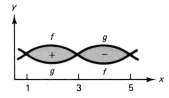

Summary

We defined the average value of a continuous function over an interval as the definite integral of the function divided by the length of the interval.

$$\begin{pmatrix} \text{Average value} \\ \text{of } f(x) \text{ on } [a, b] \end{pmatrix} = \frac{1}{b - a} \int_a^b f(x)\, dx$$

To find the area between two curves:

1. If the x-values are not given, set the functions equal to each other and solve for the points of intersection.

2. Use a test points within each interval to determine which curve is "upper" and which is "lower."

3. Integrate "upper minus lower" on each interval.

This technique provides a very powerful method for finding areas, taking us far beyond the few formulas (for rectangles, triangles, and circles) that we knew before calculus.

SOLUTIONS TO PRACTICE EXERCISES

1. $\dfrac{1}{2}\displaystyle\int_0^2 3x^2\,dx = \dfrac{1}{2}\cdot x^3\,\Big|_0^2 = \dfrac{1}{2}\cdot 2^3 - \dfrac{1}{2}\cdot 0^3 = \dfrac{1}{2}\cdot 8 = 4$

2. $\displaystyle\int_{-1}^1 [(2x^2+1)-(-x^2-1)]\,dx = \int_{-1}^1 (2x^2+1+x^2+1)\,dx$

$= \displaystyle\int_{-1}^1 (3x^2+2)\,dx = (x^3+2x)\,\Big|_{-1}^1 = (1+2)-(-1-2) = 6$ square units

3. $2x^2 - 1 = 2 - x^2$

$3x^2 - 3 = 0$

$3(x^2 - 1) = 0$

$3(x+1)(x-1) = 0$

$x = 1 \quad$ and $\quad x = -1$

Test value $x = 0$ shows that $2 - x^2$ is "upper" and $2x^2 - 1$ is "lower."

$\displaystyle\int_{-1}^1 [(2-x^2)-(2x^2-1)]\,dx = \int_{-1}^1 (3-3x^2)\,dx = (3x-x^3)\,\Big|_{-1}^1$

$= (3-1)-(-3+1) = 4$ square units

EXERCISES 5.4

AVERAGE VALUE

Find the average value of each function over the given interval.

1 $f(x) = x^2 \quad$ on $[0, 3]$

2 $f(x) = x^3 \quad$ on $[0, 2]$

3 $f(x) = 3\sqrt{x} \quad$ on $[0, 4]$

4 $f(x) = \sqrt[3]{x} \quad$ on $[0, 8]$

5 $f(x) = \dfrac{1}{x^2} \quad$ on $[1, 5]$

6 $f(x) = \dfrac{1}{x^2} \quad$ on $[1, 3]$

7 $f(x) = 2x + 1 \quad$ on $[0, 4]$

8 $f(x) = 4x - 1 \quad$ on $[0, 10]$

9 $f(x) = 36 - x^2 \quad$ on $[-2, 2]$

10 $f(x) = 9 - x^2 \cdot$ on $[-3, 3]$

11 $f(x) = 3 \quad$ on $[10, 50]$

12 $f(x) = 2 \quad$ on $[5, 100]$

13 $f(x) = e^{(1/2)x} \quad$ on $[0, 2]$

14 $f(x) = e^{(1/3)x} \quad$ on $[0, 3]$

15 $f(x) = e^{-x} \quad$ on $[0, 2]$

16 $f(x) = e^{-2x} \quad$ on $[0, 1]$

17 $f(x) = \dfrac{1}{x} \quad$ on $[1, 2]$

18 $f(x) = \dfrac{1}{x} \quad$ on $[1, 10]$

19 $f(x) = x^n \quad$ on $[0, 1]$, where n is a constant $(n > -1)$

20 $f(x) = e^{kx} \quad$ on $[0, 1]$, where k is a constant $(k \neq 0)$

21 $f(x) = ax + b \quad$ on $[0, 2]$, where a and b are constants

22 $f(x) = \dfrac{1}{x} \quad$ on $[1, c]$, where c is a constant $(c > 1)$

APPLIED EXERCISES

23–24 (*Business–Sales*) A store's sales on day x are given by the function $S(x)$ below. Find the average sales during the first 3 days (day 0 to day 3).

23 $S(x) = 200x + 6x^2$ **24** $S(x) = 400x + 3x^2$

25 (*General–Temperature*) The temperature at time t hours is $T(t) = -.3t^2 + 4t + 60$ (for $t \le 12$). Find the average temperature between time 0 and time 10.

26 (*Behavioral Science–Practice*) After x practice sessions, a person can accomplish a task in $f(x) = 12x^{-1/2}$ minutes. Find the average time required from the end of session 1 to the end of session 9.

27 (*General–Pollution*) The amount of pollution in a lake x years after the closing of a chemical plant is $P(x) = 100/x$ tons (for $x \ge 1$). Find the average amount of pollution between 1 and 10 years after the closing.

28 (*General–Population*) The population of the United States is predicted to be $P(t) = 260e^{.01t}$ million, where t is the number of years after the year 2000. Find the average population between the years 2000 and 2050.

29 (*Business–Compound Interest*) A deposit of $1000 at 5% interest compounded continuously will grow to $V(t) = 1000e^{.05t}$ dollars after t years. Find the average value during the first 40 years (that is, from time 0 to time 40).

30 (*Biomedical–Bacteria*) A colony of bacteria is of size $S(t) = 300e^{.1t}$ after t hours. Find the average size during the first 12 hours (that is, from time 0 to time 12).

AREA

31 Find the area between the curve $y = x^2 + 1$ and the straight line $y = 2x - 1$ (shown below) from $x = 0$ to $x = 3$.

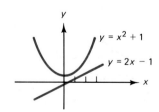

32 Find the area between the curve $y = x^2 + 3$ and the straight line $y = 2x$ (shown below) from $x = 0$ to $x = 3$.

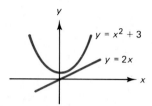

33 Find the area between the curves $y = e^x$ and $y = e^{2x}$ (shown below) from $x = 0$ to $x = 2$. (Leave the answer in its exact form.)

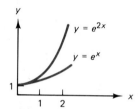

34 Find the area between the curves $y = e^x$ and $y = e^{-x}$ (shown below) from $x = 0$ to $x = 1$. (Leave the answer in its exact form.)

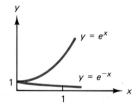

35 (a) Sketch the parabola $y = x^2 + 4$ and the straight line $y = 2x + 1$ on the same graph.
(b) Find the area between then from $x = 0$ to $x = 3$.

36 (a) Sketch the parabola $y = x^2 + 5$ and the straight line $y = 2x + 3$ on the same graph.
(b) Find the area between then from $x = 0$ to $x = 3$.

37 (a) Sketch the straight lines $y = 2x + 3$ and $y = -2x - 3$ on the same graph.
(b) Find the area between them from $x = -1$ to $x = 2$.

38 (a) Sketch the straight lines $y = 2x + 5$ and $y = -2x - 5$ on the same graph.
(b) Find the area between them from $x = -2$ to $x = 1$.

Find the area bounded by the given curves.

39 $y = x^2 - 1$ and $y = 2 - 2x^2$

40 $y = x^2 - 4$ and $y = 8 - 2x^2$

41 $y = 6x^2 - 10x - 8$ and $y = 3x^2 + 8x - 23$

42 $y = 3x^2 - x - 1$ and $y = 5x + 8$

43 $y = x^2$ and $y = x^3$

44 $y = x^3$ and $y = x^4$

45 $y = 4x^3 + 1$ and $y = 36x + 1$

46 $y = 7x^3 - 36x$ and $y = 3x^3 + 64x$

47 $y = 5x^3 + 2x$ and $y = x^3 + 6x$

48 $y = x^n$ and $y = x^{n-1}$ (for $n > 1$)

APPLIED EXERCISES

49 (*General–Population*) In an effort to reduce its growth, China instituted strict birth control methods, slowing its birth rate from $27e^{.02t}$ to $21e^{.02t}$ million births per year, where t is the number of years since 1985. Find the total decrease in the population that will result from this lower birth rate between 1985 ($t = 0$) and 2005 ($t = 20$).

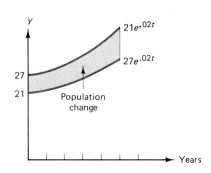

50 (*General–Population*) The birth rate in Africa has increased from $13e^{.02t}$ to $16e^{.02t}$ million births per year, where t is the number of years since 1985. Find the total increase in population that will result from this higher birth rate between 1985 ($t = 0$) and 2005 ($t = 20$).

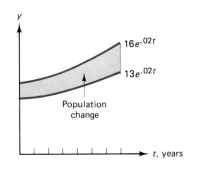

51 (*Business–Net Savings*) A factory installs new machinery which saves $S(x) = 1200 - 20x$ dollars per year, where x is the number of years since installation. However, the cost of maintaining the new machinery is $C(x) = 100x$ dollars per year.

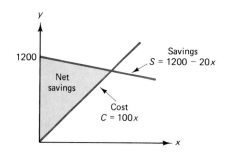

(a) Find the year x at which the maintenance costs $C(x)$ will equal the savings $S(x)$. (At this time the new machinery should be replaced.)

(b) Find the accumulated net savings [savings $S(x)$ minus cost $C(x)$] during the period $t = 0$ to the replacement time found in part (a).

52 (*General–Quilting*) A quilt pattern consists of a white square with a red shape inside, as shown below. Find the area of the red interior.

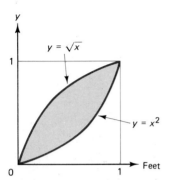

53 (*Economics–Balance of Trade*) A country's annual imports are $I(t) = 30e^{.2t}$ and its exports are $E(t) = 25e^{.1t}$, both in billions of dollars, where t is measured in years and $t = 0$ corresponds to the beginning of 1985. Find the country's accumulated trade deficit (imports minus exports) for the 10 years beginning with 1985.

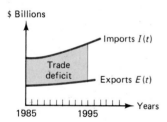

54 (*Economics–Cost of Labor Contracts*) An employer offers to pay workers at the rate of $30{,}000e^{.04t}$ dollars per year, while the union demands payment at the rate of $30{,}000e^{.08t}$ dollars per year, where $t = 0$ corresponds to the beginning of the contract. Find the accumulated difference in pay between these two rates over the 10-year life of the contract.

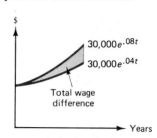

55–56 (*Business–Cumulative Profit*) A company's marginal revenue function is $MR(x) = 700x^{-1}$ and its marginal cost function is $MC(x) = 500x^{-1}$ (both in thousands of dollars), where x is the number of units ($x > 1$). Find the total profit from

55 $x = 100$ to $x = 200$ **56** $x = 200$ to $x = 300$

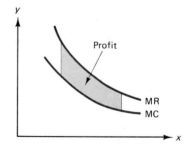

▓ REVIEW EXERCISES

These exercises review material that will be helpful in Section 5.6.

Find the derivative of each function.

57 e^{x^2+5x}

58 e^{x^3+6x}

59 $\ln(x^2 + 5x)$

60 $\ln(x^3 + 6x)$

5.5 TWO APPLICATIONS TO ECONOMICS: CONSUMERS' SURPLUS AND INCOME DISTRIBUTION

Introduction

In this section we discuss several important economic concepts—consumers' and producers' surplus, and the Gini Index of income distribution—each of which is defined as the area between two curves.

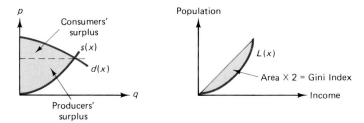

Consumers' Surplus

Imagine that you really liked pizza and were willing to pay $12 for a pizza pie. If, in fact, a pizza costs only $8, then you have, in some sense, "saved" $4, the $12 that you were willing to pay minus the $8 market price. If one were to add up this difference for all pizzas sold in a given period of time (the price that each consumer was willing to pay minus the price actually paid), the total savings would be called the *consumers' surplus* for that product. The consumers' surplus measures the benefit that consumers derive from an economy in which competition keeps prices low.

Demand Functions

Price and quantity are inversely related: if the price of an item rises, the quantity sold generally falls, and vice versa. Through market research, economists can determine the relationship between price and quantity for an item. This relationship can often be expressed as a *demand function* (or demand curve) $d(x)$, so called because it gives the price at which exactly x units will be demanded.

Demand Function

The demand function $d(x)$ for a product gives the price at which exactly x units will be sold.

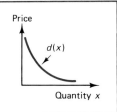

In Chapter 3 we called $d(x)$ the *price function*.

Mathematical Definition of Consumer's Surplus

The demand curve gives the price that consumers are *willing* to pay, and the *market price* is what they *do* pay, so the amount by which the demand curve is above the market price measures the benefit or "surplus" to consumers. We add up these benefits by integrating, so the area between the demand curve and the market price line gives the *total benefit* that consumers derive from being able to buy at the market price. This total benefit (the shaded area in the diagram) is called the *consumers' surplus*.

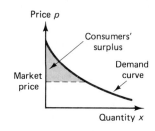

Consumers' Surplus

For a demand function $d(x)$ and a demand level A, the market price B is the demand function evaluated at $x = a$,

$$B = d(A)$$

The consumers' surplus is the area between the demand curve and the market price.

$$\left(\begin{array}{c}\text{Consumers'}\\ \text{surplus}\end{array}\right) = \int_0^A [\underbrace{d(x)}_{\substack{\text{demand}\\ \text{function}}} - \underbrace{B}_{\substack{\text{market}\\ \text{price}}}]\, dx$$

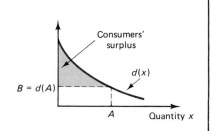

Example 1   If the demand function for electricity is $d(x) = 1100 - 10x$ dollars (where x is in millions of kilowatt-hours), find the consumers' surplus at the demand level $x = 80$.

Solution   The market price is the demand function $d(x)$ evaluated at $x = 80$.

$$\left(\begin{array}{c}\text{Market}\\ \text{price } B\end{array}\right) = d(80) = 1100 - 10\cdot 80 = 300$$

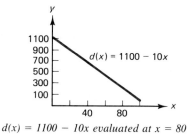

$d(x) = 1100 - 10x$ evaluated at $x = 80$

The consumers' surplus is the area between the demand curve and the market price line.

$$\left(\begin{array}{c}\text{Consumers'}\\ \text{surplus}\end{array}\right) = \int_0^{80} (\underbrace{1100 - 10x}_{\substack{\text{demand}\\ \text{function}}} - \underbrace{300}_{\substack{\text{market}\\ \text{price}}})\, dx$$

$$= \int_0^{80} (800 - 10x)\, dx = (800x - 5x^2)\Big|_0^{80}$$

$$= (64{,}000 - 32{,}000) - (0) = \underbrace{\$32{,}000}_{\substack{\text{consumers'}\\ \text{surplus}}}$$ ■

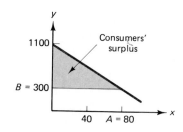

(All such answers will be in dollars unless otherwise indicated.)

How Consumers' Surplus Is Used

Example 1, at demand level $x = 80$ the consumers' surplus was \$32,000. If electricity usage were to increase to $x = 90$, the market price would then drop to $d(90) = 1100 - 10 \cdot 90 = 200$. We could then calculate the consumers' surplus at this higher demand level (the answer is \$40,500). Therefore, a price decrease from \$300 to \$200 would mean that consumers would benefit by an additional \$40,500 − \$32,000 = \$7500. This benefit would then be compared to the cost of a new generator to decide whether the expenditure would be worthwhile.

Producers' Surplus

Just as the consumers' surplus measures the total benefit to consumers, the *producers' surplus* measures the total benefit that producers derive from being able to sell at the market price. Returning to our pizza example, if a pizza producer might just be willing to remain in business if the price of pizzas dropped to \$5, the fact that he can sell them for \$8 gives him a "benefit" of \$3. The sum of all such benefits is the *producers' surplus* for a product.

Supply Functions

Clearly, as the price of an item rises, so will the quantity that producers are willing to supply. The relationship between the price of an item and the quantity that producers are willing to supply at that price can be expressed as a supply function (or supply curve) $s(x)$.

Supply Function

The supply function $s(x)$ for a product gives the price at which exactly x units will be supplied.

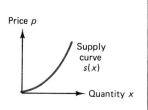

Mathematical Definition of Producers' Surplus

As before we integrate to find the total benefit, but now "upper" is the market price and "lower" is the supply curve $s(x)$.

Producers' Surplus

For supply function $s(x)$ and demand level A, the market price is $B = s(A)$. The producers' surplus is the area between the market price and the supply curve.

$$\binom{\text{Producers'}}{\text{surplus}} = \int_0^A [B - s(x)]\, dx$$

$$\underset{\substack{\uparrow \\ \text{market} \\ \text{price}}}{} \quad \underset{\substack{\uparrow \\ \text{supply} \\ \text{function}}}{}$$

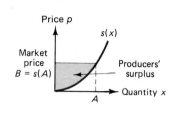

Example 2 For supply function $s(x) = .09x^2$ and demand level $x = 200$, find the producers' surplus.

Solution The market price is the supply function $s(x)$ evaluated at $x = 200$.

$$\binom{\text{Market}}{\text{price } B} = s(200) = .09(200)^2 = 3600 \qquad\qquad s(x) = .09x^2 \text{ evaluated at } x = 200$$

The producers' surplus is the area between the market price line and the supply curve.

$$\binom{\text{Producers'}}{\text{surplus}} = \int_0^{200} (\underbrace{3600}_{\substack{\text{market} \\ \text{price}}} - \underbrace{.09x^2}_{\substack{\text{supply} \\ \text{function}}})\, dx$$

$$= (3600x - .03x^3)\, \Big|_0^{200}$$

$$= (720{,}000 - 240{,}000) - (0) = \underbrace{\$480{,}000}_{\substack{\text{producers'} \\ \text{surplus}}} \qquad ■$$

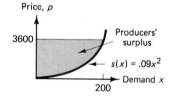

Consumers' and Producers' Surplus

The demand x at which the supply and demand curves intersect is called the *market demand*. The consumers' and the producers' surplus can be shown together on the same graph. These two areas together give a numerical measure to the total benefit that consumers and producers derive from competition, showing that both consumers and producers benefit from an open market.

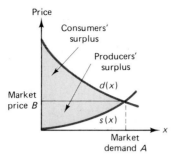

Gini Index of Income Distribution

In any society, some people make more money than others. To measure the "gap" between the rich and the poor, economists calculate the proportion of the total income that is earned by the lowest 10% of the population, and then the proportion that is earned by the lowest 20% of the population, and so on. This information (for the year 1986) is given in the table below (with percentages written as decimals), and graphed on the right.

Proportion of Population	Proportion of Income
.20	.05
.40	.16
.60	.34
.80	.58
1.00	1.00

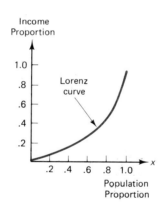

For example, the lowest 20% of the population earns only 5% of the total income, the lowest 40% earns only 16% of the total income, and so on. The curve is known as the *Lorenz curve* (after the American statistician Max Otto Lorenz).

Lorenz Curve

The Lorenz curve $L(x)$ gives the proportion of total income earned by the lowest proportion x of the population.

The Gini Index

The Lorenz curve may be compared to two extreme cases of income distribution.

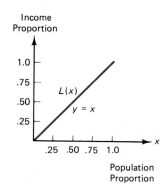

1. *Absolute equality of income* means that everyone earns exactly the same income, so the lowest 10% of the population earns exactly 10% of the total income, the lowest 20% earns exactly 20% of the income, and so on. This gives the Lorenz curve $y = x$ shown on the left below.

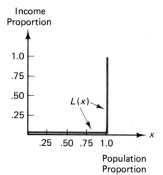

2. *Absolute inequality of income* means that nobody earns any income except one person, who earns all the income. This gives the Lorenz curve shown on the right below.

To measure how the actual distribution differs from absolute equality, we calculate the *area* between the actual distribution and the line of absolute equality $y = x$. Since this area can be at most $\frac{1}{2}$ (the area of the entire lower triangle), economists multiply the area by 2 to get a number between 0 (absolute equality) and 1 (absolute inequality). This measure is called the *Gini index*. Note that a higher Gini index means greater *in*equality (greater deviation from the line of absolute equality).

Gini Index

For a Lorenz curve $L(x)$, the Gini index is

$$\left(\begin{matrix}\text{Gini}\\\text{index}\end{matrix}\right) = 2 \cdot \int_0^1 [x - L(x)] \, dx$$

The Gini index varies from 0 (absolute equality) to 1 (absolute inequality).

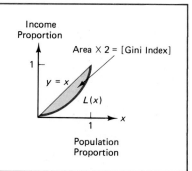

Example 3 The Lorenz curve for income distribution in the United States in 1986 is approximated by $L(x) = x^{2.2}$. Find the Gini index.

Solution First we calculate the area between the curve of absolute equality, $y = x$, and the Lorenz curve $y = x^{2.2}$.

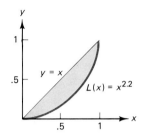

$$\int_0^1 (x - x^{2.2})\, dx = \left(\frac{1}{2}x^2 - \frac{1}{3.2}x^{3.2}\right)\Big|_0^1$$

$$= \frac{1}{2}\cdot 1^2 - \frac{1}{3.2}\cdot 1^{3.2} - 0 \approx .5 - .31 = .19$$

Multiplying by 2 gives the Gini index.

$$\left(\frac{\text{Gini}}{\text{index}}\right) = .38$$

■

Practice Exercise 1

In 1955 the Gini index for income was .405. At that time were incomes more or less equal than in 1986?   *(solution below)*

Gini Index for Total Wealth

The Gini index for total wealth can be calculated similarly.

Practice Exercise 2

The Lorenz curve for total wealth in the United States during 1985 was $L(x) = x^{5.6}$. Calculate the Gini index.   *(solution below)*

Practice Exercise 2 shows that the Gini index for wealth is greater than the Gini index for income. That is, total wealth in the United States is distributed more unequally than total income. One reason for this is that we have an income tax but no "wealth" tax.

SOLUTIONS TO PRACTICE EXERCISES

1. Less equal in 1955

2. $\int_0^1 (x - x^{5.6})\, dx = \left(\frac{1}{2}x^2 - \frac{1}{6.6}x^{6.6}\right)\Big|_0^1 = \frac{1}{2} - \frac{1}{6.6} - 0 \approx .5 - .15 = .35$

     Gini index for wealth = .70   (multiplying by 2)

For each demand function d(x) and demand level x, find the consumers' surplus.

1 $d(x) = 4000 - 12x, \ x = 100$

2 $d(x) = 500 - x, \quad x = 400$

3 $d(x) = 300 - \frac{1}{2}x, \quad x = 200$

4 $d(x) = 200 - \frac{1}{2}x, \quad x = 300$

5 $d(x) = 350 - .09x^2, \ x = 50$

6 $d(x) = 840 - .06x^2, \ x = 100$

For each supply function s(x) and demand level x, find the producers' surplus.

7 $s(x) = .2x, \quad x = 100$

8 $s(x) = .4x, \quad x = 200$

9 $s(x) = .03x^2, \ x = 200$

10 $s(x) = .06x^2, \ x = 50$

For each demand function d(x) and supply function s(x):
(a) Find the market demand (the positive value of x at which the demand function intersects the supply function).
(b) Find the consumers' surplus at the market demand found in part (a).
(c) Find the producers' surplus at the market demand found in part (a).

11 $d(x) = 300 - .4x, \quad s(x) = .2x$

12 $d(x) = 120 - .16x, \ s(x) = .08x$

13 $d(x) = 300 - .03x^2, \ s(x) = .09x^2$

14 $d(x) = 360 - .03x^2, \ s(x) = .006x^2$

Find the Gini index for the given Lorenz curve.

15 $L(x) = x^{3.2}$ (the Lorenz curve for U.S. income in 1929)

16 $L(x) = x^3$ (the Lorenz curve for U.S. income in 1935)

17 $L(x) = x^{2.1}$ (the Lorenz curve for income in Sweden)

18 $L(x) = x^{15.3}$ (the Lorenz curve for wealth in Great Britain)

19 $L(x) = .4x + .6x^2$

20 $L(x) = .2x + .8x^3$

21 $L(x) = x^n$ (for $n > 1$)

22 $L(x) = \frac{1}{2}x + \frac{1}{2}x^n$ (for $n > 1$)

REVIEW EXERCISES (These exercises are useful for Section 5.6.)

Find the derivative of each function.

23 $(x^5 - 3x^3 + x - 1)^4$

24 $(x^4 - 2x^2 - x + 1)^5$

25 $\ln(x^4 + 1)$

26 $\ln(x^3 - 1)$

27 e^{x^3}

28 e^{x^4}

5.6 INTEGRATION BY SUBSTITUTION

Introduction

The generalized power rule in Chapter 2 greatly expanded the range of functions that we could differentiate. In this section we will learn a similar technique for integration, called the substitution method, which will greatly expand the range of functions that we can integrate. First, however, we must define differentials.

Differentials

One of the notations for the derivative of a function $f(x)$ is $\dfrac{df}{dx}$. Although written as a fraction, $\dfrac{df}{dx}$ was not defined as the quotient of two quantities df and dx, but was defined as a *single* object, the *derivative*. We will now define df and dx separately (they are called *differentials*) so that their quotient $df \div dx$ is equal to the derivative $\dfrac{df}{dx}$. We begin with

$$\frac{df}{dx} = f'$$

since df/dx and f' are both notations for the derivative

$$df = f' \, dx$$

multiplying each side by dx

This leads to a definition for the differential df.

Differentials

For a differentiable function $f(x)$, the differential df is

$$df = f'(x) \, dx$$

Note that df does *not* mean d times f, but rather is defined as the derivative of the function times dx. The dx is just the notation that appears at the end of integrals. The reason for this notation will be clear later.

Example 1

(a) If $f(x) = x^2$, then $df = 2x \, dx$.

derivative 2x times dx

(b) If $f(x) = \ln x$, then $df = \dfrac{1}{x} \, dx$.

derivative 1/x times dx

(c) If $f(x) = e^{x^2}$, then $df = \underbrace{e^{x^2}(2x)}_{f'(x)} \, dx$.

(d) If $f(x) = x^4 - 5x + 2$, then $df = \underbrace{(4x^3 - 5)}_{f'(x)} dx$. ■

Practice Exercise 1

For $f(x) = x^3 - 4x - 2$ find the differential df. *(solution on page 326)*

The differential formula $df = f' \cdot dx$ is easy to remember since dividing both sides by dx gives

$$\frac{df}{dx} = f'$$

which simply says "the derivative equals the derivative." We may use other letters besides f and x.

Example 2

(a) For $u = x^3 + 1$ the differential is $du = 3x^2\, dx$.

derivative of the function ⟍   ⟋ d followed by the variable

(b) For $u = e^{2t} + 1$ the differential is $du = 2e^{2t}\, dt$. ■

Practice Exercise 2

For $u = e^{-5t}$ find the differential du. *(solution on page 326)*

Substitution Method

Using differential notation we can state three very useful integration formulas.

(A) $\displaystyle \int u^n\, du = \frac{1}{n+1} u^{n+1} + C$ $n \neq -1$

(B) $\displaystyle \int e^u\, du = e^u + C$

(C) $\displaystyle \int \frac{du}{u} = \int \frac{1}{u}\, du = \int u^{-1}\, du = \ln |u| + C$

These formulas are easy to remember since they are exactly the formulas that we learned earlier (see pages 275 and 281) except that here we use the letter u to stand for a *function*. The du is the *differential* of the function. Each of these formulas may be justified by differentiating the right-hand side (see Exercises 61–63). A few examples will illustrate their use.

Example 3 Solve $\int (x^2 + 1)^3 \, 2x \, dx$.

Solution This integral involves a function to a power,

$$\underbrace{(x^2 + 1)}_{\text{function}}{}^{\overset{\uparrow}{3}}_{\text{power}}$$

so we use formula A,

$$\int u^n \, du = \frac{1}{n + 1} u^{n+1} + C$$

since it involves u to a power. The idea is to choose some part of the given integrand and call it u, so that the entire integral can be expressed as $\int u^n \, du$. Since u (in the formula) and $x^2 + 1$ (in the problem) are both raised to powers, we take $u = x^2 + 1$. Therefore, the first part of the integrand is u^3.

$$\int \underbrace{(x^2 + 1)^3}_{u^3} \, 2x \, dx$$

The differential of $u = x^2 + 1$ is $du = 2x \, dx$, and this is exactly what appears at the end of our integral.

$$\int \underbrace{(x^2 + 1)^3}_{u^3} \, \underbrace{2x \, dx}_{du}$$

We *substitute* u^3 and du for these quantities.

$$\int u^3 \, du$$

This integral is solved by formula A with $n = 3$.

$$\frac{1}{4} u^4 + C \qquad\qquad using \int u^n \, du = \frac{1}{n + 1} u^{n+1} + C$$
$$with \ n = 3$$

For the final answer we substitute back to the original variable x, using our relationship $u = x^2 + 1$.

$$\frac{1}{4} (x^2 + 1)^4 + C \qquad\qquad \frac{1}{4} u^4 + C \ with \ u = x^2 + 1$$

This is the solution to the integral. ▪

The procedure is not as complicated as it might seem. All of these steps may be written together as follows.

$$\int \underbrace{(x^2 + 1)^3}_{u^3} \; \underbrace{2x \, dx}_{du} \;\; = \;\; \int u^3 \, du \;\; = \;\; \frac{1}{4} u^4 + C \;\; = \frac{1}{4}(x^2 + 1)^4 + C \qquad \textit{answer}$$

choosing $u = x^2 + 1$ therefore $du = 2x \, dx$	substituting $u^3 = (x^2 + 1)^3$ $du = 2x \, dx$	integrating using formula A with $n = 3$	substituting back to x using $u = x^2 + 1$

We may check this answer by differentiation (using the generalized power rule).

$$\frac{d}{dx}\underbrace{\left[\frac{1}{4}(x^2 + 1)^4 + C \right]}_{\text{answer}} = \frac{1}{4} \cdot 4(x^2 + 1)^3 \; \underbrace{2x}_{\substack{\text{derivative} \\ \text{of the inside}}} = \underbrace{(x^2 + 1)^3 \, 2x}_{\text{integrand}}$$

Since the result of the differentiation agrees with the original integrand, the integration is correct.

Multiplying Inside and Outside by Constants

To solve the integral $\int (x^2 + 1)^3 \, 2x \, dx$ we chose $u = x^2 + 1$, and the differential of this is $du = 2x \, dx$, which happened to equal, exactly, the rest of the integral. If the rest of the integral does not exactly match du, we may sometimes solve the integral anyway by multiplying by constants.

Example 4   Solve $\displaystyle\int (x^2 + 1)^3 x \, dx$.   *same as Example 3 but without the 2*

Solution   As before, we use formula A with $u = x^2 + 1$, which gives $du = 2x \, dx$. But the integral has only an $x \, dx$, not the $\underline{2}x \, dx$, which would allow us to substitute du.

$$\int \underbrace{(x^2 + 1)^3}_{u^3} \underbrace{x \, dx}_{\substack{\text{}}}$$
$$\text{—not } du, \text{ because there is no 2}$$

Therefore, the integral is *not* in the form $\int u^3 \, du$. (The integral must fit the formula exactly: *everything* in the integral must be accounted for either by the u^n or by the du.)

However, we may multiply inside the integral by 2 as long as we compensate by also multiplying by $\frac{1}{2}$, and the $\frac{1}{2}$ may be written outside the integral (since constants may be moved across the integral sign).

$$\frac{1}{2} \int \underbrace{(x^2 + 1)^3 \, 2x \, dx}_{\text{compensate}}$$

multiplying inside by 2 and outside by 1/2

This 2 completes the *du*, so now the integral *is* in the form $\int u^n \, du$ and we may solve it as before (substituting u^3 and *du*), simply carrying along the $\frac{1}{2}$ outside.

$$\frac{1}{2} \int \underbrace{(x^2 + 1)^3}_{\substack{u^3 \\ \text{with } u = x^2 + 1 \\ du = 2x \, dx}} \underbrace{2x \, dx}_{du} = \frac{1}{2} \int u^3 \, du = \underset{\substack{\text{integrating} \\ \text{by formula A}}}{\frac{1}{2} \frac{1}{4} u^4 + C} = \underset{\substack{\text{substituting back} \\ \text{to } x \text{ using } u = x^2 + 1}}{\frac{1}{8} (x^2 + 1)^4 + C}$$
$$\underset{\text{substituting}}{}$$
▪

Again, this answer can be checked by differentiation.

Multiplying inside and outside the integral by constants is a very useful trick. We first choose *u*, then calculate *du*, and multiply inside by whatever constant is necessary to complete the *du*. Finally, we multiply outside by the reciprocal of the "inside" constant. The justification is that the constants can be moved across the integral sign, and multiplying by a number and its reciprocal leaves the value of the integral unchanged. This method is equally useful with integration formulas B and C.

Example 5   Solve the integral $\int e^{x^3} x^2 \, dx$. ⌐— function

Solution   The integral involves *e* to a function, e^{x^3}, so we use formula B from page 318,

$$\int e^u \, du = e^u + C$$

since it involves e^u. The *u* (in the formula) and the x^3 (in the problem) are both exponents of *e*, so we take $u = x^3$. The differential of $u = x^3$ is $du = 3x^2 \, dx$, so to complete the *du* in $\int e^{x^3} x^2 \, dx$ we need a 3. Therefore, we multiply inside by 3 and outside by $\frac{1}{3}$.

$$\int e^{x^3} x^2 \, dx = \underset{\substack{u = x^3 \\ du = 3x^2 \, dx}}{\frac{1}{3} \int} e^{x^3} \underbrace{3x^2 \, dx}_{\substack{e^u \quad du \\ \text{multiplying} \\ \text{by 3 and 1/3}}} = \underset{\text{substituting}}{\frac{1}{3} \int e^u \, du} = \underset{\substack{\text{integrating} \\ \text{using formula B}}}{\frac{1}{3} e^u + C} = \underset{\substack{\text{substituting back} \\ \text{to } x \text{ using } u = x^3}}{\frac{1}{3} e^{x^3} + C}$$
▪

We multiplied by 3 and by 1/3, and 1/3 remained in the answer but the 3 did not. This is because the 3 became part of the *du* ($du = 3x^2 \, dx$), which was then "used up" in the integration along with the integral sign.

Application to Business

The substitution method is useful for solving integrals that appear in applications, such as recovering cost from marginal cost.

Example 6   A company's marginal cost function is $MC(x) = \dfrac{x^3}{x^4 + 1}$ and fixed costs are \$1000. Find the cost function.

Solution   Cost is the integral of marginal cost.

$$C(x) = \int \frac{x^3 \, dx}{x^4 + 1}$$

The differential of the denominator is $4x^3 \, dx$, which except for the 4 is just the numerator. This suggests formula C, $\int \dfrac{du}{u} = \ln |u| + C$ with $u = x^4 + 1$. The problem already has the differential $du = 4x^3 \, dx$ except for the 4, so we multiply inside by 4 (in the numerator) and outside by $\tfrac{1}{4}$.

$$\int \frac{x^3 \, dx}{x^4 + 1} = \frac{1}{4} \int \frac{4x^3 \, dx}{x^4 + 1} = \frac{1}{4} \int \frac{du}{u} = \frac{1}{4} \ln |u| + C = \frac{1}{4} \ln |x^4 + 1| + C$$

$u = x^4 + 1$ $du = 4x^3 \, dx$	multiplying by 4 and $\tfrac{1}{4}$	substituting	integrating by formula C	substituting back to x

Since $x^4 + 1$ is always positive, we may drop the absolute value bars.

$$C(x) = \frac{1}{4} \ln (x^4 + 1) + K \qquad \qquad \textit{using K to avoid confusion with C(x)}$$

To evaluate the constant K, we set the cost function (evaluated at $x = 0$) equal to the given fixed cost.

$$\frac{1}{4} \underbrace{\ln (1)}_{0} + K = 1000 \qquad \qquad \textit{ln } (x^4 + 1) + K \textit{ at } x = 0 \\ \textit{set equal to 1000}$$

This gives $K = 1000$. Therefore, the cost function is

$$C(x) = \frac{1}{4} \ln (x^4 + 1) + 1000 \qquad \qquad \textit{C(x)} = \textit{ln } (x^4 + 1) + K \textit{ with } K = 1000 \quad ■$$

Which Formula to Use

The three formulas apply to three different types of integrals.

(A) $\displaystyle \int u^n \, du = \frac{1}{n + 1} u^{n+1} + C \qquad (n \neq -1)$ *integrates a function to a constant power (except power −1) times the differential of the function*

(B) $\displaystyle \int e^u \, du = e^u + C$ *integrates e to a power times the differential of the exponent*

(C) $\displaystyle \int \frac{du}{u} = \int u^{-1}\, du = \ln|u| + C$

integrates a fraction whose top is the differential of the bottom, or equivalently, a function to the power −1 times the differential of the function

To solve an integral by substitution, choose the formula whose left-hand side most closely resembles the given problem.

Practice Exercise 3

For each integral below, choose the most appropriate formula: (A), (B), or (C). (Do not solve the integral.)

(a) $\displaystyle \int e^{5x^2-1} x\, dx$

(b) $\displaystyle \int \frac{x\, dx}{x^2 + 1}$

(c) $\displaystyle \int (x^4 - 12)^4 x^3\, dx$

(d) $\displaystyle \int (x^4 - 12)^{-1} x^3\, dx$ *(solutions on page 326)*

Only Constants Can Be Adjusted

We may multiply inside and outside only by *constants*, not variables, since only constants can be moved across the integral sign. Therefore, the *du* in a problem must already be "complete" except for adjusting the constant. Otherwise, the problem cannot be solved by substitution. For example, the integral

$$\int e^{x^3} x\, dx$$

$$\underbrace{\phantom{e^{x^3}}}_{\substack{e^u \\ u = x^3 \\ du = 3x^2\, dx}} \quad \uparrow\!\!\!-\!\!\! \text{not } du$$

cannot be solved by substitution since the *du* requires an x^2 and the problem only has an x.

Practice Exercise 4

Which of these integrals can be solved by substitution? (*Hint*: See whether only a constant is needed to complete the *du*.)

(a) $\displaystyle \int (x^3 + 1)^3 x^3\, dx$

(b) $\displaystyle \int e^{x^2}\, dx$ *(solution on page 326)*

Further Examples of the Substitution Method

Example 7 Solve $\displaystyle \int \sqrt{x^3 - 3x}\, (x^2 - 1)\, dx.$

Solution Since $\sqrt{x^3 - 3x} = (x^3 - 3x)^{1/2}$ is a function to a power, we use the formula for $\int u^n\, du$ with $u = x^3 - 3x$. Comparing the differential $du = 3x^2 - 3 = 3(x^2 - 1)$ with the problem shows that we need to multiply by 3.

$$\int (x^3 - 3x)^{1/2}(x^2 - 1)\, dx = \frac{1}{3} \int \underbrace{(x^3 - 3x)^{1/2}}_{u^{1/2}} \underbrace{3(x^2 - 1)\, dx}_{du}$$ *multiplying by 3 and 1/3*

$$u = x^3 - 3x$$
$$du = (3x^2 - 3)\, dx$$
$$= 3(x^2 - 1)\, dx$$

$$= \frac{1}{3} \int u^{1/2}\, du \;=\; \frac{1}{3}\frac{2}{3}\, u^{3/2} \;=\; \frac{2}{9}(x^3 - 3x)^{3/2} + C \quad \textit{answer}$$

substituting   integrating   substituting
                            using formula (A)   back to x ■

Example 8   Find $\int e^{\sqrt{x}} x^{-1/2}\, dx$.

Solution   The integral involves $e^{\sqrt{x}}$, e to a function, so we use the formula $\int e^u\, du = e^u + C$. Equating the exponents of e in the formula and in the problem gives $u = x^{1/2}$.

$$\int e^{x^{1/2}} x^{-1/2}\, dx \;=\; 2 \int \underbrace{e^{x^{1/2}}}_{e^u} \underbrace{\frac{1}{2}\, x^{-1/2}\, dx}_{du}$$ *multiplying by 1/2 and 2*

$$u = x^{1/2}$$
$$du = \frac{1}{2}\, x^{-1/2}$$

$$= 2 \int e^u\, du \;=\; 2e^u + C \;=\; 2e^{x^{1/2}} + C$$

substituting   integrating   substituting back to x ■

Solving Definite Integrals by Substitution

Sometimes a *definite* integral requires a substitution. In such cases changing from x to u also requires changing the limits of integration from x-values to u-values, using the substitution formula for u.

Example 9   Solve the definite integral $\int_4^5 \dfrac{dx}{3 - x}$.

Solution   The differential of the denominator is $-1 \cdot dx$, which except for the -1 is just the numerator. This suggests formula C, $\int \dfrac{du}{u} = \ln |u| + C$ with $u = 3 - x$ (from equating the denominators). We multiply inside and outside by -1.

$$u = 3 - 5 = -2$$

$$\int_4^5 \frac{dx}{3-x} = -\int_4^5 \frac{-dx}{3-x} = -\int_{-1}^{-2} \frac{du}{u}$$

$$u = 3 - x$$
$$du = -dx$$

$$u = 3 - 4 = -1$$

new upper and lower limits of integration for u are found by evaluating u = 3 − x at the old x limits

$$= -\ln|u| \Big|_{-1}^{-2} = -\ln|-2| - (-\ln|-1|) = -\ln 2 + \ln 1 = -\ln 2$$ ■

Further Applications

Definite integrals are used to find areas, total accumulations, and average values, and any of these applications may require a substitution.

Example 10   Pollution is being discharged into a lake at the rate of $r(t) = 400te^{t^2}$ tons per year, where t is the number of years since measurements were begun. Find the total amount of pollutant discharged into the lake during the first 2 years.

Solution   The total accumulation is the definite integral from $t = 0$ (the beginning) to $t = 2$ (2 years later). Since the integral involves e to a function, we use the formula for $\int e^u \, du$ with $u = t^2$ (from equating exponents).

$$u = 2^2 = 4$$

$$\int_0^2 400te^{t^2} \, dt = 400 \cdot \frac{1}{2} \int_0^2 2te^{t^2} \, dt = 200 \int_0^4 e^u \, du$$

changing the limits to u-values using $u = t^2$

$$e^u$$ taking out the constant

$$u = t^2$$
$$du = 2t \, dt$$

$$e^u$$
$$du$$
$$u = 0^2 = 0$$

$$= 200e^u \Big|_0^4 = 200e^4 - 200e^0 \approx 10{,}720$$

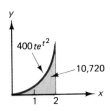

Therefore, during the first 2 years 10,720 tons of pollutant were discharged into the lake.   ■

Notice that the du does not need to be "all together," but can be in several separate pieces, as long as it is all there.

Example 11   After x months the water level in a newly built reservoir is $L(x) = 40x(x^2 + 9)^{-1/2}$ feet. Find the average depth during the first 4 months.

Solution The average value is the definite integral from $x = 0$ to $x = 4$ (the end of month 4) divided by the length of the interval.

$$\frac{1}{4}\int_0^4 40x(x^2 + 9)^{-1/2}\ dx = \frac{1}{4}\cdot 40\cdot\frac{1}{2}\int_0^4 2x(x^2 + 9)^{-1/2}\ dx = 5\int_9^{25} u^{-1/2}\ du$$

changing the limits to u-values using $u = x^2 + 9$

$u = x^2 + 9$
$du = 2x\ dx$

$$= 5\cdot 2u^{1/2}\Big|_9^{25} = 10(25)^{1/2} - 10(9)^{1/2} = 10\cdot 5 - 10\cdot 3 = 20$$

That is, the average depth of the reservoir over the last 4 months was 20 feet.

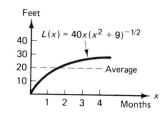

Summary

The three substitution formulas are listed on the inside back cover. Most of the work in using these formulas is making a problem "fit" one of the left-hand sides (choosing the u and adjusting constants). Once a problem fits a left-hand side, the right-hand side immediately gives the answer (except for substituting back to the original variable).

Note that the du now plays a very important role: the du must be correct if the answer is to be correct. For example, the formula $\int e^u\ du = e^u + C$ should not be thought of as the formula for integrating e^u, but as the formula for integrating $e^u\ du$. The du is just as important as the e^u.

SOLUTIONS TO PRACTICE EXERCISES

1. $df = (3x^2 - 4)\ dx$

2. $du = -5e^{-5t}\ dt$

3. **(a)** B   **(b)** C   **(c)** A   **(d)** C

4. Neither.

   (a) Use formula A with $u = x^3 + 1$, so $du = 3x^2\ dx$. The problem has an x^3 for the differential instead of the needed x^2.

   (b) Use formula B with $u = x^2$, so $du = 2x\ dx$. The problem does not have the x that is needed for the differential.

▨ EXERCISES 5.6

Solve each integral. The integration formulas A, B, and C may be found on the inside back cover, numbered 5, 6, and 7.

1 $\int (x^2 + 1)^9\ 2x\ dx$ (*Hint:* Use $u = x^2 + 1$ and formula 5.)

2 $\int (x^3 + 1)^4\ 3x^2\ dx$ (*Hint:* Use $u = x^3 + 1$ and formula 5.)

3 $\int (x^2 + 1)^9 x \, dx$ (*Hint*: Use $u = x^2 + 1$ and formula 5.)

4 $\int (x^3 + 1)^4 x^2 \, dx$ (*Hint*: Use $u = x^3 + 1$ and formula 5.)

5 $\int e^{x^5} x^4 \, dx$ (*Hint*: Use $u = x^5$ and formula 6.)

6 $\int e^{x^4} x^3 \, dx$ (*Hint*: Use $u = x^4$ and formula 6.)

7 $\int \frac{x^5 \, dx}{x^6 + 1}$ (*Hint*: Use $u = x^6 + 1$ and formula 7.)

8 $\int \frac{x^4 \, dx}{x^5 + 1}$ (*Hint*: Use $u = x^5 + 1$ and formula 7.)

Show that each integral cannot be solved by our substitution formulas.

9 $\int \sqrt{x^3 + 1} \; x \, dx$

10 $\int \sqrt{x^5 + 9} \; x^2 \, dx$

11 $\int e^{x^4} x^5 \, dx$

12 $\int e^{x^3} x^4 \, dx$

Solve each integral by the substitution method or state that it cannot be solved by our substitution formulas.

13 $\int (x^4 - 16)^5 x^3 \, dx$

14 $\int (x^5 - 25)^6 x^4 \, dx$

15 $\int e^{-x^2} x \, dx$

16 $\int e^{-x^4} x^3 \, dx$

17 $\int e^{3x} \, dx$

18 $\int e^{5x} \, dx$

19 $\int e^{x^2} x^2 \, dx$

20 $\int e^{x^3} x \, dx$

21 $\int \frac{dx}{1 + 5x}$

22 $\int \frac{dx}{1 + 3x}$

23 $\int (x^2 + 1)^9 5x \, dx$

24 $\int (x^2 - 4)^6 3x \, dx$

25 $\int \sqrt[4]{z^4 + 16} \; z^3 \, dz$

26 $\int \sqrt[3]{z^3 - 8} \; z^2 \, dz$

27 $\int \sqrt[4]{x^4 + 16} \; x^2 \, dx$

28 $\int \sqrt[3]{x^3 - 8} \; x \, dx$

29 $\int (2y^2 + 4y)^5 (y + 1) \, dy$

30 $\int (3y^2 - 6y)^3 (y - 1) \, dy$

31 $\int e^{x^2 + 2x + 5} (x + 1) \, dx$

32 $\int e^{x^3 - 3x + 7} (x^2 - 1) \, dx$

33 $\int \frac{x^3 + x^2}{3x^4 + 4x^3} \, dx$

34 $\int \frac{x^2 - x}{2x^3 - 3x^2} \, dx$

35 $\int \frac{x^3 + x^2}{(3x^4 + 4x^3)^2} \, dx$

36 $\int \frac{x^2 - x}{(2x^3 - 3x^2)^3} \, dx$

37 $\int \frac{x}{1 - x^2} \, dx$

38 $\int \frac{1}{1 - x} \, dx$

39 $\int (2x - 3)^7 \, dx$

40 $\int (5x + 9)^9 \, dx$

41 $\int \frac{e^{2x}}{e^{2x} + 1} \, dx$

42 $\int \frac{e^{3x}}{e^{3x} - 1} \, dx$

43 $\int \dfrac{\ln x}{x}\, dx$ (*Hint*: Let $u = \ln x$.)

44 $\int \dfrac{(\ln x)^2}{x}\, dx$ (*Hint*: Let $u = \ln x$.)

45 $\int \dfrac{e^{\sqrt{x}}}{\sqrt{x}}\, dx$ (*Hint*: Let $u = \sqrt{x}$.)

46 $\int \dfrac{e^{1/x}}{x^2}\, dx$ (*Hint*: Let $u = 1/x$.)

Solve each integral. (*Hint*: Try some algebra.)

47 $\int (x + 1)x^2\, dx$

48 $\int (x + 4)(x - 2)\, dx$

49 $\int (x + 1)^2 x^3\, dx$

50 $\int (x - 1)^2\sqrt{x}\, dx$

Evaluate each definite integral.

51 $\int_0^3 e^{x^2}x\, dx$

52 $\int_0^2 e^{x^3}x^2\, dx$

53 $\int_0^1 \dfrac{x}{x^2 + 1}\, dx$

54 $\int_2^3 \dfrac{x^2}{x^3 - 7}\, dx$

55 $\int_0^4 \sqrt{x^2 + 9}\, x\, dx$

56 $\int_0^3 \sqrt{x^2 + 16}\, x\, dx$

57 $\int_2^3 \dfrac{dx}{1 - x}$

58 $\int_3^4 \dfrac{dx}{2 - x}$

59 $\int_1^8 \dfrac{e^{\sqrt[3]{x}}}{\sqrt[3]{x^2}}\, dx$

60 $\int_1^4 \dfrac{e^{\sqrt{x}}}{\sqrt{x}}\, dx$

61 Prove the integration formula $\int u^n\, du = \dfrac{1}{n + 1}u^{n+1} + C$ ($n \neq -1$) as follows.

   (a) Differentiate the right-hand side of the formula with respect to x (remembering that u is a function of x).

   (b) Verify that this agrees with the integrand in the formula (after replacing du in the formula by $u'\, dx$).

62 Prove the integration formula $\int e^u\, du = e^u + C$ by following the steps in Exercise 61.

63 Prove the integration formula $\int \dfrac{du}{u} = \ln u + C$ ($u > 0$) by following the steps in Exercise 61. (Absolute value bars come from applying the same argument to $-u$ for $u < 0$.)

64 Solve $\int (x + 1)\, dx$ by

   (a) using the formula for $\int u^n\, du$ with $n = 1$,

   (b) dropping the parentheses and integrating directly.

   (c) Can you reconcile the two seemingly different answers? (*Hint*: Think of the arbitrary constant.)

APPLIED EXERCISES

65 (*Business–Cost*) A company's marginal cost function is $MC(x) = \dfrac{1}{2x + 1}$ and its fixed costs are 50. Find the cost function.

66 (*Business–Cost*) A company's marginal cost function is $MC(x) = \dfrac{1}{\sqrt{2x + 25}}$ and its fixed costs are 100. Find the cost function.

67 (*General–Average Value*) The population of a city is expected to be $P(x) = x(x^2 + 36)^{-1/2}$ million people after x years. Find the average population between year $x = 0$ and year $x = 8$.

68 (*General–Area*) Find the area between the curve $y = xe^{x^2}$ and the x-axis from $x = 1$ to $x = 3$. (Leave the answer in its exact form.)

69 (*Business–Average Sales*) A company's sales (in millions) during week x are given by $S(x) = \dfrac{1}{x + 1}$. Find the average sales from week $x = 1$ to week $x = 4$.

70 (*Behavioral Science–Repeated Tasks*) A subject can perform a task at the rate of $r(t) = \sqrt{2t + 1}$ tasks per minute at time t minutes. Find the total number of tasks performed from time $t = 0$ to time $t = 12$.

71 (*Biomedical–Cholesterol*) An experimental drug lowers a patient's body serum cholesterol at the rate of $r(t) = t\sqrt{25 - t^2}$ units per day, where t is the number of days since the drug was administered ($t \leq 5$). Find the total change during the first 3 days.

72 (*Business–Total Sales*) During an automobile sale, cars are selling at the rate of $r(x) = \dfrac{12}{x + 1}$ cars per day, where x is the number of days since the sale began. How many cars will be sold during the first 7 days of the sale?

73 (*Business–Total Sales*) A real estate office is selling condominiums at the rate of $r(x) = 100e^{-(1/4)x}$ per week after x weeks. How many condominiums will be sold during the first 8 weeks?

74 (*Business–Profit*) An aircraft company estimates its marginal revenue function for helicopters to be $MR(x) = \sqrt{x^2 + 80x}\,(x + 40)$ thousand dollars, where x is the number of helicopters sold. Find the total revenue from the sale of the first 10 helicopters.

75–76 (*General–Pollution*) A factory is discharging pollution into a lake at the rate $r(t)$ tons per year given below, where t is the number of years that the factory has been in operation. Find the total amount of pollution discharged during the first 3 years of operation.

75 $r(t) = \dfrac{t}{t^2 + 1}$ **76** $r(t) = t\sqrt{t^2 + 16}$

5.7   REVIEW OF CHAPTER FIVE

Introduction

With this chapter we began our study of integral calculus. Historically, integral calculus came before differential calculus, and was used for finding areas. At that time there was no systematic procedure for solving integrals, and each had to be solved by some special "trick." Later, Newton and Leibniz invented derivatives, and showed that reversing the process of differentiation provides a simple way of solving integrals. It was for this reason that the procedure for calculating areas by definite integration was called the "Fundamental Theorem of Integral Calculus"—the problem of calculating areas was very difficult until the "fundamental" connection with antidifferentiation was understood.

The developments of this chapter have been along two different lines— one technical (how to solve integrals) and the other conceptual (what integrals do). First we review the technical developments.

Techniques of Integration

We began by defining indefinite integration as "inverse differentiation," and we developed several formulas.

$$\int x^n \, dx = \frac{1}{n+1} \, x^{n+1} + C \qquad\qquad (n \neq -1)$$

$$\int \frac{dx}{x} = \int \frac{1}{x} \, dx = \int x^{-1} \, dx = \ln |x| + C$$

$$\int e^{ax} \, dx = \frac{1}{a} \, e^{ax} + C \qquad\qquad (a \neq 0)$$

These, together with the sum rule (the integral of sum is the sum of the integrals) and the constant multiple rule (constants may be moved across the integral sign) enable us to integrate many functions.

Many more functions can be integrated by the substitution method, using the following formulas.

$$\int u^n \, du = \frac{1}{n+1} \, u^{n+1} + C \qquad\qquad (n \neq 1)$$

$$\int \frac{du}{u} = \int \frac{1}{u} \, du = \int u^{-1} \, du = \ln |u| + C$$

$$\int e^u \, du = e^u + C$$

These formulas are similar to the previous ones, and follow immediately from them using the chain rule.

Definite integrals were defined as indefinite integrals with an added step of evaluation.

$$\int_a^b f(x) \, dx = F(b) - F(a)$$

where $F(x)$ is any antiderivative of $f(x)$.

Uses of Integration

On the conceptual side, integrals represent *continuous sums*.

$$\left(\begin{array}{l}\text{Total accumulation at rate} \\ f(x) \text{ from } x = a \text{ to } x = b\end{array}\right) = \int_a^b f(x) \, dx$$

Definite integrals give the accumulation over a particular interval (such as the total cost of units 100 to 200), while indefinite integrals give a *formula* for the total up to x (such as the total cost of x units).

Definite integrals of nonnegative functions also give areas.

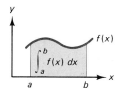 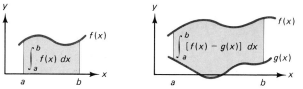

Dividing a definite integral by the length of the interval gives the *average value* of a function on the interval.

$$\left(\begin{array}{c} \text{Average value} \\ \text{of } f(x) \text{ on } [a, b] \end{array} \right) = \frac{1}{b - a} \int_a^b f(x)\, dx$$

Differentiation and Integration

Differentiation may be thought of as a process of "breaking down into pieces," and integration as a process of "combining into a whole." (In fact, the word "integrate" means "make whole.") For example, differentiation takes a cost function and gives the marginal cost (cost per unit), while integration combines all of these per-unit costs back into a total cost.

The rules of differentiation from Chapter 2 enable us to differentiate *any* function composed of sums, differences, products, quotients, roots, powers, exponential functions, and logarithms (so-called "elementary functions"). The same, however, is not true for integration: there are continuous functions (like e^{x^2}) that cannot be integrated (as finite combinations of elementary functions). We can only solve an integral if it "fits" one of our formulas. In this sense, differentiation is a purely mechanical procedure, while integration is more of an "art," often involving a clever choice of the function u. Further techniques in the "art of integration" are discussed in Chapter 6. All of these developments are reviewed in the following exercises.

Practice Test

Exercises: 3, 5, 7, 21, 25, 29, 35, 41, 47, 53, 57, 61, 67, 69

▦ EXERCISES 5.7

Solve each integral (if possible).

1 $\int (6\sqrt{x} - 5)\, dx$

2 $\int (8\sqrt[3]{x} - 2)\, dx$

3 $\int (10\sqrt[3]{x^2} - 4x)\, dx$

4 $\int (5\sqrt{x^3} - 6x)\, dx$

5 $\int x^2 \sqrt[3]{x^3 - 1}\, dx$

6 $\int x^3 \sqrt{x^4 - 1}\, dx$

7 $\int x \sqrt[3]{x^3 - 1}\, dx$

8 $\int x^2 \sqrt{x^4 - 1}\, dx$

9 $\int \dfrac{dx}{9 - 3x}$

10 $\int \dfrac{dx}{1 - 2x}$

11 $\int \dfrac{dx}{(9 - 3x)^2}$

12 $\int \dfrac{dx}{(1 - 2x)^2}$

13 $\int \dfrac{x^2}{\sqrt[3]{8 + x^3}}\, dx$

14 $\int \dfrac{x}{\sqrt{9 + x^2}}\, dx$

15 $\int \dfrac{w + 3}{(w^2 + 6w - 1)^2}\, dw$

16 $\int \dfrac{t - 2}{(t^2 - 4t + 1)^2}\, dt$

17 $\int \dfrac{(1 + \sqrt{x})^2}{\sqrt{x}}\, dx$

18 $\int \dfrac{(1 + \sqrt[3]{x})^2}{\sqrt[3]{x^2}}\, dx$

19 $\int e^{(1/2)x}\, dx$

20 $\int e^{-2x}\, dx$

21 $\int \left(6e^{3x} - \dfrac{6}{x}\right) dx$

22 $\int (x - x^{-1})\, dx$

23 $\int \dfrac{e^x}{e^x - 1}\, dx$

24 $\int \dfrac{1}{x \ln x}\, dx$

25 $\int (x + 4)(x - 4)\, dx$

26 $\int \dfrac{3x^3 + 2x^2 + 4x}{x}\, dx$

Evaluate each definite integral.

27 $\int_1^9 \left(x - \dfrac{1}{\sqrt{x}}\right) dx$

28 $\int_0^4 \dfrac{dz}{\sqrt{2z + 1}}$

29 $\int_0^4 \dfrac{w}{\sqrt{25 - w^2}}\, dw$

30 $\int_1^2 \dfrac{x + 1}{(x^2 + 2x - 2)^2}\, dx$

31 $\int_0^8 \dfrac{x}{\sqrt{x^2 + 36}}\, dx$

32 $\int_1^5 \dfrac{dx}{x}$

33 $\int_1^{e^4} \dfrac{dx}{x}$

34 $\int_4^5 \dfrac{dx}{x - 6}$

35 $\int_0^2 e^{-x}\, dx$

36 $\int_0^2 e^{(1/2)x}\, dx$

37 $\int_0^{100} (e^{.05x} - e^{.01x})\, dx$

38 $\int_0^{10} (e^{.04x} - e^{.02x})\, dx$

39 $\int_0^1 x^3 e^{x^4}\, dx$

40 $\int_0^1 x^4 e^{x^5}\, dx$

Find the area under the given curve between the two x-values. (Leave answers in exact form.)

41 $f(x) = 12e^{2x}$, $x = 0$ to $x = 3$

42 $f(x) = e^{x/2}$, $x = 0$ to $x = 4$

43 $f(x) = \dfrac{1}{x}$, $x = 1$ to $x = 100$

44 $f(x) = x^{-1}$, $x = 1$ to $x = 1000$

45 $y = \dfrac{x^2 + 6x}{\sqrt[3]{x^3 + 9x^2 + 17}}$ $x = 1$ to $x = 3$

46 $y = \dfrac{x + 6}{\sqrt{x^2 + 12x + 4}}$ $x = 0$ to $x = 3$

Find the area bounded by the each pair of curves.

47 $y = x^2 + 3x$ and $y = 3x + 1$

48 $y = 12x - 3x^2$ and $y = 6x - 24$

49 $y = x^2$ and $y = x$

50 $y = x^4$ and $y = x$

51 $y = 4x^3$ and $y = 12x^2 - 8x$

52 $y = x^3 - x^2$ and $y = x^2 + x$

Find the average value of the function on the given interval.

53 $f(x) = \dfrac{1}{x^2}$ on $[1, 4]$

54 $f(x) = 6\sqrt{x}$ on $[1, 4]$

55 $f(x) = xe^{-x^2}$ on $[0, 2]$

56 $f(x) = \dfrac{x}{x^2 - 3}$ on $[2, 4]$

▨ APPLIED EXERCISES

57 (*Business–Cost*) A company's marginal cost function is

$$\mathrm{MC}(x) = \frac{1}{\sqrt{2x + 9}}$$

and fixed costs are 100. Find the cost function.

58 (*General–Weight*) An average child of age t years gains weight at the rate of $r(t) = 1.7t^{1/2}$ kilograms per year. Find the total weight gain from age 1 to age 9.

59 (*Biomedical–Temperature*) An experimental drug changes a patient's temperature at the rate $r(t) = 3x^2/(x^3 + 1)$ degrees per milligram of the drug, where x is the amount of the drug administered. Find the total change in temperature resulting from the first 3 milligrams of the drug. (*Note*: The rate of change of temperature with respect to dosage, is called the "drug sensitivity.")

60 (*Behavioral Science–Learning*) A student can memorize foreign vocabulary words at the rate $r(t) = 2/\sqrt[3]{x}$ words per minute, where t is the number of minutes since the studying began. Find the number of words that can be memorized in the first 8 minutes.

61 (*General–Global Warming*) The temperature of the earth is rising, due to the "greenhouse effect," in which pollution prevents the escape of heat from the atmosphere. If the temperature is rising at the rate of $r(t) = .15e^{.1t}$ degrees per year, find the total rise in temperature over the next 10 years.

62 (*General–Consumption of Natural Resources*) World consumption of aluminum is running at the rate of $r(t) = 24e^{.05t}$ thousand tons per year, where t is the number of years since 1985. At this rate, when will the known world resources of 8000 thousand tons of aluminum be exhausted?

63 (*General–Total Savings*) A homeowner installs a solar heating system, which is expected to generate savings at the rate of $r(t) = 200e^{.1t}$ dollars per year, where t is the number of years since the system was installed. If the system originally cost $1500, when will it "pay for itself?"

64 (*General–Average Population*) The population of the world is predicted to be $P(t) = 5.7e^{.01t}$ billion, where t is the number of years after the year 2000. Find the average population between the years 2000 and 2100.

65 (*General–Art*) An artist wants to paint the interior shape shown below on the side of a building. If the shape is to be painted red, how much red paint (in square meters) will the artist need?

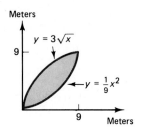

66 (*Economics–Balance of Trade*) A country's annual exports will be $E(t) = 40e^{.2t}$ and its imports will be $I(t) = 20e^{.1t}$ (both in billions of dollars per year), where t is the number of years from now. Find the accumulated trade surplus (exports minus imports) over the next 10 years.

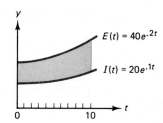

67–68 (*Economics–Consumers' Surplus*) For the demand function $d(x)$ and the demand level x given below, find the consumers' surplus.

67 $d(x) = 8000 - 24x$, $x = 200$

68 $d(x) = 1800 - .03x^2$, $x = 200$

69–70 (*Economics–Gini Index*) For the Lorenz curve given below, find the Gini Index.

69 $L(x) = x^{3.5}$ **70** $L(x) = x^{2.5}$

6 INTEGRATION TECHNIQUES AND DIFFERENTIAL EQUATIONS

The length of a cable in a suspension bridge can be calculated by integration.

6.1 INTEGRATION BY PARTS

Introduction

In this chapter we introduce further techniques for solving integrals: integration by parts, integration by tables, and numerical integration. We also discuss improper integrals (integrals over infinite intervals) and differential equations.

First we discuss the method of *integration by parts,* which comes from interpreting the product rule as an integration formula.

Integration by Parts Formula

For two differentiable functions $u(x)$ and $v(x)$, hereafter denoted simply u and v, the product rule is

$$(uv)' = u'v + uv'$$

the derivative of a product is the derivative of the first times the second, plus the first times the derivative of the second

If we integrate both sides of this equation, the integral on the left side "undoes" the differentiation.

$$uv = \int u'v \, dx + \int uv' \, dx$$
$$\quad\quad\quad \underset{du}{\diagdown\diagup} \quad\quad \underset{dv}{\underbrace{}}$$

using differential notation $du = u' \, dx$, $dv = v' \, dx$

$$uv = \int v \, du + \int u \, dv$$

the above formula in differential notation

Solving this equation for the second integral $\int u \, dv$ (writing $\int u \, dv$ on the left side with everything else on the right side) gives

$$\int u \, dv = u \cdot v - \int v \, du$$

This formula is the basis for a technique called "integration by parts."

Integration by Parts

For differentiable functions u and v,

$$\int u \, dv = uv - \int v \, du$$

We use this formula to solve integrals by a "double substitution," substituting u for part of the given integral and dv for the rest, and then expressing the integral in the form $u \cdot v - \int v \, du$. The point is to choose the u and the dv so that the resulting integral $\int v \, du$ is *simpler* than the original integral $\int u \, dv$. A few examples will make the method clear.

Integration by Parts Procedure

We solve the integral $\int xe^x \, dx$, explaining the steps on the right.

Example 1   Use integration by parts to solve $\int xe^x \, dx$.

Solution

$$\int \underbrace{x}_{u}\,\underbrace{e^x \, dx}_{dv}$$

original integral

$$\begin{bmatrix} u = x & dv = e^x \, dx \\ \downarrow & \downarrow \\ du = 1 \, dx = dx & v = \int e^x \, dx = e^x \end{bmatrix}$$

calling part of the integrand u and everything else dv, we take $u = x$ and $dv = e^x \, dx$

differentiating u to find du and integrating dv to find v (omitting the C) (the arrows show which part leads to which other part)

$$= \underbrace{xe^x}_{u\ v} - \int \underbrace{e^x}_{v}\,\underbrace{dx}_{du}$$

replacing the original integral $\int u \, dv$ by the right-hand side of the formula, $u \cdot v - \int v \, du$, with u, du, v, and dv replaced by what they are equal to

$$= xe^x - e^x + C$$

$$\text{from } \int e^x \, dx$$

solving the new integral $\int e^x \, dx = e^x$ to give the final answer (with a + C) ■

The procedure is not as complicated as it might seem. All the steps may be written together as follows.

$$\int \underbrace{x}_{u}\,\underbrace{e^x \, dx}_{dv} = \underbrace{xe^x}_{u\ v} - \int \underbrace{e^x}_{v}\,\underbrace{dx}_{du} = xe^x - e^x + C \qquad \longleftarrow \text{answer}$$

$$\text{from } \int e^x \, dx$$

$$\begin{bmatrix} u = x & dv = e^x \, dx \\ du = dx & v = \int e^x \, dx = e^x \end{bmatrix}$$

We may check this answer by differentiation.

$$\frac{d}{dx}(xe^x - e^x + C) = \underbrace{e^x + xe^x}_{\substack{\text{differentiating } xe^x \\ \text{by the product rule}}} - e^x = e^x + xe^x - e^x = xe^x$$

cancel

The result xe^x agrees with the original integrand, so the integration is correct.

REMARKS ON THE INTEGRATION BY PARTS PROCEDURE

(i)   The differentials du and dv include the dx.

(ii)   We omit the constant C when we integrate dv to get v, because one C at the end is enough.

(iii) The integration by parts formula does not give a "final answer," but rather, expresses the given integral as $uv - \int v\,du$, a product $u \cdot v$ (already integrated) and a new integral $\int v\,du$. That is, integration by parts "exchanges" the original integral $\int u\,dv$ for another integral $\int v\,du$. The hope is that the second integral will be simpler than the first. In our example we "exchanged" $\int xe^x\,dx$ for the simpler $\int e^x\,dx$, which could be integrated immediately by formula 4 (inside back cover).

(iv) Integration by parts is rather complicated, so should be used only if formulas 1–7 (inside back cover) fail to solve the integral.

Other Choices for *u* and *dv*

To solve $\int xe^x\,dx$, we chose $u = x$ and $dv = e^x\,dx$:

$$\int \underset{u}{x}\underset{dv}{e^x\,dx}$$

If we had, instead, chosen $u = e^x$ and $dv = x\,dx$, the integration by parts would have gone as follows.

$$\int xe^x\,dx = e^x \frac{1}{2}x^2 - \int \frac{1}{2}x^2 e^x\,dx = \frac{1}{2}x^2 e^x - \frac{1}{2}\int x^2 e^x\,dx$$

$$\left[\begin{array}{ll} u = e^x & dv = x\,dx \\ du = e^x\,dx & v = \int x\,dx = \frac{1}{2}x^2 \end{array} \right]$$

moving the 1/2 outside

new integral

This time we "exchanged" the original integral $\int xe^x\,dx$ for the integral $\int x^2 e^x\,dx$, which is *more* complicated than the original. Our work has been mathematically correct, but we have succeeded only in expressing a difficult integral $\int xe^x\,dx$ in terms of an even more difficult integral, $\int x^2 e^x\,dx$. Therefore, this choice of u and dv was a poor choice, and our earlier choice of $u = x$ and $dv = e^x\,dx$ was the "right" choice, since it led to a solution.

How to Choose the *u* and the *dv*

The proper choice of u and dv may involve some "trial and error." Generally, only one choice will be "best," while other choices lead to more complicated integrals. There is no "foolproof" rule that always gives the "right" choice for u and dv, but the following guidelines often help.

> **Guidelines for Choosing u and dv**
>
> 1. Choose dv to be the most complicated part of the integral that can be easily integrated.
>
> 2. Choose u so that u' is simpler than u.

Further Examples of Integration by Parts

Example 2   Solve $\int x^2 \ln x \, dx$.

Solution   None of the easier formulas (1–6 inside the back cover) solve the integral, as you may easily check. Therefore we try integration by parts. The integrand is a product, x^2 times $\ln x$. The guidelines say choose dv to be the most complicated part that can be easily integrated. We can integrate x^2 but not $\ln x$ (we know how to *differentiate* $\ln x$, but not how to *integrate* it), so we choose $dv = x^2 \, dx$, and therefore $u = \ln x$.

$$\int x^2 \ln x \, dx = \underbrace{(\ln x)}_{u}\underbrace{\left(\tfrac{1}{3} x^3\right)}_{} - \int \underbrace{\tfrac{1}{3} x^3}_{v} \, \underbrace{\tfrac{1}{x} \, dx}_{du} = \tfrac{1}{3} x^3 \ln x - \tfrac{1}{3}\underbrace{\int x^2 \, dx}_{}$$

(under the first line: u / dv; under second term: u, v; under third: v, du)

$$\text{moving the 1/3 outside and}$$
$$\text{simplifying } \tfrac{1}{x} x^3 \text{ to } x^2$$

$$\left[\begin{array}{ll} u = \ln x & dv = x^2 \, dx \\[2mm] du = \tfrac{1}{x} \, dx & v = \int x^2 \, dx = \tfrac{1}{3} x^3 \end{array} \right]$$

$$= \tfrac{1}{3} x^3 \ln x - \tfrac{1}{3}\tfrac{1}{3} x^3 + C = \tfrac{1}{3} x^3 \ln x - \tfrac{1}{9} x^3 + C \qquad \leftarrow answer$$

▪

We check this answer by differentiation.

$$\frac{d}{dx}\left[\tfrac{1}{3} x^3 \ln x - \tfrac{1}{9} x^3 + C \right] = \underbrace{x^2 \ln x + \tfrac{1}{3} x^3 \tfrac{1}{x}}_{} - \tfrac{1}{3} x^2$$

$$\text{differentiating } \tfrac{1}{3} x^3 \ln x$$
$$\text{by the product rule}$$

$$= x^2 \ln x + \tfrac{1}{3} x^2 - \tfrac{1}{3} x^2 = x^2 \ln x$$

$$\text{from } x^3 \tfrac{1}{x} \qquad \text{cancel}$$

The result agrees with the original integrand, so the integration is correct.

Practice Exercise 1

Solve using integration by parts: $\int x^3 \ln x \, dx$.

(solution on page 343)

Integration by parts is also useful for integrating products of powers of linear functions.

Example 3   Solve using integration by parts: $\int (x - 2)(x + 4)^8 \, dx$.

Solution   The guidelines recommend that dv be the most complicated part that can be integrated. Both $x - 2$ and $(x + 4)^8$ can be integrated. For example,

$$\int (x + 4)^8 \, dx = \frac{1}{9} (x + 4)^9 + C$$

by the substitution method with $u = x + 4$ (omitting the details)

Since $(x + 4)^8$ is more complicated than $(x - 2)$, we take $dv = (x + 4)^8 \, dx$.

$$\underbrace{\int}_{} \underbrace{(x - 2)}_{u}\underbrace{(x + 4)^8 \, dx}_{dv} = \underbrace{(x - 2)}_{u}\underbrace{\frac{1}{9}(x + 4)^9}_{v} - \int \underbrace{\frac{1}{9}(x + 4)^9}_{v}\underbrace{dx}_{du}$$

$$\left[\begin{array}{ll} u = x - 2 & dv = (x + 4)^8 \, dx \\ du = dx & v = \int (x + 4)^8 \, dx = \frac{1}{9}(x + 4)^9 \\ & \text{(see above)} \end{array}\right]$$

using the integration by parts formula

$$= \frac{1}{9}(x - 2)(x + 4)^9 - \frac{1}{9}\int (x + 4)^9 \, dx$$

taking out the 1/9

$$= \frac{1}{9}(x - 2)(x + 4)^9 - \frac{1}{9}\frac{1}{10}(x + 4)^{10} + C$$

integrating by the substitution method

$$= \frac{1}{9}(x - 2)(x + 4)^9 - \frac{1}{90}(x + 4)^{10} + C$$

Again we could check this answer by differentiation.   ■

Practice Exercise 2

Solve using integration by parts: $\int (x + 1)(x - 1)^3 \, dx$.

(solution on page 343)

Application: Present Value of a Continuous Stream of Income

If a business or some other asset generates income continuously at the rate $C(t)$ dollars per year, where t is the number of years from now, then $C(t)$ is called a *continuous stream of income.** In Section 4.1 we saw that to find the *present value* of a sum (the amount now that will later yield the stated sum) under continuous compounding we multiply by e^{-rt}, where r is the interest rate and t is the number of years. (We will refer to a rate with continuous compounding as a "continuous interest rate.") Therefore, the present value of the continuous stream $C(t)$ is found by multiplying by e^{-rt} and summing (integrating) over the time period.

The present value of the continuous stream of income $C(t)$ over T years at continuous interest rate r is

$$\int_0^T C(t)e^{-rt}\,dt$$

Example 4   A business generates income at the rate of $2t$ million dollars per year, where t is the number of years from now. Find the present value of this continuous stream for the next five years at the continuous interest rate of 10%.

Solution   We multiply $C(t) = 2t$ by $e^{-.1t}$ (since $r = 10\% = .1$) and integrate from 0 to 5 years.

$$\int_0^5 2te^{-.1t}\,dt$$

This is a *definite* integral, but we will ignore the limits of integration until after we have solved the *indefinite* integral. None of the formulas 1 through 7 on the inside back cover will solve this integral, so we try integration by parts with $u = 2t$ and $dv = e^{-.1t}\,dt$. (Do you see why the guidelines on page 339 suggest this choice?)

$$\int \underbrace{2t}_{u}\underbrace{e^{-.1t}\,dt}_{dv} = \underbrace{(2t)}_{u}\underbrace{(-10e^{-.1t})}_{v} - \int \underbrace{(-10e^{-.1t})}_{v}\underbrace{2\,dt}_{du}$$

$$\left[\begin{array}{ll} u = 2t & dv = e^{-.1t}\,dt \\ du = 2\,dt & v = \int e^{-.1t}\,dt = -10e^{-.1t} \end{array}\right]$$

$$\left(\text{using } \int e^{at}\,dt = \frac{1}{a}e^{at}\right)$$

$$= -20te^{-.1t} + 20\int e^{-.1t}\,dt = -20te^{-.1t} + \underbrace{20(-10)e^{-.1t} + C}$$

again using
$$\int e^{at}\,dt = \frac{1}{a}e^{at} + C$$

$$= -20te^{-.1t} - 200e^{-.1t} + C$$

---

* $C(t)$ must be continuous, meaning that the income is being paid continuously rather than in "lump-sum" payments. However, even lump-sum payments can be approximated by a continuous stream $C(t)$ if the payments are frequent enough.

For the *definite* integral, we evaluate this from 0 to 5.

$$[-20te^{-.1t} - 200e^{-.1t}]\Big|_0^5 = \underbrace{-20{\cdot}5e^{-.5} - 200e^{-.5}}_{\text{evaluation at } t = 5} - \underbrace{[-20{\cdot}\underset{0}{0}e^0 - \underset{200}{200}e^0]}_{\text{evaluation at } t = 0}$$

$$= -300e^{-.5} + 200 \approx 117.9 \quad \text{(using a calculator)}$$

This is in millions of dollars, so the present value of the stream of income over 5 years is $117,900,000. ■

This answer means that $117,900,000 at 10% interest compounded continuously would exactly generate the continuous stream $C(t) = 2t$ million dollars for 5 years. This method is often used to determine the fair value of a business or some other asset, since it gives the present value of all future income.

Practice Exercise 3

Find the present value of the continuous stream of income $C(t) = 3t$ thousand dollars per year (where t is the number of years from now) for 10 years at the continuous interest rate of 5%.   *(solution on page 343)*

Summary

The integration by parts formula

$$\int u \, dv = uv - \int v \, du$$

is simply the integration version of the product rule. The guidelines on page 339 and the examples lead to the following particular suggestions for choosing u and dv.

For Integrals of the Form:		Choose:	
$\int x^n e^{ax} \, dx$	$u = x^n$	$dv = e^{ax} \, dx$	
$\int x^n \ln x \, dx$	$u = \ln x$	$dv = x^n \, dx$	
$\int (x + a)(x + b)^n \, dx$	$u = x + a$	$dv = (x + b)^n \, dx$	

Remember, however, that you should use integration by parts only if the "easier" formulas (1 through 7 on the inside back cover) fail to solve the integral.

Practice Exercise 4

Which of the following integrals requires integration by parts, and which can be solved by the substitution formula $\int e^u \, du = e^u + C$? (Do not solve the integrals.)

(a) $\displaystyle\int xe^x \, dx$ (b) $\displaystyle\int xe^{x^2} \, dx$ *(solutions below)*

Integration by parts can be useful in any situation involving integrals, such as recovering total cost from marginal cost, calculating areas, average values (the integral divided by the length of the interval), continuous accumulations, or present values of continuous income streams.

SOLUTIONS TO PRACTICE EXERCISES

1. $\displaystyle\int x^3 \ln x \, dx = (\ln x)\left(\frac{1}{4} x^4\right) - \int \frac{1}{x}\frac{1}{4} x^4 \, dx = (\ln x)\left(\frac{1}{4} x^4\right) - \frac{1}{4}\int x^3 \, dx = (\ln x)\left(\frac{1}{4} x^4\right) - \frac{1}{16} x^4 + C$

$$\left[\begin{array}{ll} u = \ln x & dv = x^3 \, dx \\ du = \dfrac{1}{x}\,dx & v = \int x^3 \, dx = \dfrac{1}{4} x^4 \end{array}\right]$$

2. $\displaystyle\int (x + 1)(x - 1)^3 \, dx = (x + 1)\frac{1}{4}(x - 1)^4 - \int \frac{1}{4}(x - 1)^4 \, dx$

$$\left[\begin{array}{ll} u = x + 1 & dv = (x - 1)^3 \, dx \\ du = dx & v = \int (x - 1)^3 \, dx = \dfrac{1}{4}(x - 1)^4 \end{array}\right]$$

$$= \frac{1}{4}(x + 1)(x - 1)^4 - \frac{1}{4}\cdot\frac{1}{5}(x - 1)^5 + C = \frac{1}{4}(x + 1)(x - 1)^4 - \frac{1}{20}(x - 1)^5 + C$$

3. We solve the indefinite integral first.

$$\int 3te^{-.05t} \, dt = (3t)(-20e^{-.05t}) - \int (-20e^{-.05t})3 \, dt = -60te^{-.05t} + 60\int e^{-.05t} \, dt$$

$$\left[\begin{array}{ll} u = 3t & dv = e^{-.05t} \, dt \\ du = 3 \, dt & v = \int e^{-.05t} \, dt = -20e^{-.05t} \end{array}\right]$$

$$= -60te^{-.05t} + 60(-20e^{-.05t}) + C = -60te^{-.05t} - 1200e^{-.05t} + C$$

For the definite integral we evaluate this from 0 to 10.

$$-60\cdot10e^{-.5} - 1200e^{-.5} - (-60\cdot0e^{0-} - 1200e^{0}) = -1800e^{-.5} + 1200 \approx 108.245$$

Present value: $108,245

4. (a) Requires integration by parts

(b) Can be solved by the substitution $u = x^2$

EXERCISES 6.1

*Integration by parts often involves solving an integral like one of the following when integrating dv to find v. Solve the following integrals **without** using integration by parts (using formulas 1 through 7 on the inside back cover). Be ready to solve similar integrals during the integration by parts procedure.*

1 $\int e^{2x}\, dx$ **2** $\int x^5\, dx$ **3** $\int (x + 2)\, dx$ **4** $\int (x - 1)\, dx$

5 $\int \sqrt{x}\, dx$ **6** $\int e^{-.5t}\, dt$ **7** $\int (x + 3)^4\, dx$ **8** $\int (x - 5)^6\, dx$

Solve each integral using integration by parts.

9 $\int xe^{2x}\, dx$ **10** $\int xe^{3x}\, dx$ **11** $\int x^5 \ln x\, dx$ **12** $\int x^4 \ln x\, dx$

13 $\int (x + 2)e^x\, dx$ (*Hint*: Take $u = x + 2$.)   **14** $\int (x - 1)e^x\, dx$ (*Hint*: Take $u = x - 1$.)

15 $\int \sqrt{x} \ln x\, dx$ **16** $\int \sqrt[3]{x} \ln x\, dx$ **17** $\int (x - 3)(x + 4)^5\, dx$ **18** $\int (x + 2)(x - 5)^5\, dx$

19 $\int te^{-.5t}\, dt$ **20** $\int te^{-.2t}\, dt$ **21** $\int \frac{\ln t}{t^2}\, dt$ **22** $\int \frac{\ln t}{\sqrt{t}}\, dt$

23 $\int s(2s + 1)^4\, ds$ **24** $\int \frac{x + 1}{e^{3x}}\, dx$ **25** $\int \frac{x}{e^{2x}}\, dx$ **26** $\int \frac{\ln (x + 1)}{\sqrt{x + 1}}\, dx$

27 $\int \frac{x}{\sqrt{x + 1}}\, dx$ **28** $\int x\sqrt{x + 1}\, dx$

29 $\int xe^{ax}\, dx$, $a \neq 0$ **30** $\int (x + b)e^{ax}\, dx$, $a \neq 0$

31 $\int x^n \ln ax\, dx$, $a \neq 0, n \neq -1$ **32** $\int (x + a)^n \ln(x + a)\, dx$, $n \neq -1$

33 $\int \ln x\, dx$ (*Hint*: Take $u = \ln x$, $dv = dx$.)   **34** $\int \ln x^2\, dx$ (*Hint*: Take $u = \ln x^2$, $dv = dx$.)

35 $\int x^3 e^{x^2}\, dx$ (*Hint*: Take $u = x^2$, $dv = xe^{x^2}\, dx$, using a substitution to find v from dv.)

36 $\int x^3(x^2 - 1)^6\, dx$ (*Hint*: Use $u = x^2$, $dv = x(x^2 - 1)^6\, dx$, using a substitution to find v from dv.)

Solve by whatever means are necessary (integration by parts or substitution).

37 (a) $\int xe^{x^2}\, dx$ (b) $\int \frac{(\ln x)^3}{x}\, dx$ (c) $\int x^2 \ln 2x\, dx$ (d) $\int \frac{e^x}{e^x + 4}\, dx$

38 (a) $\int \sqrt{\ln x}\, \frac{1}{x}\, dx$ (b) $\int x^2 e^{x^3}\, dx$ (c) $\int x^7 \ln 3x\, dx$ (d) $\int xe^{4x}\, dx$

Solve each definite integral using integration by parts.
(Leave answers in exact form.)

39 $\int_0^2 xe^x\, dx$ **40** $\int_0^3 xe^x\, dx$ **41** $\int_1^3 x^2 \ln x\, dx$ **42** $\int_1^2 x \ln x\, dx$

43 $\int_0^2 z(z - 2)^4\, dz$ **44** $\int_0^4 z(z - 4)^6\, dz$ **45** $\int_0^{\ln 4} te^t\, dt$ **46** $\int_1^e \ln x\, dx$

Solve in two different ways.

47 $\int x(x-2)^5\,dx$

(a) Use integration by parts.

(b) Use the substitution $u = x - 2$ (so x is replaced by $u + 2$) and then multiply out the integrand.

48 $\int x(x+4)^6\,dx$

(a) Use integration by parts.

(b) Use the substitution $u = x + 4$ (so x is replaced by $u - 4$) and then multiply out the integrand.

Derive each formula by using integration by parts on the left-hand side.

49 $\int x^n e^x\,dx = x^n e^x - n \int x^{n-1} e^x\,dx, \quad n > 0$

50 $\int (\ln x)^n\,dx = x(\ln x)^n - n \int (\ln x)^{n-1}\,dx \quad n > 0$

51 Use the formula in Exercise 49 to solve the integral $\int x^2 e^x\,dx$. (*Hint*: Apply the formula twice.)

52 Use the formula in Exercise 50 to solve the integral $\int (\ln x)^2\,dx$. (*Hint*: Apply the formula twice.)

53 (a) Solve the integral $\int x^{-1}\,dx$ by integration by parts (using $u = x^{-1}$ and $dv = dx$), obtaining

$$\int x^{-1}\,dx = x^{-1}x - \int (-x^{-2})x\,dx$$

which gives

$$\int x^{-1}\,dx = 1 + \int x^{-1}\,dx$$

(b) Subtract the integral from both sides of this last equation, obtaining $0 = 1$. Explain this apparent contradiction.

54 We omit the constant of integration when we integrate dv to get v. Including the constant C in this step simply replaces v by $v + C$, giving the formula

$$\int u\,dv = u(v+C) - \int (v+C)\,du$$

Multiplying out the parentheses and expanding the last integral into two gives

$$\int u\,dv = uv + Cu - \int v\,du - C\int du$$

Show that the second and fourth terms on the right cancel, thereby obtaining the "old" integration by parts formula $\int u\,dv = u\cdot v - \int v\,du$. This shows that including the constant in the dv to v step gives the same formula. One constant of integration at the end is enough.

■ **APPLIED EXERCISES** (Most require 🖩.)

55 (*Business–Revenue*) If a company's marginal revenue function is $MR(x) = xe^{(1/4)x}$, find the revenue function. (*Hint*: Evaluate the constant C so that revenue is 0 at $x = 0$.)

56 (*Business–Cost*) A company's marginal cost function is $MC(x) = xe^{-(1/2)x}$ and fixed costs are 200. Find the cost function. (*Hint*: Evaluate the constant C so that the cost is 200 at $x = 0$.)

57 (*Business–Present Value of a Continuous Stream of Income*) An electronics company generates a continuous stream of income of $4t$ million dollars per year, where t is the number of years that the company has been in operation. Find the present value of this stream of income over the first 10 years at a continuous interest rate of 10%.

58 (*Business–Present Value of a Continuous Stream of Income*) An oil well generates a continuous stream of income of $60t$ thousand dollars per year, where t is the number of years that the rig has been in operation. Find the present value of this stream of income over the first 20 years at a continuous interest rate of 5%.

59 (*Biomedical–Drug Dosage*) A drug taken orally is absorbed into the bloodstream at the rate of $r(t) = te^{-.5t}$ milligrams per hour, where t is the number of hours since the drug was taken. Find the total amount of the drug absorbed during the first 5 hours.

60 (*General–Pollution*) Contamination is leaking from an underground waste-disposal tank at the rate of $r(t) = t\cdot\ln t$ thousand gallons per month, where t is the number of months since the leak began. Find the total leakage between months $t = 1$ and $t = 4$.

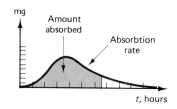

mg
Amount absorbed
Absorbtion rate
t, hours

61 (*General–Area*) Find the area under the curve

$$y = x \cdot \ln x$$

and above the *x*-axis from $x = 1$ to $x = 2$.

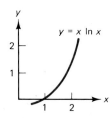

y
$y = x \ln x$
2
1
1   2
x

62 (*Political Science–Fundraising*) A politician can raise campaign funds at the rate of $r(t) = 50te^{-.1t}$ thousand dollars per week during the first *t* weeks of a campaign. Find the average amount raised during the first 5 weeks.

63 (*Business–Product Recognition*) A company begins advertising a new product and finds that after *t* months the product is gaining customer recognition at the rate of $r(t) = t^2 \ln t$ thousand customers per week (for $t \geq 1$). Find the total gain in recognition between weeks $t = 1$ and $t = 6$.

64 (*General–Population*) The population of a town is increasing at the rate of $r(x) = 400te^{.02t}$ people per year, where *t* is the number of years from now. Find the total gain in population during the next 5 years.

REPEATED INTEGRATION BY PARTS

65–70 Sometimes an integral requires two or more integration by parts. As an example, we apply integration by parts to the integral $\int x^2 e^x \, dx$.

$$\int \underbrace{x^2}_{u} \underbrace{e^x \, dx}_{dv} = \underbrace{x^2}_{u} \underbrace{e^x}_{v} - \int \underbrace{e^x}_{v} \underbrace{2x \, dx}_{du} = x^2 e^x - 2 \int x e^x \, dx$$

$$\begin{bmatrix} u = x^2 & dv = e^x \, dx \\ du = 2x \, dx & v = \int e^x \, dx = e^x \end{bmatrix}$$

The new integral $\int x e^x \, dx$ is solved by a second integration by parts. Continuing with the solution:

$$= x^2 e^x - 2 \left(\int \underbrace{x}_{u} \underbrace{e^x \, dx}_{dv} \right) = x^2 e^x - 2 \left(\underbrace{x e^x}_{u \ v} - \int \underbrace{e^x}_{v} \underbrace{dx}_{du} \right) = x^2 e^x - 2(x e^x - e^x) + C$$

$$\begin{bmatrix} u = x & dv = e^x \, dx \\ du = dx & v = \int e^x \, dx = e^x \end{bmatrix}$$

$$= x^2 e^x - 2x e^x + 2e^x + C \longleftarrow \text{final answer}$$

After reading the explanation above, solve each integral by repeated integration by parts.

65 $\int x^2 e^{-x} \, dx$ **66** $\int x^2 e^{2x} \, dx$ **67** $\int (x + 1)^2 e^x \, dx$ **68** $\int (\ln x)^2 \, dx$

69 $\int x^2 (\ln x)^2 \, dx$ **70** $\int x^3 e^x \, dx$

6.2 INTEGRATION USING TABLES

Introduction

There are many techniques of integration, and only a few of the most useful ones will be discussed in this book. Many of the advanced techniques lead to simple integration formulas, which can then be collected into a "table of integrals." In this section we will see how to solve integrals by choosing an appropriate formula from such a table.

Inside the back cover of this book is a short table of integrals which we shall use. The formulas are grouped according to the type of integrand (for example, "Forms Involving $x^2 - a^2$"). Look at the table now to see how it is organized.

Using Integral Tables

Given a particular integral, we first look for a formula that fits it exactly.

Example 1 Solve the integral $\int \dfrac{1}{x^2 - 4}\, dx$.

Solution The denominator $x^2 - 4$ is of the form $x^2 - a^2$ (with $a = 2$), so we look under "Forms Involving $x^2 - a^2$" (inside back cover). Formula 15,

$$\int \frac{1}{x^2 - a^2}\, dx = \frac{1}{2a} \ln \left| \frac{x - a}{x + a} \right| + C \qquad \text{formula 15}$$

with $a = 2$, solves our integral.

$$\int \frac{1}{x^2 - 4}\, dx = \frac{1}{4} \ln \left| \frac{x - 2}{x + 2} \right| + C \qquad \text{\textit{formula 15 with } a = 2 \textit{ substituted on both sides}}$$

Note that the expression $x^2 - a^2$ does not require that the last number be a "perfect square." For example, $x^2 - 3$ can be written $x^2 - a^2$ with $a = \sqrt{3}$. ∎

Application to Business

Integral tables are useful in many applications, such as integrating a rate to find the total accumulation.

Example 2 A company's sales rate is

$$r(x) = \frac{x}{\sqrt{x + 9}}$$

sales per week after x weeks. Find a formula for the total sales after x weeks.

Solution   To find the *total* sales $S(x)$ we integrate the *rate* of sales.

$$S(x) = \int \frac{x}{\sqrt{x+9}}\, dx$$

In the table under "Form Involving $\sqrt{ax+b}$" we find formula 13:

$$\int \frac{x}{\sqrt{ax+b}}\, dx = \frac{2ax - 4b}{3a^2}\sqrt{ax+b} + C \qquad \text{\textit{formula 13}}$$

This formula with $a = 1$ and $b = 9$ solves the integral.

$$S(x) = \int \frac{x}{\sqrt{x+9}}\, dx = \frac{2x - 36}{3}\sqrt{x+9} + C \qquad \text{\textit{formula 13 with a = 1 and b = 9}}$$

$$= \left(\frac{2}{3}x - 12\right)\sqrt{x+9} + C \qquad \text{\textit{simplifying}}$$

To evaluate the constant C we use the fact that total sales at time $x = 0$ must be zero: $S(0) = 0$.

$$(-12)\sqrt{9} + C = 0 \qquad \text{\textit{$\left(\frac{2}{3}x - 12\right)\sqrt{x+9} + C$ at x = 0 set}}$$

$$\text{\textit{equal to zero}}$$

$$-36 + C = 0 \qquad \text{\textit{simplifying}}$$

Therefore, $C = 36$. Substituting this into $S(x)$ gives the formula for the total sales in the first x months.

$$S(x) = \left(\frac{2}{3}x - 12\right)\sqrt{x+9} + 36 \qquad \text{\textit{$S(x) = \left(\frac{2}{3}x - 12\right)\sqrt{x+9} + C$ with}}$$

$$\text{\textit{C = 36}}$$

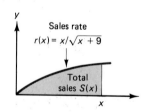

Application to Genetic Engineering

Modern techniques of biotechnology are being used to develop many new products, including powerful antibiotics, disease-resistant crops, and bacteria that literally "eat" oil spills. These "gene splicing" techniques require solving definite integrals such as the following.

Example 3   Under certain circumstances, the number of generations of bacteria needed to increase the frequency of a gene from .2 to .5 is

$$n = 2.5 \int_{.2}^{.5} \frac{1}{q^2(1 - q)}\, dq$$

Find n (rounded to the nearest integer).

Solution   Formula 12 (inside back cover) integrates a similar-looking fraction.

$$\int \frac{1}{x^2(ax + b)}\, dx = -\frac{1}{b}\left(\frac{1}{x} + \frac{a}{b}\ln\left|\frac{x}{ax + b}\right|\right) + C \qquad \text{\textit{formula 12}}$$

To make $(ax + b)$ into $(1 - x)$ we take $a = -1$ and $b = 1$, so the left-hand side of the formula becomes

$$\int \frac{1}{x^2(-x + 1)}\, dx \quad \text{or} \quad \int \frac{1}{x^2(1 - x)}\, dx \qquad \text{\textit{from formula 12 with } } a = -1, b = 1$$

Except for replacing x by q, this is the same as our integral. Therefore, the indefinite integral is solved by formula 12 with $a = -1$ and $b = 1$ (which we express in the variable q).

$$\int \frac{1}{q^2(1 - q)}\, dq = -\left(\frac{1}{q} - \ln\left|\frac{q}{1 - q}\right|\right) + C \qquad \text{\textit{formula 12 with } } a = -1 \text{ and } b = 1$$

For the *definite* integral from .2 to .5, we evaluate and subtract.

$$\underbrace{-\left(\frac{1}{.5} - \ln\left|\frac{.5}{1 - .5}\right|\right)}_{\text{evaluation at } q = .5} - \underbrace{\left[-\left(\frac{1}{.2} - \ln\left|\frac{.2}{1 - .2}\right|\right)\right]}_{\text{evaluation at } q = .2} = -(2 - \underbrace{\ln 1}_{0}) + \left(5 - \underbrace{\ln \frac{.2}{.8}}_{\ln .25}\right)$$

$$= -2 + 5 - \ln .25 \approx 4.38 \qquad \text{\textit{using a calculator}}$$

We multiply this by the 2.5 in front of the original integral.

$$(2.5)(4.38) = 10.95$$

Therefore, 11 generations are needed to raise the gene frequency from .2 to .5.   ▪

Further Examples

Sometimes a substitution is needed to transform a formula to fit a given integral. In such cases both the x and the dx must be transformed. A few examples will make the method clear.

Example 4   Solve the integral $\int \frac{x}{\sqrt{x^4 + 1}}\, dx$.

Solution   The table has no formula involving x^4. However, $x^4 = (x^2)^2$, so a formula involving x^2, along with a substitution, might work. Formula 18,

$$\int \frac{1}{\sqrt{x^2 \pm a^2}}\, dx = \ln|x + \sqrt{x^2 \pm a^2}| + C \qquad \textit{formula 18}$$

has a square root in the denominator, like our problem. (A double sign like $\pm$ in a formula means that it is actually *two* formulas, one read with the upper sign all the way across, and the other read with the lower sign all the way across.) This formula with $a = 1$ and the upper sign is

$$\int \frac{1}{\sqrt{x^2 + 1}}\, dx = \ln|x + \sqrt{x^2 + 1}| + C$$

and with the substitution

$$x = z^2$$
$$dx = 2z\, dz \qquad \textit{differential of } x = z^2$$

it becomes

$$\int \frac{1}{\sqrt{z^4 + 1}}\, 2z\, dz = \ln|z^2 + \sqrt{z^4 + 1}| + C$$

This, except for the 2 on the left-hand side (and z instead of x) is the integral that we want to solve. Therefore, dividing by 2 and replacing z by x gives the final answer.

$$\int \frac{x}{\sqrt{x^4 + 1}}\, dx = \frac{1}{2} \ln(x^2 + \sqrt{x^4 + 1}) + C$$

We drop the absolute value bars since the quantity in parentheses is always positive. ▪

Notice that the original integral had an x in the numerator, while the formula that we successfully used had only a 1 in the numerator. This illustrates the principle that in choosing a formula, try to match the *most complicated part* of the integral (in this case, the $\sqrt{x^4 + 1}$ in the denominator), to a formula, and hope that the rest works out when you substitute for the differential dx.

Practice Exercise 1

Solve the integral $\int \dfrac{t}{9t^4 - 1}\, dt$. (*Hint:* Use formula 15 with the substitution

$x = 3t^2$.)

(solution on page 353)

Example 5 Solve the integral $\int \dfrac{e^{-2t}}{e^{-t} + 1}\, dt$.

Solution Looking in the table under "Forms Involving e^{ax} and $\ln x$," none of the formulas looks anything like this integral. However, replacing e^{-t} by x would make the denominator of our integral into $x + 1$, so formula 9 (inside back cover) might help. This formula with $a = 1$ and $b = 1$ is

$$\int \frac{x}{x + 1}\, dx = x - \ln|x + 1| + C \qquad\qquad \textit{formula 9 with } a = 1 \textit{ and } b = 1$$

With the substitution

$$x = e^{-t}$$

$$dx = -e^{-t}\, dt \qquad\qquad \textit{differential of } x = e^{-t}$$

formula 9 becomes

$$\int \frac{e^{-t}}{e^{-t} + 1}\, (-e^{-t})\, dt = e^{-t} - \ln|e^{-t} + 1| + C$$

or

$$-\int \frac{e^{-2t}}{e^{-t} + 1}\, dt = e^{-t} - \ln\,(e^{-t} + 1) + C$$

Except for the negative sign this is the given integral. Multiplying through by -1 gives the final answer.

$$\int \frac{e^{-2t}}{e^{-t} + 1}\, dt = -e^{-t} + \ln\,(e^{-t} + 1) + C \qquad\qquad ▪$$

Reduction Formulas

Sometimes we must apply a formula several times to simplify an integral in stages.

Example 6 Solve the integral $\int x^3 e^{-x}\, dx$.

Solution Looking through the integral table, we see that formula 21,

$$\int x^n e^{ax}\, dx = \frac{1}{a}\, x^n e^{ax} - \frac{n}{a} \int x^{n-1} e^{ax}\, dx \qquad\qquad \textit{formula 21}$$

with $n = 3$ and $a = -1$, fits this integral. Notice that the right-hand side of this formula involves a new integral, but with a *lower* power of x. We will apply formula 21 several times, each time reducing the power of x until we eliminate it completely. Applying formula 21 with $n = 3$ and $a = -1$, we proceed as follows.

$$\int x^3 e^{-x}\, dx = -x^3 e^{-x} + 3 \int x^2 e^{-x}\, dx \qquad\qquad \textit{after 1 application}$$

the power has been reduced

$$= -x^3 e^{-x} + 3\left(-x^2 e^{-x} + 2 \int x^1 e^{-x}\, dx\right) \qquad \textit{applying formula 21 again (now with}$$
$\textit{n = 2) to the last integral above}$

$$= -x^3 e^{-x} - 3x^2 e^{-x} + 6 \int x^1 e^{-x}\, dx \qquad \textit{multiplying out}$$

$$= -x^3 e^{-x} - 3x^2 e^{-x} + 6\left(-x e^{-x} + \int \underbrace{x^0}_{1} e^{-x}\, dx\right) \qquad \textit{using formula 21 a third time (now with}$$
$\textit{n = 1)}$

$$= -x^3 e^{-x} - 3x^2 e^{-x} - 6x e^{-x} + 6 \int e^{-x}\, dx \qquad \textit{now solve this last integral with the}$$
$\textit{formula } \int e^{ax}\, dx = \dfrac{1}{a} e^{ax}$

$$= -x^3 e^{-x} - 3x^2 e^{-x} - 6x e^{-x} - 6 e^{-x} + C \qquad \textit{the solution, after three applications of}$$
$\textit{formula 21}$ ■

In the example we used formula 21 three times, reducing the x^3 in steps, first down to x^2, then to x^1, and finally to $x^0 = 1$, at which point we could solve the integral easily. If the power of x in the integral had been higher, more applications of formula 21 would have been necessary. Formulas like 21 and 22 are called *reduction formulas*, since they express an integral in terms of a similar integral but with a reduced power of x.

Summary

Finding a formula in an integral table to "fit" a given integral can be tricky, especially if a substitution is involved. Look for a formula that matches *the most complicated part* of the given integral. We have been using a very short table of integrals, but the techniques are the same with more extensive tables. Many integral tables have been published, some book length, containing several thousand formulas.*

---

* A useful table of integrals is contained in the *CRC Standard Mathematical Tables*, published by the CRC Press, Boca Raton, Fla., with more than 400 formulas.

We can perform substitutions "in either direction." That is, we could look for a substitution to change a formula into the problem, or to change the problem into a formula. It is easier to use a substitution that changes the chosen formula into the problem (as we did in the examples), since this simplifies the substitution of the differential.

SOLUTION TO PRACTICE EXERCISE

1. Formula 15 with the substitution $x = 3t^2$, $dx = 6t\,dt$, and $a = 1$ becomes

$$\int \frac{1}{9t^4 - 1}\, 6t\, dt = \frac{1}{2}\ln\left|\frac{3t^2 - 1}{3t^2 + 1}\right| + C$$

Dividing each side by 6 gives the answer.

$$\int \frac{t}{9t^4 - 1}\, dt = \frac{1}{12}\ln\left|\frac{3t^2 - 1}{3t^2 + 1}\right| + C$$

EXERCISES 6.2

For each integral, state the number of the integration formula (from the inside back cover) and the values of the constants a and b that will solve the integral. (Do not solve the integral.)

1 $\int \dfrac{1}{x^2(5x - 1)}\, dx$ **2** $\int \dfrac{x}{2x - 3}\, dx$ **3** $\int \dfrac{1}{x\sqrt{-x + 7}}\, dx$ **4** $\int \dfrac{x}{\sqrt{-2x + 1}}\, dx$

5 $\int \dfrac{x}{1 - x}\, dx$ **6** $\int \dfrac{1}{x\sqrt{1 - 4x}}\, dx$

Solve each integral by using the integral table on the inside back cover.

7 $\int \dfrac{1}{9 - x^2}\, dx$ (*Hint:* Use formula 16 with $a = 3$.) **8** $\int \dfrac{1}{x^2 - 25}\, dx$ (*Hint:* Use formula 15 with $a = 5$.)

9 $\int \dfrac{1}{x^2(2x + 1)}\, dx$ (*Hint:* Use formula 12 with $a = 2$, $b = 1$.) **10** $\int \dfrac{x}{x + 2}\, dx$ (*Hint:* Use formula 9 with $a = 1$, $b = 2$.)

11 $\int \dfrac{x}{1 - x}\, dx$ (*Hint:* Use formula 9.) **12** $\int \dfrac{x}{\sqrt{1 - x}}\, dx$ (*Hint:* Use formula 13.)

13 $\int \dfrac{1}{(2x + 1)(x + 1)}\, dx$ **14** $\int \dfrac{x}{(x + 1)(x + 2)}\, dx$ **15** $\int \sqrt{x^2 - 4}\, dx$ **16** $\int \dfrac{1}{\sqrt{x^2 - 1}}\, dx$

17 $\int \dfrac{1}{z\sqrt{1 - z^2}}\, dz$ **18** $\int \dfrac{\sqrt{4 + z^2}}{z}\, dz$ **19** $\int x^3 e^{2x}\, dx$ **20** $\int x^{99} \ln x\, dx$

21 $\int x^{-101} \ln x\, dx$ **22** $\int (\ln x)^2\, dx$ **23** $\int \dfrac{1}{x(x + 3)}\, dx$ **24** $\int \dfrac{1}{x(x - 3)}\, dx$

25 $\int \dfrac{z}{z^4 - 4}\, dz$

26 $\int \dfrac{z}{9 - z^4}\, dz$

27 $\int \sqrt{9x^2 + 16}\, dx$

28 $\int \dfrac{1}{\sqrt{16x^2 - 9}}\, dx$

29 $\int \dfrac{1}{\sqrt{4 - e^{2t}}}\, dt$

30 $\int \dfrac{e^t}{9 - e^{2t}}\, dt$

31 $\int \dfrac{e^t}{e^{2t} - 1}\, dt$

32 $\int \dfrac{e^{2t}}{1 - e^t}\, dt$

33 $\int \dfrac{x^3}{\sqrt{x^8 - 1}}\, dx$

34 $\int x^2 \sqrt{x^6 + 1}\, dx$

35 $\int \dfrac{1}{x\sqrt{x^3 + 1}}\, dx$

36 $\int \dfrac{\sqrt{1 - x^6}}{x}\, dx$

37 $\int \dfrac{e^t}{(e^t - 1)(e^t + 1)}\, dt$

38 $\int \dfrac{e^{2t}}{(e^t - 1)(e^t + 1)}\, dt$

39 $\int x e^{x/2}\, dx$

40 $\int \dfrac{x}{e^x}\, dx$

41 $\int \dfrac{1}{e^{-x} + 4}\, dx$

42 $\int \dfrac{1}{\sqrt{e^{-x} + 4}}\, dx$

Solve each definite integral using the table of integrals on the inside back cover. (Leave answers in exact form.)

43 $\int_4^5 \sqrt{x^2 - 16}\, dx$

44 $\int_0^4 \dfrac{1}{\sqrt{x^2 + 9}}\, dx$

45 $\int_2^3 \dfrac{1}{x^2 - 1}\, dx$

46 $\int_2^4 \dfrac{1}{1 - x^2}\, dx$

47 $\int_3^5 \dfrac{\sqrt{25 - x^2}}{x}\, dx$

48 $\int_3^4 \dfrac{1}{x\sqrt{25 - x^2}}\, dx$

Solve each integral by whatever means are necessary (either substitution or tables).

49 $\int \dfrac{1}{2x + 6}\, dx$

50 $\int \dfrac{x}{x^2 - 4}\, dx$

51 $\int \dfrac{x}{2x + 6}\, dx$

52 $\int \dfrac{1}{4 - x^2}\, dx$

53 $\int x\sqrt{1 - x^2}\, dx$

54 $\int \dfrac{x}{\sqrt{1 - x^2}}\, dx$

55 $\int \dfrac{\sqrt{1 - x^2}}{x}\, dx$

56 $\int \dfrac{1}{\sqrt{x^2 - 1}}\, dx$

Solve each integral. (Hint: Separate each integral into two integrals, using the fact that the numerator is a sum or difference, and solve the two integrals by two different formulas.)

57 $\int \dfrac{x - 1}{(3x + 1)(x + 1)}\, dx$

58 $\int \dfrac{x - 1}{x^2(x + 1)}\, dx$

59 $\int \dfrac{x + 1}{x\sqrt{1 + x^2}}\, dx$

60 $\int \dfrac{x - 1}{x\sqrt{x^2 + 4}}\, dx$

61 $\int \dfrac{x + 1}{x - 1}\, dx$ (*Hint:* After separating into two integrals, solve one by a formula and the other by substitution.)

62 $\int \dfrac{x + 1}{\sqrt{x^2 + 1}}\, dx$ (*Hint:* After separating into two integrals, solve one by a formula and the other by substitution.)

■ **APPLIED EXERCISES**

63 (*Business–Total Sales*) A company's sales rate is $r(x) = x^2 e^{-x}$ million sales per month after x months. Find a formula for the total sales in the first x months. (*Hint:* Integrate the sales rate to find the total sales and determine the constant C so that total sales are zero at time $x = 0$.)

64 (*General–Population*) The population of a city is expected to grow at the rate of $r(x) = x/\sqrt{x + 9}$ thousand people per year after x years. Find the total change in population from year 0 to year 27.

▦ **65** (*Biomedical–Gene Frequency*) Under certain circumstances, the number of generations necessary to increase the frequency of a gene from .1 to .3 is

$$n = 3 \int_{.1}^{.3} \frac{1}{q^2(1-q)} \, dq$$

Find n (rounded to the nearest integer).

66 (*Behavioral Science*) A subject in a psychology experiment gives responses at the rate of $r(t) = t/\sqrt{t+1}$ correct answers per minute after t minutes. Find the total number of correct responses between times $t = 0$ and $t = 15$.

67 (*Business–Cost*) The marginal cost function for a computer chip manufacturer is $MP(x) = 1/\sqrt{x^2 + 1}$, and fixed costs are \$2000. Find the cost function.

68 (*Social Science–Welfare Cases*) An urban welfare office estimates that the number of social security recipients t years from now will be $t/(2t + 4)$ million people. Find the average number of social security recipients during the period $t = 0$ to $t = 10$.

6.3 IMPROPER INTEGRALS

Introduction

In this section we define integrals over intervals that are infinite in length. Such integrals are called ''improper'' integrals and have many applications. Suppose, for example, that after you become rich and famous, you decide to commission a statue of yourself for your home town. The town, however, will accept your generous gift only if you pay for the perpetual upkeep of the statue by establishing a fund that will generate \$2000 annually for every year in the future. (Interestingly, we will see that such a fund does *not* require an infinite amount of money). Before deciding whether or not to accept this condition, you of course want to know the size of the fund that is required. In Section 4.1 we saw that \$2000 in t years requires only $\$2000e^{-rt}$ (the ''present value'') deposited in a bank now at continuous interest rate r. Therefore, the size of the fund needed to generate an annual \$2000 *forever* is found by summing (integrating) the present value $2000e^{-rt}$ over the infinite time interval from zero to infinity (∞).

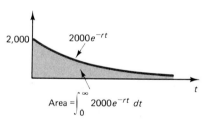

$$\left(\begin{array}{c} \text{Size of} \\ \text{fund} \end{array} \right) = \int_0^\infty 2000 e^{-rt} \, dt$$

In this section we learn how to calculate such ''improper'' integrals.

Limits as x Approaches ±∞

The notation $x \to \infty$ (''x approaches infinity'') means that x takes on arbitrary large values.

$x \to \infty$ means:	x takes values arbitrarily far to the *right* on the number line.
$x \to -\infty$ means:	x takes values arbitrarily far to the *left* on the number line.

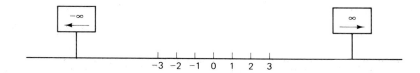

Evaluating limits as x approaches positive or negative infinity is simply a matter of thinking about large and small numbers. The reciprocal of a large number is a small number. For example:

$$\frac{1}{1,000,000} = .000001$$

As the denominator becomes larger (with the numerator staying constant), the value of the fraction approaches zero. For example, the denominators in the following two fractions grow arbitrarily large, so the limits are both zero.

$$\lim_{x \to \infty} \frac{1}{x^2} = 0 \qquad \lim_{x \to \infty} \frac{1}{e^x} = \lim_{x \to \infty} e^{-x} = 0$$

These examples illustrate the following general rules.

$$\lim_{x \to \infty} \frac{1}{x^n} = 0 \qquad (n > 0)$$

as x approaches infinity, one over x to a positive power approaches zero

$$\lim_{x \to \infty} e^{-ax} = 0 \qquad (a > 0)$$

as x approaches infinity, e to a negative number times x approaches zero

We will often use the letter b for a variable approaching infinity.

Example 1

(a) $\lim_{b \to \infty} \dfrac{1}{b^2} = 0$

using the first rule in the box above

(b) $\lim_{b \to \infty} \left(3 - \dfrac{1}{b} \right) = 3$

because the 1/b approaches zero

(c) $\lim_{b \to \infty} (e^{-2b} - 5) = -5$

because the e^{-2b} approaches zero   ▪

Some quantities, instead of approaching zero, become arbitrarily large as x approaches infinity. If a quantity becomes arbitrarily large, then it does not approach any number, so the limit does not exist. (For a limit to exist, the limit must be finite.)

The following limits do not exist

$$\lim_{x\to\infty} x^n \qquad (n > 0)$$

as x approaches infinity, x to a positive power has no limit

$$\lim_{x\to\infty} e^{ax} \qquad (a > 0)$$

as x approaches infinity, e to a positive number times x has no limit

$$\lim_{x\to\infty} \ln x$$

as x approaches infinity, the natural logarithm of x has no limit

Example 2

(a) $\lim\limits_{b\to\infty} b^3$ does not exist.

because b^3 becomes arbitrarily large as b approaches infinity

(b) $\lim\limits_{b\to\infty} (\sqrt{b} - 1)$ does not exist.

because $\sqrt{b}$ becomes arbitrarily large as b approaches infinity

■

Practice Exercise 1

Evaluate the following limits (if they exist).

(a) $\lim\limits_{b\to\infty} \left(1 - \dfrac{1}{b}\right)$

(b) $\lim\limits_{b\to\infty} (\sqrt[3]{b} + 3)$

(solutions on page 363)

Similar rules hold for x approaching negative infinity.

$$\lim_{x\to-\infty} \frac{1}{x^n} = 0 \qquad \text{(for any integer } n > 0)$$

as x approaches negative infinity, 1 over x to a positive integer approaches zero

$$\lim_{x\to-\infty} e^{ax} = 0 \qquad (a > 0)$$

as x approaches negative infinity, e to a positive number time x approaches zero (because the exponent is approaching $-\infty$)

These results will be useful for evaluating improper integrals.

Integrating to Infinity

As a first example we evaluate the improper integral $\int_1^\infty (1/x^2)\, dx$, which gives the area under the curve $y = 1/x^2$ from $x = 1$ arbitrarily far to the right.

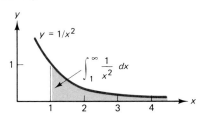

Example 3   Evaluate $\int_1^\infty \frac{1}{x^2}\, dx$.

Solution   To integrate to infinity we first integrate over a *finite* interval, from 1 to some number b (think of b as some very large number), and then take the limit as b approaches ∞. First integrate from 1 to b.

$$\underbrace{\int_1^b \frac{1}{x^2}\, dx = \int_1^b x^{-2}\, dx = (-x^{-1})\Big|_1^b = \left(-\frac{1}{x}\right)\Big|_1^b}_{\substack{\text{integrating by} \\ \text{the power rule}}} = \underbrace{-\frac{1}{b}}_{\substack{\text{evaluating} \\ \text{at } x = b}} - \underbrace{\left(-\frac{1}{1}\right)}_{\substack{\text{evaluating} \\ \text{at } x = 1}} = -\frac{1}{b} + 1$$

Then take the limit of this answer as $b \to \infty$.

$$\lim_{b \to \infty} \left(-\frac{1}{b} + 1\right) = 1 \qquad\qquad \textit{the 1/b approaches zero as } b \to \infty$$

This gives the answer:

$$\int_1^\infty \frac{1}{x^2}\, dx = 1 \qquad\qquad \textit{the integral of 1/x}^2 \textit{ from 1 to } \infty \textit{ equals 1}$$

■

Geometrically, this procedure amounts to finding the area under the curve from 1 to some number b, shown on the left below, and then letting $b \to \infty$ to find the area shown on the right.

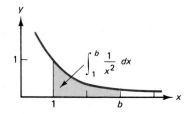
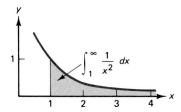

For a function f that is continuous and nonnegative for $x \geq a$,

$$\int_a^\infty f(x)\, dx = \lim_{b \to \infty} \int_a^b f(x)\, dx$$

If the limit exists, then the improper integral is *convergent*. Otherwise, the integral is *divergent*.

Therefore, the integral that we solved, $\int_1^\infty 1/x^2\, dx$, is *convergent,* since the limit was finite.

We require the function to be continuous and nonnegative on the interval. It is possible to define improper integrals for functions that take negative values, and even for discontinuous functions, but we shall not do so since most applications involve functions that are positive and continuous.

Practice Exercise 2

Solve the improper integral $\int_2^\infty \frac{1}{x^2}\, dx.$ *(solution on page 363)*

Divergent Integrals

If an improper integral diverges (that is, if the limit does not exist), the integral cannot be solved.

Example 4   Solve the improper integral $\int_1^\infty \frac{1}{\sqrt{x}}\, dx.$

Solution   As usual, we first integrate up to b.

$$\int_1^b \frac{1}{\sqrt{x}}\, dx = \int_1^b x^{-1/2}\, dx = \underbrace{2\cdot x^{1/2}\Big|_1^b}_{\substack{\text{integrating by}\\ \text{the power rule}}} = \underbrace{2\sqrt{b} - 2\sqrt{1}}_{\text{evaluating}} = 2\sqrt{b} - 2$$

Then we let b approach infinity.

$$\lim_{b\to\infty} (2\sqrt{b} - 2) \quad \text{does not exist}$$ *because $\sqrt{b}$ becomes infinite as $b \to \infty$*

Therefore,

$$\int_1^\infty \frac{1}{\sqrt{x}}\, dx \quad \text{is } \textit{divergent}$$ *the integral cannot be solved*   ∎

   Notice from Examples 3 and 4 that $\int_1^\infty (1/x^?)\, dx$, was convergent, while $\int_1^\infty (1/\sqrt{x})\, dx$ was divergent.

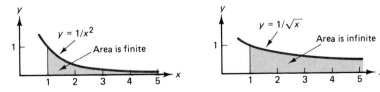

Although both areas are infinitely long, the area under $1/x^2$ is finite while the area under $1/\sqrt{x}$ is infinite. Intuitively, this is because for large values of x the curve $1/x^2$ is lower than the curve $1/\sqrt{x}$ (as may be seen by evaluating $1/x^2$ and $1/\sqrt{x}$ at $x = 100$), so has a smaller area under it.

   From now on we will combine the two steps of integrating up to b and letting $b \to \infty$.

$$\int_1^\infty x^{-2}\,dx = \lim_{b\to\infty}\int_1^b x^{-2}\,dx = \lim_{b\to\infty}(-x^{-1})\Big|_1^b = \lim_{b\to\infty}\left(-\frac{1}{x}\right)\Big|_1^b$$

integrating

$$= \lim_{b\to\infty}\left[-\frac{1}{b} - \left(-\frac{1}{1}\right)\right] = 0 + 1 = 1$$

approaches
0

Application: Permanent Endowments

At the beginning of this section we found that the size of the fund necessary to generate \$2000 annually forever is $\int_0^\infty 2000e^{-rt}\,dt$, where r is the continuous interest rate (the interest rate with continuous compounding). Such a fund that generates steady income forever is called a "permanent endowment."

Example 5 Find the size of the permanent endowment needed to generate \$2000 annually at the continuous interest rate of 10%.

Solution We must solve the following improper integral.

$$\int_0^\infty 2000e^{-.1t}\,dt$$

annual income   continuous interest rate

$$\int_0^\infty 2000e^{-.1t}\,dt = \lim_{b\to\infty}\left[2000\int_0^b e^{-.1t}\,dt\right] = \lim_{b\to\infty}\left[2000(-10)e^{-.1t}\Big|_0^b\right]$$

moved outside   integrating by $\int e^{ax}\,dx = \frac{1}{a}e^{ax}$

$$= \lim_{b\to\infty}[-20{,}000e^{-b} + 20{,}000e^0] = 20{,}000$$

approaches 0   1

Therefore, a permanent endowment that will pay the \$2000 annual maintenance forever costs \$20,000. ∎

Permanent endowments are used to estimate the ultimate cost of anything that requires steady long-term funding, from buildings to governmental agencies.

Practice Exercise 3

Find the size of the permanent endowment necessary to generate an annual $3000 at an interest rate of 5% compounded continuously.

(solution on page 363)

Solving Improper Integrals Using Substitutions

Solving an improper integral may require a substitution. In such cases we apply the substitution not only to the integrand but also to the differential and the upper and lower limits of integration.

Example 6   Solve the improper integral $\int_2^\infty \dfrac{x}{(x^2+1)^2}\,dx$.

Solution   We use the substitution $u = x^2 + 1$, so $du = 2x\,dx$, requiring multiplication by 2 and by $\frac{1}{2}$. Notice how the substitution changes the limits.

$$\text{as } x \to \infty \; u = x^2 + 1 \to \infty$$

$$\int_2^\infty \frac{x}{(x^2+1)^2}\,dx = \frac{1}{2}\int_2^\infty \frac{2x}{(x^2+1)^2}\,dx = \frac{1}{2}\int_5^\infty \frac{du}{u^2} = \frac{1}{2}\lim_{b\to\infty}\int_5^b u^{-2}\,du$$

$$\begin{bmatrix} u = x^2 + 1 \\ du = 2x\,dx \end{bmatrix} \qquad u^2 \qquad u = 2^2 + 1 = 5$$

$$= \frac{1}{2}\lim_{b\to\infty}\left[-u^{-1}\right]_5^b = \frac{1}{2}\lim_{b\to\infty}\left[-\frac{1}{b} - \left(-\frac{1}{5}\right)\right] = \frac{1}{2}\cdot\frac{1}{5} = \frac{1}{10}$$

$$\underbrace{}_{\text{integrating}} \qquad \underbrace{}_{\text{approaches 0}} \qquad \underbrace{}_{\text{final answer}}$$

■

Integrating to $-\infty$ and from $-\infty$ to ∞

To integrate over an interval that extends arbitrarily far to the *left*, we again integrate over a finite interval and then take the limit.

For a function f that is continuous and nonnegative for $x \le b$, we define

$$\int_{-\infty}^b f(x)\,dx = \lim_{a\to-\infty}\int_a^b f(x)\,dx$$

The improper integral is *convergent* if the limit is finite.

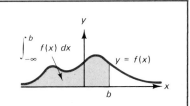

To integrate over the entire x-axis, from $-\infty$ to ∞, we use two integrals, one from $-\infty$ to 0, and the other from 0 to ∞, and then add the results.

For a function f that is continuous and nonnegative for all values of x, we define

$$\int_{-\infty}^{\infty} f(x)\ dx = \lim_{a \to -\infty} \int_a^0 f(x)\ dx + \lim_{b \to \infty} \int_0^b f(x)\ dx$$

The improper integral is *convergent* if both lmiits are finite.

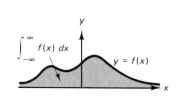

Example 7   Solve $\int_{-\infty}^3 4e^{2x}\ dx$.

Solution

$$\int_{-\infty}^3 4e^{2x}\ dx = \lim_{a \to -\infty} \int_a^3 4e^{2x}\ dx = \lim_{a \to -\infty} \underbrace{\left[4 \cdot \tfrac{1}{2} \cdot e^{2x} \Big|_a^3 \right]}_{\text{integrating}} = \lim_{a \to -\infty} [\underbrace{2e^6 - 2e^{2a}}_{\substack{\text{approaches 0} \\ \text{as } a \to -\infty}}] = 2e^6$$

■

Practice Exercise 4

Solve the improper integral $\int_{-\infty}^1 12e^{3x}\ dx$.

(solution on page 363)

Summary

A definite integral in which one or both limits of integration are infinite is called an "improper" integral. For a continuous nonnegative function the improper integral is defined as the limit of the integral over a finite interval. The integral is *convergent* if the limit is finite, and otherwise the integral is *divergent*. This idea of dealing with the infinite by "dropping back to the finite and then taking the limit" is a standard trick in mathematics.

Several particular limits are helpful in evaluating improper integrals.

x **Approaching Infinity**		x **Approaching Negative Infinity**	
$\lim\limits_{x \to \infty} \dfrac{1}{x^n} = 0$	$(n > 0)$	$\lim\limits_{x \to -\infty} \dfrac{1}{x^n} = 0$	(for any integer $n > 0$)
$\lim\limits_{x \to \infty} e^{-ax} = 0$	$(a > 0)$	$\lim\limits_{x \to -\infty} e^{ax} = 0$	$(a > 0)$

Improper integrals give continuous sums over infinite intervals. For example, the total future output of an oil well can be found by integrating the production rate out to infinity. Improper integrals are used to sum the

present value of future income, for example, calculating the cost of permanent endowments, or estimating the value of assets (like land) that lasts
forever. These and many other applications are illustrated in the exercises.

SOLUTIONS TO PRACTICE EXERCISES

1. **(a)** $\displaystyle\lim_{b\to\infty}\left(1-\frac{1}{b}\right)=1$ **(b)** $\displaystyle\lim_{b\to\infty}(\sqrt[3]{b}+3)$ does not exist.

2. $\displaystyle\int_2^b x^{-2}\,dx=(-x^{-1})\Big|_2^b=-\frac{1}{b}-\left(-\frac{1}{2}\right)=\frac{1}{2}-\frac{1}{b}$

$\displaystyle\lim_{b\to\infty}\left(\frac{1}{2}-\frac{1}{b}\right)=\frac{1}{2}$

Therefore, $\displaystyle\int_2^\infty \frac{1}{x^2}\,dx=\frac{1}{2}$

3. $\displaystyle\int_0^\infty 3000e^{-.05t}\,dt=\lim_{b\to\infty}\int_0^b 3000e^{-.05t}\,dt=\lim_{b\to\infty}\left[3000(-20)e^{-.05t}\Big|_0^b\right]$

$\displaystyle\qquad\qquad = \lim_{b\to\infty}[-60{,}000e^{-.05b}+60{,}000e^0]=\$60{,}000$

4. $\displaystyle\int_{-\infty}^1 12e^{3x}\,dx=\lim_{a\to-\infty}\int_a^1 12e^{3x}\,dx=\lim_{a\to-\infty}\left[12\cdot\frac{1}{3}\,e^{3x}\Big|_a^1\right]$

$\displaystyle\qquad\qquad = \lim_{a\to-\infty}[4e^3-4e^{3a}]=4e^3$

EXERCISES 6.3

Evaluate each limit (or state that it does not exist).

1 $\displaystyle\lim_{x\to\infty}\frac{1}{x^2}$ **2** $\displaystyle\lim_{b\to\infty}\left(\frac{1}{\sqrt{b}}-8\right)$ **3** $\displaystyle\lim_{b\to\infty}(1-2e^{-5b})$ **4** $\displaystyle\lim_{b\to\infty}(3e^{3b}-4)$

5 $\displaystyle\lim_{x\to\infty}(2-e^{x/2})$ **6** $\displaystyle\lim_{x\to\infty}(1-e^{-x/3})$ **7** $\displaystyle\lim_{b\to\infty}(3+\ln b)$ **8** $\displaystyle\lim_{b\to\infty}(2-\ln b^2)$

Find the value of each improper integral that is convergent.

9 $\displaystyle\int_1^\infty \frac{1}{x^3}\,dx$ **10** $\displaystyle\int_1^\infty \frac{1}{\sqrt[3]{x^4}}\,dx$ **11** $\displaystyle\int_2^\infty 3x^{-4}\,dx$ **12** $\displaystyle\int_0^\infty e^{-t}\,dt$

13 $\displaystyle\int_2^\infty \frac{1}{x}\,dx$ **14** $\displaystyle\int_1^\infty \frac{1}{x^{.99}}\,dx$ **15** $\displaystyle\int_1^\infty \frac{1}{x^{1.01}}\,dx$ **16** $\displaystyle\int_0^\infty 8e^{-2x}\,dx$

17 $\displaystyle\int_4^\infty e^{-x/2}\,dx$ **18** $\displaystyle\int_{10}^\infty e^{-x/5}\,dx$ **19** $\displaystyle\int_3^\infty x^2\,dx$ **20** $\displaystyle\int_1^\infty \sqrt{x}\,dx$

21 $\int_0^\infty e^{.05t}\, dt$

22 $\int_0^\infty e^{-.01t}\, dt$

23 $\int_0^\infty e^{-.05t}\, dt$

24 $\int_0^\infty e^{.01t}\, dt$

25 $\int_5^\infty \dfrac{1}{(x-4)^3}\, dx$

26 $\int_0^\infty \dfrac{x}{(x^2+1)^2}\, dx$

27 $\int_0^\infty \dfrac{x}{x^2+1}\, dx$

28 $\int_0^\infty \dfrac{x^2}{x^3+1}\, dx$

29 $\int_0^\infty x^2 e^{-x^3}\, dx$

30 $\int_e^\infty (\ln x)^{-2}\, \dfrac{1}{x}\, dx$

31 $\int_{-\infty}^0 e^{3x}\, dx$

32 $\int_{-\infty}^0 \dfrac{x^4}{(x^5-1)^2}\, dx$

33 $\int_{-\infty}^1 \dfrac{1}{2-x}\, dx$

34 $\int_{-\infty}^0 \dfrac{1}{1-x}\, dx$

35 $\int_{-\infty}^\infty \dfrac{e^x}{(1+e^x)^2}\, dx$

36 $\int_{-\infty}^\infty \dfrac{e^{-x}}{(1+e^{-x})^3}\, dx$

37 $\int_{-\infty}^\infty \dfrac{e^x}{1+e^x}\, dx$

38 $\int_{-\infty}^\infty \dfrac{e^{-x}}{1+e^{-x}}\, dx$

APPLIED EXERCISES

39 (*General–Permanent Endowments*) Find the size of the permanent endowment needed to generate an annual $12,000 forever at continuous interest rate 6%.

40 (*General–Permanent Endowments*) Show that the size of the permanent endowment needed to generate an annual C dollars forever at interest rate r compounded continuously is C/r dollars.

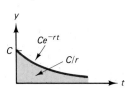

🖩 41 (*General–Permanent Endowments*)

(a) Find the size of the permanent endowment needed to generate an annual $1000 forever at a continuous interest rate of 10%.

(b) At this same interest rate, the size of the fund needed to generate an annual $1000 for precisely 100 years is $\int_0^{100} 1000 e^{-.1t}\, dt$. Evaluate this integral (it is not an improper integral), approximating your answer using a calculator.

(c) Notice that the cost for the first hundred years is almost the same as the cost forever. This illustrates the principle that in endowments, the short term is expensive, eternity is cheap.

42 (*Business–Capital Value of an Asset*) The capital value of an asset is defined as the present value of all future earnings. For an asset that may last indefinitely (like real estate or a corporation) the capital value is

$$\binom{\text{capital}}{\text{value}} = \int_0^\infty C(t) e^{-rt}\, dt$$

where $C(t)$ is the income per year and r is the continuous interest rate. Find the capital value of a piece of property that will generate an annual $8000 forever at the continuous interest rate of 5%.

43 (*Business–Oil Well Output*) An oil well is expected to produce oil at the rate $r(t) = 50 e^{-.05t}$ thousand barrels per month indefinitely, where t is the number of months that the well has been in operation. Find the total output over the lifetime of the well by integrating this rate from 0 to ∞. (*Note:* The owner will shut down the well when production falls too low, but is it convenient to estimate the total output as if production continued forever.)

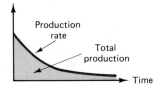

🖩 44 (*General–Duration of Telephone Calls*) Studies have shown that the proportion of telephone calls that last longer than t minutes is approximately

$$\int_t^\infty 0.3 e^{-0.3s}\, ds$$

Use this formula to find the proportion of telephone calls that last longer than 4 minutes.

45 (*Area*) Find the area between the curve $y = 1/x^{3/2}$ and the x-axis from $x = 1$ to ∞.

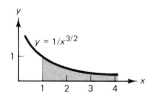

46 (*Area*) Find the area between the curve $y = e^{-4x}$ and the x-axis from $x = 0$ to ∞.

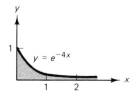

47 (*Area*) Find the area between the curve $y = e^{-ax}$ (for $a > 0$) and the x-axis from $x = 0$ to ∞.

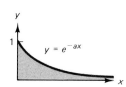

48 (*Area*) Find the area between the curve $y = 1/x^n$ (for $n > 1$) and the x-axis from $x = 1$ to ∞.

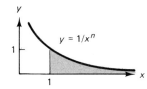

49 (*Behavioral Science–Mazes*) In a psychology experiment, rats were placed in a T-maze, and the proportion of rats who required more than t seconds to reach the end was $\int_t^\infty .05e^{-.05s}\,ds$. Use this formula to find the proportion of the rats who required more than 10 seconds.

50 (*Sociology–Prison Terms*) If the proportion of prison terms that are longer than t years is given by the improper integral $\int_t^\infty .2e^{-.2s}\,ds$, find the proportion of prison terms that are longer than 5 years.

51 (*Business–Product Reliability*) The proportion of light bulbs that last longer than t hours is predicted to be $\int_t^\infty .001e^{-.001s}\,ds$. Use this formula to find the proportion of light bulbs that will last longer than 1200 hours.

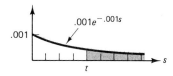

52 (*Business–Warranties*) When a company sells a product with a lifetime guarantee, the number of items returned for repair under the guarantee usually decreases with time. A company estimates that the annual rate of returns after t years will be $800e^{-.2t}$. Find the total number of returns by summing (integrating) this rate from 0 to ∞.

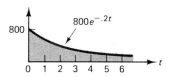

53 (*Business–Sales*) A publisher estimates that a book will sell at the rate of $16{,}000e^{-.8t}$ books per year t years from now. Find the total number of books that will be sold by summing (integrating) this rate from 0 to ∞.

54 (*Biomedical–Drug Absorption*) To determine how much of a drug is absorbed into the body, researchers measure the difference between the dosage D and the amount of the drug excreted from the body. The total amount excreted is found by integrating the excretion rate $r(t)$ from 0 to ∞. Therefore, the amount of the drug absorbed by the body is

$$D - \int_0^\infty r(t)\,dt.$$

If the initial dose is $D = 200$ mg (milligrams), and the excretion rate is $r(t) = 40e^{-.5t}$ mg per hour, find the amount of the drug absorbed by the body.

6.4 NUMERICAL INTEGRATION

Introduction

Even with all our techniques of integration, there are still some integrals that cannot be solved. For example, many statistical predictions depend upon the normal or "bell-shaped" distribution, which involves an integral of the form

$$\int e^{-\frac{1}{2}x^2}\, dx$$

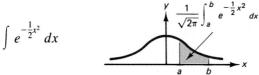

This integral cannot be solved (as a finite combination of powers, roots, logarithms, and exponentials) by *any* method.

For definite integrals there is a "method of last resort" called *numerical integration*. A definite integral is equal to a *number*, and numerical integration finds an *approximation* to this number by interpreting the integral as an area and using area formulas from elementary geometry. The approximation can be as accurate as you wish, but greater accuracy comes at the cost of more calculation.

We will discuss three methods of numerical integration, based on approximating areas by rectangles, trapezoids, and parabolas (known as "Simpson's rule," discussed in the exercises). Any of these methods may be studied independently, and each may be programmed on a calculator or computer.

Idea of Rectangular Approximation

Suppose that we want to find the area under a curve from $x = a$ to $x = b$. If we cannot integrate the function, we may approximate the area as follows.

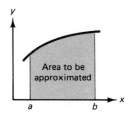

The area under the curve is approximated by the area of the two rectangles shown on the right. The distance from a to b is $b - a$, and the width of each rectangle is half of this distance, $\frac{1}{2}(b - a)$. The height of each rectangle is the height of the curve at the left-hand edge of that rectangle. Notice that this approximation underestimate the area under the curve by the size of the white spaces just above the rectangles.

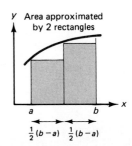

A better approximation is obtained by using *four* rectangles, each with base $\frac{1}{4}(b - a)$, as shown on the right. Again, the error in this approximation would be the total area of the white spaces just above the rectangles.

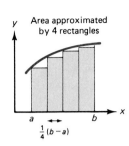

Area approximated by 4 rectangles

$\frac{1}{4}(b-a)$

Using more rectangles gives an approximation that is even closer to the area under the curve. For example, the diagram on the right shows the approximation using 16 rectangles, and the white "error" area just above the rectangles is so small as to be almost invisible.

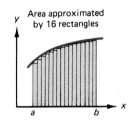

Area approximated by 16 rectangles

Example of Rectangular Approximation

We will estimate the definite integral $\int_1^2 x^2 \, dx$ by rectangular approximation. Of course, there is no need to approximate this particular integral since it can be solved exactly.

$$\int_1^2 x^2 \, dx = \frac{1}{3} x^3 \Big|_1^2 = \frac{1}{3} \cdot 8 - \frac{1}{3} \cdot 1 = \frac{8}{3} - \frac{1}{3} = \frac{7}{3} \approx 2.33$$

Nevertheless, we will approximate this integral because knowing the exact answer will enable us to judge the accuracy of the approximation.

Example 1 Approximate the definite integral $\int_1^2 x^2 \, dx$ using four rectangles.

Solution We use the four rectangles shown below, each of width $\Delta x = \frac{1}{4}$, one quarter of the distance from $a = 1$ to $b = 2$.

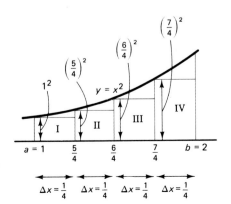

The base of each rectangle is $\Delta x = 1/4$.

The height of each rectangle is the height of the curve at the left-hand edge of that rectangle.

The curve is $f(x) = x^2$, so the heights are the squares of the x-values at the left-hand edges of the rectangles.

Height of rectangle I   is $f(1) = 1^2 = 1.$

Height of rectangle II is $f\left(\frac{5}{4}\right) = \left(\frac{5}{4}\right)^2 = \frac{25}{16}.$

Height of rectangle III is $f\left(\frac{6}{4}\right) = \left(\frac{6}{4}\right)^2 = \frac{36}{16}.$

Height of rectangle IV is $f\left(\frac{7}{4}\right) = \left(\frac{7}{4}\right)^2 = \frac{49}{16}.$

The approximation for the integral is the sum of these areas (base Δx times height).

$$\int_1^2 x^2\, dx \approx \left(\frac{1}{4}\right)(1) + \left(\frac{1}{4}\right)\left(\frac{25}{16}\right) + \left(\frac{1}{4}\right)\left(\frac{36}{16}\right) + \left(\frac{1}{4}\right)\left(\frac{49}{16}\right)$$

base times height for each rectangle

$$\underbrace{\Delta x\, f(1)}_{\substack{\text{area of}\\\text{rectangle}\\\text{I}}} \quad \underbrace{\Delta x\, f\left(\frac{5}{4}\right)}_{\substack{\text{area of}\\\text{rectangle}\\\text{II}}} \quad \underbrace{\Delta x\, f\left(\frac{6}{4}\right)}_{\substack{\text{area of}\\\text{rectangle}\\\text{III}}} \quad \underbrace{\Delta x\, f\left(\frac{7}{4}\right)}_{\substack{\text{area of}\\\text{rectangle}\\\text{IV}}}$$

$$= \frac{1}{4} + \frac{25}{64} + \frac{36}{64} + \frac{49}{64}$$

$$= \frac{16 + 25 + 36 + 49}{64} = \frac{126}{64} = 1\frac{31}{32}$$

This number, $1\frac{31}{32} \approx 1.97$, is our approximation for the definite integral.

$$\int_1^2 x^2\, dx \approx 1.97$$

■

Earlier we found that the actual value of this integral is 2.33, so our approximation 1.97 is "off" by .36. The *relative* error (the actual error .36 divided by the actual value 2.33, expressed as a percentage) is 15%. Although not astonishingly accurate, this is reasonably close considering that we used only four rectangles.

Shortly, we will develop a better method of approximation, using trapezoids. Rectangular approximation, however, is the simplest method, and it enables us to make several important remarks about definite integrals.

Riemann Sums and Integral Notation

We approximated the definite integral $\int_a^b f(x)\, dx$ by a sum of terms of the form $\Delta x\, f(x)$, or reversing the order, $f(x)\, \Delta x$.

$$\int_a^b f(x)\ dx \approx f(x_1)\ \Delta x + f(x_2)\ \Delta x + f(x_3)\ \Delta x + f(x_4)\ \Delta x$$

We used the leftmost x-value from each interval, but we could have taken *any* x-value from each rectangle. Such an approximation with n rectangles would be written

$$\int_a^b f(x)\ dx \approx f(x_1)\ \Delta x + f(x_2)\ \Delta x + \cdots + f(x_n)\ \Delta x$$

└ the dots stand for the middle terms

where the approximation becomes exact as n, the number of rectangles, approaches infinity. (Recall that the white "error" area above the rectangles became vanishingly small as the number of rectangles increased.) Sums of this form, adding terms of the form $f(x_i)\cdot\Delta x$, are called *Riemann sums* (after the nineteenth-century German mathematician Georg Bernhard Riemann). The symbol Σ (the Greek letter S) means a sum, and we may represent symbolically the fact that the sum of n terms of the form $f(x_i)\ \Delta x$ approaches the definite integral as

$$\sum_1^n f(x_i)\ \Delta x \xrightarrow[\text{as } n \to \infty]{} \int_a^b f(x)\ dx$$

This expression suggests a meaning for the integral sign $\int$ and the dx in the integral notation. It is as if n approaching infinity makes the Σ (Greek S) turn into the integral sign $\int$ (a "stretched out" S), and Δ (Greek D) turn into d:

$$\sum f(x)\ \Delta x \quad \text{as } n \to \infty \quad \int f(x)\ dx$$

Σ becomes $\int$

Δ becomes d

In other words, the integral sign $\int$ reminds us that the integral comes from a sum, with the $f(x)$ and dx playing the role of the height and base of the rectangles being added. All of this supports our interpretation of the integral as a sum: the integral is the *limit* of a sum.

Average Value of a Function

In Section 5.4 we defined the average value of a continuous function over an interval as the integral divided by the length of the interval. We could also define the average value of $f(x)$ over $[a, b]$ by choosing n numbers

$$x_1, x_2, \ldots, x_n$$

spaced evenly throughout the interval and averaging the values of the function at these numbers, and then letting $n \to \infty$.

$$\binom{\text{Average of } n}{\text{function values}} = \frac{f(x_1) + f(x_2) + \cdots + f(x_n)}{n}$$

$$= [f(x_1) + f(x_2) + \cdots + f(x_n)]\frac{1}{n} \qquad \textit{dividing by } n \textit{ is equivalent to} \\ \textit{multiplying by } 1/n$$

$$= \frac{1}{b-a}[f(x_1) + f(x_2) + \cdots + f(x_n)]\frac{b-a}{n} \qquad \textit{dividing and multiplying by } b - a$$

$$= \frac{1}{b-a}[f(x_1) + f(x_2) + \cdots + f(x_n)]\,\Delta x \qquad \textit{replacing } \frac{b-a}{n} \textit{ by } \Delta x \textit{ (since they are} \\ \textit{equal)}$$

$$= \frac{1}{b-a}[f(x_1)\,\Delta x + f(x_2)\,\Delta x + \cdots + f(x_n)\,\Delta x] \qquad \textit{multiplying the } \Delta x \textit{ inside the brackets}$$

$$= \frac{1}{b-a}\sum_{1}^{n} f(x_i)\,\Delta x \qquad \textit{using } \Sigma \textit{ for summation}$$

as $n \to \infty$ this Riemann sum approaches the definite integral $\int_{a}^{b} f(x)\,dx$

Taking the limit as $n \to \infty$ gives the same formula for average value that we used earlier:

$$\binom{\text{average value}}{\text{of } f(x) \text{ on } [a,\,b]} = \frac{1}{b-a}\int_{a}^{b} f(x)\,dx$$

That is, the formula for the average value of a function is equivalent to averaging the function at n numbers throughout the interval and then letting $n \to \infty$.

Trapezoidal Approximation

While rectangular approximation contributed to our understanding of integrals and average values, it did not provide a very accurate approximation for the integral in Example 1. We could improve the accuracy by increasing n, the number of rectangles, but this would mean more calculation (and more round-off errors). A better way to improve accuracy is to use, instead of rectangles, another shape that fits a curve more closely. We will modify rectangles by allowing their tops to slant with the curve, as shown below.

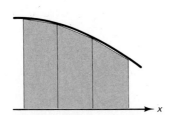

Such shapes, in which the sides are parallel but the top and bottom are not, are called *trapezoids*. Approximating an area by trapezoids is more accurate than approximating it by rectangles, as shown above on the right.

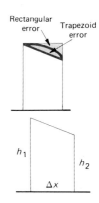

The area of a trapezoid is the width Δx times the average of the two heights.

$$\text{Area} = \underbrace{\frac{h_1 + h_2}{2}}_{\substack{\text{average} \\ \text{height}}} \overset{\text{\Large\textupharpoon}}{\underset{\text{width}}{}}\Delta x$$

If we use trapezoids that have the same width Δx, we may add up all the heights first and multiply by Δx at the end. Averaging the heights means dividing each height by 2 [since $(h_1 + h_2)/2 = h_1/2 + h_2/2$], but a side *between* two rectangles will be counted twice, once for each trapezoid on either side, thereby canceling the division by 2. Therefore, in adding up the areas, only the two outside heights a and b should be divided by two. This gives us the following procedure for trapezoidal approximation.

Trapezoidal approximation of $\int_a^b f(x)\,dx$ **with** n **trapezoids**

(i)   Calculate the trapezoid width $\Delta x = \dfrac{b - a}{n}$.

(ii)   Calculate equally spaced numbers starting with $x_1 = a$, successively adding Δx, and ending with $x_{n+1} = b$.

(iii)   The integral is approximated by the sum

$$\int_a^b f(x)\,dx \approx \left[\frac{1}{2}f(x_1) + f(x_2) + f(x_3) + \cdots + \frac{1}{2}f(x_{n+1})\right]\Delta x$$

This formula calculates the total area of the trapezoids shown below.

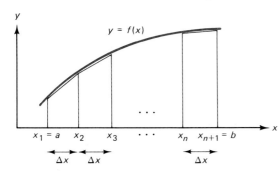

We perform this calculation in a table, as shown in the following example.

Example 2   Approximate $\int_1^2 x^2\, dx$ using four trapezoids.

Solution   The limits of integration are $a = 1$ and $b = 2$ and we are using $n = 4$ trapezoids. The method consists of the following six steps.

1.   Calculate the trapezoid width $\Delta x = \dfrac{b - a}{n} = \dfrac{2 - 1}{4} = .25$

2.   List the x-values a through b with spacing Δx

3.   Apply $f(x)$ to each x-value

x	$f(x) = x^2$	
initial point a → 1	$(1)^2$ $= \cancel{1}$ .5	
add Δx 1.25	$(1.25)^2 = 1.56$	**4.** Take half of first and last entry
add Δx 1.5	$(1.5)^2 = 2.25$	
add Δx 1.75	$(1.75)^2 = 3.06$	
final point b → 2.0	$(2)^2$ $= \cancel{4}$ 2	**5.** Sum last column

$$9.37{\cdot}(.25) \approx 2.34$$

6.   Multiply by Δx final answer

Therefore, our estimate is $\int_1^2 x^2\, dx \approx 2.34$ ◾

Earlier we found that the exact value of this integral is $7/3 \approx 2.33$. The trapezoidal approximation of 2.34 is very accurate indeed, despite the fact that we used only four trapezoids, far more accurate than the rectangular approximation of 1.97. Therefore, from now on we will use trapezoidal rather than rectangular approximation.

Example 3   Approximate $\int_3^5 \dfrac{1}{x}\, dx$ using five trapezoids.

Solution   The limits of integration are $a = 3$ and $b = 5$, and we are using $n = 5$ trapezoids.

1.   The width is

$$\Delta x = \frac{b - a}{n} = \frac{5 - 3}{5} = \frac{2}{5} = .4$$

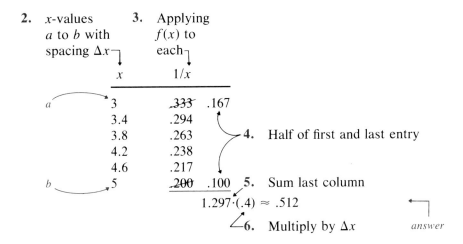

2. x-values
      a to b with
      spacing Δx

3. Applying
      f(x) to
      each

x	1/x
a 3	.333 .167
3.4	.294
3.8	.263
4.2	.238
4.6	.217
b 5	.200 .100

4. Half of first and last entry

5. Sum last column

$$1.297 \cdot (.4) \approx .512$$

6. Multiply by Δx   *answer*

Therefore, our estimate is $\int_3^5 \frac{1}{x}\, dx \approx .512$.   ▪

This integral too may be evaluated directly by antiderivatives.

$$\int_3^5 \frac{1}{x}\, dx = \ln x \Big|_3^5 = \ln 5 - \ln 3 \approx .511 \qquad \textit{using a calculator}$$

Therefore, our estimate of .512 is only off by .001, for a relative error of

$$\frac{.001}{.511} \approx .002$$

or .2% (two-tenths of one percent). This is an excellent approximation, using only five trapezoids.

Practice Exercise 1

(a) Approximate $\int_4^6 1/x\, dx$ using four trapezoids. Keep three decimal places in your calculations.

(b) Calculate the actual value of this integral, using a calculator to estimate the natural logarithms.

(c) From your answers to parts (a) and (b), find the actual error and also the relative error (the actual error divided by the actual value, expressed as a percentage).   (solutions on page 376)

Each of the integrals that we approximated so far could also be solved exactly. We did this to judge the accuracy of our estimates. Example 4 involves an integral that *cannot* be solved exactly, and which must therefore be approximated.

Application to IQ Tests

Although it is increasingly clear that human intelligence cannot be measured by a single number, IQ tests are still widely used. (IQ stands for "intelligence quotient" and is defined as mental age divided by chronological age, multiplied by 100.) The average American IQ is 100, and the distribution of IQs follows the famous "bell-shaped curve" of probability.* The proportion of Americans with IQs between two numbers A and B (with $A < B$) is given by the following integral.

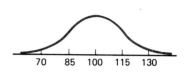

$$\begin{pmatrix} \text{Proportion of Americans} \\ \text{with IQs between } A \text{ and } B \end{pmatrix} \approx \int_{(A-100)/15}^{(B-100)/15} .4e^{-(1/2)x^2} \, dx$$

Example 4   Find the proportion of Americans who have IQs between 115 and 130. Use trapezoidal approximation with three trapezoids.

Solution   Substituting the given IQs $A = 115$ and $B = 130$ into the formulas at the top and bottom of the integral, we obtain the following definite integral.

$$\begin{pmatrix} \text{Proportion of IQs} \\ \text{between 115 and 130} \end{pmatrix} \approx \int_{1}^{2} .4e^{-(1/2)x^2} \, dx$$

$$\longleftarrow 2 \text{ from } \frac{B - 100}{15} = \frac{130 - 100}{15} = \frac{30}{15} = 2$$

$$\longleftarrow 1 \text{ from } \frac{A - 100}{15} = \frac{115 - 100}{15} = \frac{15}{15} = 1$$

This integral cannot be solved exactly, and we must resort to numerical integration. For $n = 3$ trapezoids, the trapezoidal width is

$$\Delta x = \frac{b - a}{n} = \frac{2 - 1}{3} = .33$$

This gives the following table (using a calculator for the second column).

x	$.4e^{-(1/2)x^2}$	
1	$.4e^{-(1/2)(1)^2} \approx .24$	.12
1.33	$.4e^{-(1/2)(1.33)^2} \approx .17$	
1.66	$.4e^{-(1/2)(1.66)^2} \approx .10$	
2.00	$.4e^{-(1/2)(2)^2} \approx .05$	.03

half of first and last

$$.42 \cdot (.33) = .14$$

$$\underset{\Delta x}{\uparrow}$$

---

* For those familiar with statistics, IQ scores are normally distributed with mean 100 and standard deviation 15.

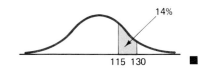

The integral is approximately .14, so about 14% of all Americans have IQs in the range 115 to 130.

The actual proportion is 13.6%, so our approximation of 14% is remarkably accurate considering that we used only three trapezoids. ■

Error in Trapezoidal Approximation

The maximum error in approximation by n trapezoids is given by the following formula.

The error in the trapezoidal approximation of $\int_a^b f(x)\, dx$ with n trapezoids does not exceed:

$$(\text{Error}) \leq \frac{(b - a)^3}{12n^2}\, [\max_{a \leq x \leq b} f''(x)]$$

This formula is very difficult to use, as it involves maximizing the second derivative. We will not make further use of it except to observe that the n^2 in the denominator means that doubling the number of trapezoids reduces the maximum error by a factor of *four*.

Computer Programs (Optional)

It is a simple matter to write a computer program to carry out the trapezoidal approximation. (Some computers and calculators come with such a program.) The inputs would be n, the numbers of trapezoids, and a and b, the upper and lower limits of integration. The function $f(x)$ could be supplied through a subroutine called by the main program, and the computer would then carry out the calculation and print the result.

As for the error, the preceding formula is useless, since the computer would not know how to maximize the second derivative. What is usually done is to choose a given accuracy (say, .001) and to repeat the trapezoidal approximation for larger and larger values of n until two successive approximations differ by less than the given amount. Although this does not guarantee the desired accuracy, it is often the best that one can do in practice.

Summary

Our discussion of rectangular approximation led to an interpretation of the integration symbols $\int$ and dx (they represent "sum" and Δx), and also to another derivation of the formula for the average value of a function.

Trapezoidal approximation (explained in Example 2) provides greater accuracy and so is generally preferred to rectangular approximation in applications. Accordingly, the exercises use trapezoidal approximation. Another method, approximating with parabolas, is known as Simpson's rule and is discussed later in the exercises. It is curious that in Chapter 5 we used integrals to evaluate areas, and now we are using areas to evaluate integrals.

SOLUTION TO PRACTICE EXERCISE

1. **(a)** $\Delta x = \dfrac{b - a}{n} = \dfrac{6 - 4}{4} = .5$

x	$1/x$	
4	~~.25~~	.125
4.5	.222	
5.0	.200	
5.5	.182	
6	~~.167~~	.084
	.813·(.5) =	.407

So $\displaystyle\int_4^6 \frac{1}{x}\, dx \approx .407$.

(b) $\displaystyle\int_4^6 \frac{1}{x}\, dx = \ln x \,\Big|_4^6 = \ln 6 - \ln 4 \approx .406$

(c) Absolute error, .001.

Relative error: $\dfrac{.001}{.406} = .0025$ or .25%.   *(one quarter of one percent)*

■ **EXERCISES 6.4** (Most require ▦) *Note:* Your answers may differ slightly, depending on the stage at which you do the rounding.

For each integral:

(a) Approximate it using trapezoidal approximation with n = 4 trapezoids.

(b) Solve the integral exactly using antiderivatives.

(c) Find the actual error (the difference between the actual value and the approximation).

(d) Find the relative error (the actual error divided by the actual value, expressed as a percent).

Round all calculations to three decimal places.

1 $\displaystyle\int_1^3 x^2\, dx$ **2** $\displaystyle\int_1^2 x^3\, dx$ **3** $\displaystyle\int_2^4 \frac{1}{x}\, dx$ **4** $\displaystyle\int_1^3 \frac{1}{x}\, dx$

Estimate each integral using trapezoidal approximation with the given value of n.
Round all calculations to three decimal places.

5 $\int_0^1 \sqrt{1 + x^2}\, dx$, $n = 3$ **6** $\int_0^1 \sqrt{1 + x^3}\, dx$, $n = 3$ **7** $\int_0^1 e^{-x^2}\, dx$, $n = 4$ **8** $\int_0^1 e^{x^2}\, dx$, $n = 4$

9 $\int_1^2 \sqrt{\ln x}\, dx$, $n = 3$ **10** $\int_0^1 \ln (x^2 + 1)\, dx$, $n = 3$ **11** $\int_{-1}^1 \sqrt{16 + 9x^2}\, dx$, $n = 4$ **12** $\int_{-1}^1 \sqrt{25 - 9x^2}\, dx$, $n = 4$

APPLIED EXERCISES

13–14 (*General–IQ Scores*) Using the formula on page 374 and trapezoidal approximation with $n = 4$, find the proportion of people in the United States with IQs between

13 100 and 130   **14** 115 and 145

15–16 (*General–Tree Growth*) If the speed with which a tree grows is e^{-t^2} feet per month (where t is the age of the tree in months), use trapezoidal approximation with $n = 3$ to estimate the height of the tree after

15 2 months   **16** 3 months

(*Hint*: Total growth is the integral of the growth rate.)

Simpson's Rule (Parabolic Approximation)

For better accuracy we could increase n, but this would mean more calculation and more round-off errors. If, instead, we modify our trapezoids by allowing their tops to be parabolic curves, we get an approximation like that shown below.

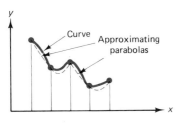

A curve approximated by two parabolas

To estimate the definite integral $\int_a^b f(x)\, dx$ using parabolic estimation, we again use equally spaced numbers from a through b, fitting a parabola to three successive points on the curve. Each parabola spans two intervals, so the number n of intervals must be even. The area under the parabolas is easy to find by integrating. The resulting formula, called *Simpson's rule*,* is

$$\int_a^b f(x)\, dx = [f(x_1) + 4f(x_2) + 2f(x_3) + \cdots + 4f(x_n) + f(x_{n+1})]\, \frac{\Delta x}{3}$$

As before, x_1 through x_{n+1} are numbers with equal spacing $\Delta x = (b - a)/n$ starting

---

* After Thomas Simpson (1701–1761), an early user, but not the discoverer, of the formula.

with $x_1 = a$ and ending with $x_{n+1} = b$. The function values are multiplied by "weights," which, written out by themselves, are

$$\underset{\substack{\uparrow \\ \text{initial } 1}}{1} \quad \underbrace{\underset{\substack{\\ \text{alternating 4s and 2s} \\ \text{beginning and ending with 4}}}{4 \quad 2 \quad 4 \quad 2 \quad 4 \quad \cdots \quad 4 \quad 2 \quad 4}} \quad \underset{\substack{\uparrow \\ \text{final } 1}}{1}$$

Example of Simpson's Rule

We illustrate Simpson's rule by approximating the integral $\int_3^5 \frac{1}{x}\, dx$ with $n = 4$ intervals (so 2 parabolas) as follows.

1. Calculate $\quad \Delta x = \dfrac{b - a}{n} = \dfrac{5 - 3}{4} = .5 \quad$ (n must be even)

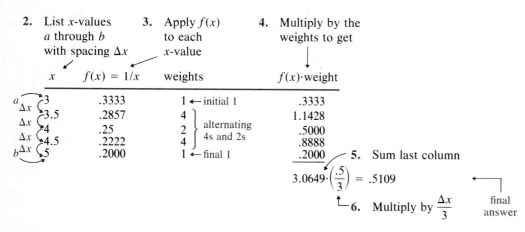

2. List x-values a through b with spacing Δx

3. Apply $f(x)$ to each x-value

4. Multiply by the weights to get

x	$f(x) = 1/x$	weights	$f(x) \cdot$ weight
3	.3333	1 ← initial 1	.3333
3.5	.2857	4	1.1428
4	.25	2 } alternating 4s and 2s	.5000
4.5	.2222	4	.8888
5	.2000	1 ← final 1	.2000

5. Sum last column

$$3.0649 \cdot \left(\frac{.5}{3}\right) = .5109$$

6. Multiply by $\dfrac{\Delta x}{3}$ — final answer

Therefore $\int_3^5 \frac{1}{x}\, dx \approx .5109.$ ■

The exact answer is $\ln 5 - \ln 3 \approx .5108$, so our error is only .0001, far smaller than the error of .01 that we had using trapezoidal approximation with $n = 5$.

Error in Simpson's Rule

The maximum error in Simpson's rule with n intervals (n even) is

$$\text{Error} \le \frac{(b - a)^5}{180 n^4}\, [\max_{a \le x \le b} f^{(4)}(x)]$$

This formula is difficult to use because it involves maximizing the fourth derivative. However, it does show that Simpson's rule is *exact* for cubics (third-degree polynomials) since these have zero fourth derivative, and that doubling n reduces the maximum error by a factor of *sixteen* (because of the n^4).

Simpson's rule is generally the "method of choice" in applications. Trapezoidal approximation, however, has the advantage of having a simpler formula for its error.

EXERCISES USING SIMPSON'S RULE

Estimate each definite integral using Simpson's rule with n = 4. Round all calculations to three decimal places. Exercises 17–20 correspond to Exercises 1–4, in which the same integrals were estimated using trapezoids. If
you did the corresponding exercise, compare your Simpson's rule answer with part (b) of the corresponding exercise in 1–4.

17 $\int_1^3 x^2\, dx$

18 $\int_1^2 x^3\, dx$

19 $\int_2^4 \frac{1}{x}\, dx$

20 $\int_1^3 \frac{1}{x}\, dx$

21 $\int_0^1 \sqrt{1 + x^2}\, dx$

22 $\int_0^1 \sqrt{1 + x^3}\, dx$

23 $\int_0^1 e^{-x^2}\, dx$

24 $\int_0^1 e^{x^2}\, dx$

25 $\int_1^2 \sqrt{\ln x}\, dx$

26 $\int_0^1 \ln(x^2 + 1)\, dx$

27 $\int_{-1}^1 \sqrt{16 + 9x^2}\, dx$

28 $\int_{-1}^1 \sqrt{25 - 9x^2}\, dx$

APPLIED EXERCISES USING SIMPSON'S RULE

29 (*General–Suspension Bridges*) The cable of a suspension bridge hangs in a parabolic curve. The equation of the cable shown is $y = x^2/2000$. Its length is given by the integral

$$\int_{-400}^{400} \sqrt{1 + \left(\frac{x}{1000}\right)^2}\, dx$$

Approximate this integral using Simpson's rule with n = 4.

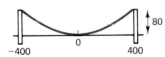

30 (*Approximation of* π) The number π is the circumference of a circle divided by its diameter (since $C = \pi D$). It can be shown that the following definite integral is equal to π.

$$\int_0^1 \frac{4}{x^2 + 1}\, dx = \pi$$

Approximate this integral using Simpson's rule with n = 4 to find the resulting estimate for π. Round all calculations to four decimal places.

6.5 DIFFERENTIAL EQUATIONS

Introduction

A differential equation is simply an equation involving derivatives. Practically any relationship involving rates of change can be expressed in differential equations, and many differential equations can be solved by a technique called "separation of variables."

We consider a function $y = f(x)$, which we will sometimes write as $y(x)$ to indicate that y depends on x. We will write the derivative of y as either y' or dy/dx.

Differential Equation $y' = f(x)$

We have actually been solving differential equations since the beginning of Chapter 5. For example, the differential equation

$$y' = 2x$$

saying that the derivative of a function is $2x$, is solved simply by integrating.

$$y = \int 2x \, dx = x^2 + C$$

In general, a differential equation of the form

$$y' = f(x)$$

is solved by integrating.

$$y = \int f(x) \, dx$$

Therefore, whenever we integrate a marginal cost function to find cost, or integrate any rate to find the total accumulation, we are solving a differential equation of the form $y' = f(x)$.

General and Particular Solutions

The solution to the differential equation $y' = 2x$ is $y = x^2 + C$, with an arbitrary constant C. We call $y = x^2 + C$ the *general solution* to $y' = 2x$ because taking all possible values of the constant C gives *all* solutions to the differential equation. If we take C to be a particular number, we get what is called a *particular solution*. Some particular solutions to the differential equation $y' = 2x$ are

$$y = x^2 + 2 \qquad \text{(taking } C = 2)$$

$$y = x^2 \qquad \text{(taking } C = 0)$$

$$y = x^2 - 2 \qquad \text{(taking } C = -2)$$

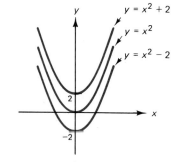

The different values of the arbitrary constant C give a "family" of curves, and the general solution $y = x^2 + C$ may be thought of as the entire family.

Verifying Solutions

Verifying that a function is a solution of a differential equation is simply a matter of calculating the necessary derivatives and substituting them into the differential equation.

Example 1 Verify that $y = e^{2x} + e^{-x} - 1$ is a solution to the differential equation

$$y'' - y' - 2y = 2.$$

Solution   The differential equation involves y, y', and y'', so we calculate

$$y = e^{2x} + e^{-x} - 1 \qquad \text{\textit{given function}}$$

$$y' = 2e^{2x} - e^{-x} \qquad \text{\textit{derivative}}$$

$$y'' = 4e^{2x} + e^{-x} \qquad \text{\textit{second derivative}}$$

Then we substitute these into the differential equation.

$$\overbrace{y'' - y' - 2y = 2}$$

$$(4e^{2x} + e^{-x}) - (2e^{2x} - e^{-x}) - 2(e^{2x} + e^{-x} - 1) \stackrel{?}{=} 2 \qquad \text{\textit{$\stackrel{?}{=}$ means the equation may not be true}}$$

$$4e^{2x} + e^{-x} - 2e^{2x} + e^{-x} - 2e^{2x} - 2e^{-x} + 2 \stackrel{?}{=} 2 \qquad \text{\textit{expanding}}$$

$$\cancel{4e^{2x}} + \cancel{e^{-x}} - \cancel{2e^{2x}} + \cancel{e^{-x}} - \cancel{2e^{2x}} - \cancel{2e^{-x}} + 2 \stackrel{?}{=} 2 \qquad \text{\textit{canceling}}$$

$$2 = 2 \qquad \text{\textit{it checks!}}$$

Since the equation checks, the given function is indeed a solution of the differential equation.   ■

If the two sides had not turned out to be exactly the same, the given function y would *not* have been a solution to the differential equation.

Practice Exercise 1

Verify that $y = e^{-x} + e^{3x}$ is a solution of the differential equation

$$y'' - 2y' - 3y = 0 \qquad \text{\textit{(solution on page 388)}}$$

We have seen that differential equations of the form $y' = f(x)$ can be solved by integrating. To solve more complicated differential equations we use a method called "separation of variables."

Separation of Variables

A differential equation is said to be "separable" if the variables can be "separated" by moving every x to one side of the equation and every y to the other side. We may then solve the differential equation by integrating both sides. Several examples will make the method clear.

Example 2   Find the general solution to the differential equation $\dfrac{dy}{dx} = 2xy^2$.

Solution

$$dy = 2xy^2\, dx \qquad \text{\textit{multiplying both sides of the differential equation by dx}}$$

$$\frac{dy}{y^2} = 2x\, dx \qquad \text{\textit{dividing each side by y^2}}$$

$$y^{-2}\, dy = 2x\, dx \qquad \text{\textit{in exponential form}}$$

The variables have been separated, with every y on the left and every x on the right. Now we integrate both sides.

$$\int y^{-2}\, dy = \int 2x\, dx$$

$$-y^{-1} = x^2 + C \qquad\qquad \text{\textit{solving each integral by the power rule}}$$

We write just one C, as the combined constant from both integrations. We now solve for y.

$$\frac{1}{y} = -x^2 - C \qquad\qquad \text{\textit{writing } } y^{-1} \text{ \textit{as } } 1/y \text{ \textit{and multiplying by } } -1$$

$$y = \frac{1}{-x^2 - C} \qquad\qquad \text{\textit{taking reciprocals of both sides}}$$

This is the general solution to the differential equation. The solution may be left in this form, but if we replace the arbitrary constant C by $-c$, another constant but with the opposite sign, this solution may be written

$$y = \frac{1}{-x^2 + c}$$

or

$$y = \frac{1}{c - x^2} \qquad\qquad \text{\textit{reversing the order}}$$

for any constant c. This is the general solution to the differential equation $y' = 2xy^2$. ■

   In the example we were asked for the *general* solution to a differential equation. Sometimes we will be given a differential equation together with some additional information that selects a *particular* solution from the general solution (that is, information that determines the value of the arbitrary constant in the general solution). This additional information is called a *boundary condition*.

Example 3 Solve the differential equation $y' = \dfrac{6x}{y^2}$ with the boundary condition $y(1) = 2$.

Solution   First we find the general solution by separating variables.

$$\frac{dy}{dx} = \frac{6x}{y^2} \qquad\qquad \text{\textit{replacing } } y' \text{ \textit{by } } dy/dx$$

$$y^2\, dy = 6x\, dx \qquad\qquad \text{\textit{multiplying both sides by dx and }} y^2$$
$$\qquad\qquad\qquad\qquad\qquad \text{\textit{(the variables are separated)}}$$

$$\int y^2\, dy = \int 6x\, dx \qquad\qquad \text{\textit{integrating both sides}}$$

$$\frac{1}{3} y^3 = 3x^2 + C$$

solving the integrals

$$y^3 = 9x^2 + \underbrace{3C}_{c}$$

(3 times a constant is
just another constant)

multiplying by 3

$$y^3 = 9x^2 + c$$

replacing 3C by c

$$y = \sqrt[3]{9x^2 + c}$$

taking cube roots

This is the general solution, with arbitrary constant c. The boundary condition $y(1) = 2$ says that $y = 2$ when $x = 1$. We substitute these into the general solution $y = \sqrt[3]{9x^2 + c}$ and solve for c.

$$2 = \sqrt[3]{9 + c}$$

$y = \sqrt[3]{9x^2 + c}$ with $x = 1$ and $y = 2$

$$8 = 9 + c$$

cubing each side

$$-1 = c$$

solving for c gives c = −1

Therefore, we replace c by -1 in the general solution.

$$y = \sqrt[3]{9x^2 - 1}$$

$y = \sqrt[3]{9x^2 + c}$ with $c = -1$

This is the solution to the given differential equation and boundary condition. ▪

This solution $y = \sqrt[3]{9x^2 - 1}$ with $x = 1$ gives $y = \sqrt[3]{9 - 1} = \sqrt[3]{8} = 2$, so the boundary condition $y(1) = 2$ is indeed satisfied. We could also verify that the differential equation is satisfied by substituting the solution into it.

In general, solving a differential equation with a boundary condition just means finding the general solution and then using the boundary condition to evaluate the constant.

Practice Exercise 2

Solve the differential equation $y' = 6x^2/y^4$ with the boundary condition $y(0) = 2$.

(solution on page 388)

Recall that to solve for y in the logarithmic equation $\ln y = f(x)$

we simply *exponentiate* the right-hand side: $y = e^{f(x)}$

This idea will be useful in the next example.

Example 4   Solve the differential equation $\frac{dy}{dx} = xy$ with the boundary condition $y(0) = 5$.

Solution

$$dy = xy\ dx \qquad\qquad\qquad\qquad\qquad \textit{multiplying by dx}$$

$$\frac{dy}{y} = x\ dx \qquad\qquad\qquad\qquad\qquad \textit{dividing by y (the variables are separated)}$$

$$\int \frac{dy}{y} = \int x\ dx \qquad\qquad\qquad\qquad \textit{integrating}$$

$$\ln y = \frac{1}{2}\,x^2 + C \qquad\qquad\qquad\qquad \textit{solving the integrals}$$

$$y = e^{(1/2)x^2 + C} \qquad\qquad\qquad\qquad \textit{solving for y by exponentiating the right-hand side}$$

$$y = e^{(1/2)x^2}\,\underbrace{e^{C}}_{c} \qquad\qquad\qquad \textit{e to a sum can be expressed as a product}$$

$$\text{(a constant to a constant}$$
$$\text{is just another constant)}$$

$$y = ce^{(1/2)x^2} \qquad\qquad\qquad\qquad \textit{replacing } e^{C} \textit{ by c (moved to the front)}$$

This is the general solution. To satisfy the boundary condition $y(0) = 5$ we substitute $y = 5$ and $x = 0$ and solve for c.

$$5 = ce^{0} \qquad\qquad\qquad\qquad\qquad \textit{y = ce}^{(1/2)x^2} \textit{ with y = 5 and x = 0}$$

$$5 = c \qquad\qquad\qquad\qquad\qquad\qquad \textit{since } e^{0} = 1$$

Substituting $c = 5$ into the general solution gives the particular solution

$$y = 5e^{(1/2)x^2} \qquad\qquad\qquad\qquad \textit{y = ce}^{(1/2)x^2} \textit{ with c = 5}$$

■

We wrote the solution to the integral $\int \frac{dy}{y}$ as $\ln y$, without absolute value bars. This is because the functions in most applications are positive, so absolute value bars are unnecessary.

Practice Exercise 3

Solve the differential equation $dy/dx = x^2 y$ with the boundary condition $y(0) = 2$.

(solution on page 389)

Example 5   Find the general solution to the differential equation

$$yy' - x = 0$$

Solution   We are asked for the general solution rather than a particular solution.

$$y \frac{dy}{dx} = x$$ *replacing y' by dy/dx and moving the x to the other side*

$$y \, dy = x \, dx$$ *multiplying by dx*

$$\int y \, dy = \int x \, dx$$ *integrating*

$$\frac{1}{2} y^2 = \frac{1}{2} x^2 + C$$ *solving the integrals by the power rule*

$$y^2 = x^2 + \overset{\uparrow}{\underset{c}{2C}}$$ *multiplying by 2*

replacing 2C by c

To solve for y we take the square root of each side. However, there are *two* square roots, one positive and one negative.

$$y = \quad \sqrt{x^2 + c}$$

$$y = -\sqrt{x^2 + c}$$

These two solutions together are the general solution to the differential equation.   ▪

Practice Exercise 4

Find the general solution of the differential equation $yy' = 1$.   *(solution on page 389)*

For some differential equations the integration step requires a substitution.

Example 6   Solve the differential equation $y' = xy - x$ with the boundary condition $y(0) = 4$.

Solution

$$\frac{dy}{dx} = xy - x$$ *replacing y' by dy/dx*

$$\frac{dy}{dx} = x(y - 1)$$ *factoring (to separate variables)*

$$\frac{dy}{y - 1} = x \, dx$$ *dividing by y − 1 (variables separated)*

$$\int \frac{dy}{y - 1} = \int x \, dx$$ *integrating*

using a substitution

$$\begin{bmatrix} u = y - 1 \\ du = dy \end{bmatrix}$$

$$\int \frac{du}{u} = \int x\, dx \qquad\qquad \textit{substituting}$$

$$\ln u = \frac{1}{2}\, x^2 + C \qquad\qquad \textit{solving the integrals}$$

$$\ln (y - 1) = \frac{1}{2}\, x^2 + C \qquad\qquad \textit{substituting back to y using u = y - 1}$$

$$y - 1 = e^{(1/2)x^2 + C} \qquad\qquad \textit{solving for y - 1 by exponentiating}$$

$$y - 1 = e^{(1/2)x^2} \cdot \underbrace{e^C = c}_{c} \cdot e^{(1/2)x^2} \qquad\qquad \textit{replacing } e^C \textit{ by c}$$

$$y = c \cdot e^{(1/2)x^2} + 1 \qquad\qquad \textit{adding 1 to each side}$$

This is the general solution. To satisfy the boundary condition $y(0) = 4$ we substitute $y = 4$ and $x = 0$.

$$4 = ce^0 + 1 \qquad\qquad \textit{y = ce}^{(1/2)x^2} + 1 \textit{ with y = 4 and x = 0}$$

$$4 = c + 1 \qquad\qquad \textit{since } e^0 = 1$$

This gives $c = 3$. Therefore, the particular solution is

$$y = 3e^{(1/2)x^2} + 1 \qquad\qquad \textit{y = ce}^{(1/2)x^2} + 1 \textit{ with c = 3}$$

■

Practice Exercise 5

Solve the differential equation $y' = 2x + xy$ with the boundary condition $y(0) = 0$.

(solution on page 389)

Application: Accumulation of Wealth

The examples so far have *given* us a differential equation to solve. In this application we will first *derive* a differential equation and then solve it.

Example 7 Suppose that you have saved $5000, and that you expect to save an additional $3000 during each year. If you deposit these savings in a bank paying 5% interest compounded continuously, find a formula for your bank balance after t years.

Solution Let $y(t)$ stand for your bank balance (in thousands of dollars) after t years. Each year $y(t)$ grows by 3 (thousand dollars) plus 5% interest. This growth can be expressed as a differential equation.

$$\underbrace{y'}_{\substack{\text{rate of}\\\text{change of } y}} \underset{\text{is}}{=} \underset{3}{3} + \underbrace{.05y}_{5\% \text{ of } y}$$

Before continuing, be sure that you understand how this differential equation models the changes due to savings and interest. We solve it by separating variables.

$$\frac{dy}{dt} = 3 + .05y$$ *replacing y' by dy/dt*

$$\int \frac{dy}{3 + .05y} = \int dt$$ *dividing by $3 + .05y$, multiplying by dt, and then integrating*

$$\begin{bmatrix} u = 3 + .05y \\ du = .05dy \end{bmatrix}$$ *using a substitution*

$$20 \int \frac{du}{u} = \int dt$$ *substituting (the 20 comes from $1/.05 = 20$)*

$$20 \ln u = t + C$$ *solving the integrals*

$$\ln (3 + .05y) = .05t + \underbrace{.05C}_{c}$$ *substituting $u = 3 + .05y$ and dividing by 20*

$$\ln (3 + .05y) = .05t + c$$ *replacing $.05C$ by c*

$$3 + .05y = e^{.05t+c} = e^{.05t} \underbrace{e^c}_{k} = ke^{.05t}$$ *exponentiating and then simplifying constants*

$$.05y = ke^{.05t} - 3$$ *subtracting 3*

$$y = \underbrace{20k}_{b} e^{.05t} - 60$$ (always simplify arbitrary constants)   *dividing by .05*

$$y = be^{.05t} - 60$$ *with arbitrary constant b*

You began at time $t = 0$ with 5 thousand dollars, which gives the boundary condition $y(0) = 5$. We substitute $y = 5$ and $t = 0$.

$$5 = be^0 - 60$$ *$y = be^{.05t} - 60$ with $y = 5$ and $t = 0$*

$$5 = b - 60$$ *since $e^0 = 1$*

Therefore, $b = 65$, which we substitute into the general solution.

$$y = 65e^{.05t} - 60$$ *$y = be^{.05t} - 60$ with $b = 65$*

This is a formula for your accumulated wealth after t years.   ■

For example, to find your wealth after 10 years we evaluate the solution at
$t = 10$.

$$y = 65e^{.5} - 60 \approx 47.167$$

Therefore, you will have \$47,167 in the bank.

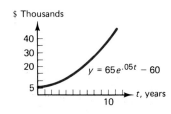

Summary

We have seen how to solve differential equations by separating variables
(moving every y to one side and every x to the other) and integrating. Given
a differential equation we can find the *general* solution (including an arbitrary constant), or given a differential equation with a boundary condition
we can find the *particular* solution (by evaluating the constant).

   We also saw how to *derive* a differential equation from information about
how a quantity changes. For example, we can "read" the differential equation as follows.

$$y' \quad = \quad ay + b$$

| rate | is | a constant times | plus a constant |
| of change | | the amount present | amount added |

The applied exercises will show how differential equations led to important
formulas in a wide variety of fields.

SOLUTIONS TO PRACTICE EXERCISES

1. $y = e^{-x} + e^{3x}$ Substituting these into the differential equation:

$y' = -e^{-x} + 3e^{3x}$ $(e^{-x} + 9e^{3x}) - 2(-e^{-x} + 3e^{3x}) - 3(e^{-x} + e^{3x}) \overset{?}{=} 0$

$y'' = e^{-x} + 9e^{3x}$ $\cancel{e^{-x}} + \cancel{9e^{3x}} + \cancel{2e^{-x}} - \cancel{6e^{3x}} - \cancel{3e^{-x}} - \cancel{3e^{3x}} \overset{?}{=} 0$

$$0 = 0 \quad \text{(it checks)}$$

2. $\dfrac{dy}{dx} = \dfrac{6x^2}{y^4}$ The boundary condition gives

$y^4 \, dy = 6x^2 \, dx$ $2 = \sqrt[5]{0 + c}$

$\displaystyle\int y^4 \, dy = \int 6x^2 \, dx$ $2 = \sqrt[5]{c}$

$\dfrac{1}{5} y^5 = 2x^3 + C$ $32 = c$ (raising each side to the fifth power)

$y^5 = 10x^3 + 5C = 10x^3 + c$ *Solution*: $y = \sqrt[5]{10x^3 + 32}$

$y = \sqrt[5]{10x^3 + c}$

3. $\dfrac{dy}{y} = x^2\, dx$ The boundary condition gives

$\displaystyle\int \dfrac{dy}{y} = \int x^2\, dx$ $2 = ce^0 = c$

$\ln y = \dfrac{1}{3}\, x^3 + C$ $Solution\colon y = 2e^{(1/3)x^3}$

$y = e^{(1/3)x^3+C} = e^{(1/3)x^3}e^C = ce^{(1/3)x^3}$

4. $y\,\dfrac{dy}{dx} = 1$ $\dfrac{1}{2}\, y^2 = x + C$

$y\, dy = dx$ $y^2 = 2x + 2C = 2x + c$

$\displaystyle\int y\, dy = \int dx$ $Solutions\colon y = \sqrt{2x + c}$ and $y = -\sqrt{2x + c}$

5. $\dfrac{dy}{dx} = 2x + xy$ $\ln\,(2 + y) = \dfrac{1}{2}\, x^2 + C$

$= x(2 + y)$ $2 + y = e^{(1/2)x^2+C} = e^{(1/2)x^2}e^C = ce^{(1/2)x^2}$

$\dfrac{dy}{2 + y} = x\, dx$ $y = ce^{(1/2)x^2} - 2$

$\displaystyle\int \dfrac{dy}{2 + y} = \int x\, dx$ The boundary condition gives

$u = 2 + y$ $0 = c - 2$
$du = dy$ $c = 2$

$\displaystyle\int \dfrac{du}{u} = \int x\, dx$ $Solution\colon y = 2e^{(1/2)x^2} - 2$

$\ln u = \dfrac{1}{2}\, x^2 + C$

EXERCISES 6.5

Verify that the function y satisfies the given differential equation.

1 $y = e^{2x} - 3e^x + 2$
   $y'' - 3y' + 2y = 4$

2 $y = e^{5x} - 4e^x + 1$
   $y'' - 6y' + 5y = 5$

3 $y = ke^{ax} - \dfrac{b}{a}$ (for constants a, b, and k)
   $y' = ay + b$

4 $y = ax^2 + bx$ (for constants a and b)
   $y' = \dfrac{y}{x} + ax$

Find the general solution of each differential equation. If the exercise says "check," verify that your answer is a solution.

5 $y^2 y' = 4x$

6 $y^4 y' = 8x$

7 $y' = 6x^2 y$ and check

8 $y' = 12x^3 y$ and check

9 $y' = \dfrac{y}{x}$ and check

10 $y' = \dfrac{y^2}{x^2}$ and check

11 $yy' = 4x$

12 $yy' = 6x^2$

13 $y' = 9x^2$

14 $y' = 6e^{-2x}$

15 $y' = \dfrac{x}{x^2 + 1}$

16 $y' = xy^2$

17 $y' = x^2 y$

18 $y' = x/y$

19 $y' = x^m y^n$ (for $m > 0$, $n \neq 1$)

20 $y' = x^m y$ (for $m > 0$)

21 $y' = 2\sqrt{y}$

22 $y' = 5 + y$

23 $y' = xy + x$

24 $y' = x - 2xy$

25 $y' = ye^x - e^x$

26 $y' = ye^x - y$

27 $y' = ay^2$ (for constant $a > 0$)

28 $y' = axy$ (for constant a)

29 $y' = ay + b$ (for constants a and b)

30 $y' = (ay + b)^2$ (for constants $a \neq 0$ and b)

Solve each differential equation and boundary condition. If the exercise says "check," verify that your answer satisfies both the differential equation and the boundary condition.

31 $y^2 y' = 2x$
$y(0) = 2$

32 $y^4 y' = 3x^2$
$y(0) = 1$

33 $y' = xy$ and
$y(0) = -1$ check

34 $y' = y^2$ and
$y(2) = -1$ check

35 $y' = 2xy^2$ and
$y(0) = 1$ check

36 $y' = 2xy^4$ and
$y(0) = 1$ check

37 $y' = \dfrac{y}{x}$ and
$y(1) = 3$ check

38 $y' = \dfrac{2y}{x}$ and
$y(1) = 2$ check

39 $y' = 2\sqrt{y}$
$y(1) = 4$

40 $y' = \sqrt{y}e^x - \sqrt{y}$
$y(0) = 1$

41 $y' = y^2 e^x + y^2$
$y(0) = 1$

42 $y' = xy - 5x$
$y(0) = 4$

43 $y' = ax^2 y$ (for constant $a > 0$)
$y(0) = 2$

44 $y' = axy$ (for constant $a > 0$)
$y(0) = 4$

APPLIED EXERCISES

45 (*Business–Elasticity*) For a demand function $D(p)$, the elasticity of demand (see page 252) is defined as $E = -pD'/D$. Find demand functions $D(p)$ that have constant elasticity by solving the differential equation: $-pD'/D = k$ (k is a constant).

46 (*Biomedical–Cell Growth*) A cell receives nutrients through its surface, and its surface area is proportional to the two-thirds power of its weight. Therefore, if $w(t)$ is the cell's weight at time t, $w(t)$ satisfies $w' = aw^{2/3}$, where a is a positive constant. Solve this differential equation with the boundary condition $w(0) = 1$ (initial weight 1 unit).

47–48 (*Business–Annuities*) An annuity is a fund into which one makes equal payments at regular intervals. If the fund earns interest at rate r compounded continuously, and deposits are made continuously at the rate of d dollars per year (a "continuous annuity"), then the value $y(t)$ of the fund after t years satisfies the differential equation $y' = d + ry$. (Do you see why?)

47 Solve this differential equation for the continuous annuity $y(t)$ with deposit rate $d = \$1000$ and continuous interest rate $r = .05$, subject to the boundary condition $y(0) = 0$ (zero initial value).

48 Solve this differential equation for the continuous annuity $y(t)$, where d and r are unknown constants, subject to the boundary condition $y(0) = 0$ (zero initial value).

49 (*General–Crime*) A medical examiner called to the scene of a murder will usually take the temperature of the body. A body cools at a rate proportional to the difference between its temperature and the temperature of the room. If $y(t)$ is the temperature of the body t hours after the murder, and if the room temperature is 70, then y satisfies

$$y' = -.32(y - 70)$$

$y(0) = 98.6$ (body temperature initially 98.6)

(a) Solve this differential equation and boundary condition.

(b) Use your answer to part (a) to estimate how long ago the murder took place if the temperature of the body when it was discovered was 80. (*Hint*: Find the value of t that makes your solution equal 80.)

50 (*Biomedical–Glucose Levels*) Hospital patients are often given glucose (blood sugar) through a tube connected to a bottle suspended over their beds. Suppose that this "drip" supplies glucose at the rate of 25 mg per minute, and each minute 10% of the accumulated glucose is consumed by the body. Then the amount $y(t)$ of glucose (in excess of the normal level) in the body after t minutes satisfies

$$y' = 25 - .1y \quad \text{(Do you see why?)}$$

$y(0) = 0$ (zero excess glucose at $t = 0$)

Solve this differential equation and boundary condition.

51 (*General*) Suppose that you meet 30 new people each year, but each year you forget 20% of all of the people that you know. If $y(t)$ is the total number of people who you remember after t years, then y satisfies the differential equation $y' = 30 - .2y$. (Do you see why?) Solve this differential equation subject to the condition $y(0) = 0$ (you knew no one at birth).

52 (*General–Pollution*) For more than 75 years the Flexfast Rubber Company in Massachusetts discharged toxic toluene solvents into the ground at a rate of 5 tons per year. Each year approximately 10% of the accumulated pollutants evaporated into the air. If $y(t)$ is the total accumulation of pollution in the ground after t years, then y satisfies:

$$y' = 5 - .1y$$

$y(0) = 0$ (initial accumulation is zero)

Solve this differential equation and boundary condition to find a formula for the accumulated pollutant after t years.

53 (*General–Accumulation of Wealth*) Suppose that you now have $6000 and that you expect to save an additional $3000 during each year, and all of this is deposited in a bank paying 10% interest compounded continuously. Let $y(t)$ be your bank balance (in thousands of dollars) after t years.

(a) Write a differential equation and boundary condition to model your bank balance. (*Hint*: See Example 7.)

(b) Solve your differential equation and boundary condition.

6.6 FURTHER APPLICATIONS OF DIFFERENTIAL EQUATIONS: THE LAWS OF GROWTH

Introduction

This section continues our study of differential equations, but with a different approach. Instead of solving individual differential equations, we will solve three important classes of differential equations (for *unlimited*, *limited*, and *logistic* growth) and remember their solutions. This will enable us to solve many problems by determining the appropriate differential equation, and then immediately writing the solution. In this section we begin to *think in terms of differential equations*.

INITIAL CONDITIONS

The variable t will usually stand for time, and boundary condition specifying the value of a solution $y(t)$ at time $t = 0$ will be called an *initial condition*.

PROPORTION

We say that one quantity is *proportional to* another quantity if the first quantity is a *constant multiple* of the second. That is, y is proportional to x if $y = ax$ for some "proportionality constant" a. For example, the formula $C = \pi D$ for the circumference of a circle shows that the circumference C is proportional to the diameter D.

The circumference of a circle is proportional to its diameter.

Unlimited Growth

In many situations the growth of a quantity is proportional to its present size. For example, a population of cells will grow in proportion to its present size, and a bank account earns interest in proportion to its size. For a quantity $y(t)$, if its rate of growth y' is proportional to its present size y, then y satisfies the differential equation

$$
\underbrace{y'}_{\substack{\text{rate of} \\ \text{growth}}} \underbrace{=}_{\text{is}} \underbrace{a}_{\substack{\text{propor-} \\ \text{tional to}}} \underbrace{y}_{\substack{\text{current} \\ \text{size}}}
$$

We solve this differential equation by separating variables.

$$\frac{dy}{dt} = ay$$

replacing y' by dy/dt

$$\int \frac{dy}{y} = \int a\, dt$$

dividing by y, multiplying by dt, and integrating

$$\ln y = at + C \qquad \text{\textit{solving the integrals}}$$

$$y = e^{at+C} = e^{at}e^{C} = ce^{at} \qquad \text{\textit{solving for y by exponentiating and}}$$
$$\text{\textit{then simplifying } } e^{C} = c$$

$$\underset{c}{\underbrace{}}$$

At time $t = 0$ this solution $y(t) = ce^{at}$ gives

$$y(0) = ce^{0} = c \qquad \qquad \text{\textit{y(t) = ce^{at} with t = 0}}$$

Therefore, the *initial value* of the solution [the value of $y(t)$ at $t = 0$] is c.

Law of Unlimited Growth

The differential equation $y' = ay$

with initial value $y(0) = c$

is solved by $y(t) = ce^{at}$

(for constants a and c). This is called the *law of unlimited growth* because the solution y grows arbitrarily large. Given this result, whenever we encounter a differential equation of the form $y' = ay$ with an initial value, we can immediately write the solution $y = ce^{at}$ with c replaced by the initial value.

Example 1   An art collection, initially worth \$25,000, continuously grows in value by 5% a year. Express this growth rate as a differential equation and find a formula for the value of the collection after t years.

Solution   Growing by 5% means that the value $y(t)$ grows by 5% *of itself.*

$$y' \quad = \quad .05y$$

rate of   is   5% of the
growth   current value

This differential equation is of the form $y' = ay$ (unlimited growth) with $a = .05$ and initial value 25,000, so we may immediately write its solution.

$$y(t) = 25{,}000e^{.05t} \qquad \qquad \text{\textit{y = ce^{at} with a = .05 and c = 25,000}}$$

This formula gives the value of the art collection after t years.   ■

   The solution ce^{at} is the same as the continuous compounding formula Pe^{rn} (except for different letters). In Chapter 4 we derived the formula Pe^{rn} rather laboriously, using the discrete interest formula $P(1 + r)^{n}$ and taking the limit as $n \to \infty$. The present derivation, using differential equations, is much simpler and shows that ce^{at} applies to *any* situation governed by the differential equation $y' = ay$.

Limited Growth

No real population can undergo unlimited growth for very long. Restrictions of food and space would soon slow its growth. If a quantity $y(t)$ cannot grow larger than a certain fixed maximum size M, and if its growth rate y' is proportional to how far it is from its upper limit, then y satisfies

$$y' \quad = \quad a(M - y) \qquad\qquad a > 0$$

rate of   is   propor-   distance below
growth   tional to   upper bound M

We solve this by separating variables.

$$\frac{dy}{dt} = a(M - y) \qquad\qquad \text{replacing } y' \text{ by } dy/dt$$

$$\int \frac{dy}{M - y} = \int a \, dt \qquad\qquad \begin{array}{l}\text{dividing by } M - y, \text{ multiplying by } dt, \\ \text{and integrating}\end{array}$$

$$\begin{bmatrix} u = M - y \\ du = -dy \end{bmatrix} \qquad\qquad \text{using a substitution}$$

$$-\int \frac{du}{u} = \int a \, dt \qquad\qquad \text{substituting}$$

$$-\ln u = at + C \qquad\qquad \text{solving each integral}$$

$$\ln (M - y) = -at - C \qquad\qquad \begin{array}{l}\text{multiplying by } -1 \text{ and replacing } u \text{ by} \\ (M - y)\end{array}$$

$$M - y = e^{-at-C} = e^{-at} \, e^{-C} = ce^{-at} \qquad \begin{array}{l}\text{solving for } M - y \text{ by exponentiating,} \\ \text{and simplifying exponents}\end{array}$$

$$\underbrace{\qquad\qquad}_{c}$$

$$y = M - ce^{-at} \qquad\qquad \text{subtracting } M \text{ and multiplying by } -1$$

We impose the initial condition $y(0) = 0$ (size zero at time $t = 0$).

$$0 = M - ce^0 = M - c \qquad\qquad y = M - ce^{-at} \text{ with } y = 0 \text{ and } t = 0$$

Therefore $c = M$, which gives the solution

$$y = M - Me^{-at} = M(1 - e^{-at}) \qquad\qquad y = M - ce^{-at} \text{ with } c = M$$

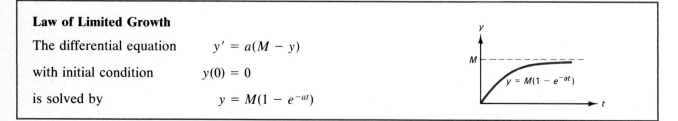

Law of Limited Growth

The differential equation $y' = a(M - y)$

with initial condition $y(0) = 0$

is solved by $y = M(1 - e^{-at})$

$y = M(1 - e^{-at})$

(for constants M and a). This is called the *law of limited growth* because the solution $y = M(1 - e^{-at})$ approaches an upper limit M as $t \to \infty$. Given this result, whenever we encounter a differential equation of the form

$$y' = a(M - y)$$

with initial value zero, we can immediately write the solution

$$y = M(1 - e^{-at}).$$

Application to Sociology: Diffusion of Information by Mass Media

If a news bulletin is repeatedly broadcast over radio and television, the news at first spreads quickly, but later more slowly when most people have already heard it. Sociologists often assume that the rate at which news spreads is proportional to the number who have not yet heard the news. Let M be the population of a city, and $y(t)$ be the number of people who have heard the news within t time units. Then y satisfies the differential equation

$$y' \quad = \quad a(M - y)$$

$$\begin{array}{cccc}
| & | & | & \searrow \\
\text{rate of} & \text{is} & \text{propor-} & \text{number who have} \\
\text{growth} & & \text{tional to} & \text{not heard the news}
\end{array}$$

We recognize this as the differential equation for limited growth, whose solution is $y = M(1 - e^{-at})$. It remains only to determine the values of the constants M and a.

Example 2 An important news bulletin is broadcast to a town of 50,000 people, and after 2 hours 30,000 people have heard the news. Find a formula for the number of people who have heard the bulletin within t hours. Then find how many people will have heard the news within 6 hours.

Solution If $y(t)$ is the number of people who have heard the news within t hours, then y satisfies

$$y' = a(\underbrace{50{,}000 - y}_{\substack{\text{number who have} \\ \text{not heard the news}}})$$

$y' = a(M - y)$ with $M = 50{,}000$

This is the differential equation for limited growth with $M = 50{,}000$, so the solution (from the preceding box) is

$$y = 50{,}000(1 - e^{-at})$$

$y' = a(M - y)$ with $M = 50{,}000$

To find the value of the constant a, we use the given information that 30,000 people have heard the news within 2 hours.

$$30,000 = 50,000(1 - e^{-a \cdot 2})$$

y = 50,000(1 − e⁻ᵃᵗ) with y = 30,000 and t = 2

$$.6 = 1 - e^{-2a}$$

dividing each side by 50,000

$$.4 = e^{-2a}$$

subtracting 1 and then multiplying by −1

$$-.916 \approx -2a$$

taking natural logs (using ln eˣ = x on the right)

$$a \approx .46$$

dividing by −2

Therefore, the number of people who have heard the news within t hours is

$$y(t) = 50,000(1 - e^{-.46t})$$

y = 50,000(1 − e⁻ᵃᵗ) with a = .46

To find the number who have heard the news within 6 hours, we evaluate this solution at $t = 6$.

$$y(6) = 50,000(1 - e^{-(.46)(6)}) = 50,000(1 - e^{-2.76}) \approx 46,835$$

using a calculator

Within 6 hours about 46,800 people have heard the news.

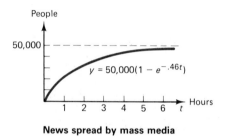

News spread by mass media   ■

We found the value of the constant a by substituting the given data into the solution, simplifying, and taking logs. Similar steps will be required in many other problems, and in the future we shall omit the details.

Application to Learning Theory

Psychologists have found that there seems to be an upper limit to the number of meaningless words that a person can memorize, and that memorizing becomes increasingly difficult approaching this upper limit. If M is this upper limit, and if $y(t)$ is the number of words that can be memorized in t minutes, then the situation is modeled by the differential equation for limited growth.

$$y' \quad = \quad a(M - y)$$

rate of	is	propor-	upper limit M minus
increase		tional to	number already memorized

This may be interpreted as saying that the rate at which new words can be memorized is proportional to the "unused memory capacity."

Example 3   Suppose that a person can memorize at most 100 meaningless words, and that after 15 minutes 10 words have been memorized. How long will it take to memorize 50 words?

Solution   The number $y(t)$ of words that can be memorized in t minutes satisfies

$$y' = a(100 - y)$$

$y' = a(M - y)$ with $M = 100$

This is the differential equation for limited growth, so the solution is

$$y = 100(1 - e^{-at})$$

$y = M(1 - e^{-at})$ with $M = 100$

To evaluate the constant a we use the given information that 10 words have been memorized in 15 minutes.

$$10 = 100(1 - e^{-a \cdot 15})$$

$y = 100(1 - e^{-at})$ with $y = 10$ and $t = 15$

Solving this for the constant a (omitting the details, which are the same as in Example 2) gives $a = .007$.

$$y(t) = 100(1 - e^{-.007t})$$

$y = 100(1 - e^{-at})$ with $a = .007$

This gives the number of words that can be memorized in t minutes. To find how long it takes to memorize 50 words, we set this solution equal to 50 and solve for t.

$$50 = 100(1 - e^{-.007t})$$

$y = 100(1 - e^{-.007t})$ with $y = 50$

Solving for t (again the details are similar to those in Example 2) gives $t = 99$ minutes. Therefore, 50 words can be memorized in about 1 hour and 39 minutes.

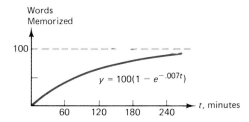

Words Memorized

100

$y = 100(1 - e^{-.007t})$

60   120   180   240

t, minutes

The slope (the rate at which additional words can be memorized) decreases near the upper limit.

■

Logistic Growth

Some quantities grow in proportion both to their present size *and* to their distance from an upper limit M.

$$y' \quad = \quad ay(M - y)$$

rate of growth | is | propor-tional to | present size | upper limit M minus present size

This differential equation can be solved by separation of variables (see Exercise 31) to give the solution

$$y = \frac{M}{1 + ce^{-aMt}}$$

where c and a are positive constants. This function is called the *logistic function*, or the *law of logistic growth*.

Law of Logistic Growth

The solution to $y' = ay(M - y)$

is $y = \dfrac{M}{1 + ce^{-aMt}}$

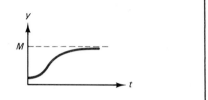

---

where M, c, and a are positive constants. As before, this result enables us to solve a differential equation "at sight," leaving only the evaluation of constants.

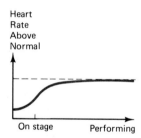

Stage fright follows a logistic curve, showing that performers do not relax during a performance, despite their claims.

This curve is called a *sigmoidal* or *S-shaped curve,* and is used to model growth that begins slowly, then becomes more rapid, and finally slows again near the upper limit.

Application to Ecology

An animal environment (like a lake or a forest) will have an upper limit to the population that it can support. This limit is called the *carrying capacity of the environment.* Ecologists often assume that an animal population grows in proportion both to its present size and to how far the population is from the carrying capacity.

$$y' \quad = \quad ay(M - y)$$

rate of growth	is	proportional to	current size	carrying capacity minus present size

Since this is the logistic differential equation, we know that the solution is the logistic function

$$y = \frac{M}{1 + ce^{-aMt}}$$

Example 4   Ecologists estimate that a man-made lake can support a maximum of 2500 fish. The lake is initially stocked with 500 fish, and after 6 months the fish population is estimated to be 1500. Find a formula for the number of fish in the lake after t months, and estimate the fish population at the end of the first year.

Solution   Letting $y(t)$ stand for the number of fish in the lake after t months, the situation is modeled by the logistic differential equation.

$$y' \quad = \quad ay(2500 - y)$$

rate of   is   propor-   current   carrying capacity
growth   tional to   size   minus present size

$y' = ay(M - y)$ with $M = 2500$

The solution is the logistic function with $M = 2500$.

$$y = \frac{2500}{1 + ce^{-a2500t}}$$

$y = \dfrac{M}{1 + ce^{-aMt}}$ with $M = 2500$

$$= \frac{2500}{1 + ce^{-bt}}$$

replacing $a \cdot 2500$ by another constant b

To evaluate the constants c and b we use the fact that the lake was originally stocked with 500 fish.

$$500 = \frac{2500}{1 + ce^{0}}$$

$y = \dfrac{2500}{1 + ce^{-bt}}$ with $\begin{array}{l} y = 500 \\ t = 0 \end{array}$

$$500 = \frac{2500}{1 + c}$$

simplifying

Solving this for c (omitting the details—the first step is to multiply both sides by $1 + c$) gives $c = 4$, so the logistic function becomes

$$y = \frac{2500}{1 + 4e^{-bt}}$$

$y = \dfrac{2500}{1 + ce^{-bt}}$ with $c = 4$

To evaluate b we substitute the information that the population is $y = 1500$ at $t = 6$.

$$1500 = \frac{2500}{1 + 4e^{-b6}}$$

$y = \dfrac{2500}{1 + 4e^{-bt}}$ with $\begin{array}{l} y = 1500 \\ t = 6 \end{array}$

Solving for b (again omitting the details—the first step is to multiply both sides by $1 + 4e^{-b6}$) gives $b = .30$, so the logistic function becomes

$$y(t) = \frac{2500}{1 + 4e^{-.3t}}$$

$y = \dfrac{2500}{1 + 4e^{-bt}}$ with $b = .3$

This is the formula for the population after t months. To find the population after a year we evaluate at $t = 12$.

$$y(12) = \frac{2500}{1 + 4e^{-.3(12)}} = \frac{2500}{1 + 4e^{-3.6}} = 2254 \quad \textit{using a calculator}$$

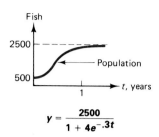

$$y = \frac{2500}{1 + 4e^{-.3t}}$$

Therefore, the population at the end of the first year is 2254, which is about 90% of the carrying capacity of the lake. ■

Biomedical Application: Epidemics

Many epidemics spread at a rate proportional both to the number of people already infected (the "carriers") and also to the number who have yet to catch the disease (the "susceptibles"). If $y(t)$ is the number of infected people at time t from a population of size M, then $y(t)$ satisfies the logistic differential equation

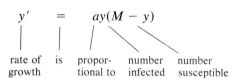

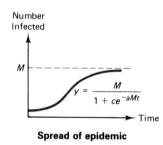

Spread of epidemic

Therefore, the size of the infected population is given by the logistic function

$$y = \frac{M}{1 + ce^{-aMt}}$$

The constants are evaluated just as in Example 4, using the (initial) number of cases reported at time $t = 0$, and also the number of cases at some later time.

Application to Sociology: Spread of Rumors

Sociologists have found that rumors spread through a population of size M at a rate proportional to the number who have heard the rumor (the "informed") and the number who have not heard the rumor (the "uninformed"). Therefore, the number $y(t)$ who have heard the rumor within t time units satisfies the logistic differential equation:

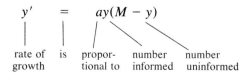

The solution $y(t)$ is then the logistic function

$$y = \frac{M}{1 + ce^{-aMt}}$$

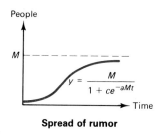

People

Spread of rumor

It remains only to evaluate the constants. The spread of an epidemic is analogous to the spread of a rumor, with an "infected" person being one who has heard the rumor.

Application to Business: Limited and Logistic Growth of Sales

Both limited and logistic growth are used to model the total sales of a product that approach an upper limit (market saturation). The difference is that limited growth begins rapidly (for a product advertised over mass media), while logistic growth begins slowly (for a product that becomes known only through "word of mouth").

Practice Exercise 1

The graphs below show the total sales through day t for two different products, A and B. Which of these products was advertised and which became known only by "word of mouth"? State an appropriate differential equation for each curve.

Product A   **Product B**

(solution on page 402)

Summary

The laws of unlimited, limited, and logistic growth are summarized in the following table. If one of these differential equations governs a particular situation, we can write the solution immediately, evaluating the constants from the given data.

The Laws of Growth

Type	Differential Equation	Solution	Examples	Graph
Unlimited (Growth is proportional to present size.)	$y' = ay$ $y(0) = c$	$y = ce^{at}$	investments bank accounts some populations	
Limited (Growth is proportional to maximum size M minus present size.)	$y' = a(M - y)$ $y(0) = 0$	$y = M(1 - e^{-at})$	information spread by mass media memorizing random information total sales (with advertising)	
Logistic (Growth is proportional to present size and to maximum size M minus present size.)	$y' = ay(M - y)$	$y = \dfrac{M}{1 + ce^{-aMt}}$	confined populations epidemics rumors total sales (without advertising)	

To decide which (if any) of the three models applies in a given situation, think of whether the growth is proportional to size, to unused capacity, or to both (as shown in the left-hand column of the chart). Notice that the differential equation gives much more insight into how the growth occurs than does the solution. This is what we meant at the beginning of the section by "thinking in terms of differential equations."

SOLUTIONS TO PRACTICE EXERCISE

1. Product A, whose growth begins slowly, was not advertised, and the differential equation is logistic: $y' = ay(M - y)$. Product B, whose growth begins rapidly, was advertised, and the differential equation is limited: $y' = a(M - y)$.

▮ EXERCISES 6.6

1 Verify that $y(t) = ce^{at}$ solves the differential equation for unlimited growth, $y' = ay$, with initial condition $y(0) = c$.

2 Verify that $y(t) = M(1 - e^{-at})$ solves the differential equation for limited growth, $y' = a(M - y)$, with initial condition $y(0) = 0$.

Determine the type of each differential equation: **unlimited** *growth,* **limited** *growth,*
logistic *growth, or* **none** *of these. (You do not need to find the solution.)*

3 $y' = .02y$

4 $y' = 5(100 - y)$

5 $y' = 30(.5 - y)$

6 $y' = .4y(.01 - y)$

7 $y' = 2y^2(.5 - y)$

8 $y' = 6y$

9 $y' = y(6 - y)$

10 $y' = .01(100 - y^2)$

11 $y' = 4y(.04 - y)$

12 $y' = 4500(1 - y)$

APPLIED EXERCISES

Write the differential equation (unlimited, limited, or logistic) that applies to the
situation described. Then use its solution to solve the problem.

13 (*General–Appreciation*) The value of a stamp collection, initially worth $1500, grows continuously by 8% per year. Find a formula for its value after t years.

14 (*General–Appreciation*) The value of a home, originally worth $25,000, grows continuously by 6% per year. Find a formula for its value after t years.

15 (*Business–Total Sales*) A manufacturer estimates that he can sell a maximum of 100,000 digital tape recorders in a city. His total sales grow at a rate proportional to the distance below this upper limit. If after 5 months total sales are 10,000, find a formula for the total sales after t months. Then use your answer to estimate the total sales at the end of the first year.

16 (*Business–Product Recognition*) Let $p(t)$ be the number of people in a city who have heard of a new product after t weeks of advertising. The city is of size 1,000,000, and $p(t)$ grows at a rate proportional to the number of people in the city who have *not* heard of the product. If after 8 weeks 250,000 people have heard of the product, find a formula for $p(t)$. Use your formula to estimate the number of people who will have heard of the product after 20 weeks of advertising.

17 (*General–Fundraising*) In a drive to raise $5000, fund-raisers estimate that the rate of contributions is proportional to the distance from the goal. If $1000 was raised in 1 week, find a formula for the amount raised in t weeks. How many weeks will it take to raise $4000?

18 (*General–Learning*) A person can memorize at most 40 two-digit numbers. If the person can memorize 15 numbers in the first 20 minutes, find a formula for the number that can be memorized in t minutes. Use your answer to estimate how long it will take to memorize 30 numbers.

19 (*Business–Sales*) A telephone company estimates the maximum market for car phones in a city to be 10,000. Total sales are proportional both to the number already sold and to the size of the remaining market. If at time $t = 0$ 100 phones have been sold and after 6 months 2000 have been sold, find a formula for the total sales after t months. Use your answer to estimate the total sales at the end of the first year.

20 (*Business–Sales*) During a flu epidemic in a city of 1,000,000, a flu vaccine sells in proportion to the number of people already inoculated and to the number not yet inoculated. If at time $t = 0$ 100 doses have been sold and after 4 weeks 2000 doses have been sold, find a formula for the total number of doses sold within t weeks. Use your formula to predict the sales after 10 weeks.

21 (*Sociology–Rumors*) One person at an airport starts a rumor that a plane has been hijacked, and within 10 minutes 200 people have heard the rumor. If there are 800 people in the airport, find a formula for the number who have heard the rumor within t minutes. Use your answer to estimate how many will have heard the rumor within 15 minutes.

22 (*General–Epidemics*) A flu epidemic on a college campus of 4000 students begins with 12 cases, and after 1 week has grown to 100 cases. Find a formula for the size of the epidemic after t weeks. Use your answer to estimate the size of the epidemic after 2 weeks.

23 (*Ecology–Deer Population*) A wildlife refuge is initially stocked with 100 deer, and can hold at most 800 deer. If 2 years later the deer population is 160, find a formula for the deer population after t years. Use your answer to estimate when the deer population will reach 400.

24 (*Political Science–Voting*) Suppose that a bill in the U.S. Senate gains votes in proportion to the number of votes that it already has and to the number of votes that it does not have. If it begins with one vote (from its sponsor) and after 3 days it has 30 votes, find a formula for the number of votes that it will have after t days. (*Note:* the number of votes in the Senate is 100.) When will the bill have "majority support" of 51 votes?

The U.S. Senate Chamber

Solve each exercise by recognizing that the differential equation is one of the three types whose solution we know.

25 (*Biomedical–Drug Absorption*) A drug injected into a vein is absorbed by the body at a rate proportional to the amount remaining in the blood. For a certain drug, the amount $y(t)$ remaining in the blood after t hours satisfies

$$y' = -.15y$$
$$y(0) = 5$$

Find $y(t)$ and use your answer to estimate the amount present after 2 hours.

26 (*Business–Stock Value*) One model for the growth of the value of stock in a corporation assumes that the stock has a limiting "market value" L, and that the value $v(t)$ of the stock on day t satisfies the differential equation

$$v' = a(L - v)$$

for some constant a. Find a formula for the value $v(t)$ of a stock if whose market value is $L = 40$ if on day $t = 10$ it was selling for $v = 30$.

Solve each differential equation by separation of variables.

27 (*General–Population*) If $y(t)$ is the size of a population at time t, then y'/y, the population growth rate divided by the size of the population, is called the *individual birthrate*. Suppose that the individual birthrate is proportional to the size of the population, $y'/y = ay$ for some constant a. Find a formula for the size of the population after t years.

28 (*General–Gompertz Curve*) Another differential equation that is used to model the growth of a population $y(t)$ is $y' = bye^{-at}$, where a and b are constants. Solve this differential equation.

29 (*General–Allometry*) Solve the differential equation of allometric growth: $y' = ay/x$ (where a is a constant). This differential equation governs the relative growth rates of different parts of the same animal.

30 (*General–Population*) Suppose that a population $y(t)$ in a certain environment grows in proportion to the *square* of the difference between the carrying capacity M and the present population, $y' = a(M - y)^2$, where a is a constant. Solve this differential equation.

THE LAW OF LOGISTIC GROWTH

31 Solve the logistic differential $y' = ay(M - y)$ as follows.

(a) Separate variables to obtain

$$\frac{dy}{y(M - y)} = a\, dt$$

(b) Integrate, using on the left-hand side the integration formula

$$\int \frac{dy}{y(M - y)} = \frac{1}{M} \ln\left(\frac{y}{M - y}\right)$$

(which may be checked by differentiation). [continued in next column.]

(c) Exponentiate to solve for $y/(M - y)$ and then solve for y.

(d) Show that the solution can be expressed

$$y = \frac{M}{1 + ce^{-aMt}}$$

32 Find the inflection point of the logistic curve

$$f(x) = \frac{M}{1 + ce^{-aMx}}$$

and show that it occurs at midheight between $y = 0$ and the upper limit $y = M$.

6.7 REVIEW OF CHAPTER SIX

Introduction

In this chapter we discussed several topics related to integration.

In **Section 6.1** we treated integration by parts, using the formula

$$\int u\, dv = uv - \int v\, du$$

which is simply an integration version of the product rule. Given an integral, we choose parts of it to be u and dv, calculate du and v, and then express the integral according to the right-hand side of the formula. For example,

$$\int xe^x\, dx = xe^x - \int e^x\, dx = xe^x - e^x + C$$

$$\begin{bmatrix} u = x & dv = e^x\, dx \\ du = dx & v = \int e^x\, dx = e^x \end{bmatrix}$$

Making the right choice for u and dv may involves some "trial and error." The guidelines on pages 339 and 342 may help with the choice. Integrals like $\int_0^T C(t)e^{-rt}\, dt$ [for the present value of a continuous stream of income $C(t)$] are often solved by integration by parts.

In **Section 6.2** we used a table of integrals (on the inside back cover) to find a formula that "fits" a given integral. Making a formula fit may be as simple as choosing values for the constants, or it may involve a substitution.

In **Section 6.3** we treated definite integrals in which one or both limits are infinite (improper integrals). We defined improper integrals for continuous nonnegative functions by "dropping back to the finite and then taking the limit."

$$\int_a^\infty f(x)\,dx = \lim_{b\to\infty} \int_a^b f(x)\,dx$$

If the limit does not give a (finite) number, then the integral is "divergent" (its value is undefined). For example,

$$\int_1^\infty x^{-2}\,dx = \lim_{b\to\infty} \int_1^b x^{-2}\,dx = \lim_{b\to\infty} \left[(-x^{-1}) \,\Big|_1^b \right] = \lim_{b\to\infty} \left[-\frac{1}{b} - \left(-\frac{1}{1} \right) \right] = 0 + 1 = 1$$

integrating   evaluating   taking the limit

The improper integral $\int_0^\infty 2000e^{-rt}\,dt$ gives the size of the permanent endowment that will generate \$2000 annually forever (at continuous interest rate r). Improper integrals of the form

$$\int_{-\infty}^b f(x)\,dx \quad \text{and} \quad \int_{-\infty}^\infty f(x)\,dx$$

are defined similarly by limits.

In **Section 6.4** we discussed numerical integration, interpreting definite integrals as areas, which we then approximated using area formulas from elementary geometry. For trapezoidal approximation, the definite integral of a continuous function is approximated by the formula

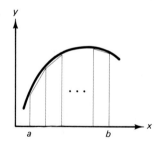

$$\int_a^b f(x)\,dx \approx \left[\frac{1}{2} f(x_1) + f(x_2) + f(x_3) + \cdots + \frac{1}{2} f(x_{n+1}) \right] \Delta x$$

The actual computation is best carried out in a table, as shown on page 372. The table for Simpson's rule (parabolic approximation) is given on page 378.

In **Section 6.5** we solved differential equations by separating variables (moving every x to one side of the equation and every y to the other) and then integrating both sides. A solution with an arbitrary constant of integration is called a *general* solution, and if the arbitrary constant is assigned a value (by a boundary condition) it is a *particular* solution.

In **Section 6.6** we solved three very important differential equations, governing unlimited, limited, and logistic growth. By remembering their solutions (see the table on page 402) we may solve problems modeled by these differential equations immediately by writing the solution and evaluating the constants.

The exercises review all these developments.

Practice Test

Exercises 3, 5, 13, 25, 27, 33, 39, 41, 57, 63, 75, 87, 89, 97, 99

▨   **EXERCISES 6.7**

─────────────────────────────── **6.1** ───────────────────────────────

Solve each integral using integration by parts.

1 $\int xe^{2x}\,dx$

2 $\int xe^{-x}\,dx$

3 $\int x^8 \ln x\,dx$

4 $\int \sqrt[4]{x}\,\ln x\,dx$

5 $\int (x-2)(x+1)^5\,dx$

6 $\int (x+3)(x-1)^4\,dx$

7 $\int \dfrac{\ln t}{\sqrt{t}}\,dt$

8 $\int x^7 e^{x^4}\,dx$

9 $\int x^2 e^x\,dx$

10 $\int (\ln x)^2\,dx$

11 $\int x(x+a)^n\,dx$ (for constants a and $n > 0$)

12 $\int x(1-x)^n\,dx$ (for constant $n > 0$)

13 $\int_0^5 xe^x\,dx$

14 $\int_1^e x \ln x\,dx$

Solve each integral by whatever means are necessary.

15 $\int \dfrac{dx}{1-x}\,dx$

16 $\int xe^{-x^2}\,dx$

17 $\int x^3 \ln 2x\,dx$

18 $\int \dfrac{dx}{(1-x)^2}$

19 $\int \dfrac{\ln x}{x}\,dx$

20 $\int \dfrac{e^{2x}}{e^{2x}+1}\,dx$

21 $\int \dfrac{e^{\sqrt{x}}}{\sqrt{x}}\,dx$

22 $\int (e^{2x}+1)^3 e^{2x}\,dx$

▨ **23** (*Business–Present Value of a Continuous Stream of Income*) A company generates a continuous stream of income of $25t$ million dollars per year, where t is the number of years that the company has been in operation. Find the present value of this stream for the first 10 years at 5% interest compounded continuously.

▨ **24** (*General–Pollution*) Radioactive waste is leaching out of cement storage vessels at the rate of $r(t) = te^{.2t}$ hundred gallons per month, where t is the number of month since the leak began. Find the total leakage during the first 3 months.

─────────────────────────────── **6.2** ───────────────────────────────

Use the integral table inside the back cover to solve each integral.

25 $\int \dfrac{1}{25-x^2}\,dx$

26 $\int \dfrac{1}{x^2-4}\,dx$

27 $\int \dfrac{x}{(x-1)(x-2)}\,dx$

28 $\int \dfrac{1}{(x-1)(x-2)}\,dx$

29 $\int \dfrac{1}{x\sqrt{x+1}}\,dx$

30 $\int \dfrac{x}{\sqrt{x+1}}\,dx$

31 $\int \dfrac{1}{\sqrt{x^2+9}}\,dx$

32 $\int \dfrac{1}{\sqrt{x^2+16}}\,dx$

33 $\int \dfrac{z^3}{\sqrt{z^2+1}}\,dz$

34 $\int \dfrac{e^{2t}}{e^t+2}\,dt$

35 $\int x^2 e^{2x}\,dx$

36 $\int (\ln x)^4\,dx$

37 (*Business–Cost*) A company's marginal cost function is

$$MC(x) = \dfrac{1}{(2x+1)(x+1)}$$

and fixed costs are 1000. Find the company's cost function.

▨ **38** (*General–Population*) The population of a town is growing at the rate of $r(t) = \sqrt{t^2+1600}$ people per year, where t is the number of years from now. Find the gain in population during the next 30 years.

─────────────────────────── **6.3** ───────────────────────────

Find the value of each improper integral that is convergent.

39 $\int_1^\infty \frac{1}{x^5} dx$

40 $\int_1^\infty \frac{1}{x^6} dx$

41 $\int_1^\infty \frac{1}{\sqrt[5]{x}} dx$

42 $\int_1^\infty \frac{1}{\sqrt[6]{x}} dx$

43 $\int_0^\infty e^{-2x} dx$

44 $\int_4^\infty e^{-.5x} dx$

45 $\int_0^\infty e^{2x} dx$

46 $\int_4^\infty e^{.5x} dx$

47 $\int_0^\infty e^{-t/5} dt$

48 $\int_{100}^\infty e^{-t/10} dt$

49 $\int_0^\infty \frac{x^3}{(x^4 + 1)^2} dx$

50 $\int_0^\infty \frac{x^4}{(x^5 + 1)^2} dx$

51 $\int_{-\infty}^0 e^{2t} dt$

52 $\int_{-\infty}^0 e^{4t} dt$

53 $\int_{-\infty}^4 \frac{1}{(5 - x)^2} dx$

54 $\int_{-\infty}^8 \frac{1}{(9 - x)^2} dx$

55 $\int_{-\infty}^\infty \frac{e^{-x}}{(1 + e^{-x})^4} dx$

56 $\int_{-\infty}^\infty \frac{e^{-x}}{(1 + e^{-x})^3} dx$

57 (*General–Permanent Endowments*) Find the size of the permanent endowment needed to generate an annual $6000 forever at an interest rate of 10% compounded continuously.

58 (*General–Automobile Age*) Insurance records indicate that the proportion of cars on the road that are more than x years old is approximated by the integral $\int_x^\infty .21e^{-.21t} dt$. Find the proportion of cars that are more than 5 years old.

59 (*Business–Book Sales*) A publisher estimates that the demand for a certain book will be $12e^{-.05t}$ thousand copies per year, where t is the number of years since its publication. Find the total number of books that will be sold from its publication onward.

60 (*General–Resource Consumption*) If the rate of consumption of a certain mineral is $300e^{-.04t}$ million tons per year (where t is the number of years from now), find the total amount of the mineral that will be consumed from now on.

─────────────────────────── **6.4** ───────────────────────────

▨ *Estimate each integral using trapezoidal approximation with the given value of n. (Round all calculations to three decimal places.)*

61 $\int_0^1 \sqrt{1 + x^4} \, dx, \quad n = 3$

62 $\int_0^1 \sqrt{1 + x^5} \, dx, \quad n = 3$

63 $\int_0^1 e^{(1/2)x^2} \, dx, \quad n = 4$

64 $\int_0^1 e^{-(1/2)x^2} \, dx, \quad n = 4$

65 $\int_{-1}^1 \ln(1 + x^2) \, dx, \quad n = 4$

66 $\int_{-1}^1 \ln(x^3 + 2) \, dx, \quad n = 4$

▨ *Estimate each integral using Simpson's rule (parabolic approximation) with the given value of n. (Round all calculations to four decimal places.)*

67 $\int_0^1 \sqrt{1 + x^4} \, dx, \quad n = 4$

68 $\int_0^1 \sqrt{1 + x^5} \, dx, \quad n = 4$

69 $\int_0^1 e^{(1/2)x^2} \, dx, \quad n = 4$

70 $\int_0^1 e^{-(1/2)x^2} \, dx, \quad n = 4$

71 $\int_{-1}^1 \ln(1 + x^2) \, dx, \quad n = 4$

72 $\int_{-1}^1 \ln(x^3 + 2) \, dx, \quad n = 4$

—————————— 6.3 and 6.4 ——————————

NUMERICAL INTEGRATION OF IMPROPER INTEGRALS

73-74 For each improper integral:

(a) Make it a "proper" integral by using the substitution $x = 1/t$ and simplifying.

(b) Approximate the proper integral using trapezoidal approximation with $n = 4$. Keep three decimal places.

73 $\int_1^\infty \frac{1}{x^2 + 1}\, dx$ **74** $\int_1^\infty \frac{x^2}{x^4 + 1}\, dx$

—————————— 6.5 ——————————

Find the general solution for each differential equation.

75 $y^2 y' = x^2$ **76** $y' = x^2 y$ **77** $y' = \dfrac{x^3}{x^4 + 1}$ **78** $y' = xe^{-x^2}$

79 $y' = y^2$ **80** $y' = y^3$ **81** $y' = 1 - y$ **82** $y' = \dfrac{1}{y}$

83 $y' = xy - y$ **84** $y' = x^2 + x^2 y$

Solve each differential equation and boundary condition.

85 $y^2 y' = 3x^2$ **86** $y' = \dfrac{y}{x^2}$ **87** $y' = \dfrac{y}{x^3}$ **88** $y' = \sqrt[3]{y}$

$y(0) = 1$ $y(1) = 1$ $y(0) = 7$ $y(1) = 0$

89 (*General–Accumulation of Wealth*) Suppose that you now have $10,000 and that you expect to save an additional $4000 during each year, and all of this is deposited in a bank paying 5% interest compounded continuously. Let $y(t)$ be your bank balance (in thousands of dollars) after t years.

(a) Write a differential equation and boundary condition to model your bank balance.

(b) Solve your differential equation and boundary condition.

(c) Use your solution to find your bank balance after 10 years.

90 (*General–Pollution*) A town discharges 4 tons of pollutant into a lake annually, and each year bacterial action removes 25% of the accumulated pollution.

(a) Write a differential equation and boundary condition for the amount of pollution in the lake.

(b) Solve your differential equation to find a formula for the amount of pollution in the lake after t years.

91 (*General–Fever Thermometers*) How long should you keep a thermometer in your mouth to take your temperature? *Newton's law of cooling* says that the thermometer reading rises at a rate proportional to the difference between your actual temperature and the present reading. For a fever of 106 degrees, the thermometer reading $y(t)$ after t minutes in your mouth satisfies

$$y' = 2.3(106 - y)$$

$$y(0) = 70 \quad \text{(initially at room temperature)}$$

(The constant 2.3 is typical for household thermometers.) Solve this differential equation and boundary value.

92 (*Continuation of Exercise 91*). Use your solution $y(t)$ to Exercise 91 to calculate $y(1)$, $y(2)$, and $y(3)$, the thermometer reading after 1, 2, and 3 minutes. Do you see why 3 minutes is the usually recommended time for keeping the thermometer in your mouth?

6.6

For each situation write an appropriate differential equation (unlimited, limited, or logistic). Then find its solution and solve the problem. (All require 🖩.)

93 (*General–Postage Stamps*) The price of a first-class postage stamp grows continuously by about 7% each year (on the average). If in 1988 the price was 25 cents, estimate the price in the year 2000.

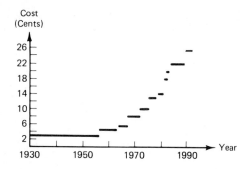

94 (*Economics–Spending on Health*) The average amount spent on health care by an American citizen is increasing continuously by about 10% per year. If in 1986 the amount was $1600, estimate the amount in the year 2000.

95 (*Biomedical–Epidemics*) A virus spreads through a university community of 8000 people at a rate proportional both to the number already infected and to the number not yet infected. If it begins with 10 cases and grows in a week to 150 cases, estimate the size of the epidemic after 2 weeks.

96 (*Social Science–Rumors*) A rumor spreads through a school of 500 students at a rate proportional both to those who know and those who do not know the rumor. If the rumor began with 2 students and within a day had spread to 75, how many students will have heard the rumor within 2 days?

97 (*Business–Total Sales*) A manufacturer estimates that he can sell a maximum of 10,000 video cassette recorders in a city. His total sales grow at a rate proportional to how far total sales are below this upper limit. If after 7 months total sales are 3000, find a formula for the total sales after t months. Then use your answer to estimate the total sales at the end of the first year.

98 (*General–Learning*) Suppose that the maximum rate at which a mail carrier can sort letters is 60 letters per minute, and that she learns at a rate proportional to her distance from this upper limit. If after 2 weeks on the route she can sort 25 letters per minute, how many weeks will it take her to sort 50 letters per minute?

99 (*Business–Advertising*) A new product is advertised extensively on television to a city of 500,000 people, and the number of people who have seen the ads increases at a rate proportional to the number who have not yet seen the ads. If within 2 weeks 200,000 have seen the ads, how long must the product be advertised to reach 400,000 people?

7 CALCULUS OF SEVERAL VARIABLES

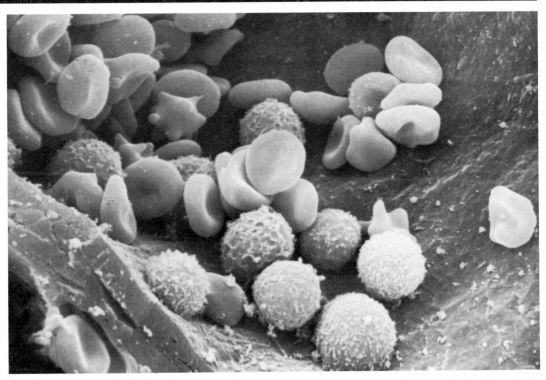

An electron microphotograph of white blood cells (rough spheres) and red blood cells (smooth disks) in an artery. Blood flow depends upon *two* variables, the radius and the length of the artery.

7.1 FUNCTIONS OF SEVERAL VARIABLES

Introduction

Many quantities depend on *several* variables. For example, the "wind chill" factor announced by the weather bureau during the winter depends on two variables: temperature and wind speed. The cost of a telephone call depends on *three* variables: distance, duration, and time of day.

In this chapter we define functions of two or more variables and learn how to differentiate and integrate them. We use derivatives for calculating rates of change and optimizing functions (for example, maximizing profit functions). We will use integrals for finding continuous sums, average values, and volumes of three-dimensional figures.

Functions of Two Variables

The notation $f(x, y)$ (read: "f of x and y") stands for a function f that depends on *two* variables, x and y. The *domain* of $f(x, y)$ is the set of all ordered pairs (x, y) for which the function is defined. The *range* is the set of all resulting values of the function. Formally:

> A function f of two variables is a rule such that to each ordered pair (x, y) in the domain of f there corresponds one and only one number $f(x, y)$.

If the domain is not stated, it will always be taken to be the largest set of ordered pairs for which the function is defined (the "natural domain").

Example 1 For $f(x, y) = \dfrac{\sqrt{x}}{y^2}$, find

(a) the domain   (b) $f(9, -1)$

Solution

(a) In $f(x, y) = \sqrt{x}/y^2$ the x cannot be negative (because of the square root) and y cannot be zero (since it is the denominator), so the domain is

$$\{(x, y) \mid x \geq 0, y \neq 0\}$$

set of all ordered pairs (x, y) such that $x \geq 0$ and $y \neq 0$

(b) $f(9, -1) = \dfrac{\sqrt{9}}{(-1)^2} = \dfrac{3}{1} = 3$

$f(x, y) = \sqrt{x}/y^2$ with $x = 9$ and $y = -1$ ∎

Example 2   For $g(u, v) = e^{uv} - \ln u$, find

(a)   the domain   (b) $g(1, 2)$

Solution

(a)   In $g(u, v) = e^{uv} - \ln u$ the u must be positive (so its logarithm will be defined), so the domain is

$$\{(u, v) \mid u > 0\}$$

set of all ordered pairs (u, v) such that $u > 0$

(b) $g(1, 2) = e^{1\cdot2} - \underbrace{\ln 1}_{0} = e^2 - 0 = e^2$

$g(u, v) = e^{uv} - \ln u$ with $u = 1$ and $v = 2$ ▪

Practice Exercise 1

For $f(x, y) = \dfrac{\ln x}{e^{\sqrt{y}}}$, find

(a)   the domain   (b) $f(e, 4)$ *(solutions on page 419)*

Functions of two variables are used in many applications.

Example 3   A company manufactures three-speed and ten-speed bicycles. It costs \$100 to make each three-speed bicycle, \$150 to make each ten-speed bicycle, and fixed costs are \$2500. Find the cost function.

Solution   Let

$$x = \text{the number of three-speed bicycles}$$
$$y = \text{the number of ten-speed bicycles}$$

The cost function is

$$C(x, y) = 100x + 150y + 2500$$

unit cost / quan-tity / unit cost / quan-tity / fixed cost

For example, the cost of producing 15 three-speed bicycles and 20 ten-speed bicycles is found by evaluating $C(x, y)$ at $x = 15$ and $y = 20$.

$$C(15, 20) = 100\cdot15 + 150\cdot20 + 2500$$
$$= 1500 + 3000 + 2500 = 7000$$

producing 15 three-speed and 20 ten-speed bicycles costs \$7000   ▪

The variables x and y in the preceding example stand for numbers of bicycles, and so should take only integer values. Instead, however, we will allow x and y to be "continuous" variables, and round to integers at the end, if necessary.

Some other "everyday" examples of functions of two variables are as follows.

$$A(l, w) = lw$$

area of a rectangle of length l and width w

$$f(w, v) = kwv^2$$

length of the skid marks for a car of weight w and velocity v skidding to a stop (k is a constant depending on the road surface)

Application to Economics: Cobb–Douglas Production Functions

A function used to model the output of a company or a nation is called a *production function,* and the most famous is the Cobb–Douglas production function*

$$P(L, K) = aL^b K^{1-b}$$

for constants a > 0 and 0 < b < 1

This function expresses the total production P as a function of L, the number of units of labor, and K, the number of units of capital. (Labor is measured in man-hours, and capital means *invested* capital, including the cost of buildings, equipment, and raw materials.)

Example 4   Cobb and Douglas modeled the output of the American economy by the function $P(L, K) = L^{.75}K^{.25}$. Find $P(150, 220)$.

Solution

$$P(150, 220) = (150)^{.75}(220)^{.25}$$

$P(L, K) = L^{.75}K^{.25}$ with $L = 150$ and $K = 220$

$$\approx (42.9)(3.85) \approx 165$$

using a calculator

That is, 150 units of labor and 220 units of capital should result in approximately 165 units of production. ■

Functions of Three or More Variables

Functions of three (or more) variables are defined analogously. Some examples are as follows.

$$V(l, w, h) = lwh$$

volume of a rectangular solid of length l, width w, and height h

$$V(P, r, n) = Pe^{rn}$$

value of P dollars invested at continuous interest rate r for n years

$$f(w, x, y, z) = \frac{w + x + y + z}{4}$$

average of four numbers

---

* First used by Charles Cobb and Paul Douglas in a landmark study of the American economy published in 1928.

Example 5   For $f(x, y, z) = \dfrac{\sqrt{x}}{y} + \ln\dfrac{1}{z}$, find

    (a)   the domain   **(b)** $f(4, -1, 1)$

Solution

    (a)   In $f(x, y, z) = \sqrt{x}/y + \ln(1/z)$ we must have $x \geq 0$ (because of the square root), $y \neq 0$ (since it is the denominator), and $z > 0$ (so that $1/z$ has a logarithm). Therefore, the domain is

$$\{(x, y, z) \mid x \geq 0, \, y \neq 0, \, z > 0\}$$

    (b) $f(4, -1, 1) = \dfrac{\sqrt{4}}{-1} + \ln\dfrac{1}{1} = \dfrac{2}{-1} + \underbrace{\ln 1}_{0} = -2$

                                                                            ▪

Example 6   An open-top box is to have a center divider, as shown in the diagram. Find formulas for the volume V of the box and for the total amount of materials M needed to construct the box.

Solution   The volume is length times width times height.

$$V = xyz$$

The box consists of a bottom, a front and back, two sides, and a divider, whose areas are shown in the diagram. Therefore, the total amount of materials (the area) is

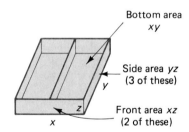

Bottom area
xy

Side area yz
(3 of these)

Front area xz
(2 of these) ▪

$$M = \underset{\substack{\text{bottom}}}{xy} + \underset{\substack{\text{back} \\ \text{and front}}}{2xz} + \underset{\substack{\text{sides and} \\ \text{divider}}}{3yz}$$

Practice Exercise 2

Find a formula for the total amount of materials M needed to construct an open-top box with three parallel dividers. Use the variables shown in the diagram.

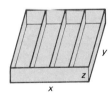

(solutions on page 419)

Three-Dimensional Coordinate Systems

Graphing a function $f(x, y)$ requires a three-dimensional coordinate system. We draw three perpendicular axes as shown on the right.* We will usually draw only the positive half of each axis, although each axis extends infinitely far in the negative direction as well. The plane at the base is called the x-y *plane*.

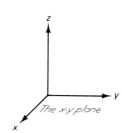

The three-dimensional ("right-handed") coordinate system

A point in a three-dimensional coordinate system is specified by three coordinates, giving its distances from the origin in the x, y, and z directions. For example, the point

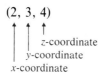

$$(2, 3, 4)$$

z-coordinate
y-coordinate
x-coordinate

is plotted by starting at the origin, moving 2 units in the x direction, 3 units in the y direction, and then 4 units in the (vertical) z direction.

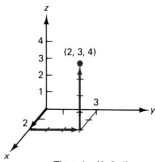

The point (2, 3, 4)

Graph of a Function of Two Variables

To graph a function $f(x, y)$ we choose values for x and y, calculate z-values from $z = f(x, y)$, and plot the points (x, y, z).

Example 7   To graph $f(x, y) = 18 - x^2 - y^2$, we set z equal to the function.

$$z = 18 - x^2 - y^2$$

z replaces f(x, y).

Then we choose values for x and y. Choosing $x = 1$ and $y = 2$ gives

$$z = 18 - 1^2 - 2^2 = 13$$

z = 18 − x² − y² with x = 1 and y = 2

for the point

$$(1, 2, 13)$$

the chosen x = 1, y = 2, and the calculated z

Choosing $x = 2$ and $y = 3$ gives

$$z = 18 - 2^2 - 3^2 = 5$$

z = 18 − x² − y² with x = 2 and y = 3

---

* This is called a "right-handed" coordinate system because the x, y, and z axes correspond to the first two fingers and thumb of the right hand.

for the point

$$(2, 3, 5)$$

*the chosen x = 2, y = 3, and the
calculated z*

These points (1, 2, 13) and (2, 3, 5) are plotted on the graph on the left
below. The completed graph of the function is shown on the right below.

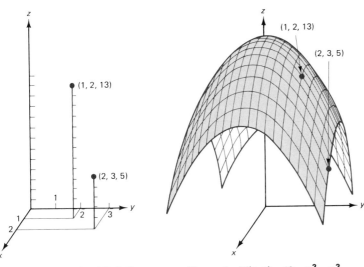

The points (1, 2, 13) and (2, 3, 5)
of the function
$f(x, y) = 18 - x^2 - y^2$.

The graph of $f(x, y) = 18 - x^2 - y^2$.

In general, the graph of a function $f(x, y)$ of *two* variables is a *surface*
above or below the *x-y* plane [just as the graph of a function $f(x)$ of *one*
variable is a *curve* above or below the *x*-axis].

Graphing functions of two variables involves drawing three-dimensional
graphs, which is very difficult. Graphing functions of *more* than two variables requires *more* than three dimensions, and is impossible. For this reason we will not graph functions of several variables. We will, however,
often speak of a function $f(x, y)$ as *surface* in three-dimensional space.

Relative Extreme Points and Saddle Points

Certain points on a surface $z = f(x, y)$ are of special importance.

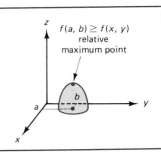

The graph of $f(x, y)$ is a surface
whose height above the point
(x, y) in the *x-y* plane is given by
$z = f(x, y)$.

A point (a, b, c) on a surface $z = f(x, y)$ is a *relative maximum point* if

$$f(a, b) \geq f(x, y)$$

for all (x, y) in some region surrounding (a, b).

A point (a, b, c) on a surface $z = f(x, y)$ is a *relative minimum point* if

$$f(a, b) \leq f(x, y)$$

for all (x, y) in some region surrounding (a, b).

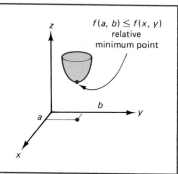

A relative maximum point is *at least as high* as the points on the surface immediately surrounding it, and a relative minimum point is *at least as low* as the points on the surface immediately surrounding it.

As before, the term *relative extreme point* means a point that is either a relative maximum or a relative minimum point. A surface may have any number of relative extreme points, even none.

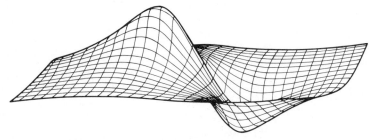

A surface with two relative extreme points: one relative maximum and one relative minimum.

The point shown on the right is called a *saddle point* (so-called because the diagram resembles a saddle).

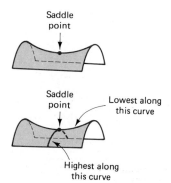

Notice that a saddle point is the highest point along one curve of the surface and the lowest point along another curve. A saddle point, however, is *not* a relative extreme point.

If we think of a surface $z = f(x, y)$ as a landscape, then relative maximum and minimum points correspond to "hilltops" and "valley bottoms" of the surface. A saddle point corresponds to a "mountain pass" between two peaks.

Gallery of Surfaces

The following are the graphs of a few functions.

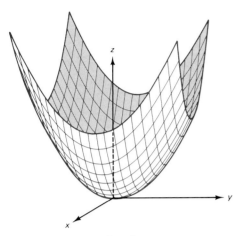

The surface $f(x, y) = x^2 + y^2$ has a relative minimum point at the origin.

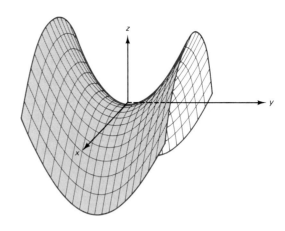

The surface $f(x, y) = y^2 - x^2$ has a saddle point at the origin.

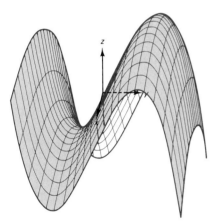

The surface $f(x, y) = 12y + 6x - x^2 - y^3$ has a saddle point and a relative maximum point.

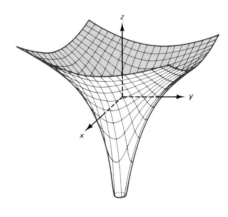

The surface $f(x, y) = \ln(x^2 + y^2)$ has no relative extreme points. It is undefined at (0, 0).

SOLUTIONS TO PRACTICE EXERCISES

1. **(a)** $\{(x, y) \mid x > 0, \ y \geq 0\}$ **(b)** $f(e, 4) = \dfrac{\ln e}{e^{\sqrt{4}}} = \dfrac{1}{e^2} = e^{-2}$

2. $M = xy + 5xz + 2yz$

■ EXERCISES 7.1

For each function find the domain.

1 $f(x, y) = \dfrac{1}{xy}$ **2** $f(x, y) = \dfrac{\sqrt{x}}{\sqrt{y}}$ **3** $f(x, y) = \dfrac{1}{x - y}$ **4** $f(x, y) = \dfrac{\sqrt[3]{x}}{\sqrt[3]{y}}$

5 $f(x, y) = \dfrac{\ln x}{y}$ 　　　　　　　　**6** $f(x, y) = \dfrac{x}{\ln y}$ 　　　　　**7** $f(x, y, z) = \dfrac{e^{1/y} \ln z}{x}$ 　　　　**8** $f(x, y, z) = \dfrac{\sqrt{x} \ln y}{z}$

For each function evaluate the given expression.

9 $f(x, y) = \sqrt{99 - x^2 - y^2}$, find $f(3, -9)$.

11 $g(x, y) = \ln (x^2 + y^4)$, find $g(0, e)$.

13 $w(u, v) = \dfrac{1 + 2u + 3v}{uv}$, find $w(-1, 1)$.

15 $h(x, y) = e^{xy + y^2 - 2}$, find $h(1, -2)$.

17 $f(x, y) = xe^y - ye^x$, find $f(1, -1)$.

19 $f(x, y, z) = xe^y + ye^z + ze^x$, find $f(1, -1, 1)$.

21 $f(x, y, z) = z \ln \sqrt{xy}$, find $f(-1, -1, 5)$.

10 $f(x, y) = \sqrt{75 - x^2 - y^2}$, find $f(5, -1)$.

12 $g(x, y) = \ln (x^3 - y^2)$, find $g(e, 0)$.

14 $w(u, v) = \dfrac{2u + 4v}{v - u}$, find $w(1, -1)$.

16 $h(x, y) = e^{x^2 - xy - 4}$, find $h(1, -2)$.

18 $f(x, y) = xe^y + ye^x$, find $f(-1, 1)$.

20 $f(x, y, z) = xe^y + ye^z + ze^x$, find $f(-1, 1, -1)$.

22 $f(x, y, z) = z\sqrt{x} \ln y$, find $f(4, e, -1)$.

▨ APPLIED EXERCISES

23 (*Business–Stock Yield*) The *yield* of a stock is defined as $Y(d, p) = d/p$, where d is the dividend per share and p is the price of a share of stock. Find the yield of a stock that sells for $140 and offers a dividend of $2.20.

24 (*Business–Price-Earnings Ratio*) The price–earnings ratio of a stock is defined as $R(P, E) = P/E$, where P is the price of a share of stock and E is its earnings. Find the price–earnings ratio of a stock that is selling for $140 with earnings of $1.70.

25 (*General–Scuba Diving*) The maximum duration of a scuba dive (in minutes) can be estimated from the

formula

$$T(v, d) = \frac{33v}{d + 33}$$

where v is the volume of air (at sea-level pressure) in the tank and d is the depth of the dive. Find $T(90, 33)$.

26 (*Social Science–Cephalic Index*) Anthropologists define the *cephalic index* to distinguish the head shapes of different races. For a head of width W and length L (measured from above), the cephalic index is

$$C(W, L) = 100 \frac{W}{L}$$

Calculate the cephalic index for a head of width 8 inches and length 10 inches.

27 (*Economics–Cobb-Douglas Functions*) A company's production is estimated to be $P(L, K) = 2L^{.6}K^{.4}$. Find $P(320, 150)$.

28 (*Biomedical–Body Area*) The surface area (in square feet) of a person of weight w pounds and height h feet is approximated by the function of two variables: $A(w, h) = .55w^{.425}h^{.725}$. Use this function to estimate the surface area of a person who weighs 160 pounds and who is 6 feet tall. (Such estimates are important in certain medical procedures.)

29 (*Economics–Cobb-Douglas Functions*) Show that the Cobb–Douglas function $P(L, K) = aL^b K^{1-b}$ satisfies the equation $P(2L, 2K) = 2 \cdot P(L, K)$. This shows that doubling the amounts of labor and capital doubles production, a property called *returns to scale*.

30 (*Economics–Cobb-Douglas Functions*) Show that the Cobb–Douglas functions $P(L, K) = aL^b K^{1-b}$ with $0 < b < 1$ satisfies

$$P(2L, K) < 2P(L, K) \quad \text{and} \quad P(L, 2K) < 2P(L, K)$$

This shows that doubling the amounts of either labor or capital alone results in *less* than double production, a property called *diminishing returns*.

31 (*General–Telephone Calls*) For two cities with populations x and y that are d miles apart, the number of telephone calls per hour between them can be estimated by the function of three variables

$$f(x, y, d) = \frac{3xy}{d^{2.4}}$$

(This is called the *gravity model*.) Use the gravity model to estimate the number of calls between two cities of populations 40,000 and 60,000 that are 600 miles apart.

32 (*Ecology–Tag and Recapture Estimates*) Ecologists estimate the size of animal populations by capturing and tagging a few animals, and then releasing them. After the first group has mixed with the population, a second group of animals is captured, and the number of tagged animals in this group is counted. If originally T animals were tagged, and the second group is of size S and contains t tagged animals, then the population is estimated by the function of three variables

$$P(T, S, t) = \frac{TS}{t}$$

Estimate the size of a deer population if 100 deer were tagged, and then a second group of 250 contained 20 tagged deer.

33 (*Business–Cost Function*) It costs an appliance company $210 to manufacture each washer and $180 to manufacture each dryer, and fixed costs are $4000. Find the company's cost function $C(x, y)$, using x and y for the number of washers and dryers, respectively.

34–35 (*General–Box Design*) For each open-top box shown below, find formulas for

(a) the volume

(b) the total amounts of materals (the area)

34

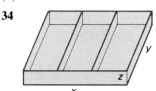

35

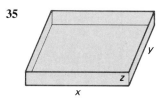

7.2 PARTIAL DERIVATIVES

Introduction

In this section we learn how to differentiate functions of several variables, and to interpret the derivatives as rates of change. Functions of several variables have several derivatives, one for each variable.

Derivatives and Constants

First we review the rules governing derivatives and constants. For a constant *standing alone*, the derivative is zero, but for a constant *multiplying a function*, we carry along the constant and differentiate the function.

$$\frac{d}{dx}\, c = 0$$

the derivative of a constant is zero

$$\frac{d}{dx}\,(cx^3) = c\cdot 3x^2$$

carry along the constant └derivative of x^3

the derivative of a constant times a function is the constant (carried along) times the derivative of the function

$$\frac{d}{dx}\,(x^3 c) = 3x^2\cdot c$$

carry along the constant

the same rule holds even if the constant appears after the function

These ideas will be very useful in this section.

Definition of Partial Derivatives

A function $f(x, y)$ has two derivatives, called *partial derivatives,* one with respect to x and the other with respect to y.

$$\frac{\partial}{\partial x} f(x, y) = \lim_{h\to 0} \frac{f(x + h, y) - f(x, y)}{h}$$

partial derivative of f with respect to x

$$\frac{\partial}{\partial y} f(x, y) = \lim_{h\to 0} \frac{f(x, y + h) - f(x, y)}{h}$$

partial derivative of f with respect to y

(provided, of course, that the limits exist). Partial derivatives are written with a "curly" ∂, $\partial/\partial x$ instead of d/dx, and are often called "partials."

To understand these definitions, notice that in the partial with respect to x, the x is increased by h, but the y is left unchanged.

─x is increased by h
─y is held constant

$$\frac{\partial}{\partial x} f(x, y) = \lim_{h\to 0} \frac{f(x + h, y) - f(x, y)}{h}.$$

If we simply omit the y in this definition, we obtain

$$\lim_{h\to 0} \frac{f(x + h) - f(x)}{h}$$

which is just the definition of the "ordinary derivative" from Chapter 2. This shows that the partial derivative with respect to x is just the ordinary derivative with respect to x with y held constant. Similarly, the partial with respect to y is just the ordinary derivative with respect to y, but now with x held constant.

$$\frac{\partial}{\partial x} f(x, y) = \begin{pmatrix} \text{the derivative of } f \text{ with respect} \\ \text{to } x, \text{ with } y \text{ held constant} \end{pmatrix}$$

$$\frac{\partial}{\partial y} f(x, y) = \begin{pmatrix} \text{the derivative of } f \text{ with respect} \\ \text{to } y, \text{ with } x \text{ held constant} \end{pmatrix}$$

Example 1 Find $\dfrac{\partial}{\partial x} x^3 y^4$.

Solution The $\partial/\partial x$ means differentiate with respect to x. Therefore, y is held constant, so y^4 is also constant (since a constant to a power is just another constant). We therefore carry along the "constant" y^4 and differentiate the x^3.

the derivative of x^3

$$\frac{\partial}{\partial x} x^3 y^4 = 3x^2 y^4$$

carry along the
"constant" y^4

∂/∂x means differentiate with respect to x, treating y like a constant

Example 2 Find $\dfrac{\partial}{\partial y} x^3 y^4$.

Solution The $\partial/\partial y$ means differentiate with respect to y, holding x (and therefore x^3) constant.

$$\frac{\partial}{\partial y} x^3 y^4 = x^3 4y^3 \quad = \quad 4x^3 y^3$$

carry along the — derivative writing the
"constant" x^3 of y^4 constant first

∂/∂y means differentiate with respect to y, treating x like a constant

Practice Exercise 1

Find

(a) $\dfrac{\partial}{\partial x} x^4 y^2$

(b) $\dfrac{\partial}{\partial y} x^4 y^2$

(solutions on page 432)

Example 3 Find $\dfrac{\partial}{\partial x} y^4$.

Solution The $\partial/\partial x$ means differentiate with respect to x, treating y^4 like a constant. But the y^4 is a "constant standing alone" (there is no x in the function), and the derivative of a constant standing alone is zero.

$$\frac{\partial}{\partial x}\, y^4 = 0$$

partial with⌐ ⌐function
respect to x　of y alone ■

Practice Exercise 2

Find $\dfrac{\partial}{\partial y}\, x^2$. *(solution on page 432)*

Example 4   Find $\dfrac{\partial}{\partial x}\,(2x^4 - 3x^3y^3 - y^2 + 4x + 1)$.

Solution

$$\frac{\partial}{\partial x}\,(2x^4 - 3x^3y^3 - y^2 + 4x + 1) = 8x^3 - 9x^2y^3 + 4$$

differentiating with respect to x, so each y is held constant ■

Practice Exercise 3

Find $\dfrac{\partial}{\partial y}\,(2x^4 - 3x^3y^3 - y^2 + 4x + 1)$. *(solution on page 432)*

Subscript Notation for Partial Derivatives

Partial derivatives are often denoted by subscripts: a subscripted x means the partial with respect to x, and a subscripted y means the partial with respect to y.*

$$f_x(x, y) = \frac{\partial}{\partial x}\, f(x, y)$$

f_x *means the partial of f with respect to x*

$$f_y(x, y) = \frac{\partial}{\partial y}\, f(x, y)$$

f_y *means the partial of f with respect to y*

Example 5   If $f(x, y) = 5x^4 - 2x^2y^3 - 4y^2$, find $f_x(x, y)$.

Solution

$$f_x(x, y) = 20x^3 - 4xy^3$$

differentiating with respect to x, holding y constant ■

---

* Sometimes subscripts 1 and 2 are used to indicate partial derivatives with respect to the first and second variables: $f_1(x, y) = f_x(x, y)$ and $f_2(x, y) = f_y(x, y)$. We will not use this notation in this book.

Example 6   If $f = xe^y$, find f_x and f_y.

Solution

$$f_x = e^y$$

the derivative of the variable x times the "constant" e^y is just the "constant" e^y

$$f_y = xe^y$$

the derivative of e^y is e^y (times the "constant" x)

■

Example 7   If $z = y \ln x$, find z_x and z_y.

Solution

$$z_x = y\frac{1}{x}$$

the derivative of ln x is 1/x (times the "constant" y)

$$z_y = \ln x$$

the derivative of the variable y times the "constant" ln x is just the "constant" ln x

■

Practice Exercise 4

If $f(x, y) = ye^x + x \ln y$, find $f_x(x, y)$ and $f_y(x, y)$.

(solution on page 432)

To differentiate a function to a power we use the generalized power rule, but now the derivative if the "inside" function means the *partial* derivative.

Example 8   If $f = (xy^2 + 1)^4$, find f_y.

Solution

$$f_y = 4(xy^2 + 1)^3(x2y)$$

the derivative of $(f)^n$ is $nf^{n-1}f'$

partial of the inside with respect to y

$$= 8xy(xy^2 + 1)^3$$

simplifying

■

Example 9   If $g = \dfrac{xy}{x^2 + y^2}$, find $\dfrac{\partial g}{\partial x}$.

Solution

partial of the top with respect to x
partial of the bottom with respect to x

$$\frac{\partial g}{\partial x} = \frac{(x^2 + y^2)y - 2x \cdot xy}{(x^2 + y^2)^2}$$

using the quotient rule

bottom squared

$$= \frac{x^2y + y^3 - 2x^2y}{(x^2 + y^2)^2}$$

simplifying

■

Example 10 If $f(x, y) = \ln (x^2 + y^2)$, find $f_x(x, y)$.

Solution

$$f_x(x, y) = \frac{2x}{x^2 + y^2}$$

partial of the bottom

the derivative of ln f is f′/f

Example 11 If $f(x, y) = e^{x^2 - y^3}$, find $f_y(x, y)$.

Solution

$$f_y(x, y) = e^{x^2 - y^3}(-3y^2)$$

the derivative of e^f is $e^f f'$

partial of the exponent with respect to y

$$= -3y^2 e^{x^2 - y^3}$$

writing the $-3y^2$ first ▪

Partials Evaluated at Numbers

An expression like $f_x(2, 5)$, which involves both differentiation and evaluation, means *first differentiate and then evaluate.**

Example 12 If $f(x, y) = 4x^3 - x^2 y^2 + 3y^4$, find $f_x(2, -1)$.

Solution

$$f_x(x, y) = 12x^2 - 2xy^2$$

differentiating with respect to x

$$f_x(2, -1) = 12(2)^2 - 2(2)(-1)^2 = 48 - 4 = 44$$

then evaluating at x = 2 and y = −1 ▪

Example 13 If $f(x, y) = e^{x^2 + y^2}$, find $f_y(1, 3)$.

Solution

$$f_y(x, y) = e^{x^2 + y^2}(2y)$$

the derivative of e^f is $e^f f'$

partial of the exponent with respect to y

$$f_y(1, 3) = e^{1^2 + 3^2}(2 \cdot 3)$$

evaluating at x = 1 and y = 3

$$= 6e^{10}$$

simplifying ▪

---

* $f_x(2, 5)$ may also be written $\dfrac{\partial f}{\partial x}(2, 5)$ or $\dfrac{\partial f}{\partial x}\Big|_{(2, 5)}$ again meaning first differentiate, then evaluate.

Practice Exercise 5

If $f(x, y) = e^{x^3+y^3}$, find $f_y(1, 2)$. (solution on page 432)

Example 14   If $f(x, y) = x^2y^2 + \ln(x^4 + y^4)$, find $f_x(1, -1)$.

Solution

$$f_x(x, y) = 2xy^2 + \frac{4x^3}{x^4 + y^4}$$

$$f_x(1, -1) = 2(1)(-1)^2 + \frac{4(1)^3}{1^4 + (-1)^4} = 2 + \frac{4}{2} = 4$$ ■

Partial Derivatives in Three or More Variables

Partial derivatives in three or more variables are defined similarly. That is, the partial derivative of $f(x, y, z)$ with respect to any one variable is the "ordinary" derivative with respect to that variable, holding all other variables constant.

Example 15

(a) $\dfrac{\partial}{\partial x}(x^3y^4z^5) = 3x^2y^4z^2$ *∂/∂x means differentiate with respect to x, holding y and z constant*

hold constant · derivative of x^3

(b) $\dfrac{\partial}{\partial y}(x^3y^4z^5) = x^3 4y^3 z^5 = 4x^3y^3z^5$ *moving the constant to the front*

hold constant · derivative of y^4

(c) $\dfrac{\partial}{\partial z}(x^3y^4z^5) = x^3y^4 5z^4 = 5x^3y^4z^4$ *moving the constant to the front*

hold constant · derivative of z^5 ■

Example 16   If $G = (w + 2x^3 + 3y^2 + 4z)^3$, find G_y.

Solution

$$G_y = 3(w + 2x^3 + 3y^2 + 4z)^2(6y)$$ *using the generalized power rule*

hold constant · derivative of the inside with respect to y

$$= 18y(w + 2x^3 + 3y^2 + 4z)^2$$ *simplifying*   ■

Practice Exercise 6

Find $\dfrac{\partial}{\partial y}\, (xy + yz)^4$. *(solution on page 432)*

Example 17 If $f(x, y, z) = e^{x^2+y^2+z^2}$, find $f_z(1, 1, 1)$.

Solution

$$f_z(x, y, z) = e^{x^2+y^2+z^2}(2z) \qquad \textit{partial with respect to z}$$

$$= 2z e^{x^2+y^2+z^2} \qquad \textit{writing the 2z first}$$

$$f_z(1, 1, 1) = 2e^{1^2+1^2+1^2} = 2e^3 \qquad \textit{evaluating}$$

■

Partials are Instantaneous Rates of Change

Now that we can calculate partials, what are they used for? Since partials are just "ordinary" derivatives with the other variable held constant, they give *instantaneous rates of change* with respect to one variable at a time.

$$f_x(x, y) = \left(\begin{array}{c}\text{the instantaneous rate of change of } f \\ \text{with respect to } x \text{ when } y \text{ is held constant}\end{array}\right)$$

$$f_y(x, y) = \left(\begin{array}{c}\text{the instantaneous rate of change of } f \\ \text{with respect to } y \text{ when } x \text{ is held constant}\end{array}\right)$$

This is why they are called *partial* derivatives: not all the variables are changed at once, only a "partial" change is made.

Application to Economics: Cobb–Douglas Production Functions

Recall that a Cobb–Douglas production function $P(L, K) = aL^b K^{1-b}$ expresses production P as a function of L and K, the number of units of labor and capital. The partial $\partial P/\partial L$ gives the rate at which production increases per additional unit of labor when capital is held constant, and is called the *marginal productivity of labor*. Similarly, $\partial P/\partial K$ gives the rate at which production increases per additional unit of capital when labor is held constant, and is called the *marginal productivity of capital*.

For production P depending on labor L and capital K,

$$\frac{\partial P}{\partial L} = \left(\begin{array}{c}\text{marginal productivity}\\ \text{of labor}\end{array}\right)$$

$$\frac{\partial P}{\partial K} = \left(\begin{array}{c}\text{marginal productivity}\\ \text{of capital}\end{array}\right)$$

Example 18   For the Cobb–Douglas function $P(L, K) = 20L^{.6}K^{.4}$, find and interpret $P_L(120, 200)$ and $P_K(120, 200)$.

Solution

$$P_L = 12L^{-.4}K^{.4} \qquad\qquad\qquad\quad \textit{partial with respect to L (the 12 is 20 times .6)}$$

$$P_L(120, 200) = 12(120)^{-.4}(200)^{.4} \approx 14.7 \qquad \textit{substituting L = 120, K = 200 evaluating using a calculator}$$

Interpretation: The marginal productivity of labor is 14.7, so production increases by about 14.7 units for each additional unit of labor (at the level $L = 120$, $K = 200$).

$$P_K = 8L^{.6}K^{-.6} \qquad\qquad\qquad\quad \textit{partial with respect to K (the 8 is 20 times .4)}$$

$$P_K(120, 200) = 8(120)^{.6}(200)^{-.6} \approx 5.9 \qquad \textit{substituting L = 120, K = 200 evaluating using a calculator}$$

Interpretation: The marginal productivity of capital is 5.9, so production increases by about 5.9 units for each additional unit of capital (at the level $L = 120$, $K = 200$).

This shows that to increase production, additional units of labor are more than twice as effective as additional units of capital.   ■

Partials are Marginals

Partial derivatives give the marginals for one product at a time.

Let $C(x, y)$ be the (total) cost function for x units of product 1 and y units of product 2. Then

$$C_x(x, y) = \left(\begin{array}{c}\text{marginal cost function for the first product}\\ \text{when production of the second is held constant}\end{array}\right)$$

$$C_y(x, y) = \left(\begin{array}{c}\text{marginal cost function for the second product}\\ \text{when production of the first is held constant}\end{array}\right)$$

Similar statements hold, of course, for revenue and profit functions: the partials give the marginals for one variable at a time when the other variables are held constant.

Example 19 A company's profit from producing x radios and y televisions per day is $P(x, y) = 4x^{3/2} + 6y^{3/2} + xy$. Find the marginal profit functions. Then find and interpret $P_y(25, 36)$.

Solution

$$P_x(x, y) = 6x^{1/2} + y$$

marginal profit for radios when television production is held constant

$$P_y(x, y) = 9y^{1/2} + x$$

marginal profit for televisions when radio production is held constant

$$P_y(25, 36) = 9\underbrace{(36)^{1/2}}_{6} + 25 = 79$$

evaluating at $x = 25$ and $y = 36$

Interpretation: Profit increases by about \$79 per additional television (when producing 25 radios and 36 televisions per day). ▪

Geometric Interpretation of Partial Derivatives

A function $f(x, y)$ represents a surface in three-dimensional space, and the partial derivatives are the slopes of certain curves on this surface: the partial with respect to x gives the slope of the surface "in the x direction," and the partial with respect to y gives the slope of the surface "in the y direction" at the point (x, y).

To put this colloquially, if you were walking on the surface $z = f(x, y)$, and if the x direction were east and the y direction were north, then $f_x(x, y)$ would be the slope or steepness of a path heading due east, and $f_y(x, y)$ would be the steepness of a path heading due north from the point (x, y).

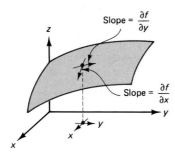

Partial derivatives are slopes.

Higher-Order Partial Derivatives

We can differentiate a function more than once to obtain "higher-order" partials. The subscript and ∂ notations for second partials are given in the two right-hand columns of the following table.

Second Partial	In Words	Subscript Notation	∂ Notation
$\dfrac{\partial}{\partial x}\left(\dfrac{\partial}{\partial x}\,f\right)$	Differentiate twice with respect to x	f_{xx}	$\dfrac{\partial^2}{\partial x^2}\,f$
$\dfrac{\partial}{\partial y}\left(\dfrac{\partial}{\partial y}\,f\right)$	Differentiate twice with respect to y	f_{yy}	$\dfrac{\partial^2}{\partial y^2}\,f$
$\dfrac{\partial}{\partial y}\left(\dfrac{\partial}{\partial x}\,f\right)$	Differentiate first with respect to x, then with respect to y	f_{xy}	$\dfrac{\partial^2}{\partial y\,\partial x}\,f$
$\dfrac{\partial}{\partial x}\left(\dfrac{\partial}{\partial y}\,f\right)$	Differentiate first with respect to y, then with respect to x	f_{yx}	$\dfrac{\partial^2}{\partial x\,\partial y}\,f$

use either of these notations

Note that we differentiate first with the letter *closest* to the f. That is, both f_{xy} and $\dfrac{\partial^2}{\partial y\,\partial x}\,f$ mean: first with respect to x and then with respect to y. Calculating a "second partial" such as f_{xy} is a two-step process: first calculate f_x, and then differentiate the *result* with respect to y.

Example 20   If $f = x^4 + 2x^2y^2 + x^3y + y^4$, find f_{xx}, f_{xy}, f_{yy}, and f_{yx}.

Solution

$$f_x = 4x^3 + 4xy^2 + 3x^2y \qquad \text{\textit{the partial with respect to }} x$$

$$f_{xx} = 12x^2 + 4y^2 + 6xy \qquad \text{\textit{differentiating }} f_x = 4x^3 + 4xy^2 + 3x^2y \text{ \textit{with respect to }} x$$

$$f_{xy} = 8xy + 3x^2 \qquad \text{\textit{differentiating }} f_x = 4x^3 + 4xy^2 + 3x^2y \text{ \textit{with respect to }} y$$

Now returning to the original function $f = x^4 + 2x^2y^2 + x^3y + y^4$, we calculate:

$$f_y = 2x^2 2y + x^3 + 4y^3 \qquad \text{\textit{the partial of }} f \text{ \textit{with respect to }} y$$

$$= 4x^2y + x^3 + 4y^3 \qquad \text{\textit{simplifying}}$$

$$f_{yy} = 4x^2 + 12y^2 \qquad \text{\textit{differentiating }} f_y = 4x^2y + x^3 + 4y^3 \text{ \textit{with respect to }} y$$

$$f_{yx} = 8xy + 3x^2 \qquad \text{\textit{differentiating }} f_y = 4x^2y + x^3 + 4y^3 \text{ \textit{with respect to }} x$$

■

The Partials f_{xy} and f_{yx}

Notice that in Example 20, $f_{xy} = f_{yx}$.

$$f_{xy} = 8xy + 3x^2$$

$$f_{yx} = 8xy + 3x^2$$

$\left.\right\}$ *same*

That is, reversing the order of differentiation (first x, then y, or first y, then x) made no difference. This is not true for all functions, but it is true for all the functions that we will encounter in this book, and it is also true for all functions that you are likely to encounter in applications.*

Practice Exercise 7

For the function $f(x, y) = x^3 - 3x^2y^4 + y^3$, find

  (a) f_x **(b)** f_{xy} **(c)** f_y **(d)** f_{yx} *(solutions on page 433)*

SOLUTIONS TO PRACTICE EXERCISES

1.   **(a)** $\dfrac{\partial}{\partial x} x^4 y^2 = 4x^3 y^2$

  (b) $\dfrac{\partial}{\partial y} x^4 y^2 = x^4 2y = 2x^4 y$

2. $\dfrac{\partial}{\partial y} x^2 = 0$

3. $\dfrac{\partial}{\partial y} (2x^4 - 3x^3 y^3 - y^2 + 4x + 1) = -3x^3 3y^2 - 2y = -9x^3 y^2 - 2y$

4. $f_x = ye^x + \ln y$

  $f_y = e^x + x\,\dfrac{1}{y} = e^x + \dfrac{x}{y}$

5. $f_y(x, y) = e^{x^3 + y^3}(3y^2) = 3y^2 e^{x^3 + y^3}$

  $f_y(1, 2) = (3 \cdot 4)e^{1^3 + 2^3} = 12e^9$

6. $\dfrac{\partial}{\partial y} (xy + yz)^4 = 4(xy + yz)^3(x + z)$

$\quad\quad\quad\quad\quad\quad\quad\quad\quad\quad \underset{\displaystyle \begin{array}{l}\text{partial of the ``inside''}\\ \text{with respect to } y\end{array}}{\underline{\quad\quad}}\!\!\uparrow$

---

* $f_{xy} = f_{yx}$ if these partials are continuous. A more detailed statement can be found in an advanced calculus book.

7. (a) $f_x = 3x^2 - 6xy^4$

(b) $f_{xy} = -6x4y^3 = -24xy^3$

(c) $f_y = -3x^24y^3 + 3y^2 = -12x^2y^3 + 3y^2$

(d) $f_{yx} = -24xy^3$

EXERCISES 7.2

For each function, find the partials (a) $f_x(x, y)$ and (b) $f_y(x, y)$.

1 $f(x, y) = x^3 + 3x^2y^2 - 2y^3 - x + y$

2 $f(x, y) = 3x^2 - 2x^2y^2 - y^3 + x - y$

3 $f(x, y) = 2x^5 - 2x^2y^3 - 18x - 3y + 2$

4 $f(x, y) = 2x^4 - 7x^3y^2 - xy + 1$

5 $f(x, y) = 12x^{1/2}y^{1/3} + 8$

6 $f(x, y) = x^{-1}y + xy^{-2}$

7 $f(x, y) = 100x^{.05}y^{.02}$

8 $f(x, y) = x/y$

9 $f(x, y) = (x + y)^{-1}$

10 $f(x, y) = (x - y)^{-1}$

11 $f(x, y) = (x^2y + 1)^5$

12 $f(x, y) = (x^2 + xy + 1)^4$

13 $f(x, y) = \ln(x^3 + y^3)$

14 $f(x, y) = x^2e^y$

15 $f(x, y) = 2x^3e^{-5y}$

16 $f(x, y) = e^{x+y}$

17 $f(x, y) = e^{xy}$

18 $f(x, y) = \ln(xy^3)$

19 $f(x, y) = \ln\sqrt{x^2 + y^2}$

20 $f(x, y) = \dfrac{xy}{x + y}$

21 $f(x, y) = \dfrac{4xy}{x^2 + y^2}$

22 $f(x, y) = \dfrac{2xy}{x^3 + y^3}$

For each function, find (a) $\partial w/\partial u$ and (b) $\partial w/\partial v$.

23 $w = (uv - 1)^3$

24 $w = (u - v)^3$

25 $w = e^{(1/2)(u^2-v^2)}$

26 $w = \ln(u^2 + v^2)$

For each function, evaluate the stated partials.

27 $f(x, y) = 4x^3 - 3x^2y^2 - 2y^2$, find $f_x(-1, 1)$ and $f_y(-1, 1)$.

28 $f(x, y) = 2x^4 - 5x^2y^3 - 4y$, find $f_x(1, -1)$ and $f_y(1, -1)$.

29 $f(x, y) = e^{x^2+y^2}$, find $f_x(0, 1)$ and $f_y(0, 1)$.

30 $g(x, y) = (xy - 1)^5$, find $g_x(1, 0)$ and $g_y(1, 0)$.

31 $h(x, y) = x^2y - \ln(x + y)$, find $h_x(1, 1)$.

32 $f(x, y) = \sqrt{x^2 + y^2}$, find $f_y(8, -6)$.

33 $f(x, y) = \sqrt{x^2 + y^2}$, find $f_x(3, -4)$.

34 $w(u, v) = \sqrt{uv}$, find $w_v(4, 1)$.

For each function, find the second order partials (a) f_{xx}, (b) f_{xy}, (c) f_{yx}, and (d) f_{yy}.

35 $f(x, y) = 5x^3 - 2x^2y^3 + 3y^4$

36 $f(x, y) = 4x^2 - 3x^3y^2 + 5y^5$

37 $f(x, y) = 9x^{1/3}y^{2/3} - 4xy^3$

39 $f(x, y) = ye^x - x \ln y$

38 $f(x, y) = 32x^{1/4}y^{3/4} - 5x^3y$

40 $f(x, y) = y \ln x + xe^y$

For each function, calculate the partials (a) f_{xxy}, (b) f_{xyx}, and (c) f_{yxx}.

41 $f(x, y) = x^4y^3 - e^{2x}$

42 $f(x, y) = x^3y^4 - e^{2y}$

For each function of three variables, find the partials (a) f_x, (b) f_y, and (c) f_z.

43 $f = xy^2z^3$

45 $f = (x^2 + y^2 + z^2)^4$

47 $f = (xy + yz + zx)^4$

49 $f = e^{x^2+y^2+z^2}$

51 $f = \ln (x^2 + y^2 + z^2)$

44 $f = x^2y^3z^4$

46 $f = (x^3 + y^3 + z^3)^2$

48 $f = (xyz + 1)^3$

50 $f = e^{x-y+2z}$

52 $f = \ln (x^2 - y^3 + z^4)$

For each function, evaluate the stated partial.

53 $f = 3x^2y - 2xz^2$, find $f_x(2, -1, 1)$.

55 $f = e^{x^2+2y^2+3z^2}$, find $f_y(-1, 1, -1)$.

54 $f = 2yz^3 - 3x^2z$, find $f_z(2, -1, 1)$.

56 $f = e^{2x^3+3y^3+4z^3}$, find $f_y(1, -1, 1)$.

APPLIED EXERCISES

57–58 (*Business–Marginal Profit*) An electronics company's profit from making x tape decks and y compact disk players per day is given by the profit function $P(x, y)$ (given below).

(a) Find the marginal profit function for tape decks.

(b) Evaluate your answer to part (a) at $x = 200$ and $y = 300$ and interpret the result.

(c) Find the marginal profit function for compact disk players.

(d) Evaluate your answer to part (c) at $x = 200$ and $y = 100$ and interpret the result.

57 $P(x, y) = 2x^2 - 3xy + 3y^2 + 150x + 75y + 200$

58 $P(x, y) = 3x^2 - 4xy + 4y^2 + 80x + 100y + 200$

59–60 (*Business–Cobb-Douglas Production Functions*) A company's production is given by the Cobb–Douglas function $P(L, K)$ (given in the next column), where L is the number of units of labor and K is the number of units of capital.

(a) Find $P_L(27, 125)$ and interpret this number.

(b) Find $P_K(27, 125)$ and interpret this number.

(c) From your answers to parts (a) and (b), which will increase production more: an additional unit of labor or an additional unit of capital?

59 $P(L, K) = 270L^{1/3}K^{2/3}$ **60** $P(L, K) = 225L^{2/3}K^{1/3}$

61 (*Business–Sales*) A store's television sales depend on x, the price of the televisions, and y, the amount spend on advertising, according to the function $S(x, y) = 200 - .1x + .2y^2$. Find and interpret the marginals S_x and S_y.

62 (*Economics–Value of an MBA*) A 1973 study found that a businessman with an MBA (Master's degree in business administration) earned an average salary of $S(x, y) = 10,990 + 1120x + 873y$ dollars, where x is the number of years of work experience before the MBA, and y is the number of years of work experience after the MBA. Find and interpret the marginals S_x and S_y.

63 (*Sociology–Status*) A study found that a person's status in a community depends on the person's income and education according to the function $S(x, y) = 7x^{1/3}y^{1/2}$, where x is income (in thousands of dollars) and y is years of education beyond high school.

(a) Find $S_x(27, 4)$ and interpret this number.

(b) Find $S_y(27, 4)$ and interpret this number.

64 (*Biomedical–Blood Flow*) The resistance of blood flowing through an artery of radius r and length L (both in centimeters) is $R(r, L) = .08Lr^{-4}$.

(a) Find $R_r(.5, 4)$ and interpret this number.

(b) Find $R_L(.5, 4)$ and interpret this number.

65 (*General–Automobile Safety*) The length of the skid marks from a truck of weight w (tons) traveling at velocity v (miles per hour) skidding to a stop on a dry road is $S(w, v) = .027wv^2$.

(a) Find $S_w(4, 60)$ and interpret this number.

(b) Find $S_v(4, 60)$ and interpret this number.

66 (*General–Wind-Chill Temperature*) The wind-chill temperature announced by the weather bureau during the cold weather measures how cold it "feels" for a given temperature and wind speed. The formula is

$$C(t, w) = (.475 + .304\sqrt{w} - .0203w)(t - 91.4)$$

where t is the temperature (Fahrenheit) and w is the wind speed (mph). Find and interpret $C_w(30, 20)$.

COMPETITIVE AND COMPLEMENTARY COMMODITIES

67 (*Economics–Competitive Commodities*) Certain commodities (like butter and margarine) are called "competitive" or "substitute" commodities because one may substitute for the other. If $B(b, m)$ gives the daily sales of butter as a function of b, the price of butter and m, the price of margarine:

(a) Give an interpretation of $B_b(b, m)$.

(b) Would you expect $B_b(b, m)$ to be positive or negative? Explain.

(c) Give an interpretation of $B_m(b, m)$.

(d) Would you expect $B_m(b, m)$ to be positive or negative? Explain.

68 (*Economics–Complementary Commodities*) Certain commodities, such as washing machines and clothes dryers, are called "complementary" commodities because they are often used together. If $D(d, w)$ gives the monthly sales of dryers as a function of d, the price of dryers and w, the price of washers:

(a) Give an interpretation of $D_d(d, w)$.

(b) Would you expect $D_d(d, w)$ to be positive or negative? Explain.

(c) Give an interpretation of $D_w(d, w)$.

(d) Would you expect $D_w(d, w)$ to be positive or negative? Explain.

7.3 OPTIMIZING FUNCTIONS OF SEVERAL VARIABLES

Introduction

In this section we optimize (maximize and minimize) functions of several variables. We do this as we did for functions of one variable, by finding "critical points" and using a two-variable version of the second derivative test. To simplify matters, we only consider functions whose first and second partials are defined everywhere.

Relative Maxima and Minima for a Function of Two Variables

In Section 7.1 we saw that a function $f(x, y)$ represents a *surface*, with relative maximum and minimum points ("hilltops" and "valley bottoms") and saddle points, as shown below. In this section we will find the relative maximum and minimum *values* of the function, that is, the values of $f(x, y)$ at the relative maximum and minimum points.

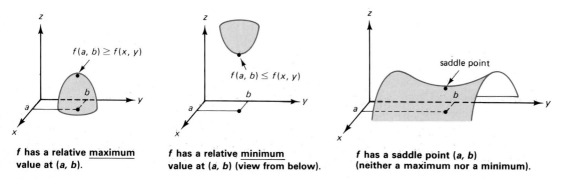

f has a relative <u>maximum</u> value at (a, b).

f has a relative <u>minimum</u> value at (a, b) (view from below).

f has a saddle point (a, b) (neither a maximum nor a minimum).

Critical Points

At the very top of a smooth hill, the slope or steepness in all directions is zero. That is, a stick will balance horizontally at the top. The partials f_x and f_y are the slopes in the x and y directions, so these partials must both be zero at a relative maximum or minimum point.

A point (a, b) at which both partials are zero is called a *critical point* of the function.

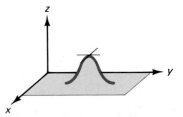

At a relative maximum or minimum point the partials must both be zero.

(a, b) is a critical point of $f(x, y)$ if

$$f_x(a, b) = 0 \quad \text{and} \quad f_y(a, b) = 0$$

Relative maximum and minimum values can only occur at critical points.

Example 1 Find all critical points of

$$f(x, y) = 3x^2 + y^2 + 3xy + 3x + y + 6$$

Solution We want all points (a, b) at which both partials are zero.

$$f_x = 6x + 3y + 3 \quad \text{and} \quad f_y = 2y + 3x + 1 \qquad \textit{partials}$$

$$6x + 3y + 3 = 0$$

$$3x + 2y + 1 = 0$$

$\Big\}$ *partials set equal to zero*

reordered so the x and y terms line up

To solve these equations simultaneously we multiply the second by -2 so that the x-terms drop out when we add.

$$6x + 3y + 3 = 0 \qquad\qquad \textit{first equation}$$

$$\underline{-6x - 4y - 2 = 0} \qquad\qquad \textit{second equation multiplied by } -2$$

$$-y + 1 = 0 \qquad\qquad \textit{adding (the x drops out)}$$

$$y = 1 \qquad\qquad \textit{from solving } -y + 1 = 0$$

Substituting $y = 1$ into either equation gives x.

$$6x + 3 + 3 = 0 \qquad\qquad \textit{substituting } y = 1 \textit{ into } 6x + 3y + 3 = 0$$

$$6x = -6 \qquad\qquad \textit{simplifying}$$

$$x = -1 \qquad\qquad \textit{solving}$$

These x- and y-values gives one critical point.

$$\text{CP:} \quad (-1, 1) \qquad\qquad \textit{from } x = -1, y = 1 \qquad\qquad ■$$

Second Derivative Test for Functions of Two Variables: The *D*-Test

To determine whether $f(x, y)$ has a relative maximum, minimum, or neither (for example, a saddle point) at a critical point, we use a two-variable version of the second derivative test. There are several second partials, f_{xx}, f_{yy}, and $f_{xy} = f_{yx}$, and the second derivative test uses all of them. The entire optimization procedure is as follows.

To find the relative maximum and minimum values of $f(x, y)$:

1. Find f_x and f_y.

2. Solve the system of equations $\begin{cases} f_x = 0 \\ f_y = 0 \end{cases}$
   Let (a, b) be a solution.

3. Find f_{xx}, f_{xy}, and f_{yy} and evaluate

$$D = f_{xx}(a, b) \cdot f_{yy}(a, b) - [f_{xy}(a, b)]^2.$$

4. Then **(i)** f has a relative *maximum* at (a, b) if $D > 0$ and $f_{xx}(a, b) < 0$;

   (ii) f has a relative *minimum* at (a, b) if $D > 0$ and $f_{xx}(a, b) > 0$;

   (iii) f has a *saddle point* at (a, b) if $D < 0$;

   (iv) no conclusion can be drawn if $D = 0$.

Step 4, checking the sign of the two numbers D and $f_{xx}(a, b)$, is known as the *D-test*, and can be understood as follows.*

1. $D > 0$ appears only in parts (i) and (ii), and so guarantees that the function has either a relative maximum or minimum at the critical point. It remains only to use the "old" second derivative test (checking the sign of f_{xx}) to determine which (maximum or minimum).

2. $D < 0$ (part iii) means a saddle point, regardless of the value of f_{xx}.

3. $D = 0$ (part iv) means that *no conclusion can be drawn:* the function may have a relative maximum, a relative minimum, or neither at the critical point.

Example 2   Find the relative extreme values of

$$f(x, y) = 3x^2 + y^2 + 3xy + 3x + y + 6$$

Solution   We find critical points by setting the two partials equal to zero and solving. But this is what we did for this same function in Example 1, finding one critical point, $(-1, 1)$. The D-test will determine whether f has a relative maximum, minimum, or neither at this critical point. The second partials are as follows.

$$f_{xx} = 6 \qquad \qquad \textit{from } f_x = 6x + 3y + 3$$

$$f_{yy} = 2 \qquad \qquad \textit{from } f_y = 2y + 3x + 1$$

$$f_{xy} = 3 \qquad \qquad \textit{from } f_x = 6x + 3y + 3$$

Since all the variables have dropped out, we do not need to substitute $x = -1$ and $y = 1$. We need only calculate $D = f_{xx}f_{yy} - (f_{xy})^2$.

$$D = 6 \cdot 2 - (3)^2 = 12 - 9 = 3 \qquad \qquad \textit{positive!}$$

$$\underset{f_{xx} \quad f_{yy} \quad f_{xy}}{\diagup \quad \diagdown \quad \diagdown}$$

D is positive and f_{xx} is positive (since $f_{xx} = 6$), so f has a *relative minimum* (part ii of the D-test) at the critical point $(-1, 1)$. (If f_{xx} had been negative,

---

* D is called the "Hessian" and, for those who have studied determinants, can be written

$$\begin{vmatrix} f_{xx} & f_{xy} \\ f_{yx} & f_{yy} \end{vmatrix}$$

The proof of this test can be found in an advanced calculus book.

there would have been a relative maximum.) The relative minimum *value* of the function is found by evaluating $f(x, y)$ at $(-1, 1)$.

$$f(-1, 1) = 3 + 1 - 3 - 3 + 1 + 6 = 5$$

$f = 3x^2 + y^2 + 3xy + 3x + y + 6$
evaluated at $x = -1, y = 1$

Answer: Relative minimum value: $f = 5$ at $x = -1, y = 1$. ▪

Note that f does not have a relative maximum value. This is because a relative maximum would have to occur at a *second* critical point, and we found only one critical point.

Example 3   Find the relative extreme values of $f(x, y) = e^{x^2 - y^2}$.

Solution

$$f_x = e^{x^2 - y^2}(2x)$$
$$f_y = e^{x^2 - y^2}(-2y)$$

} *partials*

$$e^{x^2 - y^2}(2x) = 0$$
$$e^{x^2 - y^2}(-2y) = 0$$

} *partials set equal to zero*

$e^{x^2 - y^2}$ can never be zero (since e to any power is positive), but the other factors, $(2x)$ and $(-2y)$, *can* be zero. Therefore, the solution of the first equation is $x = 0$, and the solution of the second is $y = 0$, giving one critical point.

$$\text{CP:}\quad (0, 0)$$

For the *D*-test we calculate the second partials.

$$f_{xx} = e^{x^2 - y^2}(2x)(2x) + e^{x^2 - y^2}(2)$$

from $f_x = e^{x^2 - y^2}(2x)$ *using the product rule*

$$= 4x^2 e^{x^2 - y^2} + 2e^{x^2 - y^2}$$

simplifying

$$f_{yy} = e^{x^2 - y^2}(-2y)(-2y) + e^{x^2 - y^2}(-2)$$

from $f_y = e^{x^2 - y^2}(-2y)$ *using the product rule*

$$= 4y^2 e^{x^2 - y^2} - 2e^{x^2 - y^2}$$

simplifying

$$f_{xy} = e^{x^2 - y^2}(2x)(-2y) = -4xye^{x^2 - y^2}$$

from $f_x = e^{x^2 - y^2}(2x)$ *using the product rule*

We evaluate these partials at the critical point $(0, 0)$.

$$f_{xx}(0, 0) = 0e^0 + 2e^0 = 0 + 2 = 2$$

$f_{xx} = 4x^2 e^{x^2 - y^2} + 2e^{x^2 - y^2}$ *at* $(0, 0)$

$$f_{yy}(0, 0) = 0e^0 - 2e^0 = 0 - 2 = -2$$

$f_{yy} = 4y^2 e^{x^2 - y^2} - 2e^{x^2 - y^2}$ *at* $(0, 0)$

$$f_{xy}(0, 0) = 0e^0 = 0$$

$f_{xy} = -4xye^{x^2 - y^2}$ *at* $(0, 0)$

Therefore D is

$$D = [2][-2] - [0]^2 = -4 - 0 = -4 \qquad \textit{negative!}$$

$$\underset{f_{xx}}{\diagup} \quad \underset{f_{yy}}{\diagdown} \quad \underset{f_{xy}}{\mid}$$

Since D is negative, the function has a *saddle point* (part iii of the D-test) at the critical point $(0, 0)$. A saddle point is neither a relative maximum nor a relative minimum. There are no other critical points, so there are no relative extreme values.

    Answer: f has no relative extreme values
          (it has a saddle point at $x = 0$, $y = 0$). ▪

Example 4   Find the relative extreme values of

$$f(x, y) = x^2 + y^3 - 6x - 12y$$

Solution

$$f_x = 2x - 6$$

$$f_y = 3y^2 - 12 \qquad\qquad \Big\} \textit{partials}$$

$$2x - 6 = 0$$

$$3y^2 - 12 = 0 \qquad\qquad \Big\} \textit{setting the partials equal to zero}$$

The first gives

$$x = 3 \qquad\qquad \textit{solving } 2x - 6 = 0$$

and the second gives

$$3y^2 = 12 \qquad\qquad \textit{adding 12 to each side of } 3y^2 - 12 = 0$$

$$y^2 = 4 \qquad\qquad \textit{dividing by 3}$$

$$y = \pm 2 \qquad\qquad \textit{taking square roots}$$

From $x = 3$ and $y = \pm 2$ we get *two* critical points:

$$\text{CP:} \quad (3, 2) \quad \text{and} \quad (3, -2)$$

For the D-test we calculate the second partials.

$$f_{xx} = 2 \qquad\qquad \textit{from } f_x = 2x - 6$$

$$f_{xy} = 0 \qquad\qquad \textit{also from } f_x = 2x - 6$$

$$f_{yy} = 6y \qquad\qquad \textit{from } f_y = 3y^2 - 12$$

$$D = (2)(6y) - (0)^2 = 12y \qquad\qquad D = (f_{xx})(f_{yy}) - (f_{xy})^2$$

We apply the *D*-test to the critical points one at a time.

at (3, 2): $D = 12 \cdot 2 > 0$ *D = 12y evaluated at (3, 2)*

$f_{xx} = 2 > 0$

relative minimum at $x = 3, y = 2$ *since D and f_{xx} are both positive*

at (3, −2): $D = 12 \cdot (-2) < 0$ *D = 12y evaluated at (3, 2)*

saddle point at $x = 3, y = -2$ *since D is negative*

Answer: Relative minimum value; $f = -25$ at $x = 3, y = 2$ *f = −25 from evaluating*
(saddle point at $x = 3, y = -2$)   *$f = x^2 + y^3 - 6x - 12y$ at (3, 2)*

▪

Absolute Extreme Values

The *largest* value of a function on its domain is called the *absolute maximum value* of the function. The *smallest* value of a function on its domain is called the *absolute minimum value*. As with functions of a single variable, one or both of these absolute extreme values may fail to exist.

If the absolute maximum value of a function exists, it must occur at a *relative* maximum (because the highest value on the entire domain must certainly be the highest among the nearby points). If you know that a function does have an absolute maximum, and if it has only one relative maximum, this *relative* maximum must also be the *absolute* maximum. Many applied problems are solved in this way, using relative extremes to find absolute extremes.

Business Application: Maximizing Profit

Suppose that a company produces two products, called A and B, and that the two price functions are

$$p(x) = \begin{pmatrix} \text{the price at which exactly } x \\ \text{units of product A will be sold} \end{pmatrix}$$

and

$$q(y) = \begin{pmatrix} \text{the price at which exactly } y \\ \text{units of product B will be sold} \end{pmatrix}$$

If $C(x, y)$ is the (total) cost function, then the company's profit will be the revenue for each product (price times quantity) minus cost.

$$P(x, y) = \underbrace{p(x) \cdot x}_{\substack{\text{price times} \\ \text{quantity for} \\ \text{product A}}} + \underbrace{q(y) \cdot y}_{\substack{\text{price times} \\ \text{quantity for} \\ \text{product B}}} - \underbrace{C(x, y)}_{\text{cost}}$$

$\underbrace{}_{\text{profit}}$

If the company produces too few items, the lost sales will obviously cause lower profits; if the company produces too many items and "floods the market," the depressed prices will again cause lower profits. Therefore, any realistic profit function should be maximized at some "intermediate" quantities x and y. Accordingly, we will assume that profit functions do have maximum values.

Example 5 Universal Motors makes compact and midsized cars. The price function for compacts is $p = 15 - 2x$ (for $x \le 7$), and the price function for midsized cars is $q = 20 - y$ (for $y \le 20$), both in thousands of dollars, where x and y are, respectively, the number of compact and midsized cars produced per hour. If the company's cost function is

$$C(x, y) = 13x + 16y - 2xy + 5$$

thousand dollars, find how many of each car should be produced and the prices that should be charged in order to maximize profit. Also find the maximum profit.

Solution The profit function is

$$P(x, y) = \underbrace{(15 - 2x)}_{\text{price}}\underbrace{x}_{\text{quantity}} + \underbrace{(20 - y)}_{\text{price}}\underbrace{y}_{\text{quantity}} - \underbrace{(13x + 16y - 2xy + 5)}_{\text{cost}}$$

for compacts · for mid-sized

$$= 15x - 2x^2 + 20y - y^2 - 13x - 16y + 2xy - 5 \qquad \textit{multiplying out}$$

$$= -2x^2 - y^2 + 2xy + 2x + 4y - 5 \qquad \textit{simplifying}$$

We maximize $P(x, y)$ in the usual way.

$$P_x = -4x + 2y + 2$$
$$P_y = -2y + 2x + 4$$
$\left.\right\}$ *partials*

$$-4x + 2y + 2 = 0$$
$\left.\right\}$ *partials set equal to zero*

$$\dfrac{2x - 2y + 4 = 0}{-2x \qquad + 6 = 0}$$ ← rearranged to line up xs and ys   *adding (the ys cancel)*

$$x = 3 \qquad \textit{from solving } -2x + 6 = 0$$

$$y = 5 \qquad \begin{array}{l}\textit{from substituting } x = 3 \textit{ into either}\\ \textit{equation (omitting the details)}\end{array}$$

These give one critical point.

$$\text{CP: } (3, 5)$$

For the D-test we calculate the partials.

$$P_{xx} - -4 \qquad P_{xy} = 2 \qquad P_{yy} = -2 \qquad \begin{array}{l}\textit{from } P_x = -4x + 2y + 2\\ \textit{and } P_y = -2y + 2x + 4\end{array}$$

$$D = (-4)(-2) - (2)^2 = 4 \qquad D = P_{xx}P_{yy} - (P_{xy})^2$$

D is positive and $P_{xx} = -4$ is negative, so profit is indeed *maximized* at $x = 3$ and $y = 5$. To find the prices we evaluate the price functions.

$$p = 15 - 2 \cdot 3 = 9 \text{ (thousand dollars)}$$

p = 15 − 2x evaluated at x = 3

$$q = 20 - 5 = 15 \text{ (thousand dollars)}$$

q = 20 − y evaluated at y = 5

The profit comes from the profit function.

$$P(3, 5) = -2 \cdot 3^2 - 5^2 + 2 \cdot 3 \cdot 5 + 2 \cdot 3 + 4 \cdot 5 - 5$$

P = −2x² − y² + 2xy + 2x + 4y − 5 evaluated at x = 3, y = 5

$$= 8 \text{ (thousand dollars)}$$

Answer: Profit is maximized when the company produces 3 compacts per hour, selling them for $9000 each, and 5 midsized cars per hour, selling them for $15,000 each. The maximum profit will be $8000.   ■

Application to Economics: Competition and Collusion

In 1938, the French economist Antoine Cournot published the following comparison of a monopoly (a market with only one supplier) and a duopoly (a market with two suppliers).

Monopoly. Imagine that you are selling spring water from your own spring (or any product whose cost of production is negligible). Since you are the only supplier in town, you have a "monopoly." Suppose that your price function is $p = 6 - .01x$, where p is the price at which you will sell precisely x gallons ($x \leq 600$). Your revenue is then

$$R(x) = (6 - .01x)x = 6x - .01x^2$$

price (6 − .01x) times quantity x

You maximize revenue by setting its derivative equal to zero.

$$R'(x) = 6 - .02x = 0$$

$$x = \frac{6}{.02} = 300$$

solving 6 − .02x = 0

Therefore, you should sell 300 gallons per day. (The second derivative test will verify that revenue is maximized.) The price will be

$$p = 6 - .01 \cdot 300 = 6 - 3 = 3$$

p = 6 − .01x evaluated at x = 300

or $3 dollars per gallon. Your maximum revenue will be

$$R(300) = 6 \cdot 300 - .01(300)^2 = \$900$$

R(x) = 6x − .01x² evaluated at x = 300

Duopoly. Suppose now that your neighbor opens a competing spring water business. (A market like this with two suppliers is called a "duopoly.") Now both of you must share the same market. If he sells y gallons per day (and you sell x), you must both sell at price

$$p = 6 - .01(x + y) = 6 - .01x - .01y$$

price function p = 6 − .01x with x replaced by the combined quantity x + y

Each of you calculates revenue as price times quantity.

$$\left(\begin{array}{c}\text{Your}\\\text{revenue}\end{array}\right) = p{\cdot}x = (6 - .01x - .01y)x = 6x - .01x^2 - .01xy$$

$$\left(\begin{array}{c}\text{His}\\\text{revenue}\end{array}\right) = p{\cdot}y = (6 - .01x - .01y)y = 6y - .01xy - .01y^2$$

You each want to maximize revenue, so you set the partials equal to zero.

$$6 - .02x - .01y = 0$$

partial of 6x − .01x² − .01xy with respect to x, set equal to zero

$$6 - .01x - .02y = 0$$

partial of 6y − .01xy − .01y² with respect to y, set equal to zero

These are easily solved by multiplying one of them by 2 and subtracting the other. The solution (omitting the details) is

$$x = 200 \qquad y = 200$$

so each of you sells 200 gallons per day. The selling price for both will be

$$p = 6 - .01(200 + 200) = 6 - 4 = 2$$

p = 6 − .01(x + y) evaluated at x = 200, y = 200

or $2 per gallon. Revenue is price ($2) times quantity (200), resulting in $400 for each of you.

Comparison of the Monopoly and the Duopoly

The the two systems may be compared as follows.

Monopoly		**Duopoly**	
Quantity:	300 gallons	Quantity:	200 gallons each, 400 total
Price:	$3 per gallon	Price:	$2 per gallon
Revenue:	$900	Revenue:	$400 each

Notice that the duopoly produces *more* than the monopoly (400 gallons versus only 300 in the monopoly), and does so at a *lower price* ($2 versus $3 in the monopoly). Cournot therefore concluded that consumers benefit more from a duopoly than from a monopoly.

However, the smart duopolist will notice that his revenue of $400 is less than half of the monopoly revenue of $900. Therefore, it benefits each duopolist to cooperate and share the market as a single monopoly. This is called *collusion*. It is for this reason that markets with only a few suppliers tend toward collusion rather than competition.

Optimizing Functions of Three or More Variables

Functions of *more* than two variables are optimized in the same way, by finding critical points (where all of the first partials are zero). For example, to optimize a function $f(x, y, z)$ we would set the three partials equal to zero and solve:

$$f_x = 0$$

$$f_y = 0$$

$$f_z = 0$$

The second derivative test for functions of three or more variables is very complicated, and we shall not discuss it.

EXERCISES 7.3

Find the relative extreme values of each function.

1 $f(x, y) = x^2 + 2y^2 + 2xy + 2x + 4y + 7$

2 $f(x, y) = 2x^2 + y^2 + 2xy + 4x + 2y + 5$

3 $f(x, y) = 2x^2 + 3y^2 + 2xy + 4x - 8y$

4 $f(x, y) = 3x^2 + 2y^2 + 2xy + 8x + 4y$

5 $f(x, y) = 3xy - 2x^2 - 2y^2 + 14x - 7y - 5$

6 $f(x, y) = 2xy - 2x^2 - 3y^2 + 4x - 12y + 5$

7 $f(x, y) = xy + 4x - 2y + 1$

8 $f(x, y) = 5xy - 2x^2 - 3y^2 + 5x - 7y + 10$

9 $f(x, y) = 3x - 2y - 6$

10 $f(x, y) = 5x - 4y + 5$

11 $f(x, y) = e^{(1/2)(x^2+y^2)}$

12 $f(x, y) = e^{5(x^2+y^2)}$

13 $f(x, y) = \ln(x^2 + y^2 + 1)$

14 $f(x, y) = \ln(2x^2 + 3y^2 + 1)$

15 $f(x, y) = -x^3 - y^2 + 3x - 2y$

16 $f(x, y) = x^3 - y^2 - 3x + 6y$

17 $f(x, y) = y^3 - x^2 - 2x - 12y$

18 $f(x, y) = -x^2 - y^3 - 6x + 3y + 4$

19 $f(x, y) = x^3 - 2xy + 4y$

20 $f(x, y) = y^3 - 2xy - 4x$

APPLIED EXERCISES

21 (*Business–Maximum Profit*) A company manufactures two products, and the price function for product A is $p = 12 - \frac{1}{2}x$ (for $x \leq 20$) and for product B is $q = 20 - y$ (for $y \leq 20$), both in thousands of dollars, where x and y are the amounts of products A and B, respectively. If the cost function is

$$C(x, y) = 9x + 16y - xy + 7$$

thousand dollars, find the quantities and the prices of the two products that maximize profit. Also find the maximum profit.

22 (*Business–Maximum Profit*) A company manufactures two products, and the price function for product A is $p = 16 - x$ (for $x \leq 16$) and for product B is $q = 19 - \frac{1}{2}y$ (for $y \leq 38$), both in thousands of dollars, where x and y are the amounts of products A and B, respectively. If the cost function is

$$C(x, y) = 10x + 12y - xy + 6$$

thousand dollars, find the quantities and the prices of the two products that maximize profit. Also find the maximum profit.

23 (*Business–Price Discrimination*) An automobile manufacturer sells cars in America and Europe, charging different prices in the two markets. The price function for cars sold in America is $p = 20 - .2x$ thousand dollars (for $x \leq 100$), and the price function for cars sold in Europe is $q = 16 - .1y$ thousand dollars (for $y \leq 160$), where x and y are the number of cars sold per day in America and Europe, respectively. The company's cost function is

$$C = 20 + 4(x + y) \text{ thousand dollars}$$

(a) Find the company's profit function. [*Hint:* Profit is revenue from America plus revenue from Europe, minus costs, where each revenue is price times quantity.]

(b) Find how many cars should be sold in each market to maximize profit. Also find the price for each country.

24 (*Biomedical–Drug Dosage*) In a laboratory test the combined antibiotic effect of x mg (milligrams) of medicine A and y mg of medicine B is given by the function

$$f(x, y) = xy - 2x^2 - y^2 + 110x + 60y$$

(for $x \leq 55$, $y \leq 60$). Find the amounts of the two medicines that maximize the antibiotic effect.

25 (*Psychology–Practice and Rest*) A subject in a psychology experiment who practices a skill for x hours and then rests for y hours is predicted to achieve a test score of $f(x, y) = xy - x^2 - y^2 + 11x - 4y + 120$ (for $x \leq 10$, $y \leq 4$). Find the number of hours of practice and rest that maximize the subject's score.

26 (*Sociology–Absenteeism*) The number of office workers near a beach resort who call in "sick" on a warm summer day is

$$f(x, y) = xy - x^2 - y^2 + 110x + 50y - 5200$$

where x is the air temperature ($70 \leq x \leq 100$) and y is the water temperature ($60 \leq y \leq 80$). Find the air and water temperatures that maximize the number of absentees.

27–28 (*Economics–Competition and Collusion*) Compare the output of a monopoly and a duopoly for the price function given below (repeating the analysis on page 443–4). That is:

(a) For a monopoly, calculate the quantity x that maximizes revenue. Also calculate the price p and the revenue R.

(b) For the duopoly, calculate the quantities x and y that maximize revenue for each duopolist. Calculate the price p and the two revenues.

(c) Are more goods produced under a monopoly or a duopoly?

(d) Is the price lower under a monopoly or a duopoly?

27 $p = 12 - .005x$ ($x \leq 2400$)

28 $p = a - bx$ (for positive numbers a and b with $x \leq a/b$)

OPTIMIZING FUNCTIONS OF THREE VARIABLES

29 (*Business–Price Discrimination*) An automobile manufacturer sells cars in America, Europe, and Asia, charging different prices in each of the three markets. The price function for cars sold in America is $p = 20 - .2x$ (for $x \leq 100$), the price function for cars sold in Europe is $q = 16 - .1y$ (for $y \leq 160$), and the price function for cars sold in Asia is $r = 12 - .1z$ (for $z \leq 120$), all in thousands of dollars, where x, y, and z are the number of cars sold in America, Europe, and Asia, respectively. The company's cost function is $C = 22 + 4(x + y + z)$ thousand dollars.

(a) Find the company's profit function $P(x, y, z)$. (*Hint:* The profit will be revenue from America plus revenue from Europe plus revenue from Asia, minus costs, where each revenue is price times quantity.)

(b) Find how many cars should be sold in each market to maximize profit. (*Hint:* Set the three partials P_x, P_y, and P_z, equal to zero and solve. Assuming that the maximum exists, it must occur at this point.)

30 (*Economics–Competition and Collusion*) Suppose that in the discussion of competition and collusion (pages 443–4), *two* of your neighbors began selling spring water. Use the price function $p = 36 - .01x$ (for $x \leq 3600$) and repeat the analysis, but now comparing a monopoly to competition among *three* suppliers (a "triopoly"). That is:

(a) For a monopoly, calculate the quantity x that maximizes your profit. Also calculate the price p and the revenue R.

[parts b, c, d on next page]

(b) For a triopoly, find the quantities x, y, and z for the three suppliers that maximize revenue for each. Also calculate the price p and the three revenues. (*Hint*: Find the three revenue functions, one for each supplier, and maximize each with respect to that supplier's variable.)

(c) Are more goods produced under a monopoly or under a triopoly?

(d) Is the price lower under a monopoly or under a triopoly?

Find the relative extreme values of each function.

31 $f(x, y) = x^3 + y^3 - 3xy$

32 $f(x, y) = x^5 + y^5 - 5xy$

33 $f(x, y) = 12xy - x^3 - 6y^2$

34 $f(x, y) = 6xy - x^3 - 3y^2$

35 $f(x, y) = 2x^4 + y^2 - 12xy$

36 $f(x, y) = 16xy - x^4 - 2y^2$

7.4 LEAST SQUARES

Introduction

You may have wondered how the mathematical models in this book were developed. For example, how are the constants a and b in the Cobb–Douglas production function $P = aL^bK^{1-b}$ determined for a particular company or nation? The problem of finding the function that fits a collection of data can be viewed geometrically as the problem of fitting a curve to a collection of points. The simplest case if this problem is fitting a straight line to a collection of points, and the most widely used method for doing this is called *least squares*.

Least squares lines are used extensively in forecasting and for detecting underlying trends in data. The exercises discuss the more general problem of fitting a *curve* to a collection of points.

A First Example

The graph on the right shows a company's annual sales (in millions) over a 3-year period. How can we fit a straight line to these three points? Clearly, these points do not lie exactly on a line, so rather than an "exact" fit, we want the line $y = ax + b$ that fits these three points most closely.

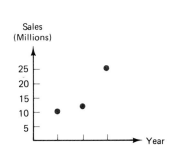

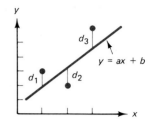

Let d_1, d_2, and d_3 stand for the vertical distances between the three points and the line $y = ax + b$. These "vertical deviations" are a measure of the distance between the points and the line. The line that minimizes the sum of the squares of these vertical deviations is called the *least squares line* or the *regression line*.*

Least Squares Lines

The least squares line for a collection of points is the line $y = ax + b$ that minimizes the sum of the squares of the vertical deviations between the points and the line.

This criterion (minimizing the sum of the squared vertical deviations) is the one most widely used for fitting a line to a collection of points. Squaring the deviations ensures that none are negative, so that a deviation below the line does not "cancel" one above the line.

Example 1 Calculate the least squares line for the points in the table below.

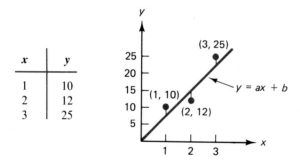

x	y
1	10
2	12
3	25

Solution The vertical deviations are the heights of the line $y = ax + b$ at each x-value, minus the y-values. These differences are then squared and summed.

$$S = (a \cdot 1 + b - 10)^2 + (a \cdot 2 + b - 12)^2 + (a \cdot 3 + b - 25)^2$$

| height of the line $y=ax+b$ at $x=1$ | y-value of the point at $x=1$ | height of the line $y=ax+b$ at $x=2$ | y-value of the point at $x=2$ | height of the line $y=ax+b$ at $x=3$ | y-value of the point at $x=3$ |

---

* The word "regression" comes from an early use of this technique to determine whether unusually tall parents have unusually tall children. It seems that tall parents do have tall offspring, but not quite as tall, with later generations exhibiting a "regression" to the average height of the population.

This sum S depends on a and b, the unknowns that determine the line $y = ax + b$. To minimize S we set its partials with respect to a and b equal to zero.

$$\frac{\partial S}{\partial a} = 2(a + b - 10) + 2(2a + b - 12)\cdot 2 + 2(3a + b - 25)\cdot 3 \qquad \textit{differentiating each part of S by the generalized power rule}$$

$$= 2a + 2b - 20 + 8a + 4b - 48 + 18a + 6b - 150 \qquad \textit{multiplying out}$$

$$= 28a + 12b - 218 \qquad \textit{combining terms}$$

$$\frac{\partial S}{\partial b} = 2(a + b - 10) + 2(2a + b - 12) + 2(3a + b - 25) \qquad \textit{differentiating each part of S by the generalized power rule}$$

$$= 2a + 2b - 20 + 4a + 2b - 24 + 6a + 2b - 50 \qquad \textit{multiplying out}$$

$$= 12a + 6b - 94 \qquad \textit{combining terms}$$

We set the two partials equal to zero and solve.

$$\begin{aligned} 28a + 12b - 218 &= 0 \\ 12a + 6b - 94 &= 0 \end{aligned} \qquad \bigg\}\ \textit{partials set equal to zero}$$

$$\begin{array}{ll} 28a + 12b - 218 = 0 & \textit{first equation} \\ \underline{-24a - 12b + 188 = 0} & \textit{second multiplied by } -2 \\ \quad 4a \qquad\qquad - 30 = 0 & \textit{adding (the bs drop out)} \end{array}$$

$$a = \frac{30}{4} = 7.5 \qquad \textit{solving } 4a - 30 = 0$$

$$b = \frac{4}{6} \approx .67 \qquad \textit{from substituting } a = 7.5 \textit{ into } 12a + 6b - 94 \textit{ and solving for } b$$

These values for a and b give the least squares line.

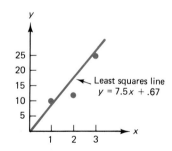

$$y = 7.5x + .67 \qquad\qquad \begin{array}{l} y = ax + b \text{ with} \\ a = 7.5 \text{ and } b = .67 \end{array}$$

The D-test will show that S has indeed been minimized (we omit the details).   ■

Interpretation of the Least Squares Line

Suppose that the data in the table of Example 1 give a company's annual sales (in millions) in each of three years. To predict the sales in year 4 we would evaluate the least squares line at $x = 4$.

$$y = 7.5(4) + .67 = 30.67 \qquad\qquad y = 7.5x + .67 \textit{ evaluated at } x = 4$$

Therefore the least squares line predicts sales of 30.67 million in year 4. The slope of the least squares line is 7.5, so that whenever x increases by 1, y increases by 7.5. This means that the *linear trend* in the company's sales is a growth of 7.5 million sales per year.

Least Squares Procedure for *n* Points

Example 1 used only three points, which is too small for most realistic applications. If we carry out the same steps for the n points

$$(x_1, \ y_1), \ (x_2, \ y_2), \ \ldots, \ (x_n, \ y_n)$$

we would obtain the formulas given in the box below. In the formulas the symbol Σ stands for summation. That is:

$\Sigma \, x$ means: add up all the x-values.

$\Sigma \, y$ means: add up all the y-values.

$\Sigma \, xy$ means: add up the products $x{\cdot}y$ for each point $(x, \ y)$

$\Sigma \, x^2$ means: add up the squares of all of the x-values.

Note that $\Sigma \, x^2$ is not the same as $(\Sigma \, x)^2$. $\Sigma \, x^2$ means the sum of the squares, first squaring the x-values and then adding, while $(\Sigma \, x)^2$ means the square of the sum, first adding the x-values and then squaring the sum.

Least Squares Line

For the n points $(x_1, \ y_1), \ (x_2, \ y_2), \ \ldots, \ (x_n, \ y_n)$ the least squares line is $y = ax + b$ with

$$a = \frac{n \, \Sigma \, xy - (\Sigma \, x)(\Sigma \, y)}{n \, \Sigma \, x^2 - (\Sigma \, x)^2} \qquad b = \frac{1}{n} \, (\Sigma \, y - a \, \Sigma \, x)$$

(n is the number of points)

From now on we will use these formulas instead of the more laborious procedure of Example 1. A derivation of these formulas by the method of Example 1 will be given following an example of how these formulas are used.

Example 2   A 1955 study compared cigarette smoking to the mortality rate for lung cancer in several countries. Find the least squares line that fits these data. Then use the line to predict lung cancer deaths if per capita cigarette consumption is 600.

	Cigarette Consumption (per capita)	Lung Cancer Deaths (per million)
Norway	250	90
Sweden	300	120
Denmark	350	170
Canada	500	150

Solution   The procedure for calculating a and b consists of six steps, beginning with the following table.

1. List the x and y values   **2.** Multiply $x \cdot y$ **3.** Square each x

x	y	xy	x^2
250	90	22,500	62,500
300	120	36,000	90,000
350	170	59,500	122,500
500	150	75,000	250,000
1400	530	193,000	525,000
$\parallel$	$\parallel$	$\parallel$	$\parallel$
Σx	Σy	Σxy	Σx^2

← **4.** Sum each column

5. Calculate a and b using the formulas in the box above.

$$a = \frac{(4)(193,000) - (1400)(530)}{(4)(525,000) - 1400^2} = \frac{30,000}{140,000} = .21$$

$$a = \frac{n\,\Sigma\,xy - (\Sigma\,x)(\Sigma\,y)}{n\,\Sigma x^2 - (\Sigma\,x)^2}$$

$$b = \frac{1}{4}\,[530 - .21(1400)] = 59$$

$$b = \frac{1}{n}\,[\Sigma\,y - a\,\Sigma\,x]$$

$n = $ *number of points*

6. Substitute these values of a and b into the least squares line $y = ax + b$.

$$y = .21x + 59$$

$y = ax + b$ *with* $a = .21$ *and* $y = 59$

This line is graphed below, showing how lung cancer mortality increases with cigarette smoking.

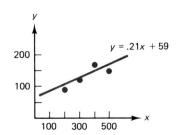

To predict the lung cancer deaths if per capita cigarette consumption reaches 600, we evaluate the least squares line at $x = 600$.

$$y = .21 \cdot 600 + 59 = 185$$

y = .21x + 59 evaluated at x = 600

The predicted annual mortality is 185 deaths per million.   ■

Derivation of the Least Squares Formulas

The formulas for a and b in the least squares line come from minimizing the squared vertical deviations, just as in Example 1, but now for the n points $(x_1, y_1), (x_2, y_2), \ldots, (x_n, y_n)$.

$$S = (ax_1 + b - y_1)^2 + (ax_2 + b - y_2)^2 + \cdots + (ax_n + b - y_n)^2$$

The partials are

$$\frac{\partial S}{\partial a} = 2(ax_1 + b - y_1)x_1 + 2(ax_2 + b - y_2)x_2 + \cdots + 2(ax_n + b - y_n)x_n$$

$$= 2ax_1^2 + 2bx_1 - 2x_1y_1 + 2ax_2^2 + 2bx_2 - 2x_2y_2 + \cdots$$
multiplying out
$$+ 2ax_n^2 + 2bx_n - 2x_ny_n$$

$$= 2a(x_1^2 + x_2^2 + \cdots + x_n^2) + 2b(x_1 + x_2 + \cdots + x_n)$$
regrouping
$$- 2(x_1y_1 + x_2y_2 + \cdots + x_ny_n)$$

$$= 2a \, \Sigma \, x^2 + 2b \, \Sigma \, x - 2 \, \Sigma \, xy$$
using Σ for sum

$$\frac{\partial S}{\partial b} = 2(ax_1 + b - y_1) + 2(ax_2 + b - y_2) + \cdots + 2(ax_n + b - y_n)$$

$$= 2ax_1 + 2b - 2y_1 + 2ax_2 + 2b - 2y_2 + \cdots + 2ax_n + 2b - 2y_n$$
multiplying out

$$= 2a(x_1 + x_2 + \cdots + x_n) + 2b(1 + 1 + \cdots + 1)$$
regrouping

$$- 2(y_1 + y_2 + \cdots + y_n)$$

$$= 2a \, \Sigma \, x + 2bn - 2 \, \Sigma \, y$$
using Σ for sum

We set the partials equal to zero.

$$a \, \Sigma \, x^2 + b \, \Sigma \, x - \Sigma \, xy = 0$$
dividing each by 2

$$a \, \Sigma \, x + bn - \Sigma \, y = 0$$

To solve for the number a we multiply the first equation by n, the second by $\Sigma \, x$, subtract, and solving the resulting equation. The result is

$$a = \frac{n \, \Sigma \, xy - (\Sigma \, x)(\Sigma \, y)}{n \, \Sigma \, x^2 - n(\Sigma \, x)^2}$$

and

$$b = \frac{1}{n} (\Sigma\, y - a\, \Sigma\, x)$$

a Σ x + nb − Σ y = 0 solved for b

These are the formulas in the box on page 450. The *D*-test would show that these values do indeed *minimize S*.

Criticism of Least Squares

Least squares is the most widely used method for fitting lines to points, but it does have one weakness: the vertical deviations from the line are squared, so one large deviation, when squared, can become enormous, and have an unexpectedly large influence on the line. For example, the graph on the right shows four points and their least squares line. The line fits the points quite closely.

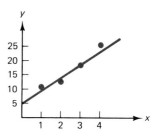

The second graph adds a fifth point, one quite out of line with the others, and shows the least squares line for the five points. The added point has an enormous effect on the line, causing it to slope downward even though all of the other points suggest an upward slope. In actual applications, one should inspect the points for such "outliers" before calculating the least squares line.

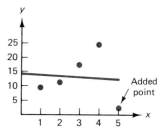

Fitting Curves by Least Squares

It is not always appropriate to fit a straight line to a set of points. Sometimes a collection of points will suggest a curve rather than a line. The exercises show how to fit an exponential curve of the form $y = ae^{bx}$ to a collection of points.

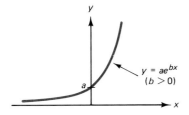
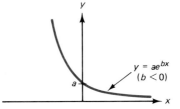

To decide whether to fit a line or a curve, you should first graph the points. For example, the points plotted on the right suggest an exponential curve.

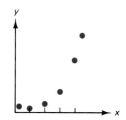

Note on Answers to the Exercises

When you check the answers to the exercises, your answers may differ slightly from those given, depending on when and how you round your calculations. Slight differences are to be expected and do not mean that your answers are wrong.

▪ **EXERCISES 7.4** (▤ useful for all exercises)

Find the least squares line for the following points.

1

x	y
1	2
2	5
3	9

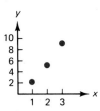

2

x	y
1	2
2	4
3	7

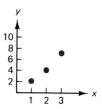

3

x	y
1	6
3	4
6	2

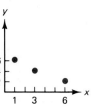

4

x	y
1	9
4	6
5	1

5

x	y
0	7
1	10
2	10
3	15

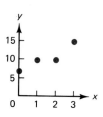

6

x	y
0	5
1	8
2	8
3	12

7

x	y
−1	10
0	8
1	5
3	0
5	−2

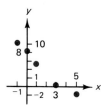

8

x	y
−2	12
0	10
2	6
4	0
5	−3

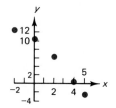

APPLIED EXERCISES

9 (*Business–Sales*) A company's annual sales are shown in the following table. Find the least squares line. Use your line to predict the sales in the next year ($x = 5$).

Year	1	2	3	4
Sales (millions)	7	10	11	14

10 (*General–Automobile Costs*) The following table gives the cost per mile of operating compact car, depending upon the number of miles driven per year. Find the least squares line for these data. Use your answer to predict the cost per mile for a car driven 25,000 miles annually ($x = 5$).

Annual Mileage	x	Cost per Mile (cents)
5,000	1	50
10,000	2	35
15,000	3	27
20,000	4	25

11 (*Sociology–Crime*) A sociologist finds the following data for the number of felony arrests per year in a town. Find the least squares line. Then use it to predict the number of felony arrests in the next year.

Year	1	2	3	4
Arrests	120	110	90	100

12 (*General–Farming*) A farmer's wheat yield (bushels per acre) depends upon the amount of fertilizer (hundreds of pounds per acre) according to the following table. Find the least squares line. Then use the line to predict the yield using 300 pounds of fertilizer per acre.

Fertilizer	1.0	1.5	2	2.5
Yield	30	35	38	40

13 (*Baseball–The Disappearance of the .400 Hitter*) Between 1901 and 1930 baseball boasted several .400 hitters (Lajoie, Cobb, Jackson, Sisler, Heilmann, Hornsby, and Terry), but only one since then (Ted Williams in 1941). The decline of the "heavy hitter" is evidenced by the following data, showing the differences between the average for the top five hitters and the major league average over several decades. Find the least squares line for these data and use it to predict the difference for the period 1981–2000 ($x = 5$).

	x	High Average Minus League Average
1901–1920	1	82
1921–1940	2	76
1941–1960	3	68
1961–1980	4	59

14 (*Economics–Consumer Price Index*) The consumer price index (CPI) is shown in the following table. Fit at least squares line to the data. Then use the line to predict the CPI in the year 2000.

	x	CPI
1970	1	116.3
1975	2	161.2
1980	3	247.0
1985	4	322.2

15–16 (*General–Percentage of Smokers*) The following tables show the percentage of the adult population who smoke (the first table is males, the second females). Find the least squares line for these data, and use your answer to predict the percentage of that sex who will smoke in the year 2000 ($x = 36$).

15

	x	Percent Males
1965	1	52.1
1976	12	41.6
1980	16	37.9
1983	19	35.4

16

	x	Percent Females
1965	1	34.2
1976	12	32.5
1980	16	29.8
1983	19	29.9

17 (*General–Smoking and Longevity*) The following data show the life expectancy of a 25-year-old male based on the number of cigarettes smoked daily. Find the least squares line for these data. The slope of the line estimates the years lost per extra cigarette per day.

Cigarettes Smoked Daily	Life Expectancy
0	73.6
5	69.0
15	68.1
30	67.4
40	65.3

18 (*General–Pollution and Absenteeism*) The following table shows the relationship between the sulfur dust content of the air (in $\mu g/m^3$) and the number of female absentees in industry. (Only absences of at least 7 days were counted.) Find the least squares line for these data. Use your answer to predict absences in a city with a sulfur dust content of 25.

	Sulfur	Absences per 1000 Employees
Cincinnati	7	19
Indianapolis	13	44
Woodbridge	14	53
Camden	17	61
Harrison	20	88

Fitting Exponential Curves by Least Squares

Least squares can be used to fit an exponential curve of the form $y = Be^{Ax}$ to a collection of points. Taking natural logs of both sides of $y = Be^{Ax}$ gives

$$\ln y = \ln (Be^{Ax}) = \ln B + \ln e^{Ax} = \ln B + Ax$$

using the properties of natural logarithms

If we introduce a new variable Y defined by $Y = \ln y$, and let $b = \ln B$, then $\ln y = \ln B + Ax$ becomes

$$Y = b + Ax$$

which, written in the other order, is the straight line

$$Y = Ax + b$$

Therefore, we fit a straight line to the *logarithms* of the y-values. (Now we are minimizing not the squared deviations, but the squared deviations of the logarithms.) The procedure consists of the eight steps shown in the following example.

Example 3 The world population since 1950 is shown in the table on the right. Fit an exponential curve to these data and predict the world population in the year 2000.

Year	Population (billion)
1950	2.51
1960	2.97
1970	3.58
1980	4.41

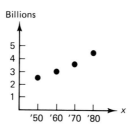

Solution   We number the years 1 through 4 (since they are evenly spaced).

1. List the x and
     y values

2. Take ln of the
     y-values (call
     them *capital Y*)

3. Multiply $x \cdot Y$
     (capital Y)

4. Square
     each x

x	y	$Y = \ln y$	xY	x^2
1	2.51	.92	.92	1
2	2.97	1.09	2.18	4
3	3.58	1.28	3.84	9
4	4.41	1.48	5.92	16
10		4.77	12.86	30
Σx		ΣY	ΣxY	Σx^2

5. Add each column
     (except y)

6. Calculate A and b
     (n = number of points)

$$A = \frac{4(12.86) - 10(4.77)}{4(30) - 10^2} = \frac{3.74}{20} = .187$$

$$b = \frac{1}{4}(4.77 - .187 \cdot 10) = .725$$

$$A = \frac{n \, \Sigma \, xY - (\Sigma \, x)(\Sigma \, Y)}{n \, \Sigma x^2 - (\Sigma \, x)^2}$$

$$b = \frac{1}{n}(\Sigma Y - A \, \Sigma \, x)$$

$$B = e^b = e^{.725} \approx 2.06 \quad \longleftarrow \quad \textbf{7.} \text{ Calculate } B = e^b$$
     (since $b = \ln B$)

The exponential curve is

$$y = 2.06e^{.187x} \quad \longleftarrow \quad \textbf{8.} \text{ The curve is } y = Be^{Ax}$$
     with the A and B found above

The population in 2000 is this evaluated at $x = 5$.

$$y = 2.06e^{.187(5)} \approx 5.25$$

or 5.25 billion people.   ■

EXERCISES

FITTING EXPONENTIAL CURVES

Use least squares to find the exponential curve $y = Be^{Ax}$ for the following points.
(Keep two decimal places in your calculations.)

19

x	y
1	2
2	4
3	7

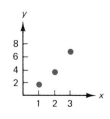

20

x	y
1	3
2	6
3	11

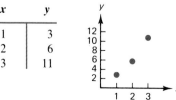

21

x	y
1	10
3	5
6	1

22

x	y
1	12
4	3
5	2

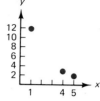

23

x	y
0	1
1	2
2	5
3	10

24

x	y
0	2
1	4
2	7
3	15

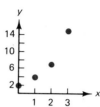

25

x	y
−1	20
0	18
1	15
3	4
5	1

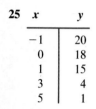

26

x	y
−2	20
0	12
2	9
4	6
5	5

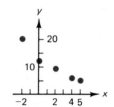

■ APPLIED EXERCISES

FITTING EXPONENTIAL CURVES (Keep two decimal places in your calculation.)

27 (*Political Science–Cost of a Congressional Victory*) The following table shows the average amount spent by the winner of a seat in the House of Representatives. Fit an exponential curve to these data and use it to predict the cost of a House seat in the year 1992.

	x	Cost (thousands of dollars)
1976	1	87
1980	2	179
1984	3	325

28 (*General–Drunk Driving*) The following table shows how a driver's blood-alcohol level (% by weight) affects the probability of being in a collision. A collision factor of 3 means that the probability of a collision is 3 times greater than normal. Fit an exponential curve to the data. Then use your curve to estimate the collision factor for a blood-alcohol level of 15.

Blood-Alcohol Level (%)	Collision Factor
0	1
6	1.1
8	3
10	6

29 (*Economics–Cost of a Stock Exchange Seat*) The following table shows the highest price paid for a seat on the New York Stock Exchange from 1979 to 1987. Fit an exponential curve to these data and use your answer to predict the price in the year 1999.

	x	Cost (thousands of dollars)
1977	1	100
1979	2	230
1981	3	300
1983	4	450
1985	5	500
1987	6	1100

30 (*General–Stamp Prices*) The following table shows the rise in the cost of a first-class postage stamp. Fit an exponential curve to these data and use your answer to predict the cost of a stamp in the year 2008.

	x	Postage (cents)
1958	1	4
1968	2	6
1978	3	15
1988	4	25

7.5 LAGRANGE MULTIPLIERS AND CONSTRAINED OPTIMIZATION

Introduction

In Section 7.3 we optimized functions of several variables. Some problems, however, involve maximizing or minimizing a function subject to a *constraint*. For example, a company might want to maximize production subject to the constraint of staying within its budget, or a beer distributor might want to design the least expensive aluminum can subject to the constraint that it hold exactly twelve ounces of beer. In this section we solve such "constrained optimization" problems by the method of Lagrange multipliers. The method was invented by the French mathematician Joseph Louis Lagrange (1736–1813).

Constraints

If a constraint can be written as an equation, such as

$$x^2 + y^2 = 100$$

then by moving all the terms to the left-hand side, it can be written with zero on the right-hand side:

$$\underbrace{x^2 + y^2 - 100}_{g(x, y)} = 0$$

In general, any equation can be written with all the terms moved to the left-hand side, taking the form

$$g(x, y) = 0$$

For the rest of this section we assume that all constraints have been written in the form $g(x, y) = 0$.

Practice Exercise 1

Write the constraint $2y^3 = 3x - 1$ in the form $g(x, y) = 0$. (*Hint*: Move everything to the left-hand side of the equation.)

(solution on page 470)

First Example of Lagrange Multipliers

We illustrate the method of Lagrange multipliers by an example. The method requires a new variable, and instead of the usual English letters a, b, c, ..., z, it is customary to use the Greek letter λ ("lambda"). Lambda is the Greek letter L (as in Lagrange).

Example 1 A cattle rancher wants to build a rectangular holding pen along an existing stone wall. If the side along the wall needs no fence, find the dimensions of the largest pen that can be enclosed using only 400 feet of fence.

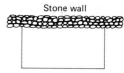

Solution We want the pen of largest *area*. There must be a largest pen, because either the length or the width being too small will make the area small, so some "intermediate" dimensions must maximize the area. Let

$$x = \text{width (perpendicular to the existing wall)}$$

$$y = \text{length (parallel to the existing wall)}$$

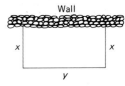

Two widths and one length must be made from the 400 feet of fence, so the constraint is

$$2x + y = 400$$

Therefore, the problem becomes:

maximize $A = xy$ *area is length times width*

subject to $2x + y - 400 = 0$ *constraint $2x + y = 400$ written with zero on the right*

We write a new function $F(x, y, \lambda)$, called the *Lagrange function,* which consists of the function to be maximized plus λ times the constraint function.

$$F(x, y, \lambda) = \underbrace{xy}_{\substack{\text{function to} \\ \text{be maximized}}} + \underbrace{\lambda(2x + y - 400)}_{\substack{\text{lambda times the} \\ \text{constraint function}}} = \underbrace{xy + \lambda 2x + \lambda y - \lambda 400}_{\text{multiplied out}}$$

The Lagrange function $F(x, y, \lambda)$ is a function of three variables, x, y, and the new variable λ. We maximize F in the usual way, setting its partials (with respect to x, y, and λ) equal to zero.

$$F_x = y + 2\lambda \qquad = 0$$

partial of $xy + \lambda 2x + \lambda y - \lambda 400$ with respect to x

$$F_y = x + \lambda \qquad = 0$$

partial of $xy + \lambda 2x + \lambda y - \lambda 400$ with respect to y

$$F_\lambda = \underbrace{2x + y - 400}_{} = 0$$

partial of $xy + \lambda 2x + \lambda y - \lambda 400$ with respect to λ

constraint $g = 0$

(Note that the last equation is just the constraint.) We solve the first two of these equations for λ.

$$\lambda = -\frac{1}{2}y$$

solving $y + 2\lambda = 0$ for λ

$$\lambda = -x$$

solving $x + \lambda = 0$ for λ

Then set these two expressions for λ equal to each other.

$$-\frac{1}{2}y = -x$$

equating $\lambda = -(1/2)y$ and $\lambda = -x$

$$y = 2x$$

multiplying by -2

We use this relationship $y = 2x$ to eliminate the y in the equation $2x + y - 400 = 0$ (which came from the third partial $F_\lambda = 0$).

$$2x + 2x - 400 = 0$$

$2x + y - 400 = 0$ with y replaced by 2x

$$4x = 400$$

simplifying

$$x = 100$$

dividing by 4

$$y = 200$$

from $y - 2x$ with $x = 100$

The maximum must occur for some dimensions x and y, so it must occur at this critical point (100, 200).

Answer: Width (perpendicular to existing wall) is 100 feet. Length (parallel to existing wall) is 200 feet. Notice that the constraint (using only 400 feet of fence) *is* satisfied.

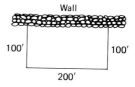

Method of Lagrange Multipliers

In general, the function to be maximized or minimized is called the *objective function,* because the "objective" of the whole procedure is to optimize it. (In Example 1 the objective function was the area function $A = xy$.) The variable λ is called the *Lagrange multiplier.* The entire method is as follows.

Lagrange Multipliers

To optimize $f(x, y)$

subject to $g(x, y) = 0$

1. Write $F(x, y, \lambda) = f(x, y) + \lambda g(x, y)$.

2. Set the partials of F equal to zero:

$$F_x = 0 \quad F_y = 0 \quad F_\lambda = 0$$

    and solve for the critical points.

3. The solution to the original problem (if it exists) will occur at one of these critical points.

objective function

constraint

objective function plus λ times the constraint function

Why Lagrange's Method Works

For the Lagrange function $F(x, y, \lambda) = f(x, y) + \lambda g(x, y)$, the partial with respect to λ is $F_\lambda = g(x, y)$, and setting this equal to zero gives $g(x, y) = 0$, which is just the constraint. Therefore, the constraint $g(x, y) = 0$ is satisfied *automatically* by setting the partials equal to zero. But if $g(x, y) = 0$, the Lagrange function

$$F(x, y, \lambda) = f(x, y) + \lambda \underbrace{g(x, y)}_{0}$$

simplifies to

$$F(x, y, \lambda) = f(x, y)$$

so that optimizing the Lagrange function F is equivalent to optimizing the original function f. This shows that optimizing the Lagrange function F is equivalent to optimizing the objective function f subject to the constraint $g = 0$.

Lagrange's Method Only Finds Critical Points

It is important to realize that Lagrange's method only finds *critical points*. It does *not* tell whether the function is maximized, minimized, or neither at the critical points. [The D-test, involving $D = f_{xx}f_{yy} - (f_{xy})^2$, is for *un*constrained optimization, and cannot be used with constraints.] In each problem we *assume* that the maximum or minimum (whichever is asked for) does exist, so it must occur at a critical point found by Lagrange multipliers.

Example 2

$$\text{Minimize} \quad f(x, y) = x^2 + y^2 \qquad \textit{objective function}$$

$$\text{subject to} \quad x + 2y = 10 \qquad \textit{constraint}$$

Solution We rewrite the constraint as $x + 2y - 10 = 0$, with zero on the right. The Lagrange function F is

$$F(x, y, \lambda) = \underbrace{x^2 + y^2}_{\substack{\text{objective} \\ \text{function}}} + \underbrace{\lambda(x + 2y - 10)}_{\substack{\text{constraint} \\ \text{function}}}$$

try keeping $\lambda(x + 2y - 10)$ in this form, multiplying it out "in your head" only when you differentiate

The partials:

$$F_x = 2x + \lambda \qquad = 0$$
$$F_y = 2y + 2\lambda \qquad = 0$$

solve these two first

$$F_\lambda = \underbrace{x + 2y - 10 = 0}_{\text{the constraint}}$$

$F_\lambda = 0$ always gives the constraint

Solving yields

$$\lambda = -2x \qquad \textit{solving } 2x + \lambda = 0 \textit{ for } \lambda$$
$$\lambda = -y \qquad \textit{solving } 2y + 2\lambda = 0 \textit{ for } \lambda$$
$$-2x = -y \qquad \textit{equating } \lambda\textit{s in } \lambda = -2x \textit{ and } \lambda = -y$$
$$y = 2x \qquad \textit{multiplying by } -1 \textit{ and reversing the order}$$

Therefore,

$$x + 2(2x) - 10 = 0 \qquad \textit{substituting } y = 2x \textit{ into } x + 2y - 10 = 0 \textit{ (from } F_\lambda = 0)$$
$$5x - 10 \qquad \textit{simplifying}$$
$$x = 2 \qquad \textit{solving}$$
$$y = 4 \qquad \textit{substituting } x = 2 \textit{ into } y = 2x$$

Answer: The minimum value of f is 20, which occurs at $x = 2$, $y = 4$.

20 is from $f = x^2 + y^2$ evaluated at $x = 2, y = 4$

∎

It is easy to check that $x = 2$ and $y = 4$ do satisfy the constraint $x + 2y = 10$.

Practice Exercise 2

$$\text{Minimize} \quad f(x, y) = x^2 + y^2$$

$$\text{subject to} \quad 2x + y = 25 \qquad \textit{(solution on page 470)}$$

Lagrange Method Versus Substitution

We could have solved Example 2 *without* Lagrange multipliers, solving the constraint $2x + y = 10$ for y and substituting the result into the objective function, as we did in Chapter 3. However, solving a more complicated constraint, such as $x^3 + y^3 = xy$, can be very difficult (or impossible), and Lagrange's method is generally easier.

How to Solve the Partial Equations

Solving the partial equations $F_x = 0$, $F_y = 0$, and $F_\lambda = 0$ can sometimes be difficult. The following strategy (used in each example in this section) may be helpful.

> First solve $F_x = 0$ and $F_y = 0$ together to eliminate λ.
> Then solve the resulting equation and $F_\lambda = 0$ for x and y.

Stated in greater detail:

1. Solve the equation $F_x = 0$ for λ. Solve the equation $F_y = 0$ for λ.

2. Set the resulting expressions for λ equal to each other. This gives an equation in x and y. Solve this equation for x or y.

3. Use the result of step 2 to eliminate one of the variables in the partial equation $F_\lambda = 0$ and solve for the remaining variable.

4. Calculate the value of the other variable using the formula from step 2.

Applications of Lagrange Multipliers

Example 3 A beer company wants to design an aluminum can requiring the least amount of aluminum, but that contains exactly 12 fluid ounces (21.3 cubic inches). Find the radius and height of the can.

Solution Minimizing the amount of aluminum means minimizing the surface area (top + bottom + sides) of the cylindrical can. Letting r and h stand for the radius and height (in inches) of the can, the area is

$$A = 2\pi r^2 + 2\pi rh$$

$$\underbrace{}_{\substack{\text{top and} \\ \text{bottom}}} \quad \underbrace{}_{\substack{\text{side} \\ \text{area}}}$$

The volume is

$$V = \pi r^2 h$$

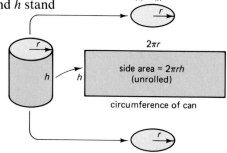

Therefore, the problem becomes:

$$\text{minimize} \qquad A = 2\pi r^2 + 2\pi rh$$ *area = top + bottom + side*

$$\text{subject to} \qquad \pi r^2 h = 21.3$$ *volume must equal 21.3*

The Lagrange function is

$$F = 2\pi r^2 + 2\pi rh + \lambda(\pi r^2 h - 21.3)$$ *objective function A plus λ times the constraint*

$$F_r = 4\pi r + 2\pi h + \lambda 2\pi rh = 0$$

$$F_h = 2\pi r + \lambda \pi r^2 \qquad\qquad = 0$$ *partials set equal to zero*

$$F_\lambda = \pi r^2 h - 21.3 \qquad = 0$$

Solving, we have

$$\lambda = -\frac{4\pi r + 2\pi h}{4\pi rh} = -\frac{2r + h}{2rh}$$ *4πr + 2πh + λ2πrh = 0 solved for λ*

└── simplified

$$\lambda = -\frac{2\pi r}{\pi r^2} = -\frac{2}{r}$$ *2πr + λπr² = 0 solved for λ*

└── simplified

Equating λs yields

$$\frac{2r + h}{rh} = \frac{2}{r}$$ *equating the two expressions for λ and multiplying each side by −1*

$$2r + h = 2h$$ *multiplying each side by rh*

$$2r = h$$ *subtracting h from each side*

Therefore,

$$\pi r^2(2r) = 21.3$$ *πr²h − 21.3 = 0 (the third partial equation) with h = 2r*

$$2\pi r^3 = 21.3$$ *simplifying*

$$r^3 = \frac{21.3}{2\pi} \approx 3.39$$ *dividing by 2π (using a calculator)*

$$r \approx \sqrt[3]{3.39} \approx 1.5$$ *taking cube roots*

$$h = 3$$ *from h = 2r with r = 1.5*

Answer: The most economical 12-fluid-ounce can has radius $r = 1.5$ inches and height $h = 3$ inches.

Notice that the height (3 inches) is twice the radius (1.5

■

inches), so the height equals the diameter. This shows that the most efficient can (least area for given volume) has "squarish" shape.

It is interesting to consider why so few cans are shaped like this. The most common 12-ounce can for beer or soft drinks is about twice as tall as it is across, requiring about 67% more aluminum than the most efficient can. This results in an enormous waste for the millions of cans manufactured each year. It seems that beverage companies prefer taller cans because they have more area for advertising, and they are easier to hold. Some products, however, are sold in "efficient" cans, with the height equal to the diameter.

Practice Exercise 3

Which of the cans pictured below are "efficiently" proportioned?

(solution on page 470)

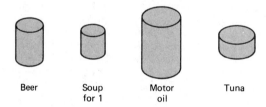

| Beer | Soup for 1 | Motor oil | Tuna |

Application to Economics

With Lagrange multipliers we can maximize a company's output subject to a budget constraint.

Example 4 A company's output given by the Cobb–Douglas production function $P = 600L^{2/3}K^{1/3}$, where L and K are the number of units of labor and capital. Each unit of labor costs the company \$40, and each unit of capital costs \$100. If the company has a total of \$3000 for labor and capital, how much of each should it use to maximize production?

Solution The budget constraint is

$$40L + 100K = 3000$$

$$\underbrace{}_{\substack{\text{cost of} \\ \text{labor}}} \quad \underbrace{}_{\substack{\text{cost of} \\ \text{capital}}} \quad \underbrace{}_{\substack{\text{total} \\ \text{budget}}}$$

unit cost times the number of units for labor and capital

Therefore, the problem becomes:

maximize $\qquad P = 600L^{2/3}K^{1/3}$

subject to $\qquad 40L + 100K - 3000 = 0$ $\qquad$ *constraint with zero on the right*

$F(L, K, \lambda) = 600L^{2/3}K^{1/3} + \lambda(40L + 100K - 3000)$ $\qquad$ *Lagrange function*

$$F_L = 400L^{-1/3}K^{1/3} + 40\lambda = 0$$

$$F_K = 200L^{2/3}K^{-2/3} + 100\lambda = 0$$

$$F_\lambda = 40L + 100K - 3000 = 0$$

partial equations

$$\lambda = -10L^{-1/3}K^{1/3}$$

solving the first for λ

$$\lambda = -2L^{2/3}K^{-2/3}$$

solving the second for λ

$$-10L^{-1/3}K^{1/3} = -2L^{2/3}K^{-2/3}$$

equating the two expressions for λ

$$5L^{-1/3}K^{1/3} = L^{2/3}K^{-2/3}$$

dividing each side by −2

$$5K = L$$

multiplying by $L^{1/3}K^{2/3}$ on each side

$$40(5K) + 100K - 3000 = 0$$

substituting L = 5K into
40L + 100K − 3000 = 0

$$300K = 3000$$

simplifying

$$K = 10$$

solving for K

$$L = 50$$

from L = 5K with K = 10

Answer: The company should use 50 units of labor and 10 units of capital.   ■

Meaning of the Lagrange Multiplier λ

The variable λ, besides being a technical device for optimizing functions, also has a useful interpretation. In Example 4 we maximized production subject to a budget constraint of $3000, and the Lagrange multiplier λ gives the *rate at which the production increases for each additional dollar* (in the jargon of economics, the "marginal productivity of money"). We can calculate the value of λ from either of the expressions for λ.

$$\lambda = \underbrace{-10L^{-1/3}K^{1/3}}_{\text{from Example 4}} = \underbrace{-10(50)^{-1/3}(10)^{1/3}}_{\substack{\text{substituting} \\ L = 50, K = 10}} \underbrace{\approx -5.8}_{\substack{\text{using a} \\ \text{calculator}}}$$

This number (ignoring the negative sign) means that production increases by about 5.8 units for each additional dollar. Therefore, an additional $100 would result in about 580 extra units of production.

In general, if the units of the objective function are called *objective units*, and if the units of the constraint function are called *constraint units*, then λ has the following interpretation.

$$|\lambda| = \binom{\text{number of additional objective units}}{\text{for each additional constraint unit}}$$

Practice Exercise 4

In Example 1 we maximized area of a pen subject to the constraint of having only 400 feet of fence.

   (a)   Interpret the meaning of λ in Example 1. (*Hint:* The objective function is area, and the constraint units are feet of fence.)

   (b)   Calculate the absolute value of λ in Example 1. (*Hint:* Use either expression for λ on page 461.)

   (c)   Use this number to approximate the additional area that could be enclosed by an additional 5 feet of fence.   *(solutions on page 471)*

Geometry of Constrained Optimization

A constrained maximization problem may be visualized by thinking of the surface as a landscape. A constraint $g(x, y) = 0$ represents a curve in the *x-y* plane, and highest point *subject to the constraint* means the highest point on a path lying directly above the curve. On the landscape shown below, the highest point is indicated by the letter M, and the highest point on the path is indicated by CM (for "constrained maximum"). In general, a constrained maximum will not be as high as an unconstrained maximum, since the highest point on a particular path may not be as high as the highest point on the entire surface.

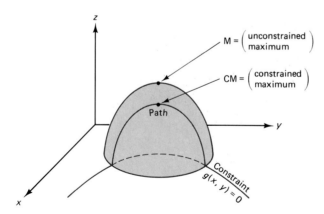

Finding Both Extreme Values

If there are several critical points, we evaluate the objective function at each. The maximum and minimum values (which we assume to exist) are the largest and smallest of the resulting values of the function. In this way we can find both the maximum and minimum values in a constrained optimization problem.

Example 5 Maximize *and* minimize $f(x, y) = 4xy$

subject to the constraint $x^2 + y^2 = 50$

Solution

$F(x, y, \lambda) = 4xy + \lambda(x^2 + y^2 - 50)$	*Lagrange function*
$F_x = 4y + \lambda 2x \quad = 0$	
$F_y = 4x + \lambda 2y \quad = 0$	*partials set equal to zero*
$F_\lambda = x^2 + y^2 - 50 = 0$	
$\lambda = -\dfrac{4y}{2x} = -\dfrac{2y}{x}$	*solving* $4y + \lambda 2x = 0$ *for* λ *(and simplifying)*
$\lambda = -\dfrac{4x}{2y} = -\dfrac{2x}{y}$	*solving* $4x + \lambda 2y = 0$ *for* λ *(and simplifying)*
$\dfrac{2y}{x} = \dfrac{2x}{y}$	*equating the two λs (multiplying by -1)*
$2y^2 = 2x^2$	*multiplying both sides by xy (or "cross-multiplying")*
$y^2 = x^2$	*dividing by 2*
$y = \pm x$	*taking square roots ($+$ and $-$)*
$x^2 + x^2 - 50 = 0$	$x^2 + y^2 - 50 = 0$ *(the third partial equation) with $y = \pm x$*
$2x^2 - 50 = 0$	*simplifying*
$\underbrace{2(x^2 - 25) = 0}_{(x + 5)(x - 5)}$	*factoring*
$x = \pm 5$	*solving*
$y = \pm 5$	*from $y = \pm x$*

These give *four* critical points (all possible combinations of $x = \pm 5$ and $y = \pm 5$) and we evaluate the objective function $f(x, y) = 4xy$ at each.

Critical Points	$f(x, y) = 4xy$	
(5, 5)	100	*100 is the highest value of $f = 4xy$,*
(−5, −5)	100	*occurring at (5, 5) and at (−5, −5)*
(5, −5)	−100	*−100 is the lowest value of $f = 4xy$,*
(−5, 5)	−100	*occurring at (−5, 5) and at (5, −5)*

Answer:
Maximum value of f is 100, occurring at $x = 5$, $y = 5$ and at $x = -5$, $y = -5$.
Minimum value of f is -100, occurring at $x = -5$, $y = 5$ and at $x = 5$, $y = -5$.

■

Constrained Optimization of Functions of Three or More Variables

To optimize a function f of any number of variables subject to a constraint $g = 0$, we proceed just as before, writing the Lagrange function $F = f + \lambda g$ and setting the partial with respect to each variable equal to zero. For example, to maximize $f(x, y, z)$ subject to $g(x, y, z) = 0$, we write

$$F(x, y, z, \lambda) = f(x, y, z) + \lambda g(x, y, z)$$ *Lagrange function*

and solve the partial equations

$$F_x = 0$$

$$F_y = 0$$

$$F_z = 0$$

$$F_\lambda = 0$$

SOLUTIONS TO PRACTICE EXERCISES

1. $2y^3 - 3x + 1 = 0$

2.
$$F(x, y, \lambda) = x^2 + y^2 + \lambda(2x + y - 25)$$

$$F_x = 2x + 2\lambda \quad = 0$$

$$F_y = 2y + \lambda \quad\;\; = 0$$

$$F_\lambda = 2x + y - 25 = 0$$

$$\lambda = -x \qquad\qquad \text{(from } 2x + 2\lambda = 0)$$

$$\lambda = -2y \qquad\qquad \text{(from } 2y + \lambda = 0)$$

$$x = 2y \qquad\qquad \text{(equating } \lambda\text{s, multiplying by } -1)$$

$$2(2y) + y - 25 = 0 \qquad (2x + y - 25 = 0 \text{ with } x = 2y)$$

$$5y = 25$$

$$y = 5$$

$$x = 10 \qquad\qquad \text{(from } x = 2y)$$

$$f(10, 5) = 10^2 + 5^2 = 125$$

The minimum value of f is 125, which occurs at $x = 10$, $y = 5$.

3. Just the "soup for one" can

4. **(a)** The approximate additional area for each additional foot of fence

   (b) $|\lambda| = 100$ (using $\lambda = -x$ with $x = 100$)

   (c) Approximately 500 more square feet of area (from $5 \cdot 100$)

▨ EXERCISES 7.5

Use Lagrange multipliers to maximize each function
f(x, y) subject to the constraint. (The maximum values
do exist.)

1 $f(x, y) = 3xy$, $\quad x + 3y = 12$

3 $f(x, y) = 6xy$, $\quad 2x + 3y = 24$

5 $f(x, y) = xy - 2x^2 - y^2$, $\quad x + y = 8$

7 $f(x, y) = x^2 - y^2 + 3$, $\quad 2x + y = 3$

9 $f(x, y) = \ln(xy)$, $\quad x + y = 2e$

2 $f(x, y) = 2xy$, $\quad 2x + y = 20$

4 $f(x, y) = 3xy$, $\quad 3x + 2y = 60$

6 $f(x, y) = 12xy - 3y^2 - x^2$, $\quad x + y = 16$

8 $f(x, y) = y^2 - x^2 - 5$, $\quad x + 2y = 9$

10 $f(x, y) = e^{xy}$, $\quad x + 2y = 8$

Use Lagrange multipliers to minimize each function
f(x, y) subject to the constraint. (The minimum values do
exist.)

11 $f(x, y) = x^2 + y^2$, $\quad 2x + y = 15$

13 $f(x, y) = xy$, $\quad y = x + 8$

15 $f(x, y) = x^2 + y^2$, $\quad 2x + 3y = 26$

17 $f(x, y) = \ln(x^2 + y^2)$, $\quad 2x + y = 25$

19 $f(x, y) = e^{x^2 + y^2}$, $\quad x + 2y = 10$

12 $f(x, y) = x^2 + y^2$, $\quad x + 2y = 30$

14 $f(x, y) = xy$, $\quad y = x + 6$

16 $f(x, y) = 5x^2 + 6y^2 - xy$, $\quad x + 2y = 24$

18 $f(x, y) = \sqrt{x^2 + y^2 + 5}$, $\quad 2x + y = 10$

20 $f(x, y) = 2x + y$, $\quad 2 \ln x + \ln y = 12$

*Use Lagrange multipliers to maximize **and** minimize each*
function subject to the constraint. (The maximum and
minimum values do exist.)

21 $f(x, y) = 2xy$, $\quad x^2 + y^2 = 8$

23 $f(x, y) = x + 2y$, $\quad 2x^2 + y^2 = 72$

22 $f(x, y) = 2xy$, $\quad x^2 + y^2 = 18$

24 $f(x, y) = 12x + 30y$, $\quad x^2 + 5y^2 = 81$

▨ APPLIED EXERCISES

Solve each using Lagrange multipliers. (The stated ex-
treme values do exist.)

25 *(General–Fences)* A parking lot, divided into two
equal parts, is to be constructed against a building, as
shown in the diagram. Only 6000 feet of fence are to
be used, and the side along the building needs no
fence.   [Continued on next page.]

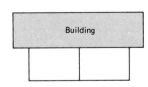

(a) What are the dimensions of the largest area that can be so enclosed?

(b) Evaluate and give an interpretation for λ.

26 (*General–Fences*) Three adjacent rectangular lots are to be fenced in, as shown in the diagram, using 12,000 feet of fence. What is the largest total area that can be so enclosed?

27–28 (*General–Container Design*) A cylindrical tank without a top is to be constructed with the least amount of materials (bottom plus side area). Find the dimensions if the volume is to be

27 160 cubic feet

28 120 cubic feet

29 (*General–Postal Regulations*) The U.S. Postal Service will accept a package if its length plus its girth is not more than 84 inches. Find the dimensions and volume of the largest package with a square end that can be mailed.

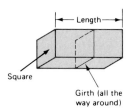

30 (*General–Postal Regulations*) Solve Exercise 29, but now for a package with a round end, so that the package is now a cylinder rather than a rectangular solid. Compare the volume with that of Exercise 29.

31 (*Business–Maximum Production*) A company's output is given by the Cobb–Douglas production function $P = 200L^{3/4}K^{1/4}$, where L and K are the number of units of labor and capital. Each unit of labor costs \$50 and each unit of capital costs \$100, and \$8000 is available to pay for labor and capital.

(a) How many units of each should be used to maximize production?

(b) Evaluate and give an interpretation for λ.

32 (*Business–Production Possibilities*) A company manufactures two products, in quantities x and y. Due to limited materials and capital, the quantities produced must satisfy the equation $2x^2 + 5y^2 = 32,500$. (This curve is called a *production possibilities curve*.) If the company's profit function is $P = 4x + 5y$ dollars, how many of each product should be made to maximize profit? Also find the maximum profit.

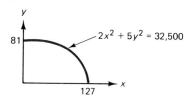

33 (*General–Package Design*) A metal box with a square base is to have a volume of 45 cubic inches. If the top and bottom cost 50 cents per square inch and the sides cost 30 cents per square inch, find the dimensions that minimize the cost. [*Hint:* The cost of the box is the area of each part (top, bottom, and sides) times the cost per square inch for that part. Minimize this subject to the volume constraint.]

34 (*General–Building Design*) A one-story building is to have 8000 square feet of floor space. The front of the building is to be made of brick, and costs \$120 per linear foot, and the back and sides are to be made of cinderblock, and cost only \$80 per linear foot. [Continued on next page.]

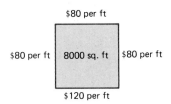

$80 per ft

$80 per ft | 8000 sq. ft | $80 per ft

$120 per ft

(a) Find the length and width that minimize the cost of the building. (*Hint*: The cost of the building is the length of the front, back, and sides, each times the cost per foot for that part. Minimize this subject to the area constraint.)

(b) Evaluate and give an interpretation for λ.

FUNCTIONS OF THREE VARIABLES

(The stated extreme values do exist.)

35 Minimize $f(x, y, z) = x^2 + y^2 + z^2$
subject to $2x + y - z = 12$

36 Minimize $f(x, y, z) = x^2 + y^2 + z^2$
subject to $x - y + 2z = 6$

37 Maximize $f(x, y, z) = x + y + z$
subject to $x^2 + y^2 + z^2 = 12$

38 Maximize $f(x, y, z) = xyz$
subject to $x^2 + y^2 + z^2 = 12$

39 (*General–Building Design*) The one-story storage building shown below is to have a volume of 250,000 cubic feet. The roof costs $32 per square foot, the walls $10 per square foot, and the floor $8 per square foot. Find the dimensions that minimize the cost of the building.

40 (*General–Container Design*) An open-top box with two parallel partitions, as in the diagram, is to have volume 64 cubic inches. Find the dimensions that require the least amount of materials.

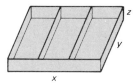

7.6 MULTIPLE INTEGRALS

Introduction

This section completes our study of functions of several variables by discussing integration. Integration is a summation process, but now we sum over *several* variables. We will use multiple integrals for calculating total accumulations and average values of functions. Multiple integrals also give volumes under surfaces, just as single integrals give areas under curves.

For simplicity, we shall restrict our attention to functions that are continuous throughout their domains (the surfaces have no holes or breaks). Most of the functions used in applications satisfy this restriction.

Review of the Integration Formulas

Multiple integration uses the same formulas that we derived in Chapter 5.

$$\int x^n \, dx = \frac{1}{n+1} x^{n+1} + C \qquad (n \neq -1)$$

$$\int e^{ax} \, dx = \frac{1}{a} e^{ax} + C \qquad (a \neq 0)$$

$$\int \frac{1}{x} \, dx = \ln |x| + C$$

Preliminary Examples

In Section 7.2 we calculated partial derivatives by differentiating with respect to one variable at a time, holding the other variable constant. Integrating a function $f(x, y)$ involves the same idea: integrating with respect to one variable at a time, treating the other variable like a constant. We begin by calculating *single* integrals of functions $f(x, y)$.

Example 1 Evaluate $\int_0^1 x^3 y^4 \, dx$.

Solution

$$\int_0^1 x^3 y^4 \, dx \quad = \quad \frac{1}{4} x^4 y^4 \Big|_{x=0}^{x=1} \quad = \quad \frac{1}{4} y^4 - 0 \quad = \quad \frac{1}{4} y^4$$

integrate with respect to x, holding y constant   integrated — held constant $\frac{1}{4}x^4y^4$ evaluated at $x = 1$ $\frac{1}{4}x^4y^4$ evaluated at $x = 0$ simplified ∎

Example 2 Evaluate $\int_{-2}^3 x^4 e^{-y} \, dy$.

Solution

$$\int_{-2}^3 x^4 e^{-y} \, dy \quad = \quad [x^4(-1)e^{-y}] \Big|_{y=-2}^{y=3} \quad = \quad -x^4 e^{-3} - (-x^4 e^2) \quad = \quad -x^4 e^{-3} + x^4 e^2$$

integrate with respect to y holding x constant   held constant   integrated   evaluated at $y = 3$ and evaluated at $y = -2$ simplified ∎

Example 3 Evaluate $\int_0^2 (3x^2 + 6xy^2) \, dx$.

Solution

$$\int_0^2 (3x^2 + 6xy^2)\, dx = \left(x^3 + 6\cdot\frac{1}{2}\, x^2 y^2\right)\Bigg|_{x=0}^{x=2} = 8 + 3\cdot 2^2 y^2 - (0) = 8 + 12y^2$$

integrate with respect to x, holding y constant   integrated   held constant   evaluated at $x = 2$ and at $x = 0$ simplified

■

Practice Exercise 1

Evaluate $\displaystyle\int_1^2 9x^2 y^3 \, dx.$

(solution on page 482)

Iterated Integrals

If we integrate a function $f(x, y)$ with respect to one variable, then integrate the result with respect to the other variable, we obtain an *iterated integral* ("iterated" means "repeated").

Example 4   Evaluate the iterated integral $\displaystyle\int_0^1 \int_0^2 (3x^2 + 6xy^2)\, dx \, dy.$

Solution   The two separate integrations will be clearer if we write the integral with parentheses.

$$\int_0^1 \left(\int_0^2 (3x^2 + 6xy^2)\, dx\right) dy$$

an inner x-integral and an outer y-integral

First we solve the inner x-integral.

$$\int_0^2 (3x^2 + 6xy^2)\, dx = \left(x^3 + 6\cdot\frac{1}{2}\, x^2 y^2\right)\Bigg|_{x=0}^{x=2} = 2^3 + 3\cdot 2^2 y^2 - (0) = 8 + 12y^2$$

integrate with respect to x, holding y constant   integrated   held constant   evaluated at $x = 2$ and at $x = 0$ simplified

We now apply the outer y-integral to this result.

$$\int_0^1 (8 + 12y^2)\, dy = (8y + 4y^3)\Bigg|_{y=0}^{y=1} = 8 + 4 - (0) = 12$$

result of solving the inner integral   integrate with respect to y $12\cdot\frac{1}{3}$ evaluated at $y=1$ and at $y=0$ final answer

Therefore, the iterated integral equals 12.

$$\int_0^1 \int_0^2 (3x^2 + 6xy^2)\, dx\, dy = 12$$

■

Always solve an iterated integral "from the inside out."

$$\int_0^1 \left(\int_0^2 (3x^2 + 6xy^2)\, dx \right) dy$$

↑   ↑   ↑   └───↑

limits   limits first integrate ─┘ └─then with
for y for x with respect to x respect to y

Reversing the Order of Integration

In Example 4 the inner integral was an x-integral and the outer one was a y-integral. The next example will show that we obtain the same answer if we *reverse* the order of integration, being careful also to reverse the x and y limits of integration. That is,

$$\int_0^1 \int_0^2 (3x^2 + 6xy^2)\, dx\, dy$$

is equal to ⤬   ⤬

$$\int_0^2 \int_0^1 (3x^2 + 6xy^2)\, dy\, dx$$

Example 5   Evaluate $\int_0^2 \int_0^1 (3x^2 + 6xy^2)\, dy\, dx$.

same as Example 4, but with the order of integration reversed

Solution   First we solve the inner y-integral.

$$\int_0^1 (3x^2 + 6xy^2)\, dy = (3x^2 y + 2xy^3) \Big|_{y=0}^{y=1} = 3x^2 + 2x - (0) = 3x^2 + 2x$$

integrate with x held └─$\tfrac{1}{3}\cdot 6$ evaluated and at
respect to y constant   at $y = 1$ $y = 0$

Then we apply the outer x-integral to this expression.

$$\int_0^2 (3x^2 + 2x)\, dx = (x^3 + x^2) \Big|_{x=0}^{x=2} = 8 + 4 - (0) = 12$$

from the inner   evaluated and at final
integration   at $x = 2$ $x = 0$ answer

This is the same answer that we obtained before, showing that either order of integration gives the same answer.

$$\int_0^1 \int_0^2 (3x^2 + 6xy^2) \; dx \; dy = \int_0^2 \int_0^1 (3x^2 + 6xy^2) \; dy \; dx$$

reversing the x and y reversing the order
limits of integration   of integration   ■

In general, for *any* continuous function $f(x, y)$ the order of integration can be reversed:

---

is equal to

$$\int_c^d \int_a^b f(x, y) \; dx \; dy$$

$$\int_a^b \int_c^d f(x, y) \; dy \; dx$$

Double Integrals Over Rectangular Regions

An x-integral from a to b and a y-integral from c to d is really an integral over the *rectangle* R shown on the right. The integral of a function over such a rectangle is called a *double integral,* and is written

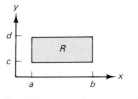

$R = \{(x, y) \mid a \le x \le b, c \le y \le d\}$

$$\iint_R f(x, y) \; dx \; dy \quad \text{with} \quad R = \{(x, y) | a \le x \le b, c \le y \le d\}$$

read: the double integral over R set of all points (x, y) such that $a \le x \le b$ and $c \le y \le d$

The rectangle R defined on the right means that x runs from a to b and y from c to d, and the integration can be done *in either order.*

The double integral

$$\iint_R f(x, y) \; dx \; dy \quad \text{with} \quad R = \{(x, y) | a \le x \le b, c \le y \le d\}$$

is evaluated by solving *either* of the iterated integrals

$$\int_c^d \int_a^b f(x, y) \; dx \; dy \quad \text{or} \quad \int_a^b \int_c^d f(x, y) \; dy \; dx$$

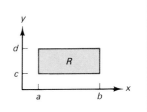

Note the distinction between iterated integrals and double integrals.

An *iterated* integral is written

$$\int_c^d \int_a^b f(x, y) \, dx \, dy \quad \text{or} \quad \int_a^b \int_c^d f(x, y) \, dy \, dx$$

A *double* integral is written

$$\iint_R f(x, y) \, dx \, dy$$

and R (which must be specified) gives the upper and lower limits of integration for x and y. The double integral is evaluated by solving *either* of the two iterated integrals—both will give the same answer. For example, the *double* integral

$$\iint_R (3x^2 + 6xy^2) \, dx \, dy \quad \text{with} \quad R = \{(x, y)|0 \le x \le 2, 0 \le y \le 1\}$$

is evaluated by solving *either* of the two iterated integrals

$$\int_0^1 \int_0^2 (3x^2 + 6xy^2) \, dx \, dy \quad \text{or} \quad \int_0^2 \int_0^1 (3x^2 + 6xy^2) \, dy \, dx$$

in which the limits of integration are taken from the x- and y-values in

$$R = \{(x, y)|0 \le x \le 2, 0 \le y \le 1\}$$

Both iterated integrals give the same answer, 12, as we saw in Examples 4 and 5.

Example 6 Evaluate $\iint_R y^2 e^{-x} \, dx \, dy$ with $R = \{(x, y)|0 \le x \le 2, -1 \le y \le 1\}$.

Solution This double integral is evaluated by solving either of the iterated integrals

$$\int_{-1}^1 \int_0^2 y^2 e^{-x} \, dx \, dy \quad \text{or} \quad \int_0^2 \int_{-1}^1 y^2 e^{-x} \, dy \, dx$$

We will solve the second one, beginning with the inner integral.

$$\int_{-1}^1 y^2 e^{-x} \, dy = \left(\frac{1}{3} y^3 e^{-x}\right)\Big|_{y=-1}^{y=1} = \frac{1}{3} e^{-x} - \left[\frac{1}{3}(-1)e^{-x}\right] = \frac{1}{3} e^{-x} + \frac{1}{3} e^{-x} = \frac{2}{3} e^{-x}$$

integrated→ ↑held constant   evaluated at $y = 1$ and at $y = -1$

Then we integrate this with respect to x.

$$\int_0^2 \frac{2}{3} e^{-x} \, dx = -\frac{2}{3} e^{-x} \Big|_{x=0}^{x=2} = -\frac{2}{3} e^{-2} - \left(-\frac{2}{3} e^0\right) = -\frac{2}{3} e^{-2} + \frac{2}{3}$$

This answer is a *number,* which could be evaluated on a calculator. ■

Practice Exercise 2

Show that evaluating this same double integral by the iterated integral in the *other* order gives the same answer. That is, solve the iterated integral

$$\int_{-1}^{1}\int_{0}^{2} y^2 e^{-x}\, dx\, dy$$

(solution on page 482)

Double Integrals Give Volumes

Just as integrals in one variable give *areas* under curves, integrals in two variables give *volumes* under surfaces.

For a nonnegative continuous function $f(x, y)$, the volume under the surface $f(x, y)$ and above a rectangular region R in the x-y plane is

$$\binom{\text{volume under}}{f(x, y) \text{ above } R} = \iint_R f(x, y)\, dx\, dy$$

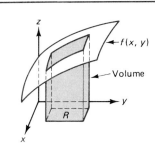

(If the surface lies *below* the x-y plane, this integral gives the *negative* of the volume). A proof of this result can be found in an advanced calculus book.

Example 7   The graph of $f(x, y) = 16 - 3x^2 - 3y^2$ is shown below. Find the volume under this surface and above the rectangle

$$R = \{(x, y)\,|\,{-2} \le x \le 2,\ -1 \le y \le 1\}$$

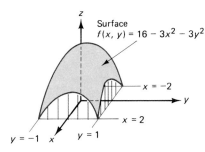

Solution   The volume is the integral of the function over the rectangle R.

$$\iint_R f(x, y)\, dx\, dy = \int_{-1}^{1}\int_{-2}^{2} (16 - 3x^2 - 3y^2)\, dx\, dy$$

limits of integration are taken from the rectangle R

We solve the inner integral first.

$$\int_{-2}^{2} (16 - 3x^2 - 3y^2)\, dx = (16x - x^3 - 3y^2x) \,\bigg|_{x=-2}^{x=2}$$

$$= \underbrace{32 - 8 - 3y^2 \cdot 2}_{\text{evaluated at } x = 2} - \underbrace{(-32 + 8 + 3y^2 \cdot 2)}_{\text{evaluated at } x = -2} = 48 - 12y^2$$

We integrate this with respect to y.

$$\int_{-1}^{1} (48 - 12y^2)\, dy = (48y - 4y^3) \,\bigg|_{-1}^{1} = 48 - 4 - (-48 + 4) = 88$$

↳ from integrating $12y^2$

Therefore, the volume under the surface is 88 cubic units. ▪

Application: Average Value

Recall that the average value of a function $f(x)$ over an interval is found by integrating over the interval and dividing by the length of the interval. For similar reasons, the average value of a function $f(x, y)$ of *two* variables over a region is defined as the *double* integral over the region divided by the *area* of the region.

$$\left(\begin{array}{c}\text{The average value}\\ \text{of } f(x, y) \text{ over } R\end{array}\right) = \frac{1}{\text{area of } R} \iint\limits_{R} f(x, y)\, dx\, dy$$

For a rectangular region R, the area of R is just length times width.

Example 8   The temperature x miles east and y miles north of a weather station is $T(x, y) = 60 + 2x - 4y$ degrees. Find the average temperature over the rectangular region R shown in the diagram.

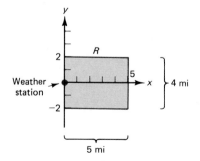

Solution   The area of the region R is $4 \cdot 5 = 20$ square miles (length times width), so to find the average temperature we integrate over the region and divide by 20.

$$\frac{1}{20} \int_{-2}^{2} \int_{0}^{5} (60 + 2x - 4y)\, dx\, dy$$

We solve the inner integral first.

$$\int_{0}^{5} (60 + 2x - 4y)\, dx = (60x + x^2 - 4yx) \,\bigg|_{x=0}^{x=5} = 300 + 25 - 20y - (0) = 325 - 20y$$

$\underbrace{\qquad\qquad\qquad\qquad}_{\text{evaluating at } x = 5}$

Then we integrate this from $y = -2$ to $y = 2$.

$$\int_{-2}^{2} (325 - 20y)\, dy = (325y - 10y^2) \Big|_{y=-2}^{y=2} = 650 - 40 - [-650 - 40]$$

— from integrating $20y$

$$= 610 + 690 = 1300$$

Finally, for the average we divide by 20 (the area of the region).

$$\frac{1300}{20} = 65$$

Therefore, the average temperature over the region is 65 degrees.   ■

Integrating Over More General Regions

We may also integrate over regions R that are bounded by *curves*.

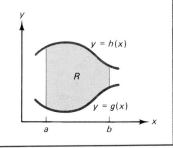

For a region R in the x-y plane bounded by an upper curve $y = h(x)$ and a lower curve $y = g(x)$ from $x = a$ to $x = b$, as shown on the right,

$$\iint_{R} f(x, y)\, dx\, dy = \int_{a}^{b} \int_{g(x)}^{h(x)} f(x, y)\, dy\, dx$$

If $f(x, y)$ is nonnegative, this integral gives the volume under the surface $f(x, y)$ above the region R.

(More precisely, f, g, and h must be continuous. A proof may be found in an advanced calculus book.)

Example 9   Find the volume under the surface $f(x, y) = 12xy$ and above the region R shown on the right.

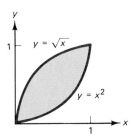

Solution   The region R is bounded by the upper curve $h(x) = \sqrt{x}$ and the lower curve $g(x) = x^2$ from $x = 0$ to $x = 1$. Therefore, the statement in the box above says that the volume is given by the following integral.

$$\int_{0}^{1} \int_{x^2}^{\sqrt{x}} 12xy\, dy\, dx$$

We solve the inner integral first.

$$\int_{x^2}^{\sqrt{x}} 12xy\, dy = 12x\, \frac{1}{2}\, y^2 \Big|_{y=x^2}^{y=\sqrt{x}} = 6xy^2 \Big|_{y=x^2}^{y=\sqrt{x}} = 6x(\sqrt{x})^2 - 6x(x^2)^2 = 6x^2 - 6x^5$$

x held constant ⌐ integrated with respect to y evaluating at $y = \sqrt{x}$ evaluating at $y = x^2$ simplified

Now we integrate the result with respect to x.

$$\int_0^1 (6x^2 - 6x^5)\, dx = \left(6{\cdot}\frac{1}{3}\, x^3 - x^6\right)\Big|_0^1 = (2x^3 - x^6)\Big|_0^1 = 2 - 1 - (0) = 1$$

Therefore, the volume under the surface and above the region R is 1 cubic unit. ■

Summary

We have defined double integrals over rectangular regions R, and over regions bounded by curves. Double integrals are evaluated by solving iterated integrals, integrating with respect to one variable at a time while holding the other variable constant.

As for the *meaning* of multiple integrals, the double integral

$$\iint_R f(x, y)\, dx\, dy$$

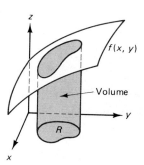

of a nonnegative function gives the *volume* under the surface $f(x, y)$ above the region R. The double integral divided by the area of R gives the *average value* of $f(x, y)$ over the region R. The exercises give illustrations of the integral as a *sum*.

The exercises discuss the "triple" integral of a function $f(x, y, z)$ of three variables. The idea is exactly the same. Just integrate successively with respect to one variable at a time, holding all else constant.

SOLUTIONS TO PRACTICE EXERCISES

1. $\displaystyle\int_1^2 9x^2 y^3\, dx = 9{\cdot}\frac{1}{3}\, x^3 y^3 \Big|_{x=1}^{x=2} = 3{\cdot}2^3 y^3 - 3y^3 = 24y^3 - 3y^3 = 21y^3$

2. $\displaystyle\int_0^2 y^2 e^{-x}\, dx = -y^2 e^{-x}\Big|_{x=0}^{x=2} = -y^2 e^{-2} + y^2 e^0 = -y^2 e^{-2} + y^2$

$\displaystyle\int_{-1}^1 (-y^2 e^{-2} + y^2)\, dy = \left(-\frac{1}{3}\, y^3 + \frac{1}{3}\, y^3\right)\Big|_{-1}^1 = -\frac{1}{3}\, e^{-2} + \frac{1}{3} - \left(\frac{1}{3}\, e^{-2} - \frac{1}{3}\right) = -\frac{2}{3}\, e^{-2} + \frac{2}{3}$

EXERCISES 7.6

Evaluate each (single) integral.

1 $\displaystyle\int_1^{x^2} 8xy^3\, dy$ **2** $\displaystyle\int_1^{y^2} 10x^4 y\, dx$

3 $\displaystyle\int_{-y}^{y} 9x^2 y\, dx$ **4** $\displaystyle\int_{-x}^{x} 6xy^2\, dy$

5 $\int_0^x (6y - x)\, dy$

6 $\int_0^y (4x - y)\, dx$

Evaluate each iterated integral.

7 $\int_0^2 \int_0^1 4xy\, dx\, dy$

8 $\int_0^2 \int_0^1 8xy\, dy\, dx$

9 $\int_0^2 \int_0^1 x\, dy\, dx$

10 $\int_0^4 \int_0^3 y\, dx\, dy$

11 $\int_0^1 \int_0^2 x^3 y^7\, dx\, dy$

12 $\int_0^1 \int_0^3 x^8 y^2\, dy\, dx$

13 $\int_1^3 \int_0^2 (x + y)\, dy\, dx$

14 $\int_1^2 \int_0^4 (x - y)\, dx\, dy$

15 $\int_{-1}^1 \int_0^3 (x^2 - 2y^2)\, dx\, dy$

16 $\int_{-1}^1 \int_0^3 (2x^2 + y^2)\, dy\, dx$

17 $\int_{-3}^3 \int_0^3 y^2 e^{-x}\, dy\, dx$

18 $\int_{-2}^2 \int_0^2 xe^{-y}\, dx\, dy$

19 $\int_{-2}^2 \int_{-1}^1 ye^{xy}\, dx\, dy$

20 $\int_{-1}^1 \int_{-1}^1 xe^{xy}\, dy\, dx$

21 $\int_0^2 \int_x^1 12xy\, dy\, dx$

22 $\int_0^1 \int_y^1 4xy\, dx\, dy$

23 $\int_3^5 \int_0^y (2x - y)\, dx\, dy$

24 $\int_2^4 \int_0^x (x - 2y)\, dy\, dx$

25 $\int_{-3}^3 \int_0^{4x} (y - x)\, dy\, dx$

26 $\int_{-1}^1 \int_0^{2y} (x + y)\, dx\, dy$

27 $\int_0^1 \int_{-y}^y (x + y^2)\, dx\, dy$

28 $\int_0^2 \int_{-x}^x (x^2 - y)\, dy\, dx$

For each double integral:
*(a) Write the **two** iterated integrals that are equal to it.*
*(b) Evaluate **both** iterated integrals (the answers should agree).*

29 $\iint\limits_R 3xy^2\, dx\, dy$ with $R = \{(x, y)|0 \le x \le 2, 1 \le y \le 3\}$

30 $\iint\limits_R 6x^2 y\, dx\, dy$ with $R = \{(x, y)|0 \le x \le 1, 1 \le y \le 2\}$

31 $\iint\limits_R ye^x\, dx\, dy$ with $R = \{(x, y)|-1 \le x \le 1, 0 \le y \le 2\}$

32 $\iint\limits_R xe^y\, dx\, dy$ with $R = \{(x, y)|0 \le x \le 1, -2 \le y \le 2\}$

Use integration to find the volume under each surface f(x, y) above the region R.

33 $f(x, y) = x + y$

$R = \{(x, y)|0 \le x \le 2, 0 \le y \le 2\}$

34 $f(x, y) = 8 - x - y$

$R = \{(x, y)|0 \le x \le 4, 0 \le y \le 4\}$

Use integration to find the volume under each surface f(x, y) above the region R.

35 $f(x, y) = 2 - x^2 - y^2$

$R = \{(x, y) | 0 \le x \le 1, 0 \le y \le 1\}$

36 $f(x, y) = x^2 + y^2$

$R = \{(x, y) | 0 \le x \le 2, 0 \le y \le 2\}$

37 $f(x, y) = 2xy$

for the region R:

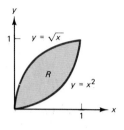

38 $f(x, y) = 3xy^2$

for the region R:

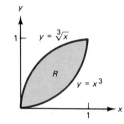

39 $f(x, y) = e^y$

for the region R:

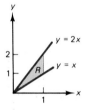

40 $f(x, y) = e^y$

for the region R:

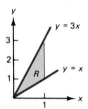

APPLIED EXERCISES

41 (*General–Average Temperature*) The temperature x miles east and y miles north of a weather station is given by the function $f(x, y) = 48 + 4x - 2y$. Find the average temperature over the region R shown below.

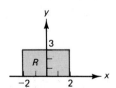

42 (*General–Average Air Pollution*) The air pollution near a chemical refinery is $f(x, y) = 20 + 6x^2y$ ppm (parts per million), where x and y are the number of miles east and north of the refinery. Find the average pollution levels for the region R shown below.

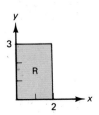

43 (*General–Total Population of a Region*) The population density (people per square mile) x miles east and y miles north of the center of a city is

$$P(x, y) = 12,000e^{-y}$$

Find the total population of the region R shown below. (*Hint*: Integrate the population density over the region R. This is an example of a double integral as a *sum*, giving a *total* population over a region.)

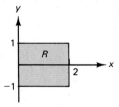

44 (*Business–Value of Mineral Deposit*) The value of an off-shore mineral deposit x miles east and y miles north of a certain point is $f(x, y) = 4x + 6y^2$ million dollars per square mile. Find the total value of the tracts shown in the next column. (*Hint*: Integrate the function over the region R. This is an example of a double integral as a *sum*, giving a *total* value over a region.)

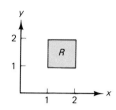

45–46 (*General–Volume of a Building*) To estimate heating and air-conditioning costs it is necessary to know the volume of a building.

45 A conference center has a curved roof, whose height is

$$f(x, y) = 40 - .006x^2 + .003y^2$$

The building sits on a rectangle extending from $x = -50$ to $x = 50$ and $y = -100$ to $y = 100$. Use integration to find the volume of the building.

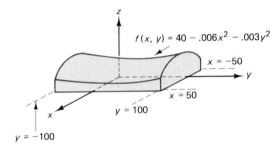

46 An airplane hanger has a curved roof, whose height is $f(x, y) = 40 - .03x^2$. The building sits on a rectangle extending from $x = -20$ to $x = 20$ and $y = -100$ to $y = 100$. Use integration to find the volume of the building.

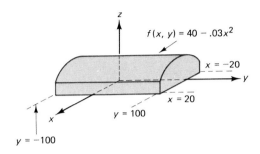

TRIPLE INTEGRALS

Evaluate each triple iterated integral. (*Hint*: Integrate with respect to one variable at a time, treating the other variables like constants, working from the inside out.)

47 $\int_1^2 \int_0^3 \int_0^1 (2x + 4y - z^2)\, dx\, dy\, dz$

48 $\int_1^2 \int_0^3 \int_0^2 (6x - 2y + z^2)\, dx\, dy\, dz$

49 $\int_1^2 \int_0^2 \int_0^1 2xy^2z^3\, dx\, dy\, dz$

50 $\int_1^3 \int_0^1 \int_0^2 12x^3y^2z\, dx\, dy\, dz$

7.7 REVIEW OF CHAPTER SEVEN

In this chapter we extended calculus to functions of two (or more) variables. The two main "themes" were using derivatives to optimize and find rates of change, and using integrals to find volumes, sums, and averages.

In **Section 7.1** we introduced functions of several variables, and showed that the graph of a function of two variables is a *surface* above or below the *x-y* plane. Such surfaces may have relative maximum and minimum points ("hilltops" and "valley bottoms") and saddle points ("mountain passes").

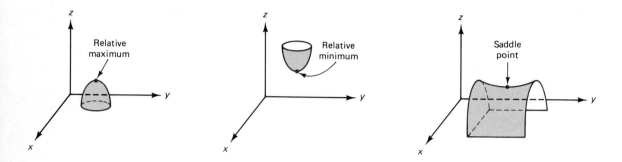

In **Section 7.2** we defined the partial derivatives of a function $f(x, y)$:

$$\frac{\partial}{\partial x} f(x, y) = f_x(x, y) = \lim_{h \to 0} \frac{f(x + h, y) - f(x, y)}{h}$$

and

$$\frac{\partial}{\partial y} f(x, y) = f_y(x, y) = \lim_{h \to 0} \frac{f(x, y + h) - f(x, y)}{h}$$

(provided that the limits exist). Only one variable at a time is increased by h, with the other remaining unchanged, so the partials are just the "ordinary" derivatives with respect to one variable with the other variable held constant. All of the "old" derivative formulas (product, quotient, and generalized power rules) apply, but always holding one variable constant. Partials may be interpreted as rates of change of the function when one variable at a time is increased.

In **Section 7.3** we optimized functions of several variables. Since at the highest (or lowest) point of a smooth surface the slope is zero in any direction, we found relative extreme points by setting the partials equal to zero and solving (finding "critical points").

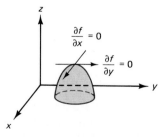

To determine whether $f(x, y)$ has a relative maximum, minimum, or neither (for example, a saddle point) at a critical point, we used a second derivative test (the D-test), checking the sign of $D = f_{xx}f_{yy} - (f_{xy})^2$ and the sign of f_{xx} at the critical point.

1. $D > 0$ and $f_{xx} > 0$ mean a relative *minimum* at the critical point.

2. $D > 0$ and $f_{xx} < 0$ mean a relative *maximum* at the critical point.

3. $D < 0$ means a *saddle point* (no relative extreme value) at the critical point.

4. $D = 0$ means that no conclusion can be drawn.

If the *absolute* extreme values of a function exist, they must occur at relative extremes found in this way.

In **Section 7.4** we found the least squares line that fits a collection of points. Minimizing the sum of the squared vertical deviations between the line and the points led to formulas for the constants a and b for the least squares line $y = ax + b$. The procedure is most easily carried out using a table such as the one on page 451. The exercises (page 457) showed how to fit an exponential curve $y = Be^{Ax}$ to a collection of points.

In **Section 7.5** we used Lagrange multipliers to optimize a function subject to a *constraint*. To optimize $f(x, y)$ subject to the constraint $g(x, y) = 0$ we combined both functions into a "Lagrange function,"

$$F(x, y, \lambda) = f(x, y) + \lambda g(x, y)$$

set the three partials equal to zero, and solved.

$$F_x = 0$$
$$F_y = 0$$
$$F_\lambda = 0$$

(The last of these equations always gives the constraint.) The maximum or minimum values (if it exists) always occurs at one of these critical points.

In **Section 7.6** we concluded our study of functions of several variables by discussing integration. We defined the double integral $\iint_R f(x, y)$ over rectangular regions, and over regions bounded by curves. Double integrals are evaluated by solving iterated integrals, integrating with respect to one variable at a time, holding the other variable constant. As before, integration is a kind of summation, which we used for finding total accumulations, average values, and volumes under surfaces.

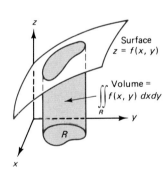

The following exercises review all these developments.

Practice Test

Exercises: 3, 7, 15, 17, 19, 33, 41, 45, 49, 57

EXERCISES 7.7

For each function, state the domain.

1 $f(x, y) = \dfrac{\sqrt{x}}{\sqrt[3]{y}}$ 　　　**2** $f(x, y) = \dfrac{\sqrt[3]{x}}{\sqrt{y}}$ 　　　**3** $f(x, y) = e^{1/x} \ln y$ 　　　**4** $f(x, y) = \dfrac{\ln y}{x}$

For each function f, calculate (a) f_x, *(b)* f_y, *(c)* f_{xy}, *and (d)* f_{yx}.

5 $f(x, y) = 2x^5 - 3x^2y^3 + y^4 - 3x + 2y + 7$

6 $f(x, y) = 3x^4 + 5x^3y^2 - y^6 - 6x + y - 9$

7 $f(x, y) = 18x^{2/3}y^{1/3}$

8 $f(x, y) = \ln(x^2 + y^3)$

9 $f(x, y) = e^{x^3 - 2y^3}$

10 $f(x, y) = 3x^2e^{-5y}$

11 $f(x, y) = ye^{-x} - x \ln y$

12 $f(x, y) = x^2e^y + y \ln x$

For each function, calculate (a) $f_x(1, -1)$ *and (b)* $f_y(1, -1)$.

13 $f(x, y) = \dfrac{x + y}{x - y}$

14 $f(x, y) = \dfrac{x}{x^2 + y^2}$

15 $f(x, y) = (x^3 + y^2)^3$

16 $f(x, y) = (2xy - 1)^4$

17 (*Business–Marginal Productivity*) A company's production is given by the Cobb–Douglas function $P(L, K) = 160L^{3/4}K^{1/4}$, where L is the number of units of labor and K is the number of units of capital.

(a) Find $P_L(81, 16)$ and interpret this number.

(b) Find $P_K(81, 16)$ and interpret this number.

(c) From your answers to parts (a) and (b), which will increase production more, an additional unit of labor or an additional unit of capital?

18 (*Business–Advertising*) A clothing designer's sales S depends on x, the amount spent on television advertising, and y, the amount spent on print advertising (both in thousands of dollars), according to the function

$$S(x, y) = 60x^2 + 90y^2 - 6xy + 200$$

Find $S_x(2, 3)$, $S_y(2, 3)$, and interpret these numbers.

For each function, find all relative extreme values.

19 $f(x, y) = 2x^2 - 2xy + y^2 - 4x + 6y - 3$

20 $f(x, y) = x^2 - 2xy + 2y^2 - 6x + 4y + 2$

21 $f(x, y) = 2xy - x^2 - 5y^2 + 2x - 10y + 3$

22 $f(x, y) = 2xy - 5x^2 - y^2 + 10x - 2y + 1$

23 $f(x, y) = 2xy + 6x - y + 1$

24 $f(x, y) = 4xy - 4x + 2y - 4$

25 $f(x, y) = e^{-(x^2 + y^2)}$

26 $f(x, y) = e^{2(x^2 + y^2)}$

27 $f(x, y) = \ln(5x^2 + 2y^2 + 1)$

28 $f(x, y) = \ln(4x^2 + 3y^2 + 10)$

29 $f(x, y) = x^3 - y^2 - 12x - 6y$

30 $f(x, y) = y^2 - x^3 + 12x - 4y$

31 (*Business–Maximum Profit*) A boatyard builds 18-foot and 22-foot sailboats. Each 18-foot boat costs $3000 to build, and each 22-foot boat costs $5000 to build, and the company's fixed costs are $6000. The price function for the 18-foot boats is $p = 7000 - 20x$, and for 22-foot boats is $q = 8000 - 30y$ (both in dollars, where x and y are the number of 18-foot and 22-foot boats respectively).

(a) Find the company's cost function $C(x, y)$.

(b) Find the company's revenue function $R(x, y)$.

(c) Find the company's profit function $P(x, y)$.

(d) Find the quantities and prices that maximize profit. Also find the maximum profit.

32 (*Business–Price Discrimination*) A company sells farm equipment in America and Europe, charging different prices in the two markets. The price function for harvesters sold in America is $p = 80 - .2x$, and the price function for harvesters sold in Europe is $q = 64 - .1y$ (both in thousands of dollars), where x and y are the number sold per day in America and Europe, respectively. The company's cost function is $C = 100 + 12(x + y)$ thousand dollars.

(a) Find the company's profit function.

(b) Find how many harvesters should be sold in each market to maximize profit. Also find the price for each country.

Find the least squares line for the points given. (Answers may vary slightly depending on rounding.)

33

x	y
1	−1
3	6
4	6
5	10

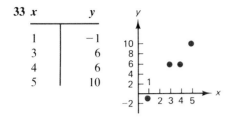

34

x	y
1	7
2	4
4	2
5	−1

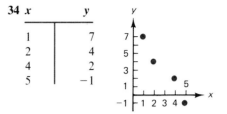

35 (*General–The Aging of America*) The population of Americans who are over 50 years old is growing faster than the population at large. Find the least squares line for the following data for the over-50 population, and use it to predict the size of the over-50 population in the year 2000 ($x = 6$).

	x	y = Population (millions)
1950	1	133
1960	2	165
1970	3	200
1980	4	232
1990	5	250

36 (*Economics–Unemployment*) The average unemployment rate has shown a general tendency to increase over recent decades. Find the least squares line for the following data and use it to predict the average unemployment rate for the year 2000 ($x = 8$).

	x	y = Percent Unemployed
1965	1	4.5
1970	2	5.0
1975	3	7.2
1980	4	7.0
1985	5	7.5

Use Lagrange multipliers to optimize each function subject to the constraint. (The stated extreme values do exist.)

37 Maximize $f(x, y) = 6x^2 − y^2 + 4$
subject to $3x + y = 12$

38 Maximize $f(x, y) = 4xy − x^2 − y^2$
subject to $x + 2y = 26$

39 Minimize $f(x, y) = 2x^2 + 3y^2 − 2xy$
subject to $2x + y = 18$

40 Minimize $f(x, y) = 12xy − 1$
subject to $y · x = 6$

41 Minimize $f(x, y) = e^{x^2+y^2}$
subject to $x + 2y = 12$

42 Maximize $f(x, y) = e^{-x^2-y^2}$
subject to $2x + y = 5$

*Use Lagrange multipliers to find the maximum **and** minimum values of each function subject to the constraint. (Both extreme values do exist.)*

43 $f(x, y) = 6x − 18y$
subject to $x^2 + y^2 = 40$

44 $f(x, y) = 4xy$
subject to $x^2 + y^2 = 32$

45 (*Business–Maximum Profit*) A company's profit is $P = 300x^{2/3}y^{1/3}$, where x and y are, respectively, the amounts spent on production and advertising. The company has a total of $60,000 to spend.

(a) Use Lagrange multipliers to find the amounts for production and advertising that maximize profit.

(b) Evaluate and give an interpretation for λ.

46 (*Biomedical–Nutrition*) A nursing home uses two vitamin supplements, and the nutritional value of x ounces of the first together with y ounces of the second is $4x + 2xy + 8y$. The first costs $2 per ounce, and the second costs $1 per ounce, and the nursing home can spend only $8 per patient per day. [Continued on next page.]

(a) Use Lagrange multipliers to find how much of each supplement should be used to maximize the nutritional value subject to the budget constraint.

(b) Evaluate and give an interpretation for λ.

47 (*Economics–Least Cost Rule*) A company's production is given by the Cobb–Douglas function $P = 60L^{2/3}K^{1/3}$, where L and K are the number of units of labor and capital. Each unit of labor costs $25 and each unit of capital costs $100. The company wants to produce exactly 1920 units.

(a) Find the number of units of labor and capital that meet the production requirements at the lowest cost.

(b) Find the marginal productivity of labor and the marginal productivity of capital. (*Hint*: This means the partials of P with respect to L and K.)

[Problem continues in next column.]

(c) Show that at the values found in part (a), the following relationship holds:

$$\frac{\text{marginal productivity of labor}}{\text{marginal productivity of capital}} = \frac{\text{price of labor}}{\text{price of capital}}$$

This is called the "least cost rule."

48 (*General–Container Design*) An open-top box with a square base and two perpendicular dividers, as in the diagram, is to have volume 108 cubic inches. Use Lagrange multipliers to find the dimensions that require the least amount of materials.

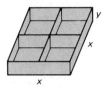

Solve each iterated integral.

49 $\int_0^4 \int_{-1}^1 2xe^{2y} \, dy \, dx$

50 $\int_{-1}^1 \int_0^3 (x^2 - 4y^2) \, dx \, dy$

51 $\int_{-1}^1 \int_{-y}^y (x + y) \, dx \, dy$

52 $\int_{-2}^2 \int_{-x}^x (x + y) \, dy \, dx$

Find the volume under the surface f(x, y) above the region R.

53 $f(x, y) = 8 - x - y$
$R = \{(x, y) \mid 0 \le x \le 2, 0 \le y \le 4\}$

54 $f(x, y) = 6 - x - y$
$R = \{(x, y) \mid 0 \le x \le 4, 0 \le y \le 2\}$

55 $f(x, y) = 12xy^3$
over the region R:

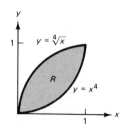

56 $f(x, y) = 15xy^4$
over the region R:

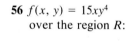

57 (*General–Average Population of a Region*) The population x miles east and y miles north of the center of a city is $P(x, y) = 12{,}000 + 100x - 200y$. Find the *average* population over the region shown on the right.

58 (*General–Total Value of a Region*) The value of land in a city x blocks east and y blocks north of the center of a town in $V(x, y) = 40 - 4x - 2y$ hundred thousand dollars per block. Find the *total* value of the parcel of land shown on the right.

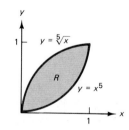

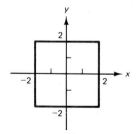

ANSWERS TO ODD-NUMBERED EXERCISES

EXERCISES 1.1 page 12

1 64   **3** 1/16   **5** 8   **7** 8/5   **9** 1/32   **11** 8/27   **13** 1

15 4/9   **17** 5   **19** 125   **21** 8   **23** 4   **25** -32 **27** 125/216

29 9/25   **31** 1/4   **33** 1/2   **35** 1/8   **37** 1/4   **39** $-1/2$ **41** 1/4

43 4/5   **45** 64/125   **47** -243 **49** 2.14   **51** 274.37   **53** x^{10}

55 z^{27} **57** x^8 **59** w^5 **61** $\dfrac{y^5}{x}$ **63** $27y^4$ **65** $u^2v^2w^2$

67 $\dfrac{1}{x^2 + 3}$ **69** $\dfrac{1}{(x + 1)^5}$ **71** $(x^2 + 4)^{-3}$ **73** $(x^2 + 1)^{-2}(x^2 + 5)^{-3}$

75 $(x^2 + 1)^{1/2}$ **77** $(x + 3)^{2/3}$ **79** $\sqrt[4]{x^3 + 1}$ **81** $\sqrt[4]{(2x + 3)^5}$ or $\left(\sqrt[4]{2x + 3}\right)^5$

83 $(x + 1)^{-1/3}$ **85** $(3x + 2)^{-7/5}$ **87** $\dfrac{1}{\sqrt[4]{x^3 + 1}}$ **89** $\dfrac{1}{\sqrt[4]{(x + 1)^3}}$ or $\dfrac{1}{\left(\sqrt[4]{x + 1}\right)^3}$

91 25.6 feet

93 Costs would be multiplied by 1.93. (Note that tripling capacity roughly doubles costs.)

95 125 beats per minute

97 About 42.6 thousand hours, or 42,600 hours, rounded to the nearest hundred hours

EXERCISES 1.2 page 27

1 $\{w \mid w \geq 1\}$, $g(10) = 3$ **3** $\{z \mid z \neq 4\}$, $h(-5) = -1$ **5** $\{x \mid x \geq 0\}$, $h(81) = 3$

7 $\{x \mid x > 0\}$, $h(81) = 1/9$ **9** $\{x \mid x \neq 0\}$, $h(27) = 1/81$ **11** Yes   **13** No

15 No   **17** No

19   **21**   **23**   **25**

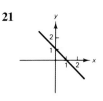 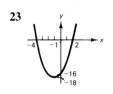 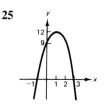

27 $(x + 3)(x + 4)$ **29** $(x - 2)(x - 4)$ **31** $(x - 5)(x + 3)$ **33** $x(x - 7)$

35 $(x + 9)(x - 9)$ **37** $3(x - 1)(x + 4)$ **39** $5(x + 3)(x - 3)$ **41** $3(x - 2)^2$

43 $-3x(x - 8)$ **45** $x = 7$ or -1 **47** $x = 3$ or -5 **49** $x = 4$ or 5

51 $x = 0$ or 10 **53** $x = 5$ or -5 **55** $x = -3$ **57** $x = 1$ or 2

59 No solutions   **61** No solutions   **63** $C(x) = 4x + 20$

65 (a) 17.7 pounds per square inch,   **67** (a) 400,   (b) 5200
(b) 15,765 pounds per square inch

69 About 208 miles per hour   **71** (a) $V(t) = 500,000 - 49,000t$,   (b) $225,000

EXERCISES 1.3 page 38

1 $\{x \mid x \neq 7\}$ **3** $\{x \mid x \neq 0 \text{ and } x \neq -3\}$ **5** $\mathbb{R}$
$f(6) = -11$ $f(-1) = -2$ $g(-1/2) = 1/2$

7 $x = 0, x = -3,$ **9** $x = 0, x = 2,$ **11** $x = 0$ and   **13** $x = 0$ and
and $x = 1$ and $x = -2$ $x = 3$ $x = -5$

15
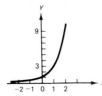

17
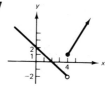

19
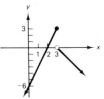

21
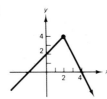

23
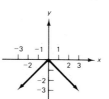

25 (a) $5x^2 + 10xh + 5h^2$ or $5(x^2 + 2xh + h^2)$
(b) $10xh + 5h^2$ or $h(10x + 5h)$ or $5h(2x + h)$

27 (a) $2x^2 + 4xh + 2h^2 - 5x - 5h + 1$ **29** $10x + 5h$ or $5(2x + h)$
(b) $4xh + 2h^2 - 5h$

31 $4x + 2h - 5$ **33** $14x + 7h - 3$ **35** $3x^2 + 3xh + h^2$ **37** $\dfrac{-2}{(x + h)x}$

39 $\dfrac{-2x - h}{x^2(x + h)^2}$ or $\dfrac{-2x - h}{x^2(x^2 + 2xh + h^2)}$ or $\dfrac{-2x - h}{x^4 + 2x^3h + x^2h^2}$

41 About 729 million people

43 (a) $300   (d)
(b) $500
(c) $2000

EXERCISES 2.1 page 52

1 Continuous   **3** Discontinuous at 1   **5** Discontinuous at -1, 1, and 3

7 Continuous   **9** Continuous   **11** Discontinuous at 1

13 Discontinuous at 3 and -4 **15** 9   **17** 2   **19** 16   **21** 15   **23** 6

25 5   **27** $4x^2$ **29** $2x^2$ **31** 7   **33** $2x$ **35** $7x^3 - 3x$ **37** $4x^2$

39 Limit exists and equals -1.

41 Limit does not exist (use x-values close to 4, such as 3.99 and 4.01).

43 Limit does not exist (use x-values close to 2, such as 1.99 and 2.01).

45 Limit exists and equals 0.   **47** 3 inches   **49** 0 feet

EXERCISES 2.2 page 63

1 (a) $f'(x) = 2x - 8$ degrees per minute
  (b) Decreasing at the rate of 4 degrees per minute (since $f'(2) = -4$)
  (c) Increasing at the rate of 2 degrees per minute (since $f'(5) = 2$)

3 (a) $f'(x) = 4x - 1$ seconds per word
  (b) When 5 words have been memorized, the memorization time is increasing at the rate of 19 seconds per word.

5 (a) $T'(x) = -2x + 5$ degrees per day
  (b) Increasing at the rate of 1 degree per day
  (c) Decreasing at the rate of 1 degree per day
  (d) Deteriorating on day 2, improving on day 3

7 (a) Slope $= f'(x) = 2x - 6$ (d)
  (b) Slope at $x = 2$ is -2
  (c) Slope at $x = 4$ is $+2$

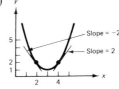

9 (a) $f'(x) = 3$, (b) The graph of $f(x) = 3x - 4$ is a straight line with slope 3.

11 (a) $f'(x) = 0$, (b) The graph of $f(x) = 5$ is a horizontal straight line with slope 0.

13 (a) $f'(x) = m$, (b) The graph of $f(x) = mx + b$ is a straight line with slope m.

15 $f'(x) = 2ax + b$ **17** $f'(x) = 3x^2$ **19** $f'(x) = \dfrac{-2}{x^2}$ **21** $f'(x) = \dfrac{-2}{x^3}$

23 $f'(x) = \dfrac{1}{2\sqrt{x}}$

EXERCISES 2.3 page 78

1 $4x^3$ **3** $500x^{499}$ **5** $(1/2)x^{-1/2}$ **7** $2x^3$ **9** $2w^{-2/3}$ **11** $-6x^{-3}$

13 $8x - 3$ **15** $(1/2)x^{-1/2} + x^{-2}$ **17** $4x^{-1/3} + 4x^{-4/3}$ **19** $-5x^{-3/2} - 15x^{2/3}$

21 (a) $f'(x) = 0$,
  (b) The graph of the constant function $f(x) = 2$ is a horizontal line and therefore has slope 0.
  (c) Since $f(x)$ is a constant function, its rate of change is zero.

23 80   **25** 4   **27** -1 **29** 3   **31** 27   **33** 6   **35** 7   **37** 1

39 (a) $MP(x) = -\dfrac{1}{2}x^3 + 3x^2 - 12x + 28$,

(b) MP(2) = 12, about \$12 profit on each additional unit
(c) MP(4) = −4, a loss of about \$4 on each additional unit

41 (a) $P'(x) = -12{,}000 + 1200x + 300x^2$
(b) Decreasing by about 10,500 per year.
(c) Increasing by about 30,000 per year.

43 Increasing by about 8000 people per additional day.

45 Increasing by about .08 square centimeter per hour

47 Increasing by about 6 phrases per hour

49 (a) $MU(x) = 50x^{-1/2}$
(b) 50   (c) .05

EXERCISES 2.4 page 90

1 $10x^9$ **3** $9x^8 + 4x^3$ **5** $f'g' = 4x^3 \cdot 6x^5 = 24x^8$ **7** $f'g' = 4x^3 \cdot 5x^4 = 20x^7$

9 $5x^4 + 2x$ **11** $4x^3$ **13** $9x^2 + 8x + 1$ **15** 1   **17** $6x^5$

19 $-\dfrac{3}{x^4}$ or $-3x^{-4}$ **21** $\dfrac{f'}{g'} = \dfrac{8x^7}{2x} = 4x^6$ **23** $\dfrac{f'}{g'} = \dfrac{0}{3x^2} = 0$ **25** $\dfrac{x^4 - 3}{x^4}$

27 $-\dfrac{2}{(x-1)^2}$ **29** $\dfrac{4t}{(t^2+1)^2}$ **31** 0   **33** 1/4

35 $\dfrac{d}{dx}[fgh] = f'[gh] + f[gh]' = f'gh + f[g'h + gh'] = f'gh + fg'h + fgh'$

37 $2f(x)f'(x)$ **39** $MAR(x) = \dfrac{xR'(x) - R(x)}{x^2}$

41 (a) $C'(x) = \dfrac{100}{(100-x)^2}$
(b) Increasing by 4 cents per additional percent of purity
(c) Increasing by 25 cents per additional percent of purity

43 (a) $AP(x) = \dfrac{12x - 1800}{x}$
(b) $MAP(x) = \dfrac{1800}{x^2}$ or $1800x^{-2}$
(c) MAP(300) = 2/100, average profit is increasing by 2 cents per additional unit

45 Increasing at the rate of 7 degrees per hour

47 $3x^2 \dfrac{x^2+1}{x+1} + (x^3+2)\dfrac{x^2+2x-1}{(x+1)^2}$ **49** $\dfrac{3x^6 + 13x^4 + 18x^2 - 2x}{(x^2+2)^2}$

51 $\dfrac{x^{-1/2}}{(x^{1/2}+1)^2}$ or $\dfrac{1}{\sqrt{x}(\sqrt{x}+1)^2}$

EXERCISES 2.5 page 101

1 (a) $12x^2 - 12x - 6$ (b) 18   (c) $24x - 12$ (d) 36   (e) 24
(f) 0

3 (a) $1 + x + \dfrac{1}{2}x^2 + \dfrac{1}{6}x^3$ (b) $6\frac{2}{3}$ (c) $1 + x + \dfrac{1}{2}x^2$ (d) 5
(e) $1 + x$ (f) 1

5 (a) $\dfrac{15}{4}x^{1/2}$ (b) $15/2$ (c) $\dfrac{15}{8}x^{-1/2}$ (d) $\dfrac{15}{16}$

   (e) $-\dfrac{15}{16}x^{-3/2}$ (f) $\dfrac{45}{32}x^{-5/2}$

7 (a) $-\dfrac{2}{x^3}$ or $-2x^{-3}$ (b) $-2/27$ **9** (a) $\dfrac{1}{x^3} = x^{-3}$ (b) $1/27$

11 (a) x^{-4} (b) $\dfrac{1}{81}$ **13** $12x^2 + 2$ **15** $12x^{-7/3}$

17 $\dfrac{2x-2}{(x^2-2x+1)^2} = \dfrac{2}{(x-1)^3}$ **19** 2π **21** 90 **23** -720 **25** 3

27 0 **29** $\dfrac{d^2}{dx^2}[fg] = \dfrac{d}{dx}[f'g + fg'] = f''g + f'g' + f'g' + fg'' = f''g + 2f'g' + fg''$

31 (a) 54 mph   (b) -42 mph or 42 mph south   (c) 24 mi/hr^2

33 310 ft/sec, 61 ft/sec^2 **35** (a) 160 ft/sec   (b) 32 ft/sec^2

37 (a) $-32t + 1280$ (b) 40 seconds   (c) $25,600$ feet

39 $D'(8) = 24$:   After 8 years the debt is growing at \$24 billion per year.
     $D''(8) = 1$:   After 8 years the debt will be growing increasingly rapidly, with the rate of growth growing by about \$1 billion per year per year.

41 $L'(4) = 1/4$:   After 4 years the sea level will be rising by 1/4 foot per year.
     $L''(4) = -3/32$:   After 4 years the rate of growth will be slowing by about 3/32 feet per year per year.

43 $20x^3 - 12x^2 + 6x - 2$ **45** $\dfrac{2x^5 + 8x^3 - 6x}{(x^4 + 2x^2 + 1)^2} = \dfrac{2x^3 - 6x}{(x^2 + 1)^3}$

47 $\dfrac{-32x - 16}{(4x^2 + 4x + 1)^2} = \dfrac{-16}{(2x + 1)^3}$

EXERCISES 2.6 page 112   **1** (a) $(7x - 1)^5$ (b) $7x^5 - 1$ **3** (a) $\dfrac{1}{x^2 + 1}$ (b) $\left(\dfrac{1}{x}\right)^2 + 1$

5 (a) $(\sqrt{x} - 1)^3 - (\sqrt{x} - 1)^2$ (b) $\sqrt{x^3 - x^2} - 1$

7 (a) $\dfrac{(x^2 - x)^3 - 1}{(x^2 - x)^3 + 1}$ (b) $\left(\dfrac{x^3 - 1}{x^3 + 1}\right)^2 - \dfrac{x^3 - 1}{x^3 + 1}$

9 $f(x) = \sqrt{x}, \;\; g(x) = x^2 - 3x + 1$ **11** $f(x) = x^{-3}, \;\; g(x) = x^2 - x$

13 $f(x) = \dfrac{x + 1}{x - 1}, \;\; g(x) = x^3$ **15** $f(x) = x^4, \;\; g(x) = \dfrac{x + 1}{x - 1}$

17 $f(x) = \sqrt{x} + 5, \;\; g(x) = x^2 - 9$ **19** $6x(x^2 + 1)^2$ **21** $45(9x - 4)^4$

23 $4(3z^2 - 5z + 2)^3(6z - 5)$ **25** $\dfrac{1}{2}(x^4 - 5x + 1)^{-1/2}(4x^3 - 5)$

27 $\dfrac{1}{3}(9z - 1)^{-2/3}(9) = 3(9z - 1)^{-2/3}$ **29** $400x(x^2 + 9)^{199}$ **31** $-8x(4 - x^2)^3$

33 $-40(2 - x)^{39}$ **35** $-12x^2(x^3 - 1)^{-5}$ **37** $4x^3 - 4(1 - x)^3$

39 $8x(x^2 + 1)^3 - 6x(x^2 + 1)^2$

41 $\frac{1}{2}(2w^3 + 3w^2)^{-1/2}(6w^2 + 6w) = (2w^3 + 3w^2)^{-1/2}(3w^2 + 3w)$

43 $-\frac{2}{3}(9x + 1)^{-5/3}(9) = -6(9x + 1)^{-5/3}$ **45** $3[(x^2 + 1)^3 + x]^2[6x(x^2 + 1)^2 + 1]$

47 $6(2x + 1)^2(2x - 1)^4 + 8(2x + 1)^3(2x - 1)^3$ **49** $\frac{-2x - 2}{x^3}$ **51** $-6\frac{(x + 1)^2}{(x - 1)^4}$

53 $\frac{x^{-1/2}}{(x^{1/2} + 1)^2} = \frac{1}{\sqrt{x}(\sqrt{x} + 1)^2}$ **55** $2x(1 + x^2)^{1/2} + x^3(1 + x^2)^{-1/2}$

57 $\frac{1}{4}x^{-1/2}(1 + x^{1/2})^{-1/2}$ **59** (a) $4x(x^2 + 1)$,   (b) $4x^3 + 4x$

61 $-\frac{3}{(3x + 1)^2}$ or $-3(3x + 1)^{-2}$ **63** $20(x^2 + 1)^9 + 360x^2(x^2 + 1)^8$

65 $MC(x) = 4x(4x^2 + 900)^{-1/2}$
$MC(20) = \frac{8}{5} = 1.60$

67 $S'(25) \approx 2.1$, at income level $25,000 status increases by about 2.1 units for each additional $1000 of income

69 $R'(50) = 32\frac{1}{3}$

71 $P'(2) = .24$, pollution is increasing by about .24 ppm per year.

EXERCISES 2.7 page 118   **1** 26   **3** $2x^2$ **5** Does not exist   **7** $4x + 3$ **9** $-\frac{3}{x^2} = -3x^{-2}$

11 $10x^{2/3} + 2x^{-3/2}$ **13** $2x(x^2 - 5) + (x^2 + 5)2x = 4x^3$

15 $(4x^3 + 2x)(x^5 - x^3 + x) + (x^4 + x^2 + 1)(5x^4 - 3x^2 + 1)$ **17** $\frac{2}{(x + 1)^2}$
$= 9x^8 + 5x^4 + 1$

19 $\frac{-10x^4}{(x^5 - 1)^2}$ **21** $3(4z^2 - 3z + 1)^2(8z - 3)$ **23** $-5(100 - x)^4$

25 $\frac{1}{2}(x^2 - x + 2)^{-1/2}(2x - 1)$

27 $\frac{1}{3}(2y^3 - 3y^2)^{-2/3}(6y^2 - 6y) = (2y^3 - 3y^2)^{-1/2}(2y^2 - 2y)$

29 $2(6z - 1)^{-2/3}$ **31** $-2(5x + 1)^{-7/5}$ **33** $10(1 - x)^{-11}$

35 $3x^2(x^3 + 1)^{1/3} + x^5(x^3 + 1)^{-2/3}$ **37** $3[(2x^2 + 1)^4 + x^4]^2[16x(2x^2 + 1)^3 + 4x^3]$

39 $\frac{1}{2}[(x^2 + 1)^4 - x^4]^{-1/2}[8x(x^2 + 1)^3 - 4x^3]$

41 $12(3x + 1)^3(4x + 1)^3 + 12(3x + 1)^4(4x + 1)^2$

43 $8x(x^2 + 1)^3(x^2 - 1)^3 + 6x(x^2 + 1)^4(x^2 - 1)^2$ **45** $\frac{3x(x + 4)^2 - 2(x + 4)^3}{x^3}$

47 $\dfrac{-20}{x^2}\left(\dfrac{x+5}{x}\right)^3 = \dfrac{-20(x+5)^3}{x^5}$ **49** $9x^{-1/2} + 2x^{-5/3}$ **51** $2x^{-4}$

53 $20(2w^2 - 4)^4 + 320w^2(2w^2 - 4)^3$ **55** $6z(z+1)^3 + 18z^2(z+1)^2 + 6z^3(z+1)$

57 -16 **59** 1   **61** -24 **63** 480   **65** 15

67 (a) $6x^2(x^3 - 1)$ (b) $6x^5 - 6x^2$ **69** $x = -3, x = 1, x = 3$

71 $x = 0, x = 3.5$

73 $T'(10) = 1/10$, temperature is increasing by about 1/10 of a degree per year.
$T''(10) = -.02$, the rate of increase is slowing by about .02 degree per year per year.

75 60 feet   **77** 48,405 feet   **79** $MC(x) = 3 - 20x^{-3}$

81 (a) $\dfrac{5x + 100}{x}$ (b) $\dfrac{-100}{x^2}$ **83** $N'(5) = 20$, so after 5 minutes it is spreading at the rate of about 20 people per minute.

85 (a) $V' = (4/3)\pi r^2 \cdot 3 = 4\pi r^2$ **87** .08   **89** $N'(96) = -250$ At age 96,
(b) As radius increases, volume "grows by a surface area."   the number of survivors is decreasing by about 250 people per year.

EXERCISES 3.1 page 134   **1** 4 and -4 **3** 0, -4, and 1   **5** 3   **7** $-3, -1,$ and 1   **9** No CVs

11 $f' < 0$ $f' = 0$ $f' > 0$ $f' = 0$ $f' < 0$ $f' = 0$ $f' > 0$

	$x = -4$		$x = 0$		$x = 1$	
↘	→	↗	→	↘	→	↗
	rel min		rel max		rel min	
	$(-4, -64)$		$(0, 64)$		$(1, 61)$	

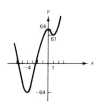

13 $f' > 0$ $f' = 0$ $f' < 0$ $f' = 0$ $f' > 0$ $f' = 0$ $f' < 0$

	$x = 0$		$x = 1$		$x = 2$	
↗	→	↘	→	↗	→	↘
	rel max		rel min		rel max	
	$(0, 1)$		$(1, 0)$		$(2, 1)$	

15 $f' < 0$ $f' = 0$ $f' > 0$ $f' = 0$ $f' > 0$

	$x = 0$		$x = 1$	
↘	→	↗	→	↗
	rel min		neither	
	$(0, 0)$		$(1, 1)$	

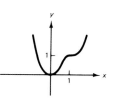

17 $f' < 0$ $f' = 0$ $f' > 0$

	$x = 1$	
↘	→	↗
	rel min	
	$(1, 0)$	

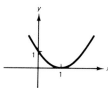

19

$f' > 0$	$f' = 0$	$f' > 0$
	$x = -1$	
↗	→	↗
	neither	
	$(-1, 0)$	

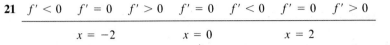

21

$f' < 0$	$f' = 0$	$f' > 0$	$f' = 0$	$f' < 0$	$f' = 0$	$f' > 0$
	$x = -2$		$x = 0$		$x = 2$	
↘	→	↗	→	↘	→	↗
	rel min		rel max		rel min	
	$(-2, 0)$		$(0, 16)$		$(2, 0)$	

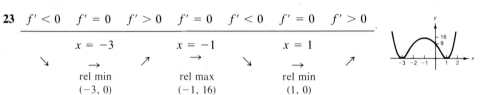

23

$f' < 0$	$f' = 0$	$f' > 0$	$f' = 0$	$f' < 0$	$f' = 0$	$f' > 0$
	$x = -3$		$x = -1$		$x = 1$	
↘	→	↗	→	↘	→	↗
	rel min		rel max		rel min	
	$(-3, 0)$		$(-1, 16)$		$(1, 0)$	

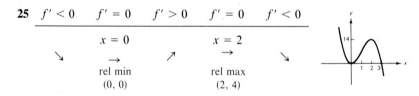

25

$f' < 0$	$f' = 0$	$f' > 0$	$f' = 0$	$f' < 0$
	$x = 0$		$x = 2$	
↘	→	↗	→	↘
	rel min		rel max	
	$(0, 0)$		$(2, 4)$	

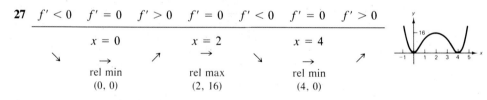

27

$f' < 0$	$f' = 0$	$f' > 0$	$f' = 0$	$f' < 0$	$f' = 0$	$f' > 0$
	$x = 0$		$x = 2$		$x = 4$	
↘	→	↗	→	↘	→	↗
	rel min		rel max		rel min	
	$(0, 0)$		$(2, 16)$		$(4, 0)$	

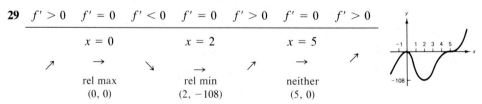

29

$f' > 0$	$f' = 0$	$f' < 0$	$f' = 0$	$f' > 0$	$f' = 0$	$f' > 0$
	$x = 0$		$x = 2$		$x = 5$	
↗	→	↘	→	↗	→	↗
	rel max		rel min		neither	
	$(0, 0)$		$(2, -108)$		$(5, 0)$	

31 $f'(x) = 2ax + b = 0$ at $x = -b/2a$

33 $MC(x) = 3x^2 - 12x + 14$

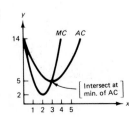

35

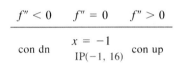

EXERCISES 3.2 page 148

1 Speeds up   **3** Points 3 and 5   **5** Points 4 and 6

7

$f' > 0$	$f' = 0$	$f' < 0$	$f' = 0$	$f' > 0$
	$x = -3$		$x = 1$	
↗	→	↘	→	↗
	rel max		rel min	
	$(-3, 32)$		$(1, 0)$	

$f'' < 0$	$f'' = 0$	$f'' > 0$
	$x = -1$	
con dn	IP$(-1, 16)$	con up

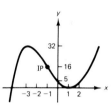

9

$f' > 0$	$f' = 0$	$f' < 0$	$f' = 0$	$f' > 0$
	$x = 1$		$x = 3$	
↗	→	↘	→	↗
	rel max		rel min	
	$(1, 28)$		$(3, 24)$	

$f'' < 0$	$f'' = 0$	$f'' > 0$
	$x = 2$	
con dn	IP$(2, 26)$	con up

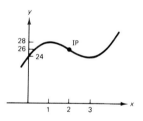

11

$f' > 0$	$f' = 0$	$f' > 0$
	$x = 1$	
↗	→	↗
	neither	
	$(1, 5)$	

$f'' < 0$	$f'' = 0$	$f'' > 0$
	$x = 1$	
con dn	IP$(1, 5)$	con up

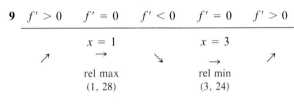
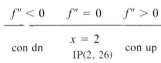
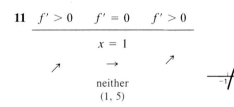
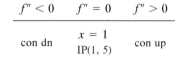

13

$f' < 0$	$f' = 0$	$f' > 0$	$f' = 0$	$f' > 0$
	$x = 0$		$x = 3$	
↘	→	↗	→	↗
	rel min (0, 2)		neither (3, 29)	

$f'' > 0$	$f'' = 0$	$f'' < 0$	$f'' = 0$	$f'' > 0$
con up	$x = 1$ IP(1, 13)	con dn	$x = 3$ IP(3, 29)	con up

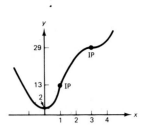

15

$f' < 0$	$f' = 0$	$f' > 0$	$f' = 0$	$f' > 0$
	$x = -3$		$x = 0$	
↘	→	↗	→	↗
	rel min (−3, −12)		neither (0, 15)	

$f'' > 0$	$f'' = 0$	$f'' < 0$	$f'' = 0$	$f'' > 0$
con up	$x = -2$ IP(−2, −1)	con dn	$x = 0$ IP(0, 15)	con up

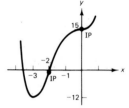

17

$f' < 0$	$f' = 0$	$f' > 0$	$f' = 0$	$f' < 0$
	$x = 0$		$x = 4$	
↘	→	↗	→	↘
	rel min (0, 0)		rel max (4, 256)	

$f'' > 0$	$f'' = 0$	$f'' > 0$	$f'' = 0$	$f'' < 0$
con up	$x = 1$	con up	$x = 3$ IP(3, 162)	con dn

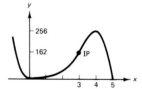

19

$f' > 0$	$f' = 0$	$f' > 0$
	$x = 1$	
↗	→	↗
	neither (1, 1)	

$f'' < 0$	$f'' = 0$	$f'' > 0$
con dn	$x = 1$ IP(1, 1)	con up

21

$f' > 0$	$f' = 0$	$f' > 0$
	$x = -2$	
↗	→	↗
	neither	
	$(-2, 0)$	

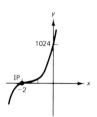

$f'' < 0$	$f'' = 0$	$f'' > 0$
	$x = -2$	
con dn	IP$(-2, 0)$	con up

23

$f' < 0$	$f' = 0$	$f' > 0$
	$x = 2$	
↘	→	↗
	rel min	
	$(2, 2)$	

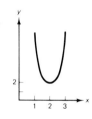

$f'' > 0$	$f'' = 0$	$f'' > 0$
	$x = 2$	
con up		con up

25

$f' > 0$	$f' = 0$	$f' < 0$	$f' = 0$	$f' > 0$
	$x = 1$		$x = 3$	
↗	→	↘	→	↗
	rel max		rel min	
	$(1, 4)$		$(3, 0)$	

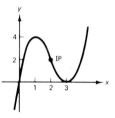

$f'' < 0$	$f'' = 0$	$f'' > 0$
	$x = 2$	
con dn	IP$(2, 2)$	con up

27

$f' > 0$	$f' = 0$	$f' < 0$	$f' = 0$	$f' > 0$
	$x = 0$		$x = 2$	
↗	→	↘	→	↗
	rel max		rel min	
	$(0, 0)$		$(2, -4)$	

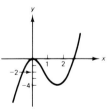

$f'' < 0$	$f'' = 0$	$f'' > 0$
	$x = 1$	
con dn	IP$(1, -2)$	con up

29

$f' > 0$	f' und	$f' > 0$
	$x = 0$	
↗		↗
	neither $(0, 0)$	

$f'' > 0$	$f'' = 0$	$f'' < 0$
con up	$x = 0$ IP$(0, 0)$	con dn

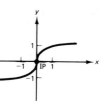

31

$f' < 0$	f' und	$f' > 0$
	$x = 0$	
↘		↗
	rel min $(0, 2)$	

$f'' < 0$	f'' und	$f'' < 0$
con dn	$x = 1$	con dn

33

$f' > 0$	f' und	$f' > 0$
	$x = 0$	
↗		↗
	neither $(0, -1)$	

$f'' > 0$	f'' und	$f'' < 0$
con up	$x = 0$ IP$(0, -1)$	con dn

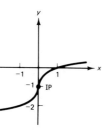

35

f' und	$f' > 0$
$x = 0$	
	↗

f'' und	$f'' < 0$
$x = 0$	con dn

37

$f' = 0$	$f' > 0$
$x = 0$	
	↗

f'' und	$f'' > 0$
$x = 0$	con up

39

$f' < 0$	f' und	$f' > 0$
	$x = 1$	
↘		↗
	rel min $(1, 0)$	

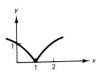

$f'' < 0$	f'' und	$f'' < 0$
	$x = 1$	
con dn		con dn

41

$f' > 0$	f' und	$f' > 0$
	$x = -1$	
↗		↗
	neither $(-1, 1)$	

$f'' > 0$	f'' und	$f'' < 0$
	$x = -1$ IP$(-1, 1)$	
con up		con dn

43

$f' < 0$	$f' = 0$	$f' > 0$
	$x = 1$	
↘	→	↗
	rel min $(1, 0)$	

$f'' > 0$	f'' und	$f'' > 0$
	$x = 1$	
con up		con up

45

$f' = 0$	$f' > 0$
$x = 3$	
→	↗

f'' und	$f'' > 0$
$x = 3$	con up

47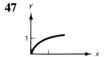

49

51 $f''(x) = 2a$, therefore: for $a > 0$: $f''(x) > 0$ and $f(x)$ is concave up,
for $a < 0$: $f''(x) < 0$ and $f(x)$ is concave down.

53 (a)

$f' > 0$	$f' = 0$	$f' < 0$	$f' = 0$	$f' > 0$
	$x = 1$		$x = 5$	
↗	→	↘	→	↗
	rel max (1, 32)		rel min (5, 0)	

$f'' < 0$	$f'' = 0$	$f'' > 0$
	$x = 3$	
con dn	IP(3, 16)	con up

(b)

55

$f' < 0$	$f' = 0$	$f' > 0$
	$x = 1$	
↘	→	↗
	rel min (1, 109)	

$f'' > 0$	$f'' = 0$	$f'' > 0$
	$x = 0$	
con up		con up

57

f' und	$f' > 0$
$x = 0$	
	↗
rel min (0, 0)	

f'' und	$f'' < 0$
$x = 0$	
	con dn

59

f' und	$f' > 0$
$x = 0$	
	↗

f'' und	$f'' < 0$
$x = 0$	
	con dn

(a)

(b) Status increases more slowly at higher income levels.

61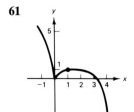

63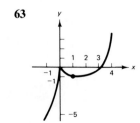

65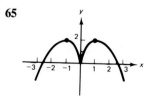

EXERCISES 3.3 page 161

1 $\lim\limits_{x\to\pm\infty} f(x) = 0$

3 $\lim\limits_{x\to\pm\infty} f(x) = 0$

5 $\lim\limits_{x\to\pm\infty} f(x) = 1$

7 $\lim\limits_{x\to\pm\infty} f(x) = 1$

9 $\lim\limits_{x\to\infty} f(x) = \infty,\quad \lim\limits_{x\to-\infty} f(x) = -\infty$

11 $\lim\limits_{x\to\pm\infty} f(x) = \infty$

13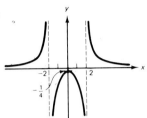

15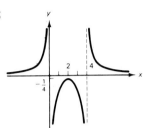

17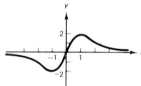

19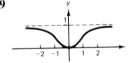

21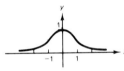

23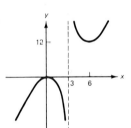

25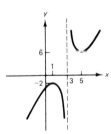

27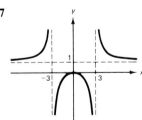

29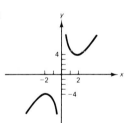

31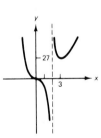

33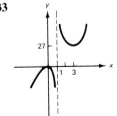

35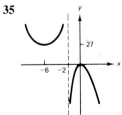

37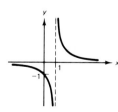

39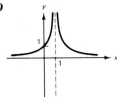

41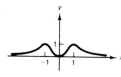

43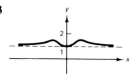

45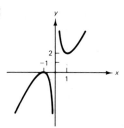

47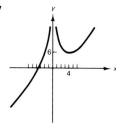

49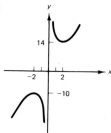

51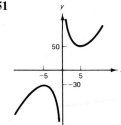

53 (a) $MC(x) = 2x + 2$

(b)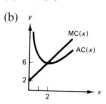

55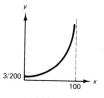

57 (a)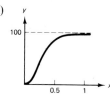

(b) $f(.65) \approx 95.5$ (more than 95% effective)

EXERCISES 3.4 page 173

1 Max f is 22 (at $x = 5$),
Min f is -48 (at $x = -2$).

3 Max f is 16 (at $x = -2$),
Min f is -16 (at $x = 2$).

5 Max f is 125 (at $x = -3$),
Min f is -100 (at $x = 0$).

7 Max f is 117 (at $x = 3$),
Min f is -99 (at $x = -3$).

9 Max f is 11 (at $x = -2$),
Min f is -16 (at $x = 1$).

11 Max f is 5 (at $x = 0$),
Min f is 0 (at $x = 5$).

13 Max f is 100 (at $x = 10$),
Min f is 0 (at $x = 0$ and at $x = 20$).

15 Max f is 1 (at $x = 0$),
Min f is 0 (at $x = 1$ and at $x = -1$).

17 Max f is 4 (at $x = 8$),
Min f is 0 (at $x = 0$).

19 Max f is 1 (at $x = 1$ and at $x = -1$),
Min f is 0 (at $x = 0$).

21 Max f is $1/2$ (at $x = 1$),
Min f is $-1/2$ (at $x = -1$).

23 On the 20th day

25 31 mph

27 24 hours

29 36 years

31 12 miles from A toward B

33 Sell 40 per day, price = $400, max profit = $6500

35 400 feet along the building and 200 feet perpendicular to the building

37 Each is 200 feet along river and 150 feet parallel to river.

39 3 inches high with a base 12 inches by 12 inches, volume: 432 cubic inches

41 The numbers are 25 and 25.

43 $r = 2$ cm

45 $r = 110/\pi \approx 35$ yards, $x = 110$ yards

EXERCISES 3.5 page 182

1 Price: $14,400, sell 16 cars per day (from $x = 2$ price reductions)

3 Ticket price: $150, number sold: 450 (from $x = 5$ price reductions)

5 Rent the cars for $90, and expect to rent 54 cars (from $x = 2$ price increases)

7 25 trees per acre (from $x = 5$ extra trees per acre)

9 Base: 2 feet by 2 feet, height: 1 foot

11 Base 14 inches by 14 inches, height: 28 inches, volume: 5488 cubic inches

13 50 feet along driveway and 100 feet perpendicular to driveway, cost: $800

15 6.4%

17 16 years

19 (*Hint*: If area is A (a constant) and one side is x, then show that the perimeter is $P = 2x + 2A/x$, which is minimized at $x = \sqrt{A}$. Then show that this means the rectangle is a square.)

21 The page should be 8 inches wide and 12 inches tall.

23 $R' = 2cpx - 3cx^2 = xc(2p - 3x)$, which is zero when $x = \frac{2}{3}p$. The second derivative test will show that R is maximized.

EXERCISES 3.6 page 190

1 Lot size: 400 bags, with 10 orders during the year

3 Lot size: 500 bottles, with 20 orders during the year

5 Lot size: 40 cars per order, with 20 orders during the year

7 Produce 1000 games per run, with 2 production runs during the year

9 Produce 40,000 records per run, with 25 runs for the year

11 Population: 20,000, yield: 40,000 (from $p = 200$)

13 Population: 75,000, yield: 2250 (from $p = 75$)

15 Population: 625,000, yield: 625,000 (from $p = 625$)

EXERCISES 3.7 page 197

1 $y' = \dfrac{2x}{3y^2}$

3 $y' = -\dfrac{x}{y}$

5 $y' = \dfrac{3x^2 + 2}{4y^3}$

7 $y' = -\dfrac{x + 1}{y + 1}$

9 $y' = -\dfrac{2y}{x}$ (after simplification)

11 $y' = \dfrac{-y + 1}{x}$

13 $y' = -\dfrac{y - 1}{2x}$ (after simplification)

15 $y' = \dfrac{1}{3y^2 - 2y + 1}$

17 $y' = -\dfrac{y^2}{x^2}$ (after simplification)

19 $y' = \dfrac{3x^2}{2(y - 2)}$

21 $y' = 2$ **23** $y' = -3$ **25** $y' = -1$ **27** $y' = -4$

29 $p' = -\dfrac{2}{2p + 1}$

31 $p' = \dfrac{1}{24p + 4}$

33 $p' = -\dfrac{p}{3x}$ (after simplification)

35 $p' = -\dfrac{p + 5}{x + 2}$

37 $dp/dx = -4$ *Interpretation*: the rate of change of price with respect to quantity is -4, so price decreases by about \$4 when quantity increases by 1.

39 (a) $s' = \dfrac{3r^2}{2s} = 8$ (b) $r' = \dfrac{2s}{3r^2} = \dfrac{1}{8}$

(c) $ds/dr = 8$ means that the rate of change of sales with respect to research expenditures is 8, so that increasing research by \$1 million will increase sales by about \$8 million (at these levels of r and s).

$dr/ds = 1/8$ means that the rate of change of research expenditures with respect to sales is $1/8$, so that increasing sales by \$1 million will increase research by about $1/8$ million dollars (at these levels of r and s).

41 $3x^2x' + 2yy' = 0$

43 $2xx'y + x^2y' = 0$

45 $6xx' - 7x'y - 7xy' = 0$

47 $2xx' + x'y + xy' = 2yy'$

49 Decreasing by 72π ≈ 226 in^3/hr

51 $32\pi \approx 101$ cm^3/week

53 Increasing by \$16,000 per day

55 Increasing by 400 cases per year

57 Slowing by 1/2 mm/sec per year

EXERCISES 3.8 page 202

1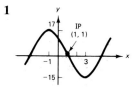

3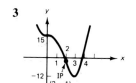

5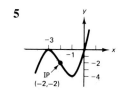

7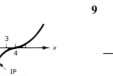

9

11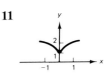

13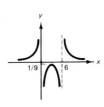

15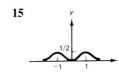

17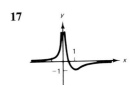

19 Max f is 220 (at $x = 5$), min f is -4 (at $x = 1$).

21 Max f is 64 (at $x = 0$), min f is -64 (at $x = 4$).

23 Max h is 4 (at $x = 9$), min h is 0 (at $x = 1$).

25 Max g is 25 (at $w = 3$ and at $w = -3$), min g is 0 (at $w = 2$ and at $w = -2$).

27 Max f is 1/2 (at $x = 1$), min f is $-1/2$ (at $x = -1$).

29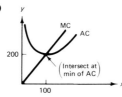

31 $v = (4/3)c$, which means that the tugboat should travel through the water at $1\frac{1}{3}$ the speed of the current. This explains why tugboats sometimes move so slowly.

33 1800 ft^2

35 Base: 10 inches by 10 inches, height: 5 inches

37 Radius = 2 inches, height = 2 inches

39 Price: $2400 each, quantity: 9 per week

41 $x = 3/4$ mile

43 Lot size: 50 per order, requiring 10 orders

45 Population: $p = 900$ (thousand), yield: 900 (thousand)

47 $y' = \dfrac{-y^2}{2xy - 1}$ (after simplification)

49 $y' = \dfrac{y^{1/2}}{x^{1/2}}$ (after simplification)

51 $y' = 1/7$

53 $y' = 1$

55 Increasing by $4200 per day

EXERCISES 4.1 page 217

1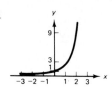

3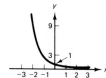

5 5.697 (rounded to 3 decimal places)

7 0.951 (rounded to 3 decimal places)

9 (a) $2144
(b) $2204
(c) $2226

11 (a) $2196
(b) $2096

13 The annual yield should be 9.69% (based on the nominal rate of 9.25%).

15 $8629

17 10% compounded quarterly (yielding 10.38%, better than 10.30%)

19 (a) $2678   (b) $12,093

21 Approximately 10 billion

23 (a) .53 (the chances are better than 50–50)
(b) .70 (quite likely)

25 (a) .267 or 26.7%
(b) .012 or 1.2%

27 208   **29** (a) About 153 degrees
(b) About 123 degrees

31 38

EXERCISES 4.2 page 232

1 (a) 2   (b) 4   (c) −1   (d) −2   (e) 1/2   (f) −1/2

3 (a) 10   (b) 1/2   (c) 4/3   (d) 0   (e) 1   (f) −3

5 $\ln x$ **7** $2 \ln x$ or $\ln x^2$ **9** $\ln x$ **11** $3x$ **13** $3 \ln x$ or $\ln x^3$

15 $7x$ **17** (a) 2.9 years (from 35 months)   (b) 1.7 years (from 20.5 months)

19 (a) 15.7 years   (b) 3.2 years   **21** 1.9 years   **23** 27 years

25 77 days   **27** .58 or 58%   **29** About 4 weeks   **31** About 13,400 years

33 1.7 million years   **35** About 138 days

EXERCISES 4.3 page 244

1 $2x \ln x + x$ **3** $\dfrac{1 - 3 \ln x}{x^4}$ **5** $\dfrac{2}{x}$ **7** $\dfrac{3x^2}{x^3 + 1}$ **9** $\dfrac{1}{2} x^{-1}$

11 $\dfrac{6x}{x^2 + 1}$ **13** $\dfrac{1}{x}$ **15** $2xe^x + x^2 e^x$ or $xe^x(2 + x)$

17 $\dfrac{xe^x - 2e^x}{x^3}$ or $\dfrac{e^x(x - 2)}{x^3}$ **19** $(3x^2 + 2)e^{x^3+2x}$ **21** $x^2 e^{x^3/3}$ **23** $5e^{x/2}$

25 $1 + e^{-x}$ **27** 2   **29** $e^{1+e^x} e^x$ or e^{1+x+e^x} **31** ex^{e-1} **33** 0

35 $\dfrac{4x^3}{x^4 + 1} - 2e^{(1/2)x} - 1$ **37** $2x \ln x + 2xe^x$ **39** (a) $\dfrac{1 - 5 \ln x}{x^6}$ (b) 1

41 (a) $\dfrac{4x^3}{x^4 + 48}$ (b) 1/2 **43** (a) $3x^2e^x + x^3e^x$ (b) $4e$

45 (a) $\dfrac{e^x - 3}{e^x - 3x}$ (b) -2 **47** (a) $5 \ln x + 5$
(b) $5 \ln 2 + 5 \approx 8.466$

49 (a) $\dfrac{xe^x - e^x}{x^2}$ (b) $\dfrac{2e^3}{9} \approx 4.463$

51 $-4x^3e^{-x^5/5} + x^8e^{-x^5/5}$ or $x^3e^{-x^5/5}(x^5 - 4)$ **53** $f^{(n)}(x) = k^n e^{kx}$

55 Rel max pt: (0, 1)
IPs: $(1/2, e^{-1/2}) \approx (1/2, .6)$
$(-1/2, e^{-1/2}) \approx (-1/2, .6)$

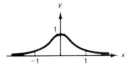

57 Rel min pt: (0, 0)
IP: $(1, \ln 2) \approx (1, .7)$
$(-1, \ln 2) \approx (-1, .7)$

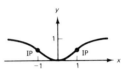

59 (a) Increasing by about \$50 per year
(b) Increasing by about \$82.44 per year

61 Increasing by about .053 billion (= 53 million) per year

63 (a) Decreasing by .06 cc/hr
(b) Decreasing by .054 cc/hr

65 (a) Increasing by about 81.4 (thousand) sales per week
(b) Increasing by about 33 (thousand) sales per week

67 $p = \$100$ **69** (a) $R(x) = 400xe^{-.20x}$ **71** 2.03 hours
(b) Quantity: $x = 5$ (thousand),
price: $p = \$147.15$

73 (a) $(\ln 10)10^x$ (b) $(\ln 3)(2x)3^{x^2+1}$ (c) $(\ln 2)3\cdot 2^{3x}$
(d) $(\ln 5)6x\cdot 5^{3x^2}$ (e) $-(\ln 2)2^{4-x}$

75 (a) $\dfrac{1}{(\ln 2)x}$ (b) $\dfrac{2x}{(\ln 10)(x^2 - 1)}$ (c) $\dfrac{4x^3 - 2}{(\ln 3)(x^4 - 2x)}$

EXERCISES 4.4 page 255 **1** (a) $\dfrac{2}{t}$ (b) 2 and .2 **3** (a) .2 (b) .2 **5** (a) $2t$ (b) 20

7 (a) $-2t$ (b) -20 **9** (a) $\dfrac{1}{2(t - 1)}$ (b) $\dfrac{1}{10}$ **11** .0071 or .71%

13 (a) $E(p) = \dfrac{5p}{200 - 5p}$ **15** (a) $E(p) = \dfrac{2p^2}{300 - p^2}$
(b) inelastic ($E = 1/3$) (b) unitary elastic ($E = 1$)

17 (a) $E(p) = 1$ **19** (a) $E(p) = \dfrac{3p}{2(175 - 3p)}$
(b) unitary elastic ($E = 1$) (b) elastic ($E = 3$)

21 (a) $E(p) = 2$
(b) elastic $(E = 2)$

23 (a) $E(p) = .01p$
(b) elastic $(E = 2)$

25 Lower prices $(E = 8)$

27 No $(E = 3/2)$

29 Yes $(E = 3/8 = .375)$

31 Lower prices $(E = 1.2)$

33 $E(p) = \dfrac{-pa(-c)e^{-cp}}{ae^{-cp}} = cp$

35 $E_s(p) = n$

EXERCISES 4.5 page 258

1 (a) \$18,845.41
(b) \$18,964.81

3 (a) $V(t) = 800,000(.8)^t$
(b) \$327,680

5 (a) 7.1 years (from 14.2 half-years)
(b) 4.2 years (from 8.3 half-years)

7 50.7 million years

9 2.3 hours   **11** (a) 12 years   (b) 11.9 years

13 $\dfrac{1}{x}$ **15** $\dfrac{-1}{1-x}$ or $\dfrac{1}{x-1}$ **17** $\dfrac{1}{3x}$ **19** $\dfrac{2}{x}$

21 $-2xe^{-x^2}$ **23** $2x$ **25** $10x + 2\ln x + 2$ **27** $6x^2 - 3e^{2x} - 6xe^{2x}$

29 Rel min: $(0, \ln 4) \approx (0, 1.4)$
IPs: $(2, \ln 8) \approx (2, 2.1)$
$\quad\ (-2, \ln 8) \approx (-2, 2.1)$

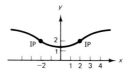

31 (a) Increasing by 136 thousand per week
(b) Increasing by 55 thousand per week

33 Decreasing by 3.3% per second

35 (a) Increasing by 6667 per hour
(b) Increasing by 816 per hour

37 (a) $R(x) = 200xe^{-.25x}$
(b) Quantity $x = 4$ (thousand),
price $p = \$73.58$

39 Price $= \$50$

41 .0031   or   .31%

43 Raise prices $(E = .7)$

EXERCISES 5.1 page 272

1 $\dfrac{1}{5}x^5 + C$ **3** $\dfrac{3}{5}x^{5/3} + C$ **5** $\dfrac{2}{3}u^{3/2} + C$ **7** $-\dfrac{1}{3}w^{-3} + C$

9 $2\sqrt{z} + C$ **11** $x^6 + C$ **13** $2x^4 - x^3 + 2x + C$ **15** $4x^{3/2} + \dfrac{3}{2}x^{2/3} + C$

17 $6x^{8/3} + 24x^{-2/3} + C$ **19** $6t^{5/3} + 3t^{1/3} + C$ **21** $\dfrac{1}{3}x^3 - x^2 + x + C$

23 $\dfrac{2}{3}w^{3/2} + 4w^{5/2} + C$ **25** $2x^3 - 3x^2 + x + C$ **27** $\dfrac{1}{3}x^3 + x^2 - 8x + C$

29 $\dfrac{1}{3}r^3 - r + C$ **31** $\dfrac{1}{2}x^2 - x + C$ **33** $\dfrac{1}{4}t^4 + t^3 + \dfrac{3}{2}t^2 + t + C$

35 (a) $-\dfrac{1}{2}x^{-2} + C$ (b) $\dfrac{x + C}{(1/4)x^4 + C_1}$ (where C_1 is another arbitrary constant)

37 $C(x) = 8x^{5/2} - 9x^{5/3} + x + 4000$　　　**39** $R(x) = 9x^{4/3} + 2x^{3/2}$

41 (a) $D(t) = -.03t^3 + 4t^2$　　　(b) 370 feet　　　**43** (a) $6t^{1/2}$　　　(b) 30 words

45 (a) $P(t) = 16t^{5/2}$　　　　　　**47** $\dfrac{1}{x}$
　　(b) 512 tons　　　(c) No

EXERCISES 5.2 page 281　　**1** $\dfrac{1}{3} e^{3x} + C$　　**3** $4e^{(1/4)x} + C$　　**5** $20e^{.05x} + C$　　**7** $-\dfrac{1}{2} e^{-2y} + C$

9 $-2e^{-.5x} + C$　　**11** $9e^{(2/3)x} + C$　　**13** $-24e^{-(3/4)u} + C$　　**15** $2 \ln |x| + C$

17 $-5 \ln |x| + C$　　**19** $3 \ln |x| + C$　　**21** $-3x^{-1} + C$　　**23** $\dfrac{1}{3} \ln |v| + C$

25 $\dfrac{3}{2} \ln |x| + C$　　**27** $\dfrac{1}{3} e^{3x} - 3 \ln |x| + C$　　**29** $6e^{.5t} - 2 \ln |t| + C$

31 $\dfrac{1}{3} x^3 + \dfrac{1}{2} x^2 + x + \ln |x| - x^{-1} + C$　　**33** $250e^{.02t} - 200e^{.01t} + C$

35 (a) $360e^{.05t} - 355$
　　(b) About 626

37 (a) $50 \ln t$　(since $t > 1$, absolute value bars are not needed)
　　(b) No (about 170 sold)

39 (a) $700e^{.02t} - 700$　　　　　　**41** (a) $500e^{.4x} - 500$
　　(b) In about 2022 (37 years from 1985)　　(b) About $3195

43 (a) $60e^{-.2t} + 10$　　**45** (a) $-4000e^{-.2t} + 4000$　　**47** $\dfrac{1}{2} x^2 + 2x + \ln |x| + C$
　　(b) In about 5 hours　　　(b) About $3\frac{1}{2}$ years

49 $t - 2 \ln |t| + 3t^{-1} + C$　　**51** $\dfrac{1}{3} x^3 - 3x^2 + 12x - 8\ln |x| + C$

EXERCISES 5.3 page 293　　**1** 16　　**3** $2e^2$　2　　**5** $\ln 5$　　**7** 19　　**9** $2\frac{1}{4}$ or $\frac{9}{4}$　　**11** 2　　**13** 16

15 13　　**17** $-e^{-1} + e$ or $e - \dfrac{1}{e}$　　**19** $-2e^{-2} + 2e^{-1}$　　**21** $e - 2$

23 $9 \ln 4$　　**25** $\dfrac{2}{3} - \ln 3$　　**27** $\dfrac{3}{2} - \ln 2$　　**29** 1　　**31** 2

33 $\dfrac{7}{2} + \ln 2$　　**35** 21 square units　　**37** $\dfrac{15}{32}$ square unit

39 27 square units　　**41** 16 square units　　**43** $3e^8 - 3e^2$ square units

45 $2e - 2$ square units　　**47** $\ln 5$ square units

49 (a) 2 square units　　　　　　**51** $\ln a$ square units (this area
　　(b)　　　　　　　　　　　　　　　is sometimes taken to
　　　　　　　　　　　　　　　　　　be the *definition* of $\ln a$)

53 132 units   **55** 411 checks

57 $-300e^{-2} + 300 \approx \$259.40$ **59** $75e^{1.2} - 75 \approx \$1.74$

61 $27e^{.15} - 27e^{.05} \approx 3$ million tons   **63** $-30e^{-2} + 30 \approx 26$ words

65 $\dfrac{a}{-b+1} B^{-b+1} - \dfrac{a}{-b+1} A^{-b+1} = \dfrac{a}{1-b}(B^{1-b} - A^{1-b})$

67 $-4,000,000e^{-.6} + 4,000,000 \approx \$1,804,754$

EXERCISES 5.4 page 305   **1** 3   **3** 4   **5** $\dfrac{1}{5}$ **7** 5   **9** $\dfrac{104}{3}$ or $34\dfrac{2}{3}$ **11** 3   **13** $e - 1$

15 $\dfrac{1}{2}(1 - e^{-2})$ **17** $\ln 2$ **19** $\dfrac{1}{n+1}$ **21** $a + b$ **23** 318   **25** 70

27 About 25.6 tons   **29** $3194.53   **31** 6 square units

33 $\dfrac{1}{2} e^4 - e^2 + \dfrac{1}{2}$ square units   **35** (a)   **37** (a)

                                                (b) 9 square units

                                                             (b) 24 square units

39 4 square units   **41** 32 square units   **43** 1/12 square unit

45 81 square units   **47** 2 square units   **49** About 148 million

51 (a) 10   **53** About $529 billion   **55** About $139 thousand
    (b) $6000

57 $(2x + 5)e^{x^2 + 5x}$ **59** $\dfrac{2x + 5}{x^2 + 5x}$

EXERCISES 5.5 page 316   **1** $60,000   **3** $10,000   **5** $7500   **7** $1000   **9** $160,000

11 (a) $x = 500$ **13** (a) $x = 50$ **15** .52   **17** .35   **19** .2
    (b) $50,000   (b) $2500
    (c) $25,000   (c) $7500

21 $1 - \dfrac{2}{n+1} = \dfrac{n-1}{n+1}$ **23** $4(x^5 - 3x^3 + x - 1)^3(5x^4 - 9x^2 + 1)$

25 $\dfrac{4x^3}{x^4 + 1}$ **27** $3x^2 e^{x^3}$

EXERCISES 5.6 page 326   **1** $\dfrac{1}{10}(x^2 + 1)^{10} + C$ **3** $\dfrac{1}{20}(x^2 + 1)^{10} + C$ **5** $\dfrac{1}{5} e^{x^5} + C$ **7** $\dfrac{1}{6} \ln(x^6 + 1) + C$

9 $u = x^3 + 1$, $du = 3x^2 \, dx$ **11** $u = x^4$, $du = 4x^3 \, dx$
    the powers in the integral   the powers in the integral
    and the du do not match.   and the du do not match.

13 $\frac{1}{24}(x^4 - 16)^6 + C$　　　**15** $-(1/2)e^{-x^2} + C$　　　**17** $(1/3)e^{3x} + C$

19 Cannot be solved by our substitution formulas　　　**21** $\frac{1}{5}\ln|1 + 5x| + C$　　　**23** $\frac{1}{4}(x^2 + 1)^{10} + C$

25 $\frac{1}{5}(z^4 + 16)^{5/4} + C$　　　**27** Cannot be solved by our substitution formulas　　　**29** $\frac{1}{24}(2y^2 + 4y)^6 + C$

31 $\frac{1}{2}e^{x^2+2x+5} + C$　　　**33** $\frac{1}{12}\ln|3x^4 + 4x^3| + C$　　　**35** $-\frac{1}{12}(3x^4 + 4x^3)^{-1} + C$

37 $-\frac{1}{2}\ln|1 - x^2| + C$　　　**39** $\frac{1}{16}(2x - 3)^8 + C$　　　**41** $\frac{1}{2}\ln(e^{2x} + 1) + C$

43 $\frac{1}{2}(\ln x)^2 + C$　　　**45** $2e^{x^{1/2}} + C$　　　**47** $\frac{1}{4}x^4 + \frac{1}{3}x^3 + C$

49 $\frac{1}{6}x^6 + \frac{2}{5}x^5 + \frac{1}{4}x^4 + C$　　　**51** $\frac{1}{2}e^9 - \frac{1}{2}$　　　**53** $\frac{1}{2}\ln 2$　　　**55** $32\frac{2}{3}$

57 $-\ln 2$　　　**59** $3e^2 - 3e$　　　**61** (a) $u^n u'$　　　(b) $\int u^n u'\, dx$

agree

63 (a) $\dfrac{u'}{u}$　　　(b) $\int \dfrac{u'}{u}\, dx$　　　**65** $\frac{1}{2}\ln(2x + 1) + 50$　　　**67** $\frac{1}{2}$ million

agree

69 $\frac{1}{3}\ln 5 - \frac{1}{3}\ln 2 \approx .305$ million　　　**71** $20\frac{1}{3}$ units　　　**73** About 346

75 $\frac{1}{2}\ln 10 \approx 1.15$ tons

EXERCISES 5.7 page 331　　　**1** $4x^{3/2} - 5x + C$　　　**3** $6x^{5/3} - 2x^2 + C$　　　**5** $\frac{1}{4}(x^3 - 1)^{4/3} + C$

7 Cannot be integrated by our methods　　　**9** $-\frac{1}{3}\ln|9 - 3x| + C$　　　**11** $\frac{1}{3}(9 - 3x)^{-1} + C$

13 $\frac{1}{2}(8 + x^3)^{2/3} + C$　　　**15** $-\frac{1}{2}(w^2 + 6w - 1)^{-1} + C$　　　**17** $\frac{2}{3}(1 + \sqrt{x})^3 + C$

19 $2e^{(1/2)x} + C$　　　**21** $2e^{3x} - 6\ln|x| + C$　　　**23** $\ln|e^x - 1| + C$

25 $\frac{1}{3}x^3 - 16x + C$　　　**27** 36　　　**29** 2　　　**31** 4　　　**33** 4　　　**35** $1 - e^{-2}$

37 $20e^5 - 100e + 80$　　　**39** $\frac{1}{4}e - \frac{1}{4}$　　　**41** $6e^6 - 6$ square units

43 $\ln 100$ square units　　　**45** 8 square units　　　**47** 4/3 square units

49 1/6 square unit   **51** 2 square units   **53** $\frac{1}{4}$

55 $\frac{1}{4} - \frac{1}{4} e^{-4} = \frac{1}{4}(1 - e^{-4})$ **57** $C(x) = (2x + 9)^{1/2} + 97$

59 ln 28 ≈ 3.33 degrees   **61** $1.5e - 1.5 \approx 2.6$ degrees   **63** About 5.6 years

65 27 square meters   **67** $480,000   **69** G.I. ≈ .56

EXERCISES 6.1 page 344

1 $\frac{1}{2} e^{2x} + C$ **3** $\frac{1}{2} x^2 + 2x + C$ **5** $\frac{2}{3} x^{3/2} + C$ **7** $\frac{1}{5}(x + 3)^5 + C$

9 $\frac{1}{2} xe^{2x} - \frac{1}{4} e^{2x} + C$ **11** $\frac{1}{6} x^6 \ln x - \frac{1}{36} x^6 + C$ **13** $(x + 2)e^x - e^x + C$

15 $\frac{2}{3} x^{3/2} \ln x - \frac{4}{9} x^{3/2} + C$ **17** $\frac{1}{6}(x - 3)(x + 4)^6 - \frac{1}{42}(x + 4)^7 + C$

19 $-2te^{-.5t} - 4e^{-.5t} + C$ **21** $-t^{-1} \ln t - t^{-1} + C$

23 $\frac{1}{10} s(2s + 1)^5 - \frac{1}{120}(2s + 1)^6 + C$ **25** $-\frac{1}{2} xe^{-2x} - \frac{1}{4} e^{-2x} + C$

27 $2x(x + 1)^{1/2} - \frac{4}{3}(x + 1)^{3/2} + C$ **29** $\frac{1}{a} xe^{ax} - \frac{1}{a^2} e^{ax} + C$

31 $\frac{1}{n + 1} x^{n+1} \ln ax - \frac{1}{(n + 1)^2} x^{n+1} + C$ **33** $x \ln x - x + C$

35 $\frac{1}{2} x^2 e^{x^2} - \frac{1}{2} e^{x^2} + C$

37 (a) $\frac{1}{2} e^{x^2} + C$ (b) $\frac{1}{4}(\ln x)^4 + C$
     (by substitution)   (by substitution)

(c) $\frac{1}{3} x^3 \ln 2x - \frac{1}{9} x^3 + C$ (d) $\ln(e^x + 4) + C$
(by parts)   (by substitution)

39 $2e^2 - e^2 + 1$ **41** $9 \ln 3 - 3 + \frac{1}{9}$ **43** $\frac{2^6}{30} = \frac{32}{15}$ **45** $4 \ln 4 - 3$

47 (a) $\frac{1}{6} x(x - 2)^6 - \frac{1}{42}(x - 2)^7 + C$ (b) $\frac{1}{7}(x - 2)^7 - \frac{1}{3}(x - 2)^6 + C$

49 Using $u = x^n$ and $dv = e^x\,dx$ **51** $x^2 e^x - 2xe^x + 2e^x + C$
the result follows
immediately.

53 (a) The result follows   **55** $R(x) = 4xe^{(1/4)x} - 16e^{(1/4)x} + 16$
     immediately
(b) (*Hint*: Think of the C.)

57 $105.7 million   **59** $-14e^{-2.5} + 4 \approx 2.85$ mg   **61** $2 \ln 2 - 1 + 1/4$
                                                                    ≈ .64 square unit

63 $72 \ln 6 - 24 + (1/9)$ **65** $-x^2e^{-x} - 2xe^{-x} - 2e^{-x} + C$
≈ 105 thousand customers

67 $(x + 1)^2e^x - 2(x + 1)e^x - 2e^x + C$ **69** $\frac{1}{3}x^3(\ln x)^2 - \frac{2}{9}x^3 \ln x + \frac{2}{27}x^3 + C$

EXERCISES 6.2 page 353

1 Formula 12   **3** Formula 14   **5** Formula 9
$a = 5, \quad b = -1$ $a = -1, \quad b = 7$ $a = -1, \quad b = 1$

7 $\frac{1}{6} \ln \left| \frac{3 + x}{3 - x} \right| + C$ **9** $-\frac{1}{x} - 2 \ln \left| \frac{x}{2x + 1} \right| + C$ **11** $-x - \ln |1 - x| + C$

13 $\ln \left| \frac{2x + 1}{x + 1} \right| + C$ **15** $\frac{x}{2} \sqrt{x^2 - 4} - 2 \ln |x + \sqrt{x^2 - 4}| + C$

17 $-\ln \left| \frac{1 + \sqrt{1 - z^2}}{z} \right| + C$ **19** $\frac{1}{2} x^3 e^{2x} - \frac{3}{4} x^2 e^{2x} + \frac{3}{4} xe^{2x} - \frac{3}{8} e^{2x} + C$

21 $-\frac{1}{100} x^{-100} \ln x - \frac{1}{10,000} x^{-100} + C$ **23** $\frac{1}{3} \ln \left| \frac{x}{x + 3} \right| + C$

25 $\frac{1}{8} \ln \left| \frac{z^2 - 2}{z^2 + 2} \right| + C$ **27** $\frac{x}{2} \sqrt{9x^2 + 16} + \frac{8}{3} \ln |3x + \sqrt{9x^2 + 16}| + C$

29 $-\frac{1}{2} \ln \left| \frac{2 + \sqrt{4 - e^{2t}}}{e^t} \right| + C$ **31** $\frac{1}{2} \ln \left| \frac{e^t - 1}{e^t + 1} \right| + C$

33 $\frac{1}{4} \ln |x^4 + \sqrt{x^8 - 1}| + C$ **35** $\frac{1}{3} \ln \left| \frac{\sqrt{x^3 + 1} - 1}{\sqrt{x^3 + 1} + 1} \right| + C$

37 $\frac{1}{2} \ln \left| \frac{e^t - 1}{e^t + 1} \right| + C$ **39** $2xe^{(1/2)x} - 4e^{(1/2)x} + C$

41 $\frac{1}{4} \ln \left| \frac{e^{-x} + 4}{e^{-x}} \right| + C = \frac{1}{4} \ln (1 + 4e^x) + C$ **43** $\frac{15}{2} - 8 \ln 8 + 8 \cdot \ln 4$

45 $\frac{1}{2} \ln \frac{1}{2} - \frac{1}{2} \ln \frac{1}{3}$ **47** $-4 + 5 \ln 3$ **49** $\frac{1}{2} \ln |2x + 6| + C$

51 $\frac{x}{2} - \frac{3}{2} \ln |2x + 6| + C$ **53** $-\frac{1}{3} (1 - x^2)^{3/2} + C$

55 $\sqrt{1 - x^2} - \ln \left| \frac{1 + \sqrt{1 - x^2}}{x} \right| + C$

57 $\frac{1}{2} \left[\ln |x + 1| - \frac{1}{3} \ln |3x + 1| \right] - \frac{1}{2} \ln \left| \frac{3x + 1}{x} \right| + C$

59 $\ln |x + \sqrt{x^2 + 1}| - \ln \left| \frac{1 + \sqrt{x^2 + 1}}{x} \right| + C$ **61** $x + 2 \ln |x - 1| + C$

63 $-x^2e^{-x} - 2xe^{-x} - 2e^{-x} + 2$ million sales   **65** 24 generations

67 $C(x) = \ln [x + \sqrt{x^2 + 1}] + 2000$

EXERCISES 6.3 page 363   **1** 0   **3** 1   **5** Does not exist   **7** Does not exist   **9** 1/2   **11** 1/8

13 Divergent   **15** 100   **17** $2e^{-2}$ **19** Divergent

21 Divergent   **23** 20   **25** 1/2   **27** Divergent   **29** 1/3

31 1/3   **33** Divergent   **35** 1   **37** Divergent   **39** $200,000

41 (a) $10,000   (b) $9999.55   **43** 1,000,000 barrels (from 1000 thousand)

45 2 square units   **47** $\dfrac{1}{a}$ square units   **49** .61 or 61%   **51** .30 or 30%

53 20,000

EXERCISES 6.4 page 376   *Note*: Your answers may differ slightly, depending on the stage at which you do the rounding.

1 (a) 8.75   **3** (a) .697   **5** 1.153   **7** .743
  (b) 8.667   (b) .693
  (c) .083   (c) .004   **9** .556   **11** 8.772
  (d) 1%   (d) 0.6%

13 .477 or about 48%   **15** .879 feet   **17** 8.667   **19** .693   **21** 1.148

23 .747   **25** .582   **27** 8.696   **29** About 821 feet

EXERCISES 6.5 page 389   **1** Check that $(4e^{2x} - 3e^x) - 3(2e^{2x} - 3e^x) + 2(e^{2x} - 3e^x + 2) \overset{?}{=} 4$

3 Check that $kae^{ax} \overset{?}{=} a(ke^{ax} - b/a) + b$ **5** $y = \sqrt[3]{6x^2 + c}$

7 $y = ce^{2x^3}$ Check that $c6x^2e^{2x^3} \overset{?}{=} 6x^2(ce^{2x^3})$

9 $y = cx$ (since $e^{\ln x} = x$) Check that $c \overset{?}{=} cx/x$

11 $y = \sqrt{4x^2 + c}$ and $y = -\sqrt{4x^2 + c}$ **13** $y = 3x^3 + C$

15 $y = \dfrac{1}{2}\ln(x^2 + 1) + C$ **17** $y = ce^{(1/3)x^3}$ **19** $y = \left(\dfrac{1-n}{m+1}x^{m+1} + c\right)^{1/(1-n)}$

21 $y = (x + c)^2$ **23** $y = ce^{(1/2)x^2} - 1$ **25** $y = ce^{e^x} + 1$

27 $y = \dfrac{1}{c - ax}$ **29** $y = ce^{ax} - \dfrac{b}{a}$ **31** $y = \sqrt[3]{3x^2 + 8}$

33 $y = -e^{(1/2)x^2}$ Check that $-xe^{(1/2)x^2} \overset{?}{=} x(-e^{(1/2)x^2})$ and $y(0) = -e^0 = -1$

35 $y = (1 - x^2)^{-1}$ Check that $-(1 - x^2)^{-2}(-2x) \overset{?}{=} 2x[(1 - x^2)^{-1}]^2$ and $y(0) = (1 - 0)^{-1} = 1$

37 $y = 3x$ (using $e^{\ln x} = x$) Check that $3 \overset{?}{=} 3x/x$ **39** $y = (x + 1)^2$

41 $y = \dfrac{1}{2 - e^x - x}$ **43** $y = 2e^{(1/3)ax^3}$ **45** $D(p) = cp^{-k}$ (for any constant c)

47 $y = 20{,}000e^{.05t} - 20{,}000$ **49** (a) $y = 28.6e^{-.32t} + 70$
                                              (b) About 3.28 hours earlier

51 $y = 150 - 150e^{-.2t}$ **53** (a) $y' = 4 + .1y$ and $y(0) = 0$ (b) $y = 46e^{.1t} - 40$

EXERCISES 6.6 page 406   **1** $y' = cae^{at} = a(ce^{at}) = ay$ **3** Unlimited   **5** Limited   **7** Neither
          $y(0) = ce^0 = c$

9 Logistic **11** Logistic **13** $y' = .08y$ **15** $y' = a(100,000 - y)$
$y = 1500e^{.08t}$ $y = 100,000(1 - e^{-.021t})$
about 22,276

17 $y' = a(5000 - y)$ **19** $y' = ay(10,000 - y)$ **21** $y' = ay(800 - y)$
$y = 5000(1 - e^{-.223t})$ $y = \dfrac{10,000}{1 + 99e^{-.535t}}$ $y = \dfrac{800}{1 + 799e^{-.558t}}$
about 7.2 weeks
about 8612   about 675 people

23 $y' = ay(800 - y)$ **25** $y = 5e^{-.15t}$ **27** $y = \dfrac{1}{c - at}$ **29** $y = ce^{a \ln x} = cx^a$
$y = \dfrac{800}{1 + 7e^{-.280t}}$ about 3.7
about 6.9 years

31 The solution follows from the indicated steps.

EXERCISES 6.7 page 407

1 $\dfrac{1}{2} xe^{2x} - \dfrac{1}{4} e^{2x} + C$ **3** $\dfrac{1}{9} x^9 \ln x - \dfrac{1}{81} x^9 + C$

5 $\dfrac{1}{6} (x - 2)(x + 1)^6 - \dfrac{1}{42} (x + 1)^7 + C$ **7** $2t^{1/2} \ln t - 4t^{1/2} + C$

9 $x^2 e^x - 2xe^x + 2e^x + C$ **11** $\dfrac{1}{n + 1} x(x + a)^{n+1} - \dfrac{1}{(n + 1)(n + 2)} (x + a)^{n+2} + C$

13 $4e^5 + 1$ **15** $-\ln |1 - x| + C$ **17** $\dfrac{1}{4} x^4 \ln 2x - \dfrac{1}{16} x^4 + C$

19 $\dfrac{1}{2} (\ln x)^2 + C$ **21** $2e^{\sqrt{x}} + C$ **23** \$902 million
(from $10,000 - 15,000e^{-.5}$)

25 $\dfrac{1}{10} \ln \left| \dfrac{5 + x}{5 - x} \right| + C$ **27** $2 \ln |x - 2| - \ln |x - 1| + C$

29 $\ln \left| \dfrac{\sqrt{x + 1} - 1}{\sqrt{x + 1} + 1} \right| + C$ **31** $\ln |x + \sqrt{x^2 + 9}| + C$

33 $\dfrac{z^2 - 2}{3} \sqrt{z^2 + 1} + C$ (from formula 13)   **35** $\dfrac{1}{2} x^2 e^{2x} - \dfrac{1}{2} xe^{2x} + \dfrac{1}{4} e^{2x} + C$

37 $\ln \left| \dfrac{2x + 1}{x + 1} \right| + 1000$ **39** $\dfrac{1}{4}$ **41** Divergent   **43** $\dfrac{1}{2}$ **45** Divergent

47 5   **49** 1/4   **51** 1/2   **53** 1   **55** 1/3   **57** \$60,000

59 240 thousand   **61** 1.102   **63** 1.204   **65** .57   **67** 1.0894

69 1.1951   **71** .5285   **73** (a) $-\displaystyle\int_1^0 \dfrac{1}{1 + t^2} \, dt = \int_0^1 \dfrac{1}{1 + t^2} \, dt$ (b) .783

75 $y = \sqrt[3]{x^3 + c}$ **77** $y = \dfrac{1}{4} \ln |x^4 + 1| + C$ **79** $y = \dfrac{1}{c - x}$

81 $y = 1 - e^{c-x}$ **83** $y = ce^{(1/2)x^2 - x}$ **85** $y = \sqrt[3]{3x^3 + 1}$ **87** $y = 7e^{-(1/2)x^{-2}}$

89 (a) $y' = 4 + .05y$ and $y(0) = 10$ **91** $y = 106 - 36e^{-2.3t}$ **93** $y' = .07y$
(b) $90e^{.05t} - 80$ $y = 25e^{.07t}$
(c) 68.385 thousand or \$68,385   58¢

95 $y' = ay(8000 - y)$ **97** $y' = a(10,000 - y)$
$y = \dfrac{8000}{1 + 799e^{-2.73t}}$ (t in weeks) $y = 10,000(1 - e^{-.051t})$
about 1819 cases   about 4577

99 $y' = a(500,000 - y)$
$y = 500,000(1 - e^{-.255t})$ (t in weeks)
about 6.3 weeks

EXERCISES 7.1 page 419

1 $\{(x, y) \mid x \neq 0, y \neq 0\}$ **3** $\{(x, y) \mid x \neq y\}$ **5** $\{(x, y) \mid x > 0, y \neq 0\}$

7 $\{(x, y, z) \mid x \neq 0, y \neq 0, z > 0\}$ **9** 3   **11** 4   **13** -2 **15** 1

17 $e^{-1} + e$ **19** e^{-1} **21** 0   **23** .0157   **25** 45 minutes   **27** 472.7

29 $P(2L, 2K) = a(2L)^b(2K)^{1-b} = a2^bL^b2^{1-b}K^{1-b} = 2aL^bK^{1-b} = 2P(L, K)$

31 1548 calls   **33** $C(x, y) = 210x + 180y + 4000$

35 (a) $V = xyz$ (b) $M = xy + 2xz + 2yz$

EXERCISES 7.2 page 433

1 (a) $3x^2 + 6xy^2 - 1$ **3** (a) $10x^4 - 4xy^3 - 18$ **5** (a) $6x^{-1/2}y^{1/3}$
(b) $6x^2y - 6y^2 + 1$ (b) $-6x^2y^2 - 3$ (b) $4x^{-1/2}y^{-2/3}$

7 (a) $5x^{-.95}y^{.02}$ **9** (a) $-(x + y)^{-2}$ **11** (a) $10xy(x^2y + 1)^4$
(b) $2x^{.05}y^{-.98}$ (b) $-(x + y)^{-2}$ (b) $5x^2(x^2y + 1)^4$

13 (a) $\dfrac{3x^2}{x^3 + y^3}$ **15** (a) $6x^2e^{-5y}$ **17** (a) ye^{xy}
                                       (b) $-10x^3e^{-5y}$ (b) xe^{xy}
(b) $\dfrac{3y^2}{x^3 + y^3}$

19 (a) $\dfrac{x}{x^2 + y^2}$ or $x(x^2 + y^2)^{-1}$ **21** (a) $\dfrac{-4x^2y + 4y^3}{(x^2 + y^2)^2}$ **23** (a) $3v(uv - 1)^2$
                                                                                                                   (b) $3u(uv - 1)^2$
(b) $\dfrac{y}{(x^2 + y^2)^2}$ or $y(x^2 + y^2)^{-1}$ (b) $\dfrac{4x^3 - 4xy^2}{(x^2 + y^2)^2}$

25 (a) $ue^{(1/2)(u^2-v^2)}$ **27** 18, -10 **29** 0, $2e$ **31** $1\frac{1}{2}$ **33** 3/5
(b) $-ve^{(1/2)(u^2-v^2)}$

35 (a) $30x - 4y^3$ **37** (a) $-2x^{-5/3}y^{2/3}$
(b) and (c) $-12xy^2$ (b) and (c) $2x^{-2/3}y^{-1/3} - 12y^2$
(d) $-12x^2y + 36y^2$ (d) $-2x^{1/3}y^{-4/3} - 24xy$

39 (a) ye^x **41** All three are $36x^2y^2$.   **43** (a) y^2z^3
(b) and (c) $e^x - \dfrac{1}{y}$ (b) $2xyz^3$
(d) xy^{-2} (c) $3xy^2z^2$

45 (a) $8x(x^2 + y^2 + z^2)^3$ **47** (a) $4(xy + yz + zx)^3(y + z)$ **49** (a) $2xe^{x^2+y^2+z^2}$
(b) $8y(x^2 + y^2 + z^2)^3$ (b) $4(xy + yz + zx)^3(x + z)$ (b) $2ye^{x^2+y^2+z^2}$
(c) $8z(x^2 + y^2 + z^2)^3$ (c) $4(xy + yz + zx)^3(x + y)$ (c) $2ze^{x^2+y^2+z^2}$

51 (a) $\dfrac{2x}{x^2 + y^2 + z^2}$

(b) $\dfrac{2y}{x^2 + y^2 + z^2}$

(c) $\dfrac{2z}{x^2 + y^2 + z^2}$

53 -14

55 $4e^6$

57 (a) $P_x = 4x - 3y + 150$
(b) \$50 (profit per additional tape deck)
(c) $P_y = -3x + 6y + 75$
(d) \$75 (profit per additional CD player)

59 (a) 250 (the marginal productivity of labor is 250, so production increases by about 250 for each additional unit of labor)
(b) 108 (the marginal productivity of capital is 108, so production increases by about 108 for each additional unit of capital)
(c) Labor

61 $S_x = -.1$ (sales fall by .1 for each dollar price increase)
$S_y = .4y$ (sales rise by .4y for each additional advertising dollar above the level y)

63 (a) .52 (status increases by .52 unit for each additional \$1000 of income)
(b) 5.25 (status increases by 5.25 units for each additional year of education)

65 (a) 97 (skid distance increases by about 97 feet for each additional ton)
(b) 12.96 (skid length increases by about 13 feet for each additional mph)

67 (a) Rate at which butter sales change as butter prices rise
(b) Negative: as prices rise, sales will fall.
(c) Rate at which butter sales change as margarine prices rise
(d) Positive: as margarine prices rise, people will switch to butter, so butter sales will rise.

EXERCISES 7.3 page 445

1 Rel min value: $f = 5$ at $x = 0$, $y = -1$

3 Rel min value: $f = -12$ at $x = -2$, $y = 2$

5 Rel max value: $f = 23$ at $x = 5$, $y = 2$

7 No rel extreme values
[saddle point at $(2, -4)$]

9 No rel extreme values

11 Rel min value: $f = 1$ at $x = 0$, $y = 0$

13 Rel min value: $f = 0$ at $x = 0$, $y = 0$

15 Rel max value: $f = 3$ at $x = 1$, $y = -1$
[saddle point at $(-1, 1)$]

17 Rel max value: $f = 17$ at $x = -1$, $y = -2$
[saddle point at $(-1, 2)$]

19 No rel extreme values
[saddle point at $(2, 6)$]

21 10 units of product A, sell for \$7000 each
7 units of product B, sell for \$13000 each
Maximum profit: \$22,000

23 (a) $P = -.2x^2 + 16x - .1y^2 + 12y - 20$
(b) 40 cars in America, sell for \$12,000;
60 cars in Europe, sell for \$10,000

25 6 hours of practice
and 1 hour of rest

27 (a) $x = 1200$, $p = \$6$, $R = \$7200$
(b) $x = 800$, $y = 800$, $p = \$4$, revenue $= \$3200$ for each
(c) Duopoly (1600 versus 1200)
(d) Duopoly

29 (a) $P = -.2x^2 + 16x - .1y^2 + 12y - .1z^2 + 8z - 22$
(b) 40 in America, 60 in Europe, 40 in Asia

31 Rel min value: $f = -1$ at $x = 1$, $y = 1$
[saddle point at $(0, 0)$]

33 Rel max value: $f = 32$ at $x = 4$, $y = 4$
[saddle point at $(0, 0)$]

35 Rel min value: $f = -162$ at $x = 3$, $y = 18$
[saddle point at $(0, 0)$]

EXERCISES 7.4 page 454

Note: Your answers may differ slightly depending on the stage at which you do the rounding.

1 $y = 3.5x - 1.67$ **3** $y = -.79x + 6.6$ **5** $y = 2.4x + 6.9$

7 $y = -1.5x + 7.4$ **9** $y = 2.2x + 5$ **11** $y = -8x + 125$
prediction: 16 million   prediction: 85

13 $y = -7.7x + 90.5$,   **15** $y = -.93x + 53$,   **17** $y = -.16x + 71.6$
prediction: 52   prediction: 34.5%

19 $y = 1.08e^{.63x}$ **21** $y = 17.46e^{-.47x}$ **23** $y = 1.02e^{.78x}$ **25** $y = 16.95e^{-.52x}$

27 $y = 46.53e^{.655x}$,   **29** $y = 79.84e^{.42x}$,
prediction: \$1,230,000   prediction: \$12,333,000
(from 1230 thousands)   (from 12,333 thousand, using $x = 12$)

EXERCISES 7.5 page 471

1 Max $f = 36$ **3** Max $f = 144$ **5** Max $f = -28$
at $x = 6$, $y = 2$ at $x = 6$, $y = 4$ at $x = 3$, $y = 5$

7 Max $f = 6$ **9** Max $f = 2$ (from ln e^2)   **11** Min $f = 45$
at $x = 2$, $y = -1$ at $x = e$, $y = e$ at $x = 6$, $y = 3$

13 Min $f = -16$ **15** Min $f = 52$ **17** Min $f = \ln 125$
at $x = -4$, $y = 4$ at $x = 4$, $y = 6$ at $x = 10$, $y = 5$

19 Min $f = e^{20}$ **21** Max $f = 8$ at $x = 2$, $y = 2$ and at $x = -2$, $y = -2$;
at $x = 2$, $y = 4$ Min $f = -8$ at $x = 2$, $y = -2$ and at $x = 2$, $y = -2$

23 Max $f = 18$ at $x = 2$, $y = 8$;
Min $f = -18$ at $x = -2$, $y = -8$

25 (a) 1000 feet perpendicular to building, 3000 feet parallel to building
(b) $|\lambda| = 1000$: each additional foot of fence adds about 1000 square feet of area

27 $r \approx 3.7$ feet,   **29** End: 14 inches by 14 inches,
$h \approx 3.7$ feet   length = 28 inches,
volume = 5488 cubic inches

31 (a) $L = 120$, $K = 20$ **33** Base: 3 inches by 3 inches,
(b) $|\lambda| \approx 1.9$, output increases by   height: 5 inches
about 1.9 for each additional dollar

35 Min $f = 24$ at   **37** Max $f = 8$ at   **39** Base: 50 feet by 50 feet
$x = 4, y = 2, z = -2$ $x = 2, y = 2, z = 2$ height: 100 feet

EXERCISES 7.6 page 482

1 $2x^9 - 2x$ **3** $6y^4$ **5** $2x^2$ **7** 4   **9** 2   **11** 1/2   **13** 12

15 14   **17** $-9e^{-3} + 9e^3$ **19** 0   **21** -12 **23** 0   **25** 72   **27** 1/2

29 (a) $\int_1^3 \int_0^2 3xy^2 \, dx \, dy$ and $\int_0^2 \int_1^3 3xy^2 \, dy \, dx$ (b) Both equal 52

31 (a) $\int_0^2 \int_{-1}^1 ye^x \, dx \, dy$ and $\int_{-1}^1 \int_0^2 ye^x \, dy \, dx$ (b) Both equal $2e - 2e^{-1}$

33 8 cubic units   **35** 4/3 cubic units   **37** 1/6 cubic unit

39 $\frac{1}{2} e^2 - e + \frac{1}{2}$ cubic units   **41** 45 degrees   **43** About 56,410 people
(from 540/12)

45 900,000 cubic feet   **47** 14   **49** 10

EXERCISES 7.7 page 487

1 $\{(x, y) \mid x \geq 0, y \neq 0\}$ **3** $\{(x, y) \mid x \neq 0, y > 0\}$ **5** (a) $10x^4 - 6xy^3 - 3$
(b) $-9x^2y^2 + 4y^3 + 2$
(c) and (d) $-18xy^2$

7 (a) $12x^{-1/3}y^{1/3}$ **9** (a) $3x^2e^{x^3 - 2y^3}$
(b) $6x^{2/3}y^{-2/3}$ (b) $-6y^2e^{x^3 - 2y^3}$
(c) and (d) $4x^{-1/3}y^{-2/3}$ (c) and (d) $-18x^2y^2e^{x^3 - 2y^3}$

11 (a) $-ye^{-x} - \ln y$ **13** (a) 1/2   (b) 1/2   **15** (a) 36   (b) -24
(b) $e^{-x} - \dfrac{x}{y}$

(c) and (d) $-e^{-x} - \dfrac{1}{y}$

17 (a) 80: rate at which production increases for each additional unit of labor
(b) 135: rate at which production increases for each additional unit of capital
(c) Capital

19 Min f is -13 (at $x = -1, y = -4$).   **21** Max f is 8 (at $x = 0, y = -1$).

23 No Rel Extr Vals   **25** Max f is 1 (at $x = 0, y = 0$).
(saddle point at $x = 1/2, y = -3$)

27 Min f is 0 (at $x = 0, y = 0$).   **29** Min f is 25 (at $x = -2, y = -3$)
(saddle point at $x = 2, y = -3$).

31 (a) $C(x, y) = 3000x + 5000y + 6000$ **33** $y = 2.6x - 3.2$
(b) $R(x, y) = 7000x - 20x^2 + 8000y - 30y^2$
(c) $P(x, y) = -20x^2 + 4000x - 30y^2 + 3000y - 6000$ **35** $y = 30.1x + 105.7$
(d) Make 100 18-foot boats, sell for $5000 each $286.3 million
and 50 22-foot boats, sell for $6500 each,
max profit: $269,000

37 Max f is 288 (at $x = 12, y = -24$).   **39** Min f is 90 (at $x = 7, y = 4$).

41 Min f is e^{45} (at $x = 3$, $y = 6$).   **43** Max f is 120 (at $x = 2$, $y = -6$),
Min f is -120 (at $x = -2$, $y = 6$).

45 (a) $40,000 for production, $20,000 for advertising
(b) $|\lambda| \approx 159$, production increases by about 159 units for each additional dollar

47 (a) $L = 64$, $K = 8$ (c) $\dfrac{40L^{-1/3}K^{1/3}}{20L^{2/3}K^{-2/3}} \stackrel{?}{=} \dfrac{25}{100}$ (and now simplify and
(b) $40L^{-1/3}K^{1/3}$, $20L^{2/3}K^{-2/3}$ substitute $L = 64$, $K = 8$)

49 $8e^2 - 8e^{-2}$ **51** $4/3$ **53** 40 cubic units   **55** 5/6 cubic unit

57 12,000 (from 192,000/16)

INDEX

Normal distribution, 242, 366
Nuclear Regulatory Commission, 218
Numerical integration (*see* Integration, numerical)
Nutcracker man, 259

O

Objective function, 461
Oil:
  conservation of, 300
  well output, 364
Only critical point in town test, 166
Optimization, 162–91, 200
  constrained (Lagrange multipliers), 459–70
    geometrical interpretation of, 468
    of function of three or more variables, 470
    of function of several variables, 435–8, 441–2, 445
Order size (*see* Lot size)
Oxygen sag, 79

P

Page design, 184
Parabola (*see* Function, quadratic)
Pareto's law, 295
Partial derivative (*see* Derivative, partial)
Penicillin, 227
People, number remembered, 391
Permanent endowments, 360, 364
Pi (π), approximation of, 379
Piecewise linear function (*see* Function, piecewise linear)
Plato, 162
Point of diminishing returns, 140
Point six, rule of, 14
Poiseuille's law (*see* Blood flow)
Pole vaulting, 239
Pollution, 391, 409
  and absenteeism, 456
Polynomial function (*see* Function, polynomial)
Population:
  animal, 187–91, 208, 398–400, 403, 404, 421
  of Africa, China, India, 218, 307
  of Detroit, 232
  of the U.S., 297–8, 306
  of the world, 39, 218, 245, 334, 456–7

Porsche, 273
Postage stamps, 410, 459
Postal package, largest, 183, 472
Potassium 40 dating, 228, 232
Power cable, 204
Power rule:
  for differentiation, 67–8
    verification of, 75–6, 242
  for integration, 263, 280
Present value, 210, 213–4, 217–8
  of a continuous stream of income, 340–2
  under continuous compounding, 313–4
Price function (*see also* Demand function), 168
Price discrimination, 446
Price-earnings ratio, 420
Principal, 208
Principal root, 6
Producers' surplus, 310–13
Product of derivatives, 110
Product recognition, 403
Product rule, 80–1, 88
  for three functions, 91
Production possibilities curve, 472
Production runs, 186–7
Productivity, total, 289
Profit, 72–3, 167–70
  average and marginal average, 84–5
  cumulative, 308
  maximizing, 167–70, 176–8, 441–3
Proportional, definition of, 392
Psychophysics, 141–2, 149

Q

Quadratic equations, solving, 24–6
Quadratic formula, 25–6
Quilting, 308
Quotient rule, 82–3, 89

R

$\mathbb{R}$, 15
Rabbits, population of, 208
Radioactive waste, 233
Rate of change, 55, 61–2, 269–70
Rational function (*see* Function, rational)
Rectangle, formulas for, 171
Rectangular solid, volume of, 171
Reduction formulas, 351–3

Reed-Frost model of contagion, 219
Region, value and average value over a, 480–2, 484, 490
Regression (*see also* Least squares), 448
Related rates, 195–6
Relative:
  extreme point and value, 123–5, 199–200
    for function of two variables, 417–9, 435–8
  maximum and minimum point and value, 123–8, 145–7
    for function of two variables, 417–9, 435–8
  rate of change, 248–50, 258
Relativity, theory of, 58
Reorder costs (*see* Lot size)
Reproduction function, 187–8
Resources, consumption of natural (*see* Natural resources)
Returns to scale, 421
Revenue, 72–3, 167–70
  and elasticity of demand, 253–4
  average and marginal average, 84–5
  function, construction of, 176–80
Reynolds number, 246
Riemann sums, 368–9
Riemann, Georg Bernhard, 369
Riots, 174, 202
Ripples in pond, 195
Rockets, 199
Roots, 5–6, 9
Rule:
  chain (*see* Chain rule)
  generalized power (*see* Generalized power rule)
  of .6, 14
  of 72, 259
  power (*see* Power rule)
  product (*see* Product rule)
  quotient (*see* Quotient rule)
Rumors, spread of, 121, 400–1, 403
Ryan, Lynn Nolan, 120

S

S-shaped curve (*see* Sigmoidal curve)
Saddle point, 418–9, 435–8
Sales, growth of, 401, 403
Scuba diving, 420
Sea level, 103
Seatbelts, 262, 299–300